STUDENT'S SOLUTIONS MANUAL

CARRIE GREEN

INTERMEDIATE ALGEBRA WITH APPLICATIONS AND VISUALIZATION

THIRD EDITION

Gary K. Rockswold
Minnesota State University, Mankato

Terry A. Krieger
Rochester Community and Technical College

PEARSON

Addison
Wesley

Boston San Francisco New York
London Toronto Sydney Tokyo Singapore Madrid
Mexico City Munich Paris Cape Town Hong Kong Montreal

Copyright © 2009 Pearson Education, Inc.
Publishing as Pearson Addison-Wesley, 75 Arlington Street, Boston, MA 02116.

ISBN-13: 978-0-321-52350-1
ISBN-10: 0-321-52350-4

6 7 8 EBM 14 13 12

Contents

Chapter 1: Real Numbers and Algebra

1.1: Describing Data with Sets of Numbers

Concepts

1. The natural numbers are the counting numbers and do not include zero. $N = \{1,\ 2,\ 3,\ ...\}$

   The whole numbers are the counting numbers along with zero. $W = \{0,\ 1,\ 2,\ 3,\ ...\}$

2. The integers are given by the set $I = \{...,\ -3,\ -2,\ -1,\ 0,\ 1,\ 2,\ 3,\ ...\}$. An example is –5. *Answers may vary.*

3. The rational numbers are numbers which can be written in the form $\frac{p}{q}$ where p and $q \neq 0$ are integers. An example is $\frac{3}{4}$. *Answers may vary.*

4. The real numbers are any numbers which can be expressed as a decimal. A number which is real but is not rational is called irrational. An example of an irrational number is $\sqrt{2}$. *Answers may vary.*

5. An example of the commutative property for addition is $3 + 2 = 2 + 3$. *Answers may vary.*

6. An example of the commutative property for multiplication is $4 \cdot 3 = 3 \cdot 4$. *Answers may vary.*

7. The number 1 is called the multiplicative identity since $1 \cdot a = a$ for any number a.

8. The number 0 is called the additive identity since $0 + a = a$ for any number a.

9. An example of the associative property for multiplication is $2 \cdot (3 \cdot 4) = (2 \cdot 3) \cdot 4$. *Answers may vary.*

10. An example of the distributive property is $2(3 + 4) = 2 \cdot 3 + 2 \cdot 4$. *Answers may vary.*

11. The property illustrated is the commutative property for addition, because the result of adding the amounts in a different order is the same.

12. The property illustrated is the commutative property for multiplication, because to find the number of people in each marching band the number of rows is multiplied by the number of people in each row, and the results are the same.

Classifying Numbers

13. The number 26,461 is a natural number, integer, rational number and a real number.

15. The number $\frac{89}{3687}$ is a rational number and a real number.

17. The number 7.5 is a rational number and a real number.

19. The number $90\sqrt{2}$ is a real number.

21. Natural numbers: 6

    Whole numbers: 6

    Integers: –5, 6

Rational Numbers: $-5, 6, \dfrac{1}{7}, 0.2$

Irrational Numbers: $\sqrt{7}$

23.   Natural numbers: $\dfrac{3}{1}$

Whole numbers: $\dfrac{3}{1}$

Integers: $\dfrac{3}{1}$

Rational Numbers: $\dfrac{3}{1}, -\dfrac{5}{8}, 0.\overline{45}$

Irrational Numbers: $\sqrt{5}, \pi$

25.   Natural numbers: $\sqrt{9} = 3$

Whole numbers: $\sqrt{9} = 3$

Integers: $-2, \sqrt{9} = 3$

Rational numbers: $-2, \dfrac{1}{2}, \sqrt{9} = 3, 0.\overline{26}$

Irrational numbers: none

27.   Rational numbers. A shoe size could be a fraction such as $8\dfrac{1}{2}$.

29.   Rational numbers. Gasoline could be measured to any fraction of a gallon.

31.   Integers. Winter temperatures in Montana could be either positive or negative but are not given as fractions.

Properties of Real Numbers

33.   Identity

35.   Commutative

37.   Distributive

39.   Associative

41.   Commutative

43.   Commutative

45.   Distributive

47. $a + 4$

49. $\dfrac{1}{3}a$

51. $x + 1$

53. xy

55. $(4 + 5) + b = 9 + b$

57. $(5 \cdot 10)x = 50x$

59. $x + (y + z)$

61. $x \cdot (3 \cdot 4) = x \cdot 12$

63. $4x + 4y$

65. $5x - 35$

67. $-x - 1$

69. $a(x - y)$

71. $3(4 + x)$

73. $(3 + 7)x = 10x$

75. $(8 - 2)t = 6t$

77. $\left(1 - \dfrac{3}{4}\right)x = \dfrac{1}{4}x$

79. $3 - 1 + 2z = 2 + 2z$

81. $(3 + 4 + 1)z = 8z$

83. $(10 - 1 - 4)t = 5t$

85. $7(11 - 3) = 7 \cdot 8 = 56$ or $7(11 - 3) = 7 \cdot 11 - 7 \cdot 3 = 77 - 21 = 56$

87. $13(16 + 23) = 13 \cdot 39 = 507$ or $13(16 + 23) = 13 \cdot 16 + 13 \cdot 23 = 208 + 299 = 507$

89. $5(19 - 7) = 5 \cdot 12 = 60$ or $5(19 - 7) = 5 \cdot 19 - 5 \cdot 7 = 95 - 35 = 60$

91. $(3 + 4 + 5 + 8) \div 4 = 5$; natural number, integer, and rational number

93. $(45 + 33 + 52) \div 3 = 43.\overline{3}$; rational number

95. $(121.5 + 45.7 + 99.3 + 45.9) \div 4 = 78.1$; rational number

97. $8 + 3 + 2 + 7 = (8 + 2) + (3 + 7) = 10 + 10 = 20$

99. $\dfrac{2}{9} \cdot 8 \cdot 9 \cdot \dfrac{3}{8} = \left(\dfrac{2}{9} \cdot 9\right) \cdot \left(8 \cdot \dfrac{3}{8}\right) = 2 \cdot 3 = 6$

101. $4 \cdot 16 - 4 \cdot 6 = 4(16 - 6) = 4 \cdot 10 = 40$

103. $\dfrac{4}{5}\left(\dfrac{5}{2} + 5 + \dfrac{15}{4}\right) = \left(\dfrac{4}{5} \cdot \dfrac{5}{2}\right) + \left(\dfrac{4}{5} \cdot 5\right) + \left(\dfrac{4}{5} \cdot \dfrac{15}{4}\right)$

$$= \dfrac{4}{2} + 4 + \dfrac{15}{5} = 2 + 4 + 3 = 9$$

105. The property used is the identity property of 1, because $\dfrac{a}{a} = 1$ and $1 \cdot \dfrac{b}{c} = \dfrac{b}{c}$. *Answers may vary.*

Applications

107.   No; $8 \cdot 4$ GB $= 4 \cdot 8$ GB $= 32$ GB. The commutative property for multiplication applies.

109.   (a) From the table, enrollment in 2003 was 16.3 million.

     (b) The average of the 4 numbers is about 15.7 million. *Answers may vary.*

     (c) $(14.5 + 15.4 + 16.3 + 16.5) \div 4 = 15.675$ million; Yes, the answers nearly agree.

111.   (a) Valid; $2r\pi = 2\pi r$

     (b) Valid; $2(\pi r) = 2\pi r$

     (c) Not valid; $2\pi + 2r \neq 2\pi r$

1.2: Operations on Real Numbers

Concepts

1.   right

2.   left

3.   positive

4.   negative

5.   negative

6.   negative

7. $-a$

8. $\dfrac{b}{a}$

9.   positive

10. $-a - b$

The Real Number Line and Absolute Value

11.

13.

15.

17. $|-6.1| = 6.1$

19. $|8 - 11| = |-3| = 3$

21. $|x| = x$ when $x > 0$

23. $|x - y| = x - y$ when $x > 0$ and $y < 0$

25.   Positive. Area can never be negative.

27.   Both. Temperature can be both above and below zero.

29.   Positive. A car's mileage can never be negative.

Arithmetic Operations

31. The additive inverse of 56 is $-(56) = -56$.

33. The additive inverse of -6.9 is $-(-6.9) = 6.9$.

35. The additive inverse of $-\pi + 2$ is $-(-\pi + 2) = \pi - 2$.

37. The additive inverse of $a - b$ is $-(a - b) = -a + b$.

39. The additive inverse of $-(x - y)$ is $-(-(x - y)) = x - y$.

41. The additive inverse of $z - (2 - z)$ is $-(z - (2 - z)) = -(2z - 2) = -2z + 2$.

43. $\dfrac{1}{3}$

45. $-\dfrac{3}{2}$

47. $\dfrac{1}{\pi}$

49. $\dfrac{1}{a+3}$

51. $-\dfrac{b}{2a}$

53. $x - 7$

55. $4 + (-6) = -2$

57. $-7.4 + (-9.2) = -16.6$

59. $-\dfrac{3}{4} - \left(-\dfrac{1}{4}\right) = -\dfrac{3}{4} + \dfrac{1}{4} = -\dfrac{2}{4} = -\dfrac{1}{2}$

61. $-9 + 1 + (-2) + 5 = -8 + (-2) + 5 = -10 + 5 = -5$

63. $-6 \cdot -12 = 72$

65. $-\dfrac{1}{2} \cdot \dfrac{5}{7} \cdot \dfrac{1}{3} = -\dfrac{5}{14} \cdot \dfrac{1}{3} = -\dfrac{5}{42}$

67. $-\dfrac{1}{2} \div -\dfrac{3}{4} = -\dfrac{1}{2} \cdot -\dfrac{4}{3} = \dfrac{4}{6} = \dfrac{2}{3}$

69. $\dfrac{3}{4} \div (-2) = \dfrac{3}{4} \cdot -\dfrac{1}{2} = -\dfrac{3}{8}$

71. $-\dfrac{8}{2} = -4$

73. $-25 \div -5 = 5$

75. $\dfrac{1}{0}$ is undefined because division by zero is not possible.

77. $-4 \cdot 7 \cdot (-5) = -28 \cdot (-5) = 140$

79. $6 \cdot \dfrac{2}{3} \cdot 3 \cdot \left(-\dfrac{1}{6}\right) = \left[6 \cdot \left(-\dfrac{1}{6}\right)\right] \cdot \left(\dfrac{2}{3} \cdot 3\right) = -1 \cdot 2 = -2$

81. $4x - 9x = (4 - 9)x = -5x$

83. $z - 4z + 2z = (1 - 4 + 2)z = -z$

85.   Entering $-23.1 + 45.7 - (-34.6)$ into a graphing calculator gives 57.2.

87.   Entering $(1/2) + (2/3) - (5/7)$ ▶Frac into a graphing calculator gives $\dfrac{19}{42}$.

89.   Entering $(-3/4)(4/5)/(5/3)$ ▶Frac into a graphing calculator gives $-\dfrac{9}{25}$.

91.   Entering $(1/2)\big((4/11) + (2/5)\big)$ ▶Frac into a graphing calculator gives $\dfrac{21}{55}$.

Data and Number Sense

93. $\big[9 + 5 + 15 + (-9)\big] \div 4 = 5$

95. $\big[(-2) + 12 + 7 + (-17)\big] \div 4 = 0$

97. $\big[101 + 99 + (-42) + 82\big] \div 4 = 60$

99. $\big[(a + b) + (a + b) + (a + b)\big] \div 6 = \dfrac{a + b}{2}$

101. $-2 + 8 + 3 + 2 - 11 = (8 + 2 + 3) + \big[(-2) + (-11)\big] = 13 - 13 = 0$

103. $103 - 44 + 97 - 56 = (103 + 97) + \big[(-44) + (-56)\big] = 200 - 100 = 100$

105. $\dfrac{1}{5} \cdot \dfrac{2}{3} \cdot \dfrac{1}{7} \cdot \dfrac{1}{9} \cdot 5 \cdot \dfrac{3}{2} \cdot 7 \cdot 9 = \left(\dfrac{1}{5} \cdot 5\right)\left(\dfrac{2}{3} \cdot \dfrac{3}{2}\right)\left(\dfrac{1}{7} \cdot 7\right)\left(\dfrac{1}{9} \cdot 9\right) = 1 \cdot 1 \cdot 1 \cdot 1 = 1$

107. $\left(\dfrac{1}{2} - \dfrac{1}{3}\right) + \left(\dfrac{1}{3} - \dfrac{1}{4}\right) + \left(\dfrac{1}{4} - \dfrac{1}{5}\right) + \left(\dfrac{1}{5} - \dfrac{1}{6}\right) = \dfrac{1}{2} + \left(\dfrac{1}{3} - \dfrac{1}{3}\right) + \left(\dfrac{1}{4} - \dfrac{1}{4}\right) + \left(\dfrac{1}{5} - \dfrac{1}{5}\right) - \dfrac{1}{6} = \dfrac{1}{2} - \dfrac{1}{6} = \dfrac{1}{3}$

109. $a \cdot b \cdot c \cdot \frac{1}{a} \cdot \frac{1}{b} \cdot \frac{1}{c} = \left(a \cdot \frac{1}{a}\right) \cdot \left(b \cdot \frac{1}{b}\right) \cdot \left(c \cdot \frac{1}{c}\right) = 1 \cdot 1 \cdot 1 = 1$

Applications

111. The box contains 1200 cubic inches because if $200 = \text{Length} \times \text{Width} \times \text{Height}$ then

$2 \cdot 3 \cdot 200 = 2 \cdot \text{Length} \times 3 \cdot \text{Width} \times \text{Height}$ and $2 \cdot 3 \cdot 200 = 1200$.

113. Since $\text{Volume} = \text{Length} \times \text{Width} \times \text{Height}$, the larger pool has a volume $4 \cdot 2 = 8$ times greater than the smaller pool. It will take 8 times longer to fill it, or $8 \cdot 2 \text{ days} = 16 \text{ days}$.

115. We may estimate the cost to be $2600 since $\$200 + \$100 \cdot 24 = \$2600$. *Answers may vary.*

The actual cost is $\$202 + \$98.99 \cdot 24 = \$2577.76$.

117. (a) If we note that the increase from 1998 to 1999 was 6125 and the increase from 2001 to 2002 was 29,436, we see that the number of hacking incidents has not increased by a fixed amount each year.

   (b) $(3734 + 9859 + 21,756 + 52,658 + 82,094) \div 5 = 34,020.2$

   (c) One possible estimate is 120,000. Since 2003 lies outside of the given data and the number of hackings does not increase by a fixed amount, it is difficult to make this estimate. *Answers may vary.*

119. (a) The number of cable modem subscribers is increasing.

   (b) We may estimate the number of subscribers in 2010 to be about 26.1 million. *Answers may vary.*

Checking Basic Concepts for Sections 1.1 & 1.2

1. (a) The number -9 is an integer, a rational number and a real number.

   (b) The number $\dfrac{8}{4}$ is a natural number, integer, rational number and a real number.

   (c) The number $\sqrt{5}$ is a real number.

   (d) The number 0.5 is a rational number and a real number.

2. (a) Commutative property; $a + b = b + a$

   (b) Associative property; $a \cdot (b \cdot c) = (a \cdot b) \cdot c$

   (c) Distributive property; $a(b + c) = ab + ac$

   (d) Distributive property; $a(b - c) = ab - ac$ with $a = -1$

3. (a) $-3 + 4 + (-5) = 1 + (-5) = -4$

   (b) $5.1 \cdot (-4) \cdot 2 = -20.4 \cdot 2 = -40.8$

   (c) $-\dfrac{2}{3} \cdot \left(\dfrac{1}{4} \div \dfrac{2}{5}\right) = -\dfrac{2}{3} \cdot \left(\dfrac{1}{4} \cdot \dfrac{5}{2}\right) = -\dfrac{2}{3} \cdot \dfrac{5}{8} = -\dfrac{10}{24} = -\dfrac{5}{12}$

4. Since there are 3 times as many acres and the water is 2 times deeper, we multiply the total gallons by 6. The larger lake has about 12 billion gallons.

1.3: Integer Exponents

Concepts

1. The base is 8 and the exponent is 3.

2. $97^0 = 1$ and $2^{-1} = \dfrac{1}{2}$

3. 7^3

4. 5^2

5. No. $2^3 = 8$ and $3^2 = 9$

6. No. $-4^2 = -16$ and $(-4)^2 = 16$

7. $\dfrac{1}{7^n}$

8. 6^{m+n}

9. 5^{m-n}

10. $3^k x^k$

11. 2^{mk}

12. $\dfrac{x^m}{y^m}$

13. x^n

14. $\dfrac{b^m}{a^n}$

15. $\left(\dfrac{z}{y}\right)^n$ or $\dfrac{z^n}{y^n}$

16. 500

Properties of Exponents

17. $8 = 2 \times 2 \times 2 = 2^3$

19. $256 = 4 \times 4 \times 4 \times 4 = 4^4$

21. $1 = 6^0$

23. (a) $4^3 = 4 \cdot 4 \cdot 4 = 64$

  (b) $3^4 = 3 \cdot 3 \cdot 3 \cdot 3 = 81$

25. (a) $\dfrac{1}{4^{-2}} = 4^2 = 4 \cdot 4 = 16$

  (b) $\dfrac{1}{3^{-3}} = 3^3 = 3 \cdot 3 \cdot 3 = 27$

27. (a) $\left(\dfrac{2}{3}\right)^3 = \dfrac{2}{3} \cdot \dfrac{2}{3} \cdot \dfrac{2}{3} = \dfrac{8}{27}$

  (b) $\left(-\dfrac{2}{3}\right)^{-3} = \left(-\dfrac{3}{2}\right)^3 = \left(-\dfrac{3}{2}\right) \cdot \left(-\dfrac{3}{2}\right) \cdot \left(-\dfrac{3}{2}\right) = -\dfrac{27}{8}$

29. (a) $\dfrac{3^{-2}}{2^{-4}} = \dfrac{2^4}{3^2} = \dfrac{2 \cdot 2 \cdot 2 \cdot 2}{3 \cdot 3} = \dfrac{16}{9}$

  (b) $\dfrac{4^{-3}}{5^{-2}} = \dfrac{5^2}{4^3} = \dfrac{5 \cdot 5}{4 \cdot 4 \cdot 4} = \dfrac{25}{64}$

31. (a) $\dfrac{1}{2x^{-3}} = \dfrac{x^3}{2}$

   (b) $\dfrac{1}{(ab)^{-1}} = ab$

33. (a) $3^5 \cdot 3^{-3} = 3^{5+(-3)} = 3^2 = 9$

   (b) $x^2 x^5 = x^{2+5} = x^7$

35. (a) $\left(-3x^{-2}\right)\left(5x^5\right) = (-3)(5)\,x^{-2+5} = -15x^3$

   (b) $(ab)\left(a^2 b^{-3}\right) = a^{1+2} b^{1+(-3)} = a^3 b^{-2} = \dfrac{a^3}{b^2}$

37. (a) $5^{-2} \cdot 5^3 \cdot 2^{-4} \cdot 2^3 = 5^{-2+3} \cdot 2^{-4+3} = 5^1 \cdot 2^{-1} = 5 \cdot \dfrac{1}{2} = \dfrac{5}{2}$

   (b) $2a^3 \cdot b^2 \cdot 4a^{-4} \cdot b^{-5} = 2 \cdot 4 \cdot a^{3+(-4)} \cdot b^{2+(-5)} = 8a^{-1} b^{-3} = \dfrac{8}{ab^3}$

39. (a) $\dfrac{4^3}{4^2} = 4^{3-2} = 4^1 = 4$

   (b) $\dfrac{10^{-3}}{10^{-5}} = 10^{-3-(-5)} = 10^2 = 100$

41. (a) $\dfrac{b^{-3}}{b^2} = b^{-3-2} = b^{-5} = \dfrac{1}{b^5}$

   (b) $\dfrac{24x^3}{6x} = \dfrac{24}{6} x^{3-1} = 4x^2$

43. (a) $\dfrac{12a^2 b^3}{18a^4 b^2} = \dfrac{12}{18} a^{2-4} \cdot b^{3-2} = \dfrac{2}{3} a^{-2} b^1 = \dfrac{2b}{3a^2}$

   (b) $\dfrac{21x^{-3} y^4}{7x^4 y^{-2}} = \dfrac{21}{7} x^{-3-4} \cdot y^{4-(-2)} = 3x^{-7} y^6 = \dfrac{3y^6}{x^7}$

45. (a) $\left(3^2\right)^4 = 3^{2 \cdot 4} = 3^8 = 6561$

   (b) $\left(x^3\right)^{-2} = x^{3 \cdot (-2)} = x^{-6} = \dfrac{1}{x^6}$

47. (a) $\left(4y^2\right)^3 = 4^3 \cdot y^{2 \cdot 3} = 64y^6$

   (b) $\left(-2xy^3\right)^{-4} = \dfrac{1}{\left(2xy^3\right)^4} = \dfrac{1}{2^4 \cdot x^4 \cdot y^{3 \cdot 4}} = \dfrac{1}{16x^4 y^{12}}$

49. (a) $\left(\dfrac{4}{x}\right)^3 = \dfrac{4^3}{x^3} = \dfrac{64}{x^3}$

(b) $\left(\dfrac{2x}{z^4}\right)^{-5} = \left(\dfrac{z^4}{2x}\right)^5 = \dfrac{\left(z^4\right)^5}{(2x)^5} = \dfrac{z^{4\cdot 5}}{2^5 \cdot x^5} = \dfrac{z^{20}}{32x^5}$

51. $\dfrac{12m^2n^{-5}}{8mn^{-2}} = \dfrac{12}{8}\, m^{2-1}n^{-5-(-2)} = \dfrac{3}{2}\, mn^{-3} = \dfrac{3m}{2n^3}$

53. $\left(2x^3y^{-2}\right)^{-2} = 2^{-2}\left(x^3\right)^{-2}\left(y^{-2}\right)^{-2} = \dfrac{1}{2^2}\, x^{3\cdot(-2)}y^{(-2)\cdot(-2)} = \dfrac{1}{4}x^{-6}y^4 = \dfrac{y^4}{4x^6}$

55. $\dfrac{\left(b^2\right)^3}{\left(b^{-1}\right)^2} = \dfrac{b^{2\cdot 3}}{b^{-1\cdot 2}} = \dfrac{b^6}{b^{-2}} = b^{6-(-2)} = b^8$

57. $\dfrac{\left(-3ab^2\right)^3}{\left(a^2b\right)^2} = \dfrac{(-3)^3\, a^3\left(b^2\right)^3}{\left(a^2\right)^2 b^2} = \dfrac{-27a^3b^{2\cdot 3}}{a^{2\cdot 2}b^2} = \dfrac{-27a^3b^6}{a^4b^2} = -27a^{3-4}b^{6-2} = -27a^{-1}b^4 = -\dfrac{27b^4}{a}$

59. $\dfrac{\left(-m^2n^{-1}\right)^{-2}}{(mn)^{-1}} = \dfrac{(mn)^1}{\left(-m^2n^{-1}\right)^2} = \dfrac{mn}{\left(-m^2\right)^2\left(n^{-1}\right)^2} = \dfrac{mn}{m^{2\cdot 2}n^{-1\cdot 2}} = \dfrac{mn}{m^4n^{-2}} = m^{1-4}n^{1-(-2)} = m^{-3}n^3 = \dfrac{n^3}{m^3}$

61. $\left(\dfrac{2a^3}{6b}\right)^4 = \left(\dfrac{a^3}{3b}\right)^4 = \dfrac{\left(a^3\right)^4}{3^4b^4} = \dfrac{a^{3\cdot 4}}{81b^4} = \dfrac{a^{12}}{81b^4}$

63. $\left(\dfrac{t^{-3}}{t^{-4}}\right)^2 = \left(t^{-3-(-4)}\right)^2 = \left(t^1\right)^2 = t^2$

65. $\dfrac{8x^{-3}y^{-2}}{4x^{-2}y^{-4}} = \dfrac{8}{4}x^{-3-(-2)}y^{-2-(-4)} = 2x^{-1}y^2 = \dfrac{2y^2}{x}$

67. $\left(\dfrac{2t}{-r^2}\right)^{-3} = \left(\dfrac{-r^2}{2t}\right)^3 = \dfrac{\left(-r^2\right)^3}{2^3t^3} = \dfrac{-r^{2\cdot 3}}{8t^3} = -\dfrac{r^6}{8t^3}$

69. $\dfrac{\left(r^2t^2\right)^{-2}}{\left(r^3t\right)^{-1}} = \dfrac{r^3t}{\left(r^2t^2\right)^2} = \dfrac{r^3t}{\left(r^2\right)^2\left(t^2\right)^2} = \dfrac{r^3t}{r^{2\cdot 2}t^{2\cdot 2}} = \dfrac{r^3t}{r^4t^4} = r^{3-4}t^{1-4} = r^{-1}t^{-3} = \dfrac{1}{rt^3}$

71. $\dfrac{4x^{-2}y^3}{\left(2x^{-1}y\right)^2} = \dfrac{4x^{-2}y^3}{2^2\left(x^{-1}\right)^2y^2} = \dfrac{4x^{-2}y^3}{4x^{-1\cdot 2}y^2} = \dfrac{4x^{-2}y^3}{4x^{-2}y^2} = \left(\dfrac{4x^{-2}}{4x^{-2}}\right)y^{3-2} = 1y = y$

73. $\left(\dfrac{-15r^2t}{3r^{-3}t^4}\right)^3 = \left(\dfrac{-15}{3}\, r^{2-(-3)}t^{1-4}\right)^3 = \left(-5r^5t^{-3}\right)^3 = (-5)^3\left(r^5\right)^3\left(t^{-3}\right)^3 = -125r^{5\cdot 3}t^{-3\cdot 3} = -125r^{15}t^{-9} = -\dfrac{125r^{15}}{t^9}$

Order of Operations

75. $4 + 5 \cdot 6 = 4 + 30 = 34$

77. $2(4 + (-8)) = 2 \cdot (-4) = -8$

79. $5 \cdot 2^3 = 5 \cdot 8 = 40$

81. $\dfrac{-2^4 - 3^2}{4} + \dfrac{1+2}{4} = \dfrac{-16-9}{4} + \dfrac{3}{4} = \dfrac{-25}{4} + \dfrac{3}{4} = \dfrac{-22}{4} = -\dfrac{11}{2}$

83. $\dfrac{1 - 2 \cdot 4^2}{5^{-1}} = \dfrac{1 - 2 \cdot 16}{\frac{1}{5}} = \dfrac{1 - 32}{\frac{1}{5}} = \dfrac{-31}{\frac{1}{5}} = \dfrac{-31}{1} \cdot \dfrac{5}{1} = -155$

85. $4 + 6 - 3 \cdot 5 \div 3 = 4 + 6 - 15 \div 3 = 4 + 6 - 5 = 10 - 5 = 5$

87. $\dfrac{-3^2 + 3}{3} = \dfrac{-9+3}{3} = \dfrac{-6}{3} = -2$

89. $-4^2 + \dfrac{15+2}{8-7} = -16 + \dfrac{17}{1} = 1$

91. $\left| 7 - 2^2 \cdot 3 \right| = \left| 7 - 4 \cdot 3 \right| = \left| 7 - 12 \right| = \left| -5 \right| = 5$

93. $\sqrt{4^2 + 3^2} = \sqrt{16 + 9} = \sqrt{25} = 5$

Scientific Notation

95. 2.447×10^6

97. 2.69×10^{10}

99. 5.1×10^{-2}

101. 1.0×10^{-6}

103. $500,000$

105. $9,300,000$

107. -0.006

109. 0.00005876

111. $\left(2 \times 10^4 \right)\left(3 \times 10^2 \right) = (2 \cdot 3) \times 10^{4+2} = 6 \times 10^6 = 6,000,000$

113. $\left(4 \times 10^{-4} \right)\left(2 \times 10^{-2} \right) = (4 \cdot 2) \times 10^{-4+(-2)} = 8 \times 10^{-6} = 0.000008$

115. $\dfrac{6.2 \times 10^3}{3.1 \times 10^{-2}} = \dfrac{6.2}{3.1} \times 10^{3-(-2)} = 2 \times 10^5 = 200,000$

Applications

117. (a) By trial and error $k = 10$

   (b) $\dfrac{1024}{52} \approx 20$ years

119. $V = (2a)^3 = 2^3 a^3 = 8a^3$

121. $A = (3ab)^2 = 3^2 a^2 b^2 = 9a^2 b^2$

123.   (a) $A = 500(1.05)^2 = \$551.25$

       (b) $A = 1000(1.05)^4 \approx \$1215.51$

125. $\left(2.19 \times 10^{12}\right) \div \left(2.49 \times 10^8\right) \approx \$8795$ per person

127. $256 \times 2^{20} = 268,435,456$ bytes

129.   See Figure 129.

Country	1996	2025
China	1.2551×10^9	1.48×10^9
Germany	8.24×10^7	8.09×10^7
India	9.758×10^8	1.3302×10^9
Mexico	9.58×10^7	1.302×10^8
U.S.	2.65×10^8	3.325×10^8

Figure 129

1.4: Variables, Equations, and Formulas

Concepts

1.   variable

2.   equation

3.   equals

4.   formula

5. x and y

6. $3 + 7 = 10$ *Answers may vary.*

7. $3x = 15$ *Answers may vary.*

8. $3y + x = 5$ *Answers may vary.*

9.   3

10. -2

11. b

12. b

13. $y = 2x = 2(3) = 6$

14. $A = s^2 = (-2)^2 = 4$

Writing Formulas

15. $y = 5280x$

17. $A = s^2$

19. $y = 3600x$

21. $A = \dfrac{1}{2}bh$

23. $r = \dfrac{1}{2}d$ and $A = \pi r^2$, so

$$A = \pi\left(\tfrac{1}{2}d\right)^2 = \pi\left(\tfrac{1}{4}d^2\right) = \tfrac{1}{4}\pi d^2.$$

Using Data, Variables, and Formulas

25. $y = 5(6) = 30$

27. $y = -3.1 + 5 = 1.9$

29. $d = (-3)^2 + 1 = 9 + 1 = 10$

31. $z = \sqrt{2(18)} = \sqrt{36} = 6$

33. $y = -\dfrac{1}{2}\sqrt[3]{\dfrac{1}{8}} = -\dfrac{1}{2}\cdot\dfrac{1}{2} = -\dfrac{1}{4}$

35. $N = 3\left(\dfrac{1}{3}\right)^3 - 1 = 3\left(\dfrac{1}{27}\right) - 1 = \dfrac{1}{9} - 1 = \dfrac{1}{9} - \dfrac{9}{9} = -\dfrac{8}{9}$

37. $P = |5 - 4.7| = |0.3| = 0.3$

39. $A = \dfrac{1}{2}(3)(6) = \dfrac{1}{2}(18) = 9$

41. $V = \pi\left(\dfrac{1}{2}\right)^2(5) = \pi\left(\dfrac{1}{4}\right)(5) = \dfrac{5}{4}\pi$

43.   (ii)

45.   (iii)

47. $a = -3$

49. $a = 2$

51.   See Figure 51.

x	0	2	4	6	8
y	−0.5	4.5	9.5	14.5	19.5

Figure 51

53.   See Figure 53.

x	−3	−1	1	3
y	15	5	5	15

Figure 53

55.   See Figure 55.

x	−1	0	1	8
y	−3	−2	−1	0

Figure 55

57.   See Figure 57.

x	a	$a-1$	a^2-1
y	$\sqrt{a+1}$	$\sqrt{a}$	a

Figure 57

59.

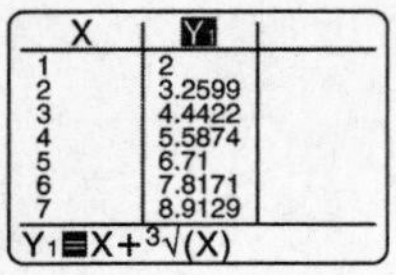

X	Y1	
1	2	
2	3.2599	
3	4.4422	
4	5.5874	
5	6.71	
6	7.8171	
7	8.9129	

$Y_1 \blacksquare X + {}^3\sqrt{(X)}$

61.

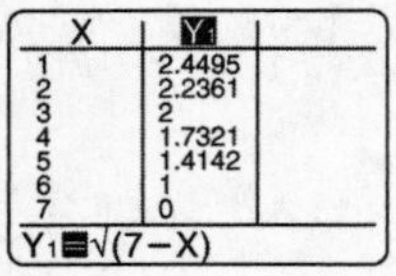

X	Y1	
1	2.4495	
2	2.2361	
3	2	
4	1.7321	
5	1.4142	
6	1	
7	0	

$Y_1 \blacksquare \sqrt{(7-X)}$

Applications

63. $y = 60t$

65.   (a) See Figure 65.

(b) $d = \dfrac{40^2}{12} = \dfrac{1600}{12} = 133.\overline{3}$ ft

(c) It quadruples.

(d) $\dfrac{60^2}{9} - \dfrac{60^2}{12} = 400 - 300 = 100$ ft

Speed (mph)	10	20	30	40	50	60	70
Braking Distance (ft)	8.3	33.3	75	133.3	208.3	300	408.3

Figure 65

67. $M = \dfrac{15}{22} \cdot 223.3 = 152.25$ mph

69.   (a) $W = 1.1(0.75)^3 \approx 0.464$ kg

(b) $W = 1.1(1.5)^3 = 3.7125$ kg

(c) It increases 8 times.

71.   If the shorter animal has a shoulder height of h, then its stepping frequency is $F_1 = \dfrac{0.87}{\sqrt{h}}$.

For the taller animal with shoulder height $4h$, $F_2 = \dfrac{0.87}{\sqrt{4h}} = \dfrac{0.87}{2\sqrt{h}} = \dfrac{1}{2}F_1$.

The stepping frequency of the taller animal is half the stepping frequency of the shorter animal.

73.   Venus: $E = \sqrt{2} \cdot 16,260 \approx 22,995$ mph; Earth: $C = \dfrac{25,040}{\sqrt{2}} \approx 17,706$ mph

Moon: $C = \dfrac{5360}{\sqrt{2}} \approx 3790$ mph; Mars: $E = \sqrt{2} \cdot 8050 \approx 11,384$ mph

75. $E = \sqrt{2} \cdot 57,000 \approx 80,610$ mph

77.   (a) $N = \dfrac{885}{\sqrt{25}} = 177$ bpm

     (b) $N = \dfrac{885}{\sqrt{1600}} \approx 22$ bpm

79.   (a) $A = \dfrac{\sqrt{3}}{4}(2)^2 = \dfrac{\sqrt{3}}{4} \cdot 4 = \sqrt{3} \approx 1.73$ ft^2

     (b) $A = \dfrac{\sqrt{3}}{4}(4)^2 = \dfrac{\sqrt{3}}{4} \cdot 16 = 4\sqrt{3} \approx 6.93$ m^2

81.   (a) $C = 2\pi(14) = 28\pi \approx 88$ inches

     (b) $C = 2\pi(1.3) = 2.6\pi \approx 8.2$ miles

83.   If $s^2 = 81$, then $s = 9$ inches.

85.   If $s^3 = 27$, then $s = 3$ meters.

87.   (a) $C = 336x$

     (b) $C = 336 \cdot 30 = 10,080$ calories

Checking Basic Concepts for Sections 1.3 & 1.4

1.   (a) $2^4 = 2 \cdot 2 \cdot 2 \cdot 2 = 16$

     (b) $3^{-2} \cdot 2^0 = \dfrac{1}{3^2} \cdot 1 = \dfrac{1}{9}$

     (c) $\dfrac{2^4}{2^2 \cdot 2^{-3}} = \dfrac{2^4}{2^{2+(-3)}} = \dfrac{2^4}{2^{-1}} = 2^{4-(-1)} = 2^5 = 32$

     (d) $x^3 x^{-4} x^2 = x^{3+(-4)+2} = x^1 = x$

     (e) $\left(\dfrac{2x^3}{y^{-4}}\right)^2 = \dfrac{2^2 \cdot x^{3 \cdot 2}}{y^{-4 \cdot 2}} = \dfrac{4x^6}{y^{-8}} = 4x^6 y^8$

2.   (a) $4 + 5 \cdot (-2) = 4 + (-10) = -6$

     (b) $\dfrac{1+3}{-4+3} = \dfrac{4}{-1} = -4$

     (c) $2^3 - 5(2 - 3 \cdot 4) = 8 - 5(2 - 12) = 8 - 5(-10) = 8 + 50 = 58$

3.   (a) 1.03×10^5

     (b) 5.23×10^{-4}

     (c) 6.7×10^0

4.   (a) 5,430,000

      (b) 0.0098

5.   See Figure 5. Each person requires 900 ft^3 of ventilation per hour.

People	10	20	30	40
Ventilation (ft³/hr)	9000	18,000	27,000	36,000

Figure 5

1.5: Introduction to Graphing

Concepts

1.   A relation is a set of ordered pairs $(x,\ y)$.

2.   The domain is the set of all *x*-values for the relation and the range is the set of all *y*-values for the relation.

3.   See Figure 3.

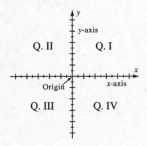   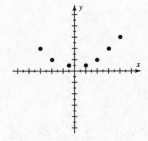   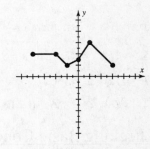

Figure 3   Figure 4a   Figure 4b

4.   A example of a scatterplot is shown in Figure 4a. An example of a line graph is shown in Figure 4b.

5.   No

6.   Quadrant III

Relations and the Rectangular Coordinate System

7. $D = \{1, 3, 5\};\ R = \{-4, 2, 6\}$

9. $D = \{-2,\ -1,\ 0,\ 1,\ 2\};\ R = \{0,\ 1,\ 2,\ 3\}$

11. $D = \{41,\ 87,\ 96\};\ R = \{24,\ 53,\ 67,\ 88\}$

13. $S = \{(1,\ 3),\ (3,\ 7),\ (5,\ 11),\ (7,\ 15),\ (9,\ 19)\};\ D = \{1,\ 3,\ 5,\ 7,\ 9\};\ R = \{3,\ 7,\ 11,\ 15,\ 19\}$

15. $S = \{(2000,\ 4.0),\ (2001, 4.2),\ (2002, 5.8),\ (2003, 6.0),\ (2004, 5.5)\};$

      $D = \{2000,\ 2001, 2002, 2003, 2004\};\ R = \{4.0,\ 4.2, 5.5, 5.8, 6.0\}$

17.   See Figure 17. (1, 2) is in QI, (–3, 0) is on the *x*-axis, (0, –2) is on the *y*-axis, (–1, 3) is in QII.

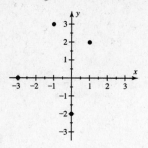

Figure 17

19.   See Figure 19. (10, 50) is in QI, (–30, 20) is in QII, (50, –25) is in QIV, (–20, –35) is in QIII.

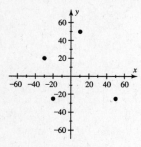

Figure 19

21.   Substituting 0, 1, 2, 3 for *x* gives the ordered pairs $(0, 0)$, $(1, 4)$, $(2, 8)$, $(3, 12)$. *Answers may vary.*

23.   Substituting 0, 1, 2, 3 for *t* gives the ordered pairs $(0, 4)$, $(1, 3)$, $(2, 0)$, $(3, -5)$. *Answers may vary.*

25.   Substituting 0, 1, 2, 3 for *r* gives the ordered pairs $(0, 1)$, $\left(1, \dfrac{1}{2}\right)$, $\left(2, \dfrac{1}{5}\right)$, $\left(3, \dfrac{1}{10}\right)$. *Answers may vary.*

27. $S = \{(-3, 2), (-2, 1), (2, -3), (3, 3)\}$; $D = \{-3, -2, 2, 3\}$; $R = \{-3, 1, 2, 3\}$

29. $S = \{(-4, 4), (-3, 2), (-2, 0), (0, -3), (2, 4), (4, 4)\}$;

$D = \{-4, -3, -2, 0, 2, 4\}$; $R = \{-3, 0, 2, 4\}$

31. $S = \{(1970, 29), (1980, 41), (1990, 79), (2000, 64)\}$

$D = \{1970, 1980, 1990, 2000\}$; $R = \{29, 41, 64, 79\}$. *Answers may vary.*

33. $a = b$

Scatterplots and Line Graphs

35. (a) $D = \{-3, -2, 0, 1\}$; $R = \{-4, -3, 0, 2, 4\}$

    (b) *x*-min: -3, *x*-max: 1; *y*-min: -4, *y*-max: 4

    (c) & (d) See Figure 35.

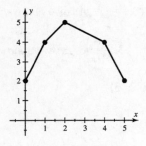

Figure 35

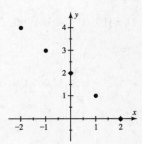

Figure 37

37. (a) $D = \{-30, 10, 20, 30\}$; $R = \{-50, 20, 40, 50\}$

    (b) *x*-min: -30, *x*-max: 30; *y*-min: -50, *y*-max: 50

    (c) & (d) See Figure 37.

39. See Figure 39.

41. See Figure 41.

43. See Figure 43.

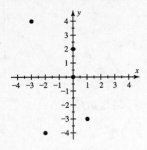

Figure 39

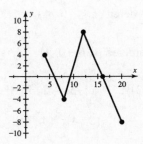

Figure 41

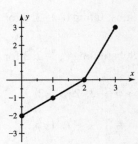

Figure 43

45. See Figure 45.

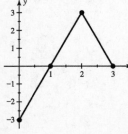

Figure 45

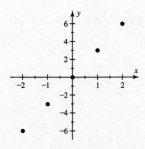

Figure 47

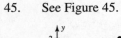

Figure 49

47. $(-2, -6)$, $(-1, -3)$, $(0, 0)$, $(1, 3)$, $(2, 6)$. See Figure 47.

49. $(-2, 4)$, $(-1, 3)$, $(0, 2)$, $(1, 1)$, $(2, 0)$. See Figure 49.

51. $(-2, 1)$, $\left(-1, \frac{1}{2}\right)$, $(0, 0)$, $\left(1, -\frac{1}{2}\right)$, $(2, -1)$. See Figure 51.

53.   (−2, 3), (−1, 0), (0, −1), (1, 0), (2, 3). See Figure 53.

55.   (−2, −4), (−1, −1), (0, 0), (1, −1), (2, −4). See Figure 55.

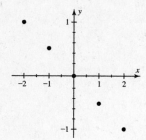

Figure 51

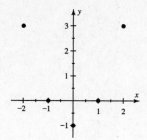

Figure 53

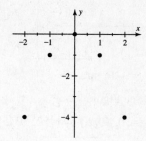

Figure 55

57.   (−2, 2), (−1, 1), (0, 0), (1, 1), (2, 2). See Figure 57.

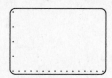

Figure 57

Graphing Calculators

59.   See Figure 59. There are 10 tick marks on the positive *x*-axis and 10 tick marks on the positive *y*-axis.

[−10, 10, 1] by [−10, 10, 1]

[0, 100, 10] by [−50, 50, 10]

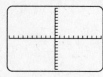

Figure 59

Figure 61

61.   See Figure 61. There are 10 tick marks on the positive *x*-axis and 5 tick marks on the positive *y*-axis.

63.   See Figure 63. There are 16 tick marks on the positive *x*-axis and 5 tick marks on the positive *y*-axis.

[1980, 1995, 1] by [12,000, 16,000, 1000]

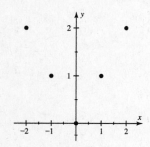

Figure 63

65. $S = \{(−2, 1), (−1, 0), (1, −1), (2, 1)\}$

67. $S = \{(−4, −2), (−2, −2), (0, 2), (2, 1), (4, −1)\}$

69.   See Figure 69.

[–6, 6, 1] by [–6, 6, 1]

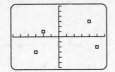

Figure 69

71.   See Figure 71.

[–30, 30, 5] by [–50, 50, 5]

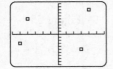

Figure 71

73.   See Figure 73.

[–200, 200, 50] by [–250, 250, 50]

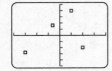

Figure 73

Applications

75.   (a) See Figure 75.

(b) Participation in Head Start first decreased and then increased.

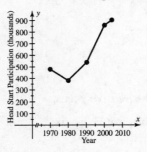

Figure 75

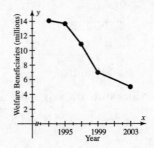

Figure 77

77.   (a) See Figure 77.

(b) The number of welfare beneficiaries first increased and then decreased.

79.   (a) See Figure 79.

    (b) The Asian-American population is increasing steadily.

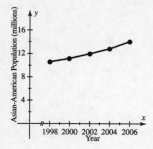

Figure 79

Checking Basic Concepts for Section 1.5

1. $D = \{-5,\ 1,\ 2\};\ R = \{-1,\ 3,\ 4\}$

2.   See Figure 2. (1, 4) is in QI, (0, –3) is on the y-axis, (2, –2) is in QIV, (–2, 3) is in QII.

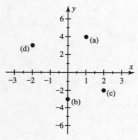   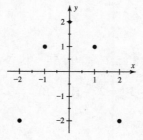   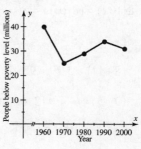

Figure 2   Figure 3   Figure 4

3. $(-2,\ -2),\ (-1,\ 1),\ (0,\ 2),\ (1,\ 1),\ (2,\ -2).$ See Figure 3.

4.   See Figure 4. The number of people below the poverty level decreased, then began to increase. then decreased.

Chapter 1 Review Exercises

Section 1.1

1.   Natural numbers: 9

    Whole numbers: 9

    Integers: –2, 9

    Rational Numbers: $-2, 9, \dfrac{2}{5},\ 2.68$

    Irrational Numbers: $\sqrt{11},\ \pi$

2.   Natural numbers: $\dfrac{6}{2}$

    Whole numbers: $\dfrac{6}{2},\ \dfrac{0}{4}$

    Integers: $\dfrac{6}{2},\ \dfrac{0}{4}$

Rational Numbers: $\dfrac{6}{2}, -\dfrac{2}{7}, 0.\overline{3}, \dfrac{0}{4}$

Irrational Numbers: $\sqrt{6}$

3.   Identity

4.   Commutative property

5.   Associative property

6.   Distributive property

7.   Distributive property

8.   Distributive property

9. $1 \cdot (a + 0) = 1 \cdot a = a$

10. $x \cdot \dfrac{1}{4} = \dfrac{1}{4}x$

11. $8(10x) = (8 \cdot 10)x = 80x$

12. $5z - 3z = (5 - 3)z = 2z$

13. $5(8 + 11) = 5 \cdot 19 = 95$ or $5(8 + 11) = 5 \cdot 8 + 5 \cdot 11 = 40 + 55 = 95$

14. $3(9 - 5) = 3 \cdot 4 = 12$ or $3(9 - 5) = 3 \cdot 9 - 3 \cdot 5 = 27 - 15 = 12$

15. $(6 + 9 + 3 + 11 + 5 + 20) \div 6 = 9$

16. $(3.2 + 6.8 + 6.1 + 10.8 + 1.7) \div 5 = 5.72$

17. $12 + 23 + (-2) + 7 = \big(12 + (-2)\big) + (23 + 7) = 10 + 30 = 40$

18. $\dfrac{3}{2} \cdot \dfrac{2}{5} \cdot \dfrac{5}{7} \cdot \dfrac{7}{3} = \left(\dfrac{3}{2} \cdot \dfrac{2}{5}\right)\left(\dfrac{5}{7} \cdot \dfrac{7}{3}\right) = \dfrac{3}{5} \cdot \dfrac{5}{3} = \dfrac{3 \cdot 5}{5 \cdot 3} = \dfrac{15}{15} = 1$

19. $5 \cdot 23 + 5 \cdot 7 = 5(23 + 7) = 5 \cdot 30 = 150$

20. $45 - 34 + 55 - 66 = (45 + 55) + (-34 - 66) = 100 - 100 = 0$

Section 1.2

21.

22. $|-7.2 + 4| = |-3.2| = 3.2$

23. $\dfrac{2}{3}$

24. $\dfrac{5}{4}$

25.   The opposite of $-2x + 3$ is $-(-2x + 3) = 2x - 3$.

26.   The reciprocal of $-\dfrac{1}{a + b}$ is $-\dfrac{a + b}{1} = -(a + b) = -a - b.$

27. $-5+(-7)+8 = -12+8 = -4$

28. $-9+11 = 2$

29. $-12-(-8) = -4$

30. $\dfrac{1}{2}+(-2)+\dfrac{3}{4} = \dfrac{2}{4}+\left(-\dfrac{8}{4}\right)+\dfrac{3}{4} = -\dfrac{6}{4}+\dfrac{3}{4} = -\dfrac{3}{4}$

31. $\dfrac{2}{3}\div(-4)-\dfrac{1}{3} = \dfrac{2}{3}\cdot\left(-\dfrac{1}{4}\right)-\dfrac{1}{3} = -\dfrac{2}{12}-\dfrac{4}{12} = -\dfrac{6}{12} = -\dfrac{1}{2}$

32. $-\dfrac{7}{11}+\dfrac{\frac{1}{5}}{\frac{2}{9}} = -\dfrac{7}{11}+\dfrac{1}{5}\cdot\dfrac{9}{2} = -\dfrac{7}{11}+\dfrac{9}{10} = -\dfrac{70}{110}+\dfrac{99}{110} = \dfrac{29}{110}$

33. $-7\cdot4\cdot-\dfrac{1}{7} = -7\cdot-\dfrac{1}{7}\cdot4 = 1\cdot4 = 4$

34. $\dfrac{3}{4}\cdot\dfrac{4}{7}+1 = \dfrac{3}{7}+1 = \dfrac{3}{7}+\dfrac{7}{7} = \dfrac{10}{7}$

35. $4x+5x = (4+5)x = 9x$

36. $2z-3z+8z = (2-3+8)z = 7z$

37. Entering $1/5-3/7+(2/13)/(4/5)$ ▶Frac into a graphing calculator gives $-\dfrac{33}{910}$.

38. $-4+9+4+11-6+16 = 30$; the average is $30\div6 = 5$.

Section 1.3

39. Base: 4; exponent: –2

40. They are not equal since $3^{\pi}\approx 31.54$ but $\pi^{3}\approx 31.01$.

41. $-2^{4} = -(2\cdot2\cdot2\cdot2) = -16$

42. $(-2)^{4} = (-2)\cdot(-2)\cdot(-2)\cdot(-2) = 16$

43. $9^{0} = 1$

44. $\left(\dfrac{2}{3}\right)^{-3} = \left(\dfrac{3}{2}\right)^{3} = \dfrac{3^{3}}{2^{3}} = \dfrac{3\cdot3\cdot3}{2\cdot2\cdot2} = \dfrac{27}{8}$

45. $4^{-3} = \dfrac{1}{4^{3}} = \dfrac{1}{4\cdot4\cdot4} = \dfrac{1}{64}$

46. $\dfrac{1}{5^{-2}} = 5^{2} = 5\cdot5 = 25$

47. $\dfrac{5^{-3}}{3^{-2}} = \dfrac{3^{2}}{5^{3}} = \dfrac{3\cdot3}{5\cdot5\cdot5} = \dfrac{9}{125}$

48. $\dfrac{1}{2 \cdot 4^{-2}} = \dfrac{4^2}{2} = \dfrac{4 \cdot 4}{2} = \dfrac{16}{2} = 8$

49. $4^3 \cdot 4^{-5} = 4^{3+(-5)} = 4^{-2} = \dfrac{1}{4^2} = \dfrac{1}{16}$

50. $10^4 \cdot 10^{-2} = 10^{4+(-2)} = 10^2 = 100$

51. $x^7 \cdot x^{-2} = x^{7+(-2)} = x^5$

52. $\dfrac{3^4}{3^{-7}} = 3^{4-(-7)} = 3^{11} = 177{,}147$

53. $\dfrac{5a^{-4}}{10a^2} = \left(\dfrac{5}{10}\right) \cdot a^{-4-2} = \left(\dfrac{1}{2}\right) \cdot a^{-6} = \left(\dfrac{1}{2}\right) \cdot \left(\dfrac{1}{a^6}\right) = \dfrac{1}{2a^6}$

54. $\dfrac{15a^4b^3}{3a^2b^6} = \left(\dfrac{15}{3}\right) \cdot a^{4-2} \cdot b^{3-6} = 5a^2b^{-3} = \dfrac{5a^2}{b^3}$

55. $\left(2^2\right)^4 = 2^{2 \cdot 4} = 2^8 = 256$

56. $\left(x^{-3}\right)^5 = x^{-3 \cdot 5} = x^{-15} = \dfrac{1}{x^{15}}$

57. $\left(4x^{-2}y^3\right)^2 = 4^2 \cdot x^{-2 \cdot 2} \cdot y^{3 \cdot 2} = 16x^{-4}y^6 = \dfrac{16y^6}{x^4}$

58. $(4a)^5 = 4^5 \cdot a^5 = 1024a^5$

59. $\left(\dfrac{5x^3}{3z^4}\right)^3 = \dfrac{5^3 \cdot x^{3 \cdot 3}}{3^3 \cdot z^{4 \cdot 3}} = \dfrac{125x^9}{27z^{12}}$

60. $\left(\dfrac{-3x^4y^3}{z}\right)^{-2} = \left(\dfrac{z}{-3x^4y^3}\right)^2 = \dfrac{z^2}{(-3)^2 \cdot x^{4 \cdot 2} \cdot y^{3 \cdot 2}} = \dfrac{z^2}{9x^8y^6}$

61. $\left(\dfrac{3a^{-4}}{4b^{-7}}\right)^2 = \dfrac{3^2\left(a^{-4}\right)^2}{4^2\left(b^{-7}\right)^2} = \dfrac{9a^{-4 \cdot 2}}{16b^{-7 \cdot 2}} = \dfrac{9a^{-8}}{16b^{-14}} = \dfrac{9b^{14}}{16a^8}$

62. $\left(\dfrac{3m^2n^{-4}}{9m^3n}\right)^{-1} = \dfrac{9m^3n}{3m^2n^{-4}} = \dfrac{9}{3}m^{3-2}n^{1-(-4)} = 3mn^5$

63. $\left(\dfrac{rt}{2r^3t^{-1}}\right)^{-3} = \left(\dfrac{2r^3t^{-1}}{rt}\right)^3 = \dfrac{2^3\left(r^3\right)^3\left(t^{-1}\right)^3}{r^3t^3} = \dfrac{8r^{3 \cdot 3}t^{-1 \cdot 3}}{r^3t^3} = \dfrac{8r^9t^{-3}}{r^3t^3} = 8r^{9-3}t^{-3-3} = 8r^6t^{-6} = \dfrac{8r^6}{t^6}$

64. $\left(\dfrac{3r^2}{4t^{-3}}\right)^2 = \dfrac{3^2\left(r^2\right)^2}{4^2\left(t^{-3}\right)^2} = \dfrac{9r^{2 \cdot 2}}{16t^{-3 \cdot 2}} = \dfrac{9r^4}{16t^{-6}} = \dfrac{9r^4t^6}{16}$

65. $2 + 3 \cdot 9 = 2 + 27 = 29$

66. $4 - 1 - 6 = 3 - 6 = -3$

67. $5 \cdot 2^3 = 5 \cdot 8 = 40$

68. $\dfrac{2+4}{2} + \dfrac{3-1}{3} = \dfrac{6}{2} + \dfrac{2}{3} = 3 + \dfrac{2}{3} = \dfrac{9}{3} + \dfrac{2}{3} = \dfrac{11}{3}$

69. $20 \div 4 \div 2 = 5 \div 2 = \dfrac{5}{2}$

70. $\dfrac{3^3 - 2^4}{4 - 3} = \dfrac{27 - 16}{1} = \dfrac{11}{1} = 11$

71. 1.86×10^5

72. 3.4×10^{-4}

73.   45,000

74.   0.00923

Section 1.4

75.   Because there are 12 inches in one foot, the formula is $y = 12x$.

76.   Because the area formula for one circle is $A = \pi r^2$, the formula for the area of six circles is $A = 6\pi r^2$.

77. $y = 12(3) = 36$

78. $d = \sqrt{67 - 3} = \sqrt{64} = 8$

79. $N = \left(\dfrac{3}{2}\right)^2 - \dfrac{3}{4} = \dfrac{9}{4} - \dfrac{3}{4} = \dfrac{6}{4} = \dfrac{3}{2}$

80. $P = (-2)^3 - 2 = -8 - 2 = -10$

81. $A = \dfrac{1}{2}(4)(5) = 2(5) = 10$

82. $V = (3)^2(3) = 9(3) = 27$

83.   (iii)

84. $a = \dfrac{3}{2}$

85.   See Figure 85.

x	-2	-1	0	1	2
y	-7	0	1	2	9

Figure 85

86.   See Figure 86.

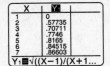

   Figure 86

Section 1.5

87. $D = \{-1, 2, 3\}$; $R = \{-6, 1, 3, 7\}$

88. $S = \{(-8, 4), (-4, -4), (4, 0), (8, 4)\}$; $D = \{-8, -4, 4, 8\}$; $R = \{-4, 0, 4\}$

89. $(-2, 6), (-1, 3), (0, 0), (1, -3), (2, -6)$. See Figure 89.

90. $(-2, -2), (-1, -1.5), (0, -1), (1, -0.5), (2, 0)$. See Figure 90.

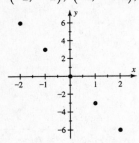

   Figure 89

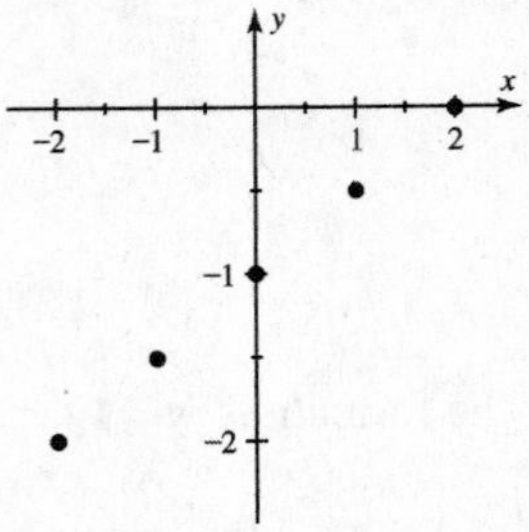

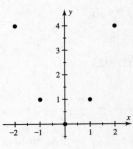

   Figure 90

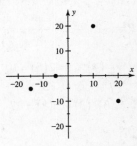

   Figure 91

91. $(-2, 4), (-1, 1), (0, 0), (1, 1), (2, 4)$. See Figure 91.

92. $(-2, 1), (-1, 2.5), (0, 5), (1, 2.5), (2, 1)$. See Figure 92.

93.   See Figure 93. $(-2, 2)$ is in QII, $(-1, -3)$ is in QIII, $(0, 1)$ is on the *y*-axis, $(2, -1)$ is in QIV.

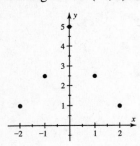

   Figure 92

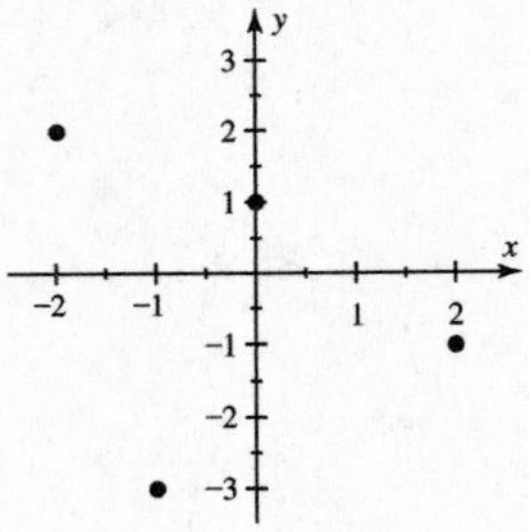

   Figure 93

   Figure 94

94.   See Figure 94. $(-15, -5)$ is in QIII, $(-5, 0)$ is on the *x*-axis, $(10, 20)$ is in QI, $(20, -10)$ is in QIV.

95.   See Figure 95. There are 9 tick marks on the positive *x*-axis and 2 tick marks on the positive *y*-axis.

   $[-9, 9, 1]$ by $[-6, 6, 3]$ $[-20, 20, 5]$ by $[-12, 12, 4]$

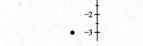

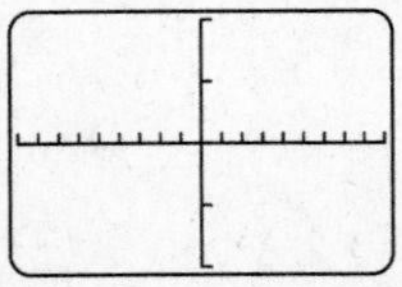

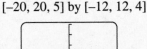

   Figure 95   Figure 96

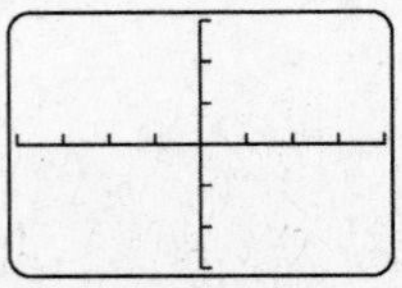

96.   See Figure 96. There are 4 tick marks on the positive *x*-axis and 3 tick marks on the positive *y*-axis.

97.   See Figure 97. $D = \{-1,\ 0,\ 2,\ 3,\ 4\}$; $R = \{-1,\ 0,\ 2,\ 3,\ 4\}$.

98.   See Figure 98. $D = \{-20,\ -10,\ 45,\ 50\}$; $R = \{-30,\ -25,\ 10,\ 20\}$.

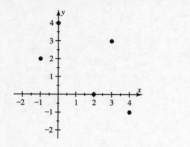

Figure 97   Figure 98

Applications

99.   (a)   The best estimate is $163,200.

      (b) $(170,000 + 165,000 + 160,500 + 157,300) \div 4 = \$163,200$

100. (a) $A = 2500(1.01)^1 = \$2525$

      (b) $A = 800(1.01)^2 = 800(1.01)(1.01) = \$816.08$

101. (a) $C = 2\pi \cdot \left(9.3 \times 10^7\right) \approx 5.8 \times 10^8$ miles

      (b) $\dfrac{5.84 \times 10^8 \text{ mi}}{1 \text{ year}} \cdot \dfrac{1 \text{ year}}{365 \text{ days}} \cdot \dfrac{1 \text{ day}}{24 \text{ hr}} \approx 66,700$ mph

102. $d = 40t$

103. For a 16-pound cat the pulse rate is $N = \dfrac{885}{\sqrt{16}} = \dfrac{885}{4} \approx 221$ bpm.

      For a 144-pound person the pulse rate is $N = \dfrac{885}{\sqrt{144}} = \dfrac{885}{12} \approx 74$ bpm.

104. See Figure 104. The poverty threshold has increased during this time period.

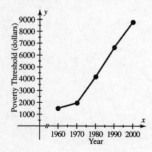

Figure 104

105. $A = (4ab)^2 = 4^2 a^2 b^2 = 16a^2 b^2$

106. $V = (5z)^3 = 5^3 z^3 = 125z^3$

Chapter 1 Test

1. Natural numbers: $\sqrt{9}$

   Whole numbers: $\sqrt{9}$

   Integers: -5, $\sqrt{9}$

   Rational Numbers: -5, $\dfrac{2}{3}$, $\sqrt{9}$, -1.83

   Irrational Numbers: $-\dfrac{1}{\sqrt{5}}$, π

2. (a) Identity

   (b) Associative property

   (c) Commutative property

   (d) Distributive property

3. (a) $0 + (a \cdot 1) \cdot 1 = 0 + a \cdot 1 = 0 + a = a$

   (b) $2x - 5x + 4x = (2 - 5 + 4)x = 1x = x$

   (c) $2(x \cdot 3) + 2x = 2(3 \cdot x) + 2x = (2 \cdot 3)x + 2x = 6x + 2x = (6 + 2)x = 8x$

   (d) $(a + b)a - ab = a \cdot a + ab - ab = a^2 + 0 = a^2$

4. $(34 + 15 + 96 + 11 + 0) \div 5 = 31.2$

5.

6. (a) $\left| \dfrac{1}{2} + \dfrac{2}{3} - \dfrac{8}{3} \right| = \left| \dfrac{3}{6} + \dfrac{4}{6} - \dfrac{16}{6} \right| = \left| -\dfrac{9}{6} \right| = \left| -\dfrac{3}{2} \right| = \dfrac{3}{2}$

   (b) $12 + (-3) + 18 + (-7) = 9 + 18 + (-7) = 27 + (-7) = 20$

   (c) $\dfrac{1}{2} \cdot \dfrac{2}{3} \cdot \dfrac{3}{4} \cdot \dfrac{3}{2} \cdot 2 = \dfrac{3}{4}$

   (d) $\dfrac{1}{2} + \dfrac{2}{3} - \dfrac{3}{2} + \dfrac{1}{3} = \dfrac{3}{6} + \dfrac{4}{6} - \dfrac{9}{6} + \dfrac{2}{6} = 0$

7. (a) $-\dfrac{4}{5}$ (b) $\dfrac{b+1}{2}$

8. (a) $-\dfrac{1}{2} + \dfrac{2}{3} \div 3 = -\dfrac{1}{2} + \dfrac{2}{3} \cdot \dfrac{1}{3} = -\dfrac{1}{2} + \dfrac{2}{9} = -\dfrac{9}{18} + \dfrac{4}{18} = -\dfrac{5}{18}$

   (b) $-4 + \dfrac{\frac{2}{3}}{-\frac{1}{4}} = -4 + \dfrac{2}{3} \cdot \left(-\dfrac{4}{1} \right) = -4 - \dfrac{8}{3} = -\dfrac{12}{3} - \dfrac{8}{3} = -\dfrac{20}{3}$

   (c) $5 - 2 \cdot 5^2 \div 5 = 5 - 2 \cdot 25 \div 5 = 5 - 50 \div 5 = 5 - 10 = -5$

   (d) $(6 - 4 \cdot 5) \div (-7) = (6 - 20) \div (-7) = (-14) \div (-7) = 2$

9.   (a) $5^{-2} = \dfrac{1}{5^2} = \dfrac{1}{25}$

(b) $\pi^0 = 1$

(c) $\left(-\dfrac{2}{5}\right)^4 = \left(-\dfrac{2}{5}\right)\cdot\left(-\dfrac{2}{5}\right)\cdot\left(-\dfrac{2}{5}\right)\cdot\left(-\dfrac{2}{5}\right) = \dfrac{16}{625}$

(d) $\dfrac{6^{-2}}{2^{-4}} = \dfrac{2^4}{6^2} = \dfrac{2\cdot 2\cdot 2\cdot 2}{6\cdot 6} = \dfrac{16}{36} = \dfrac{4}{9}$

(e) $\dfrac{1}{5^{-3}} = 5^3 = 5\cdot 5\cdot 5 = 125$

(f) $2^2 \cdot 2^{-4} = 2^{\left[2+(-4)\right]} = 2^{-2} = \dfrac{1}{2^2} = \dfrac{1}{4}$

10.   (a) $x^6 \cdot x^{-4} \cdot y^3 = x^{6+(-4)} \cdot y^3 = x^2 y^3$

(b) $\dfrac{16x^{-2}y^8}{6xy^{-7}} = \left(\dfrac{16}{6}\right)\cdot x^{-2-1} \cdot y^{8-(-7)} = \dfrac{8}{3}x^{-3}y^{15} = \dfrac{8y^{15}}{3x^3}$

(c) $\left(2yz^{-2}\right)^3 = 2^3 \cdot y^3 \cdot z^{-2\cdot 3} = 8y^3 z^{-6} = \dfrac{8y^3}{z^6}$

(d) $\left(\dfrac{15x^4}{10xy^{-2}}\right)^{-2} = \left(\dfrac{10xy^{-2}}{15x^4}\right)^2 = \dfrac{10^2 \cdot x^2 \cdot y^{-2\cdot 2}}{15^2 \cdot x^{4\cdot 2}} = \dfrac{100}{225}\cdot x^{2-8} \cdot y^{-4} = \dfrac{4}{9}x^{-6}y^{-4} = \dfrac{4}{9x^6 y^4}$

11. 0.00052

12. 3.4×10^6

13. $H = \dfrac{x}{60}$

14. $y = \sqrt{3+1} + 3 = \sqrt{4} + 3 = 2 + 3 = 5$;

$y = \sqrt{-1+1} + (-1) = \sqrt{0} + (-1) = 0 + (-1) = -1$

15. $D = \{-3,\ -1,\ 2\},\ R = \{-4,\ 2,\ 3\}$

16. $(-2, 3)$: QII; $(-1, -2)$: QIII; $(0, 2)$: y-axis;

$(1, -1)$: QIV; $(2, 1)$: QI

See Figure 16.

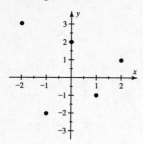

Figure 16

17. $S = \{(-30, 20), (-20, 20), (-10, 10), (10, 30), (20, 10), (30, -20)\}$

$D = \{-30, -20, -10, 10, 20, 30\}$; $R = \{-20, 10, 20, 30\}$

18.   See Figure 18. The equation $y = 1.25x$ models these data.

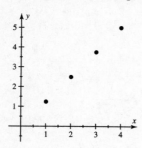

Figure 18

19.   See Figure 19.

$[-20, 20, 5]$ by $[-5, 40, 5]$

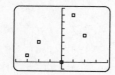

Figure 19

20. $C = \dfrac{5}{9}(5-32) = \dfrac{5}{9}(-27) = -15°C$

21. $\left(2.11 \times 10^8\right) \cdot 2.21 = 4.6631 \times 10^8$ gal

22. $r = \sqrt{\dfrac{25}{\pi}} \approx 2.82$ ft

23. $M = 61t$, because $244 \div 4 = 366 \div 6 = 610 \div 10 = 61$.

Chapter 1 Extended and Discovery Exercises

1. (a) The surface area of the earth is $A = 4\pi \cdot 3960^2 \approx 1.97 \times 10^8$ mi^2.

    (b) The surface area of the oceans is $A = \left(1.97 \times 10^8 \text{ mi}^2\right) \cdot 0.71 \approx 1.40 \times 10^8$ mi^2.

    (c) If the Arctic ice cap were to melt, the rise in sea level would be $\dfrac{6.8 \times 10^5 \text{ mi}^3}{1.40 \times 10^8 \text{ mi}^2} \approx 0.00486$ mi.

    That is $0.00486 \text{ mi} \cdot \dfrac{5280 \text{ ft}}{1 \text{ mi}} \approx 25.7$ ft.

    (d) If the Arctic ice cap melted, cities such as Boston, New Orleans and San Diego would be flooded.

    (e) If the Antarctic ice cap were to melt, the rise in sea level would be $\dfrac{6.3 \times 10^6 \text{ mi}^3}{1.40 \times 10^8 \text{ mi}^2} = 0.045$ mi.

    That is $0.045 \text{ mi} \cdot \dfrac{5280 \text{ ft}}{1 \text{ mi}} \approx 238$ ft.

2. (a) See Figure 2. The injury rate has decreased over this period of time.

    (b) The values in the table decrease by about 1 for each 3-year period, so a reasonable estimate might be about 9.3.

    (c) The values in the table decrease by about 1 for each 3-year period, so a reasonable estimate might be about 4.4.

    (d) It is not valid to try to estimate when the injury rate might reach 0 because the apparent trend cannot continue. There will always be injuries of this type.

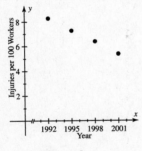

Figure 2

3. $y = \dfrac{0.455W}{\left(0.0254H\right)^2} \approx \dfrac{705W}{H^2}$

4. Noting that 5 feet 9 inches = 69 inches, Steffi Graf's BMI is $y = \dfrac{0.455(119)}{\left(0.0254(69)\right)^2} \approx 17.6$

5. Noting that 6 feet 1 inch = 73 inches, Venus Williams' BMI is $y = \dfrac{0.455(160)}{\left(0.0254(73)\right)^2} \approx 21.2.$

6. Noting that 7 feet 1 inch = 85 inches, Shaquille O'Neal's BMI is $y = \dfrac{0.455(300)}{\left(0.0254(85)\right)^2} \approx 29.3.$

7.   When x increases by 2, y decreases by 3, so $a = -\dfrac{3}{2}$. By using substitution and trial-and-error, $b = \dfrac{1}{2}$.

8.   By using substitution we find that $a + b = 1$ and $2a + b = -1$. By trial-and-error, $a = -2$ and $b = 3$.

9.   By using substitution we find that $a + b = 1$ and $4a + 2b = 3$. By trial-and-error, $a = \dfrac{1}{2}$ and $b = \dfrac{1}{2}$.

10.   Substituting $x = 0$ and $y = 3$ gives $b = 3$. Then by substituting $x = 1$ and $y = 6$ we find that $a = 2$.

Critical Thinking Solutions for Chapter 1

Section 1.1

- Yes, for example $-\sqrt{2} + \sqrt{2} = 0$.

Section 1.3

- Each value in the *value* row can be obtained by dividing the value to the left of it by 10 or by multiplying the value to the right of it by 10. Hence $10^0 = 10 \div 10 = 1$ and $10^0 = 10^{-1} \cdot 10 = 1$.

- A student who is 19 years, 4 months and 3 days old is 19 years 123 days old or about 7058 days old.

  That is $7058 \text{ days} \cdot \dfrac{24 \text{ hr}}{1 \text{ day}} \cdot \dfrac{60 \text{ min}}{1 \text{ hr}} \cdot \dfrac{60 \text{ sec}}{1 \text{ min}} \approx 610{,}000{,}000$ seconds.

  A student who is exactly 18 years old has been alive for about $568{,}000{,}000$ seconds. *Answers may vary.*

Section 1.4

- The formula is $T = \dfrac{100}{x}$.

Section 1.5

- Quadrant III, because both coordinates are negative.

Chapter 2: Linear Functions and Models

2.1: Functions and Their Representations

Concepts

1.　function

2.　*y* equals *f* of *x*

3.　symbolic

4.　numerical

5.　domain

6.　range

7.　one

8.　We use the vertical line test to identify graphs of functions.

9.　The four types of representations for a function are verbal, numerical, symbolic and graphical or diagrammatic.

10.　$(3, 4)$; 3; 6

11.　(a, b)

12.　*d*

13.　1

14.　equal

15.　Yes, there is only one output for each input.

16.　Yes, there is only one age for each person.

17.　No, one exam can have many students who pass.

18.　No, the parent may have more than one child.

Representing and Evaluating Functions

19.　$f(-1) = 4(-1) - 2 = -6$; $f(0) = 4(0) - 2 = -2$

21.　$f(0) = \sqrt{0} = 0$; $f\left(\frac{9}{4}\right) = \sqrt{\frac{9}{4}} = \frac{3}{2}$

23.　$f(-5) = (-5)^2 = 25$; $f\left(\frac{3}{2}\right) = \left(\frac{3}{2}\right)^2 = \frac{9}{4}$

25.　$f(-8) = 3$; $f\left(\frac{7}{3}\right) = 3$

27.　$f(-2) = 5 - (-2)^3 = 5 - (-8) = 13$;

　　$f(3) = 5 - 3^3 = 5 - 27 = -22$

29.　$f(-5) = \frac{2}{-5+1} = \frac{2}{-4} = -\frac{1}{2}$; $f(4) = \frac{2}{4+1} = \frac{2}{5}$

31.   (a) Because there are 36 inches in 1 yard, the formula is $I(x) = 36x$.

(b) $I(10) = 36(10) = 360$. There are 360 inches in 10 yards.

33.   (a) The area formula for a circle is $A(r) = \pi r^2$.

(b) $A(10) = \pi(10)^2 = 100\pi \approx 314.2$. The area of a circle with radius 10 is about 314.2.

35.   (a) Because there are 43,560 square feet in 1 acre, the formula is $A(x) = 43,560x$.

(b) $A(10) = 43,560(10) = 435,600$. There are 435,600 square feet in 10 acres.

37. $f = \{(1,3),(2,-4),(3,0)\}; D = \{1,2,3\}, R = \{-4,0,3\}$

39. $f = \{(a,b),(c,d),(e,a),(d,b)\}; D = \{a,c,d,e\}, R = \{a,b,d\}$

41.   See Figure 41.

43.   See Figure 43.

45.   See Figure 45.

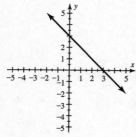
Figure 41

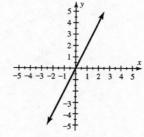
Figure 43

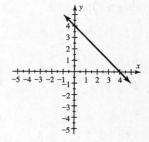
Figure 45

47.   See Figure 47.

49.   See Figure 49.

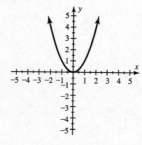
Figure 47

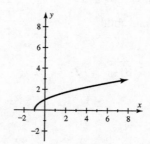
Figure 49

51. $f(0) = 3; f(2) = -1$

53. $f(-2) = 0; f(1) = 2$

55. $f(1) = -4; f(2) = -3$

57. $f(0) = 5.5; f(2) = 3.7$

59. $f(1990) = 26.9$ mpg ; In 1990 average fuel efficiency was 26.9 mpg.

61.   Symbolic: $y = x + 5$.   Numerical: See the table in Figure 61a.   Graphical: See the graph in Figure 61b.

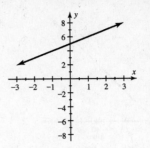

x	−3	−2	−1	0	1	2	3
$y = f(x)$	2	3	4	5	6	7	8

Figure 61a

Figure 61b

63.   Symbolic: $y = 5x - 2$.   Numerical: See the table in Figure 63a.   Graphical: See the graph in Figure 63b.

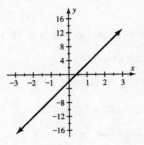

x	−3	−2	−1	0	1	2	3
$y = f(x)$	−17	−12	−7	−2	3	8	13

Figure 63a

Figure 63b

65.   Numerical   Graphical

[−10, 10, 1] by [−10, 10 1]

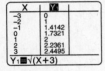

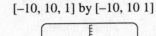

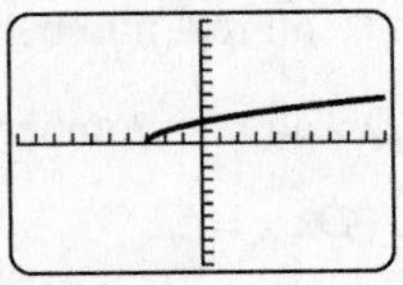

$Y_1 = \sqrt{(X+3)}$

67.   Numerical   Graphical (Dot Mode)

[−10, 10, 1] by [−10, 10, 1]

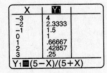

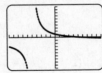

$Y_1 = (5-X)/(5+X)$

69.   Subtract $\dfrac{1}{2}$ from the input x to obtain the output y.

71.   Divide the input x by 3 to obtain the output y.

73.   Subtract 1 from x and then take the square root to obtain the output y.

75.   Symbolic: $f(x) = 0.50x$. Graphical: See the graph in Figure 75a. Numerical: See the table in Figure 75b.

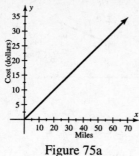

Figure 75a

Miles	10	20	30	40	50	60	70
Cost	$5	$10	$15	$20	$25	$30	$35

Figure 75b

77. $P(1995) = 5.7(1995 - 1990) + 150 = 5.7(5) + 150 = 28.5 + 150$

$$= 178.5$$

In 1995 the average price was $178,500.

Identifying Domains and Ranges

79. $D: -2 \le x \le 2; \ R: 0 \le y \le 2$

81. $D: -2 \le x \le 4; \ R: -2 \le y \le 2$

83. D: all real numbers; $R: y \ge -1$

85. $D: -3 \le x \le 3; \ R: -3 \le y \le 2$

87. $D = \{1, 2, 3, 4\}; R = \{5, 6, 7\}$

89.   Any real number is a valid input for this function. $D:$ all real numbers.

91.   Any real number is a valid input for this function. $D:$ all real numbers.

93.   The denominator of this function cannot equal zero. $D: x \ne 5$.

95.   The denominator of this function will never equal zero because the variable is squared. $D:$ all real numbers.

97.   The radicand must be greater than or equal to zero. $D: x \ge 1$.

99.   Any real number is a valid input for this function.

D: all real numbers.

101.   The denominator of this function cannot equal zero.

$D: x \ne 0$.

103.   (a) $f(1950) = 60.3$; In 1950 there were 60.3 accidental deaths per 100,000 people.

(b) $D = \{1910, 1930, 1950, 1970, 1990, 2000\}; \ R = \{35.5, 36.9, 56.2, 60.3, 80.5, 84.4\}$

(c)   The number of accidental deaths per 100,000 people decreased over this time period.

Identifying a Function

105.   No. The value 1 in the domain corresponds to more than one value in the range.

107.   Yes. Each value in the domain corresponds to exactly one value in the range.

109.   (a)   May is month number 5. The corresponding value for P is 0.2.

(b)   Yes. Each month has exactly one average precipitation.

(c)   Months 2, 3, 7 and 11.

111. Yes. The graph passes the vertical line test.; D : all real numbers; R : all real numbers

113. No. The graph does not pass the vertical line test.

115. Yes. The graph passes the vertical line test.; $D : -4 \le x \le 4$; $R : 0 \le y \le 4$

117. Yes. The graph passes the vertical line test.; D : all real numbers; $R : y = 3$

119. No. The graph does not pass the vertical line test when $x = -2$.

121. Yes. Each value in the domain corresponds to exactly one value in the range.

123. No. The value 5 in the domain corresponds to more than one value in the range.

125. Walks away from home, then turns around and walks back a little slower.

2.2: Linear Functions

Concepts

1. $ax + b$

2. b

3. line

4. horizontal line

5. 7

6. 0

7. Carpet costs \$2 per square foot. Ten square feet of carpet costs \$20.

8. The rate at which water is leaving the tank is 4 gallons per minute. After 5 minutes the tank contains 80 gallons of water.

Identifying Linear Functions

9. Yes; $a = \dfrac{1}{2}$, $b = -6$

11. No

13. Yes; $a = 0$, $b = -9$

15. Yes; $a = -9$, $b = 0$

17. Yes. The graph is a straight line.

19. No. The graph is not a straight line.

21. Yes. For each unit increase in x, the values of $f(x)$ increase by 3 units, so $a = \dfrac{3}{1} = 3$.

Because $f(x) = -6$ when $x = 0$, the y -intercept is $b = -6$. The function can be written $f(x) = 3x - 6$.

23. Yes. For each 2-unit increase in x, the values of $f(x)$ decrease by 3 units, so $a = -\dfrac{3}{2}$.

Because $f(x) = 3$ when $x = 0$, the y -intercept is $b = 3$. The function can be written $f(x) = -\dfrac{3}{2}x + 3$.

25. No. For each unit increase in x, the values of $f(x)$ do not increase by a constant amount.

27. Yes. For each unit increase in x, the values of $f(x)$ increase by 2 units, so $a = \dfrac{2}{1} = 2$.

Because $f(x) = -2$ when $x = 0$, the y-intercept is $b = -2$. The function can be written $f(x) = 2x - 2$.

Evaluating Linear Functions

29. $f(-4) = 4(-4) = -16$; $f(5) = 4(5) = 20$

31. $f\left(-\dfrac{2}{3}\right) = 5 - \left(-\dfrac{2}{3}\right) = \dfrac{15}{3} + \dfrac{2}{3} = \dfrac{17}{3}$; $f(3) = 5 - 3 = 2$

33. $f\left(-\dfrac{3}{4}\right) = -22$; $f(13) = -22$

35. $f(-1) = -2$; $f(0) = 0$

37. $f(-2) = -1$; $f(4) = -4$

39. $f(-3) = 1$; $f(1) = 1$

41. $f(x) = 6x$; $f(3) = 6(3) = 18$

43. $f(x) = \dfrac{x}{6} - \dfrac{1}{2}$; $f(3) = \dfrac{(3)}{6} - \dfrac{1}{2} = \dfrac{1}{2} - \dfrac{1}{2} = 0$

Representing Linear Functions

45. The graph should have a positive slope and pass through (0, 0). d

47. The graph should have a positive slope and pass through (0, –2). b

49. See Figure 49.

51. See Figure 51.

53. See Figure 53.

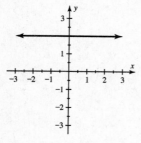

Figure 49

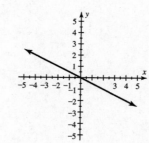

Figure 51

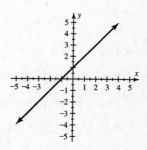

Figure 53

55.   See Figure 55.

57.   See Figure 57.

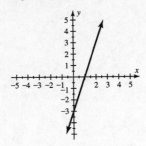

Figure 55

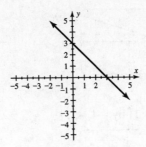

Figure 57

59.   Since each pound is divided into 16 ounces: $f(x) = \dfrac{1}{16}x$

61.   Since the car travels 65 miles each hour: $f(t) = 65t$

63.   Since every day has 24 hours: $f(x) = 24$

65.   *a*

67.   (a) *f* multiplies the input *x* by –2 and then adds 1 to obtain the output *y*.

(b)

x	–2	0	2
$y = f(x)$	5	1	–3

(c)

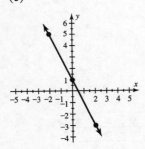

69.   (a) *f* multiplies the input *x* by $\frac{1}{2}$ and then subtracts 1 to obtain the output *y*.

(b)

x	–2	0	2
$y = f(x)$	–2	–1	0

(c)

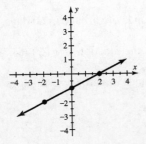

71.   (a)   (b)

[−6, 6, 1] by [−4, 4, 1]

    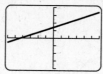

73.   (a)   (b)

[−6, 6, 1] by [−4, 4, 1]

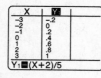   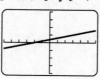

Modeling

75.   The graph should increase since the cost of tuition has been rising but it should not start at zero. b

77.   The graph should be a horizontal line since this distance has not changed over the past 10 years. c

Applications

79.   (a) Symbolic: $f(x) = 70$

Graphical: A graph of the function is shown in Figure 79a.

(b) A table of the function is shown in Figure 79b.

(c) The function f is a constant function.

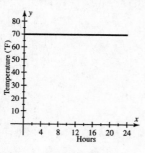

Hours	0	4	8	12	16	20	24
Temp. (°F)	70	70	70	70	70	70	70

Figure 79a   Figure 79b

81.   For each 2-hour increase in t, the distance increases by 120 miles, so $a = \dfrac{120}{2} = 60$.

Because $D = 50$ when $t = 0$, the y-intercept is $b = 50$. The function can be written $D(t) = 60t + 50$.

83.   (a) Because the average person disposed of 2.7 pounds of garbage each day, the function is $f(x) = 2.7x$.

$f(60) = 2.7(60) = 162$; In 1960 the average person disposed of 162 pounds of garbage in 60 days.

(b) Because the average person disposed of 4.3 pounds of garbage each day, the function is $g(x) = 4.3x$.

$g(60) = 4.3(60) = 258$; In 2003 the average person disposed of 258 pounds of garbage in 60 days.

85. (a) See Figure 85a.

[1, 7, 1] by [0, 300, 100]

Figure 85a

(b) Malware increased.

(c) $N(3) = 16(3) + 136 = 48 + 136 = 184$;

In March there were 184 malware viruses.

(d) The number of malware viruses is increasing at a rate of 16 per month.

87. (a) Because each 1°C increase in temperature results in a 0.5 cubic centimeter increase in volume, $a = 0.5$.

Because the volume is 137 cubic centimeters when the temperature is 0°C, the y-intercept is $b = 137$.

The formula is $V(T) = 0.5T + 137$.

(b) $V(50) = 0.5(50) + 137 = 25 + 137 = 162$ cubic centimeters.

89. $f(x) = 40x$

In 30 days each additional pound of muscle will burn $40(30) = 1200$ calories. Then the amount of muscle

necessary to lose 1 pound of fat is $3500 \div 1200 \approx 2.92$ lb.

91. (a) From the table, the average length of a baseball game in 2000 was 180 minutes.

(b) Each year the average length of a game decreased by 4 minutes.

(c) Because the constant change is a decrease of 4 minutes, $a = -4$. Because the initial length is 180,

$b = 180$. The formula is $f(x) = -4x + 180$.

(d) The year 2004 corresponds to $x = 4$. $f(4) = -4(4) + 180 = -16 + 180 = 164$ minutes.

Checking Basic Concepts for Sections 2.1 & 2.2

1. Symbolic: $f(x) = x^2 - 1$

Graphical: The graph is shown in Figure 1.

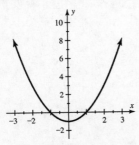

Figure 1

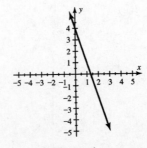

Figure 4

2. (a) $D: -3 \le x \le 3; \ R: -4 \le y \le 4$

(b) $f(0) = 0;\ f(2) = 4$

(c) No. The graph is not a straight line.

3.   (a) Yes. The function is of the form $f(x) = ax + b$.

(b) No. The function can not be written in the form $f(x) = ax + b$.

(c) Yes. The function could be written $f(x) = 0x - 7$ which is of the form $f(x) = ax + b$.

(d) Yes. The function could be written $f(x) = 3x + 9$ which is of the form $f(x) = ax + b$.

4.   The graph is shown in Figure 4. $f(-2) = 4 - 3(-2) = 4 + 6 = 10$.

5.   For each 2-unit increase in x, the values of $f(x)$ increase by 1 unit, so $a = \dfrac{1}{2}$.

Because $f(x) = -1$ when $x = 0$, the y-intercept is $b = -1$. The function can be written $f(x) = \dfrac{1}{2}x - 1$.

6.   (a) $f(20) = 0.264(20) + 27.7 = 5.28 + 27.7 = 32.98$; In 1990 the median age was about 33 years.

(b) The number 0.264 means that the median age increased by 0.264 year each year. The number 27.7 means that the initial median age in 1970 was 27.7 years.

2.3: The Slope of a Line

Concepts

1. $y;\ x$

2.   0

3.   rises

4.   horizontal

5.   vertical

6.   slope-intercept

Slope

7.   2. The *y*-value increases 2 units for every 1 unit increase in the *x*-value.

9. $-\dfrac{2}{3}$. The *y*-value decreases 2 units for every 3 units of increase in the *x*-value.

11.   0. This is a horizontal line.

13. $m = \dfrac{4-2}{2-1} = \dfrac{2}{1} = 2$

15. $m = \dfrac{3-1}{-1-2} = \dfrac{2}{-3} = -\dfrac{2}{3}$

17. $m = \dfrac{6-6}{4-(-3)} = \dfrac{0}{7} = 0$

19. $m = \dfrac{\frac{13}{17} - \left(-\frac{2}{7}\right)}{\frac{1}{2} - \frac{1}{2}} = \dfrac{\frac{125}{119}}{0} \Rightarrow$ Undefined

21. $m = \dfrac{\frac{1}{3} - \left(-\frac{4}{3}\right)}{\frac{1}{6} - \frac{1}{3}} = \dfrac{\frac{1}{3} + \frac{4}{3}}{\frac{1}{6} - \frac{2}{6}} = \dfrac{\frac{5}{3}}{-\frac{1}{6}} = \dfrac{5}{3} \cdot \left(-\dfrac{6}{1}\right) = -\dfrac{30}{3} = -10$

23. $m = \dfrac{16 - 10}{1999 - 1989} = \dfrac{6}{10} = \dfrac{3}{5}$

25. $m = \dfrac{4.3 - 3.6}{-1.2 - 2.1} = \dfrac{0.7}{-3.3} = -0.\overline{21}$

27. $m = \dfrac{4b - b}{3a - 2a} = \dfrac{3b}{a}$

29. $m = \dfrac{b - 0}{a - (a + b)} = \dfrac{b}{-b} = -1$

31.   (a) $m = \dfrac{-1 - 2}{3 - (-1)} = \dfrac{-3}{4}$

     (b)   See Figure 31b.

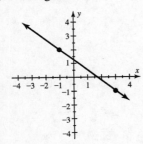

Figure 31b

     (c)   The graph falls 3 units for every 4-unit increase in x.

33.   (a) $m = \dfrac{0 - 2}{-3 - 0} = \dfrac{2}{3}$

     (b)   See Figure 33b.

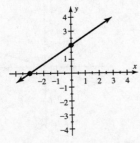

Figure 33b

     (c)   The graph rises 2 units for every 3-unit increase in x.

35. See Figure 35.

37. See Figure 37.

39. See Figure 39.

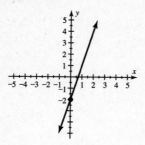

Figure 35

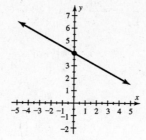

Figure 37

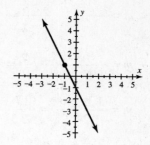

Figure 39

41. See Figure 41.

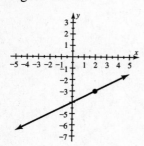

Figure 41

Slope-Intercept Form

43. (a) $m = 1$; $b = 2$

    (b) See Figure 43.

45. (a) $m = -3$; $b = 2$

    (b) See Figure 45.

47. (a) $m = \dfrac{1}{3}$; $b = 0$

    (b) See Figure 47.

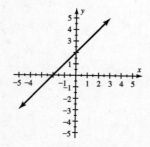

Figure 43

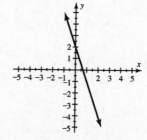

Figure 45

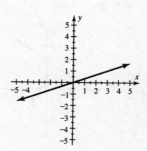

Figure 47

49.   (a) $m = 0;\ b = 2$

(b)   See Figure 49.

51.   (a) $m = -1;\ b = 3$

(b)   See Figure 51.

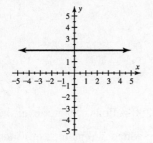

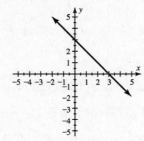

Figure 49   Figure 51

53.   The slope is -1 since the y-value decreases 1 unit for every 1–unit increase in the x-value.

The y-intercept is 4 since the graph crosses the y-axis at $(0, 4)$. The equation is $y = -x + 4$.

55.   The slope is 0 since this is a horizontal line.

The y-intercept is 2 since the graph crosses the y-axis at $(0, 2)$. The equation is $y = 2$.

57.   The slope is 1 since the y-value increases 1 unit for every 1–unit increase in the x-value.

The y-intercept is -2 since the graph passes through $(0, -2)$. The equation is $y = x - 2$.

59. $y = 3x - 5$

61.   The slope is $m = \dfrac{-\frac{3}{2} - 0}{0 - 1} = \dfrac{-\frac{3}{2}}{-1} = \dfrac{3}{2}$. The y-intercept is $-\dfrac{3}{2}$ since the graph crosses the y-axis at $\left(0, -\dfrac{3}{2}\right)$.

The equation is $y = \dfrac{3}{2}x - \dfrac{3}{2}$.

63. $m = \dfrac{0 - (-2)}{1 - 0} = \dfrac{2}{1} = 2;\ b = -2$

65. $m = \dfrac{-10 - (-3)}{2 - 1} = \dfrac{-7}{1} = -7;$ By substituting $x = 1$ and $y = -3$ into $y = -7x + b$, we find that $b = 4$.

67.   (a)   The missing value is 3 since the y-value increases 2 units for every 1–unit increase in the x-value

$(m = 2)$.

(b)   The y-intercept is -1 since the graph passes through $(0, -1)$. The equation is $f(x) = 2x - 1$.

69.   (a)   The missing number is 5 since the y-value increases 9 units for every 6 units of increase in the x-value

$\left(m = \dfrac{3}{2}\right)$.

(b)   The y-intercept is 5 since the graph passes through $(0, 5)$. The equation is $f(x) = \dfrac{3}{2}x + 5$.

Interpreting Slope

71.   The graph should be increasing since a person's pay increases with time and it should start at $(0, 0)$. c

73.   The graph should be increasing since the world's population is increasing but it should not start at $(0, 0)$. b

75.   (a) $m_1 = 50$, $m_2 = 0$, $m_3 = 150$ $\left(\text{i.e. } m_1 = \dfrac{300-100}{4-0} = \dfrac{200}{4} = 50 \right)$

       (b) m_1: the pump is adding water at the rate of 50 gallons per hour.

            m_2: the pump is turned off

            m_3: the pump is adding water at the rate of 150 gallons per hour.

       (c) Initially the pool contained 100 gallons of water. The pump added 200 gallons of water over the first 4 hours. Then the pump was turned off for 2 hours. Finally, the pump added 300 gallons of water over the last 2 hours.

77.   (a) $m_1 = 100$, $m_2 = 25$, $m_3 = -100$ $\left(\text{i.e. } m_1 = \dfrac{300-100}{2-0} = \dfrac{200}{2} = 100 \right)$

       (b) m_1: the pump is adding water at the rate of 100 gallons per hour.

            m_2: the pump is adding water at the rate of 25 gallons per hour.

            m_3: the pump is removing water at the rate of 100 gallons per hour.

       (c) Initially the pool contained 100 gallons of water. The pump added 200 gallons of water over the first 2 hours. Then the pump added 100 gallons of water over the next 4 hours. Finally, the pump removed 200 gallons of water over the last 2 hours.

79.   (a) $m_1 = 50$, $m_2 = 0$, $m_3 = -20$, $m_4 = 0$ $\left(\text{i.e. } m_1 = \dfrac{50-0}{1-0} = \dfrac{50}{1} = 50 \right)$

       (b) m_1: the car is moving away from home at a rate of 50 miles per hour.

            m_2: the car is not moving.

            m_3: the car is moving toward home at a rate of 20 miles per hour.

            m_4: the car is not moving.

       (c) The car started at home and moved away from home at 50 mph for 1 hour to a location 50 miles from home.

            The car was then parked for 1 hour. Next the car moved toward home at 20 mph for 2 hours to a location 10 miles from home. Finally, the car was parked for 1 hour.

81.   See Figure 81.

83.   See Figure 83.

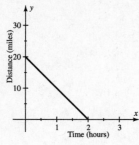

Figure 81

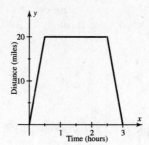

Figure 83

Applications

85.   (a) $f(1993) = 4(1993) - 7910 = 62$ thousand

(b) See Figure 85. The graph is increasing.

(c) $m = 4$

(d) The number of children born to older mothers is increasing by about 4000 each year.

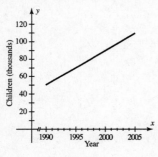

Figure 85

87.   (a) The year 2002 corresponds to $t = 15$. $N(15) = -358.4(15) + 13,723 = 8347$ banks.

(b) The slope is -358.4.   The number of federally insured banks decreased by 358.4 banks per year on average.

(c) The y-intercept is 13,723. Initially there were 13,723 federally insured banks in 1987.

89.   Since the graph passes through the points (0, 0) and (1, 10), the slope is $m = \dfrac{10-0}{1-0} = \dfrac{10}{1} = 10.$

The carpet costs $10 per square yard.

91.   At the beginning of the 5-hour period, the pool contained $\dfrac{1}{4} \cdot 20,000 = 5000$ gallons of water. At the end of

the 5-hour period, the pool contained $\dfrac{5}{8} \cdot 20,000 = 12,500$ gallons of water. That is, the amount of water in

the pool increased by $12,500 - 5000 = 7500$ gallons in 5 hours. The rate is $\dfrac{7500}{5} = 1500$ gallons per hour.

93.   (a) The year 1987 corresponds to $x = 0$ and the year 2004 corresponds to $x = 17$.

$$\text{slope} = \frac{12.1 - 3.8}{17 - 0} = \frac{8.3}{17} = \frac{83}{170}, \text{ so } a = \frac{83}{170}.$$

The graph of the line $y = \frac{83}{170}x + b$ goes through $(0, 3.8)$, so $b = 3.8$.

(b) The year 2000 corresponds to $x = 13$.

$$f(13) = \frac{83}{170}(13) + 3.8 \approx 10.1$$

Using the model, the percentage for 2000 is about 10.1%.

This estimate is 0.4% too high.

95. (a) The value of a is equal to the slope of the line passing through the points $(0, 37,000)$ and $(15, 60,000)$.

$$a = \frac{60,000 - 37,000}{15} = \frac{23,000}{15} \approx 1533.3. \text{ The value of } b \text{ is the initial value in 1990, } b = 37,000.$$

(b) The year 2000 corresponds to $x = 10$. $f(10) = 1533.3(10) + 37,000 = 15,333 + 37,000 = \$52,333$

2.4: Equations of Lines and Linear Models

Concepts

1.   One

2.   One

3. $y = mx + b$

4. $y = m(x - x_1) + y_1$ or $y - y_1 = m(x - x_1)$

5. $y = b$

6. $x = h$

7. $m_1 = m_2$

8. -1

9.   The line must have slope $-\frac{3}{4}$. One example is $y = -\frac{3}{4}x - 3$. *Answers may vary.*

10. The line must have slope $\frac{4}{3}$. One example is $y = \frac{4}{3}x + 2$. *Answers may vary.*

11. The line must be vertical and have an equation of the form $x = h$. For example, $x = 1$. *Answers may vary.*

12. The line must be horizontal and have an equation of the form $y = b$. For example, $y = -5$. *Answers may vary.*

13. Yes. By substituting for x in the equation we see that when $x = -3$, $y = 4$.

14. No. By substituting for x in the equation we see that when $x = 4$, $y = 0 \neq 2$.

15. No. By substituting for x in the equation we see that when $x = -4$, $y = 2 \neq 3$.

16. Yes. By substituting for x in the equation we see that when $x = 1$, $y = -13$.

17. Since $m > 0$ the graph must be increasing. d

18. Since $m < 0$ and $b \neq 0$ the graph must be decreasing but does not pass through the origin. **b**

19. Since $m < 0$ and $b = 0$ the graph must be decreasing and it must pass through the origin. **a**

20. This is the equation of a horizontal line. **c**

21. Since $m > 0$ and $b = 0$ the graph must be increasing and it must pass through the origin. **f**

22. This is the equation of a vertical line. **e**

Equations of Lines

23. From the given point, count 3 units down and 4 units to the right to return to the line. Thus $m = \dfrac{-3}{4} = -\dfrac{3}{4}$.

    Using the given point in the point-slope form gives $y = -\dfrac{3}{4}\left(x - (-3)\right) + 2$ or $y = -\dfrac{3}{4}x - \dfrac{1}{4}$.

25. From the given point, count 1 unit down and 3 units to the left to return to the line. Thus $m = \dfrac{-1}{-3} = \dfrac{1}{3}$.

    Using the given point in the point-slope form gives $y = \dfrac{1}{3}(x-1) + 3$ or $y = \dfrac{1}{3}x + \dfrac{8}{3}$.

27. Here $m = -2$, $x_1 = 2$ and $y_1 = -3$. The equation is $y = -2(x-2) - 3$.

29. Here $m = 1.3$, $x_1 = 1990$ and $y_1 = 25$. The equation is $y = 1.3(x - 1990) + 25$.

31. Here $m = \dfrac{-1-3}{-5-1} = \dfrac{-4}{-6} = \dfrac{2}{3}$. Then either $x_1 = 1$ and $y_1 = 3$ or $x_1 = -5$ and $y_1 = -1$.

    The equation can be written as either $y = \dfrac{2}{3}(x-1) + 3$ or $y = \dfrac{2}{3}(x+5) - 1$.

33. Here $m = \dfrac{45-5}{2000-1980} = \dfrac{40}{20} = 2$. Then either $x_1 = 1980$ and $y_1 = 5$ or $x_1 = 2000$ and $y_1 = 45$.

    The equation can be written as either $y = 2(x-1980) + 5$ or $y = 2(x-2000) + 45$.

35. Here $m = \dfrac{4-0}{0-6} = \dfrac{4}{-6} = -\dfrac{2}{3}$. Then either $x_1 = 6$ and $y_1 = 0$ or $x_1 = 0$ and $y_1 = 4$.

    The equation can be written as either $y = -\dfrac{2}{3}(x-6) + 0$ or $y = -\dfrac{2}{3}(x-0) + 4$.

37. $y = 2(x-1) - 2 \Rightarrow y = 2x - 2 - 2 \Rightarrow y = 2x - 4$

39. $y = \dfrac{1}{2}(x+4) + 1 \Rightarrow y = \dfrac{1}{2}x + 2 + 1 \Rightarrow y = \dfrac{1}{2}x + 3$

41. $y = 22(x - 1.5) - 10 \Rightarrow y = 22x - 33 - 10 \Rightarrow y = 22x - 43$

43. $m = \dfrac{-8-7}{2-(-3)} = \dfrac{-15}{5} = -3$

    $y = mx + b$

    $7 = -3(-3) + b$

$7 = 9 + b$

$-2 = b$

$\Rightarrow y = -3x - 2$

45. $m = \dfrac{-5-1}{4-(-4)} = \dfrac{-6}{8} = -\dfrac{3}{4}$

$y = mx + b$

$1 = -\dfrac{3}{4}(-4) + b$

$1 = 3 + b$

$-2 = b$

$\Rightarrow y = -\dfrac{3}{4}x - 2$

47. $y = -\dfrac{1}{3}(x-0) - 5 \Rightarrow y = -\dfrac{1}{3}x + 0 - 5 \Rightarrow y = -\dfrac{1}{3}x - 5$

49. $m = \dfrac{-1-(-2)}{2-3} = \dfrac{1}{-1} = -1 \Rightarrow y = -(x-3) - 2 \Rightarrow y = -x + 3 - 2 \Rightarrow y = -x + 1$

51. The *y*-intercept is given. The point-slope form is not needed. $m = \dfrac{-\frac{2}{3}-0}{0-2} = \dfrac{-\frac{2}{3}}{-2} = \dfrac{1}{3} \Rightarrow y = \dfrac{1}{3}x - \dfrac{2}{3}$

53. The *y*-intercept is given. The point-slope form is not needed. $m = \dfrac{3-0}{0-1} = \dfrac{3}{-1} = -3 \Rightarrow y = -3x + 3$

55. Parallel lines have the same slope. $m = 4 \Rightarrow y = 4(x-1) + 3 \Rightarrow y = 4x - 4 + 3 \Rightarrow y = 4x - 1$

57. Parallel lines have the same slope. $m = \dfrac{-2-3}{1-(-2)} = -\dfrac{5}{3}$

$y = mx + b$

$2 = -\dfrac{5}{3}(-3) + b$

$2 = 5 + b$

$-3 = b$

$\Rightarrow y = -\dfrac{5}{3}x - 3$

59. The product of the slopes of two perpendicular lines is –1.

$m = 3 \Rightarrow y = 3(x+3) + 5 \Rightarrow y = 3x + 9 + 5 \Rightarrow y = 3x + 14$

61. The line passing through (–1, 6) and (8, –4) has slope $\dfrac{-4-6}{8-(-1)} = -\dfrac{10}{9}$. The product of the slopes of two

perpendicular lines is –1.

$$m = \frac{9}{10} \Rightarrow y = \frac{9}{10}\left(x + \frac{1}{2}\right) - 2 \Rightarrow y = \frac{9}{10}x + \frac{9}{20} - \frac{40}{20} \Rightarrow y = \frac{9}{10}x - \frac{31}{20}$$

63.   The product of the slopes of two perpendicular lines is -1.

$$m = -\frac{1}{c} \Rightarrow y = -\frac{1}{c}x + b$$

65.   (a)   The slope of the perpendicular line must be the negative reciprocal of $\frac{1}{2}$. That is $m = -2$.

   Here $x_1 = 0$ and $y_1 = 2$. The equation is $y = -2(x - 0) + 2$ or $y = -2x + 2$.

   (b)   See Figure 65.

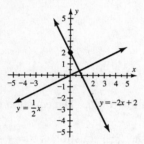

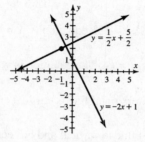

   Figure 65   Figure 67

67.   (a)   The slope of the perpendicular line must be the negative reciprocal of -2. That is $m = \frac{1}{2}$.

   Here $x_1 = -1$ and $y_1 = 2$. The equation is $y = \frac{1}{2}(x + 1) + 2$ or $y = \frac{1}{2}x + \frac{5}{2}$.

   (b)   See Figure 67.

69.   (a)   The slope of the perpendicular line must be the negative reciprocal of $-\frac{1}{3}$. That is $m = 3$.

   Here $x_1 = 1$ and $y_1 = 1$. The equation is $y = 3(x - 1) + 1$ or $y = 3x - 2$.

   (b)   See Figure 69.

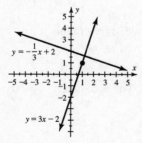

   Figure 69

71.   (a)   The slope of y_1 is $m_1 = \dfrac{2 - 0}{1 - 0} = \dfrac{2}{1} = 2$ and the y-intercept is $b = 0$. The equation is $y_1 = 2x$.

   The slope of y_2 is $m_2 = \dfrac{1 - 0}{-2 - 0} = \dfrac{1}{-2} = -\dfrac{1}{2}$ and the y-intercept is $b = 0$. The equation is $y_2 = -\dfrac{1}{2}x$.

(b) The two lines are perpendicular since $m_1 m_2 = 2\left(-\dfrac{1}{2}\right) = -1$.

73. (a) The slope of y_1 is $m_1 = \dfrac{2-(-1)}{-2-0} = \dfrac{3}{-2} = -\dfrac{3}{2}$ and the y-intercept is $b = -1$. The equation is

$$y_1 = -\dfrac{3}{2}x - 1.$$

The slope of y_2 is $m_2 = \dfrac{1-(-1)}{3-0} = \dfrac{2}{3}$ and the y-intercept is $b = -1$. The equation is $y_2 = \dfrac{2}{3}x - 1$.

(b) The two lines are perpendicular since $m_1 m_2 = -\dfrac{3}{2}\left(\dfrac{2}{3}\right) = -1$.

75. The equation of a vertical line is $x = h$. The equation is $x = -1$.

77. The equation of a horizontal line is $y = k$. The equation is $y = -\dfrac{5}{6}$.

79. A line which is perpendicular to a horizontal line is vertical and has equation $x = h$. The equation is $x = 4$.

81. A line which is parallel to a vertical line is vertical and has equation $x = h$. The equation is $x = -\dfrac{2}{3}$.

83. Yes. $m = \dfrac{0-(-4)}{2-1} = \dfrac{4}{1} = 4 \;\Rightarrow\; y = 4(x-2)+0 \;\Rightarrow\; y = 4x-8$

85. No. The change in the y-values does not remain constant as the x-values increase.

Graphical Interpretation

87. (a) Since the graph is increasing, the person is traveling away from home.

(b) $(1, 35)$: After 1 hour the person is 35 miles from home.

$(3, 95)$: After 3 hours the person is 95 miles from home.

(c) $m = \dfrac{95-35}{3-1} = \dfrac{60}{2} = 30 \;\Rightarrow\; y = 30(x-1)+35$

A slope of 30 means that the person is traveling 30 miles per hour.

89. (a) Two acres of land have 100 people and 4 acres of land have 200 people, on average.

(b) A zero-acre parcel has no people.

(c) The slope is $m = \dfrac{200-100}{4-2} = \dfrac{100}{2} = 50$, and the y-intercept is $b = 0$. The equation is $y = 50x$.

(d) The land had 50 people per acre, on average.

(e) $P(x) = 50x$

Applications

91. (a) The slope is $m = \dfrac{37,000-25,000}{2007-2003} = \dfrac{12,000}{4} = 3000$. Using the point $(2003, 25000)$ the equation is

$$y = 3000(x-2003)+25,000 \text{ or } y = 3000x - 5,984,000.$$

(b) The cost is increasing by \$3000 per year, on average.

(c) In 2005 the cost is estimated to be $y = 3000(2005) - 5,984,000 = \$31,000$.

93. First find a linear function that can be used to model these data. Then use the function to find the value in 2008.

The slope is $m = \dfrac{60 - 57}{2010 - 2005} = \dfrac{3}{5} = 0.6$. Using the point (2005, 57) the equation is

$f(x) = 0.6(x - 2005) + 57$ or $f(x) = 0.6x - 1146$.

The estimated chicken consumption in 2008 is $f(2008) = 0.6(2008) - 1146 = 58.8$ pounds.

95. First find a linear function that can be used to model these data. Then use the function to find the value in 1970.

The slope is $m = \dfrac{77 - 48}{2000 - 1900} = \dfrac{29}{100} = 0.29$. Using the point (1900, 48) the equation is

$f(x) = 0.29(x - 1900) + 48$ or $f(x) = 0.29x - 503$.

The estimated life expectancy of a baby born in 1970 is $f(1970) = 0.29(1970) - 503 = 68.3$ years.

97. Using the points (1, 1) and (4, 13) the slope is $m = \dfrac{13 - 1}{4 - 1} = \dfrac{12}{3} = 4$.

Then using the point (1, 1) the equation is $y = 4(x - 1) + 1$ or $y = 4x - 3$.

99. Using the points (1, 8) and (5, –6) the slope is $m = \dfrac{-6 - 8}{5 - 1} = \dfrac{-14}{4} = -3.5$.

Then using the point (1, 8) the equation is $y = -3.5(x - 1) + 8$ or $y = -3.5x + 11.5$.

101. Using the points (1998, 1.4) and (2002, 1.8) the slope is $m = \dfrac{1.8 - 1.4}{2002 - 1998} = \dfrac{0.4}{4} = 0.1$.

Then using the point (1998, 1.4) the equation is $y = 0.1(x - 1998) + 1.4$ or $y = 0.1x - 198.4$.

103. Using the points (1950, 1.7) and (1980, 4.4) the slope is $m = \dfrac{4.4 - 1.7}{1980 - 1950} = \dfrac{2.7}{30} = 0.09$.

Then using the point (1950, 1.7) the equation is $y = 0.09(x - 1950) + 1.7$ or $y = 0.09x - 173.8$.

105. (a) The cost per mile is $a = 0.30$. The fixed cost is $b = 189.20$.

(b) The y-intercept represents the fixed cost of owning the car for one month.

107. (a) $m = \dfrac{7 - 5}{5.4 - 4.9} = \dfrac{2}{0.5} = 4$

$\Rightarrow R(c) = 4(c - 4.9) + 5 \Rightarrow R(c) = 4c - 19.6 + 5 \Rightarrow R(c) = 4c - 14.6$

(b) $R(6.16) = 4(6.16) - 14.6 = 10.04$; the ring size is about 10.

109. (a) See Figure 109.

(b) Using the points (1970, 203) and (2000, 281) the slope is $m = \dfrac{281-203}{2000-1970} = \dfrac{78}{30} = 2.6.$

Then using the point (1970, 203), $x_1 = 1970$ and $y_1 = 203.$ *Answers may vary.*

(c) The population in 2010 is estimated to be $f(2010) = 2.6(2010-1970) + 203 = 307$ million.

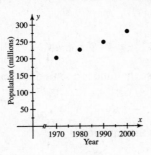

Figure 109

Checking Basic Concepts for Sections 2.3 & 2.4

1. (a) The equation is in point-slope form, $m = -3.$ One point on this line is (5, 7). *Answers may vary.*

(b) This is a horizontal line, $m = 0.$ One point on this line is (1, 10). *Answers may vary.*

(c) This is a vertical line, m is undefined. One point on this line is (–5, 0). *Answers may vary.*

(d) The equation is in slope-intercept form, $m = 5.$ One point on this line is (0, 3). *Answers may vary.*

2. (a) $m = \dfrac{2-(-4)}{5-2} = \dfrac{6}{3} = 2$

(b) For the point-slope form of the equation either $y = 2(x-2)-4$ or $y = 2(x-5)+2.$

The slope-intercept form is $y = 2(x-2)-4 \implies y = 2x-4-4 \implies y = 2x-8.$

(c) x-intercept: $0 = 2x-8 \implies 8 = 2x \implies x = 4;$ y-intercept: $y = 2(0)-8 \implies y = -8$

3. A vertical line has an equation of the form $x = h.$ The equation is $x = -2.$

A horizontal line has an equation of the form $y = k.$ The equation is $y = 5.$

4. The given line has slope $m = -\dfrac{1}{2};$ thus a parallel line has slope $m = -\dfrac{1}{2}$ and a perpendicular line has slope

$m = 2.$

Perpendicular: $y = 2(x-2)+(-4) \implies y = 2x-4-4 \implies y = 2x-8$

Parallel: $y = -\dfrac{1}{2}(x-2)+(-4) \implies y = -\dfrac{1}{2}x+1-4 \implies y = -\dfrac{1}{2}x-3$

5. (a) Since the line is decreasing, the distance from home is decreasing. The car is moving toward home.

(b) $m = -50$ since the y-value decreases 50 units for every unit increase in the x-value. The car is moving

toward home at 50 mph.

(c) x-intercept: 5; after 5 hours the car is home. y-intercept: 250; The car is initially 250 miles from home.

(d) The slope is $a = -50$ and the y-intercept is $b = 250$.

(e) $D: 0 \le x \le 5;\ R: 0 \le y \le 250$

Chapter 2 Review Exercises

Section 2.1

1. $f(-2) = 3(-2) - 1 = -7;\ f\left(\dfrac{1}{3}\right) = 3\left(\dfrac{1}{3}\right) - 1 = 0$

2. $f(-3) = 5 - 3(-3)^2 = 5 - 27 = -22;\ f(1) = 5 - 3(1)^2 = 5 - 3 = 2$

3. $f(0) = \sqrt{0} - 2 = -2;\ f(9) = \sqrt{9} - 2 = 3 - 2 = 1$

4. $f(-5) = 5;\ f\left(\dfrac{7}{5}\right) = 5$

5. (a) Since there are 2 pints in a quart, $P(q) = 2q$.

    (b) $P(5) = 2(5) = 10$. There are 10 pints in 5 quarts.

6. (a) Three less than four times a number is written $f(x) = 4x - 3$.

    (b) $f(5) = 4(5) - 3 = 17$. Three less than four times five is 17.

7. $(3, -2)$

8. $f(4) = -6$; the answers are 4 and –6.

9. See Figure 9.

10. See Figure 10.

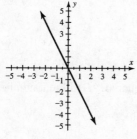

Figure 9

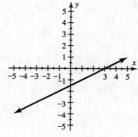

Figure 10

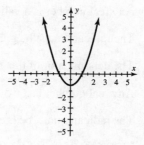

Figure 11

11. See Figure 11.

12. See Figure 12.

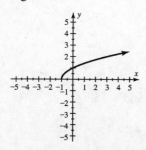

Figure 12

13. $f(0) = 1;\ f(-3) = 4$

14. $f(-2) = 1;\ f(1) = -2$

15. $f(-1) = 7;\ f(3) = -1$

16.   Numerical: The table is shown in Figure 16a.

Symbolic: $f(x) = 3x - 2$

Graphical: The graph is shown in Figure 16b.

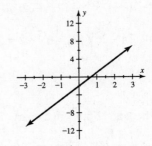

x	-3	-2	-1	0	1	2	3
$y = f(x)$	-11	-8	-5	-2	1	4	7

Figure 16a

Figure 16b

17. D: all real numbers; R: $y \leq 4$

18. $D: -4 \leq x \leq 4;\ R: -4 \leq y \leq 0$

19.   Yes. The graph passes the vertical line test.

20.   No. The graph does not pass the vertical line test.

21. $D = \{-3, -1, 2, 4\};\ R = \{-1, 3, 4\}$; Yes, S is a function since each input has exactly one output.

22. $D = \{-1, 0, 1, 2\};\ R = \{-2, 2, 3, 4, 5\}$; No, S is not a function because the input -1 has more than one output.

23.   Any real number is a valid input for this function. D: all real numbers.

24.   The radicand must be greater than or equal to zero. $D: x \geq 0$.

25.   The denominator of this function cannot equal zero. $D: x \neq 0$.

26.   Any real number is a valid input for this function because the variable is squared. D: all real numbers.

27.   The radicand must be greater than or equal to zero. $D: x \leq 5$

28.   The denominator of this function cannot equal zero. $D: x \neq -2$

29.   Any real number is a valid input for this function. D: all real numbers

30.   Any real number is a valid input for this function. D: all real numbers

Section 2.2

31.   No. The graph is not a straight line.

32.   Yes. The graph is a straight line.

33.   This function is linear because it is in the form $f(x) = ax + b$ with $a = -4$ and $b = 5$.

34.   This function is linear because it can be written in the form $f(x) = ax + b$ with $a = -1$ and $b = 7$.

35.   This function is not linear because it contains a square root.

36.   This function is linear because it can be written in the form $f(x) = ax + b$ with $a = 0$ and $b = 6$.

37.   Yes. For each 2-unit increase in x, the values of $f(x)$ increase by 3 units, so $a = \frac{3}{2}$.

Because $f(x) = -3$ when $x = 0$, the y-intercept is $b = -3$. The function can be written $f(x) = \frac{3}{2}x - 3$.

38.   No. For each unit increase in x, the values of $f(x)$ do not increase by a constant amount.

39. $f(-4) = \frac{1}{2}(-4) + 3 = -2 + 3 = 1$

40. $f(-2) = -3$ and $f(1) = 0$

41.   See Figure 41.

42.   See Figure 42.

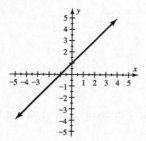
Figure 41

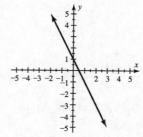
Figure 42

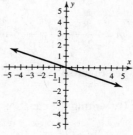
Figure 43

43.   See Figure 43.

44.   See Figure 44.

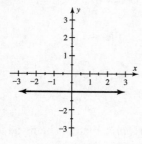
Figure 44

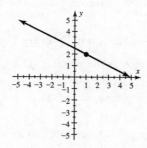
Figure 55

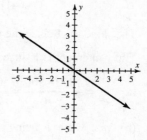
Figure 56

45.   There are 24 hours in each day, so $H(x) = 24x$.

$H(2) = 24(2) = 48$; there are 48 hours in 2 days.

46.   (a) See Figure 46a.   (b) See Figure 46b. Domain: $x \geq -2$

[–10, 10, 1] by [–10, 10, 1]

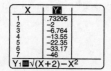
Figure 46a

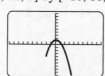
Figure 46b

Section 2.3

47.   –2. The y-value decreases 2 units for every 1–unit increase in the x-value.

48.   0. This is a horizontal line.

49. $\dfrac{1}{3}$. The y-value increases 1 unit for every 3 units of increase in the x-value.

50.   Undefined. This is a vertical line.

51. $m = \dfrac{8-2}{3-(-1)} = \dfrac{6}{4} = \dfrac{3}{2}$

52. $m = \dfrac{-\frac{1}{2}-\frac{5}{2}}{1-(-3)} = \dfrac{-\frac{6}{2}}{4} = \dfrac{-3}{4} = -\dfrac{3}{4}$

53. $m = \dfrac{-4-(-4)}{5-3} = \dfrac{0}{2} = 0$

54. $m = \dfrac{8-6}{-2-(-2)} = \dfrac{2}{0} \Rightarrow$ undefined

55.   See Figure 55.

56.   By writing the equation in the form $y = -\dfrac{2}{3}x+0$, the slope is $-\dfrac{2}{3}$ and the y-intercept is 0. See Figure 56.

57.   The slope is –3 since the y-value decreases 3 units for every 1–unit increase in the x-value.
       The y-intercept is 1 since the graph crosses the y-axis at (0, 1). The equation is $y = -3x+1$.

58.   The slope is $m = \dfrac{2-0}{0-(-1)} = \dfrac{2}{1} = 2$. The y-intercept is 2. The equation is $y = 2x+2$.

59.   The slope is –1 since the y-value decreases 1 unit for every 1–unit increase in the x-value.
       The y-intercept is 1 since the graph crosses the y-axis at (0, 1).

60.   (a) $m_1 = \dfrac{1000-500}{2-0} = \dfrac{500}{2} = 250$; Similarly, $m_2 = 500$, $m_3 = 0$, $m_4 = -750$

       (b) m_1: Water is being added to the pool at a rate of 250 gallons per hour. m_2: Water is being added to the pool at a rate of 500 gallons per hour. m_3: No water is being added or removed. m_4: Water is being removed from the pool at a rate of 750 gallons per hour.

       (c) Initially the pool contains 500 gallons of water. For the first two hours water is added to the pool at a rate of 250 gallons per hour until there are 1000 gallons in the pool. Then water is added at a rate of 500 gallons per hour for 1 hour until there are 1500 gallons in the pool. For the next hour, no water is added or removed. Finally, water is removed from the pool at a rate of 750 gallons per hour for 1 hour until it contains 750 gallons.

61.   See Figure 61.

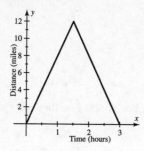

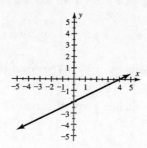

Figure 61   Figure 62

62.   (a) The equation is in the form $f(x) = mx + b$. The slope is $\frac{1}{2}$ and the y-intercept is -2.

(b) See Figure 62.

Section 2.4

63.   Yes, the point $(2, 1)$ lies on the graph because when $x = 2$, $y = \frac{3}{2}(2) - 2 = 3 - 2 = 1$.

64.   The slope is $m = \dfrac{4 - 6}{0 - (-1)} = \dfrac{-2}{1} = -2$. From the table, the x-intercept is 2 and the y-intercept is 4.

65. $y = -3(x - 2) + 1 \;\Rightarrow\; y = -3x + 6 + 1 \;\Rightarrow\; y = -3x + 7$

66. $y = -3(x + 2) + 3$; $y = -3x - 6 + 3 \;\Rightarrow\; y = -3x - 3$

67.   Here $m = \dfrac{-3 - 0}{0 - 2} = \dfrac{-3}{-2} = \dfrac{3}{2}$. Use $x_1 = 2$ and $y_1 = 0$ in the point-slope form to get the slope-intercept form.

The slope-intercept equation is $y = \dfrac{3}{2}(x - 2) + 0 \;\Rightarrow\; y = \dfrac{3}{2}x - 3$.

68.   Here $m = \dfrac{-2 - 4}{2 - (-1)} = \dfrac{-6}{3} = -2$. Use $x_1 = -1$ and $y_1 = 4$ in the point-slope form to get the slope-intercept

form.

The slope-intercept equation is $y = -2(x + 1) + 4 \;\Rightarrow\; y = -2x + 2$.

69.   Parallel lines have the same slope, thus $m = 4$. Use $x_1 = -\dfrac{3}{5}$ and $y_1 = \dfrac{1}{5}$ in the point-slope form to get the

slope-intercept form. The slope-intercept equation is $y = 4\left(x + \dfrac{3}{5}\right) + \dfrac{1}{5} \;\Rightarrow\; y = 4x + \dfrac{12}{5} + \dfrac{1}{5} \;\Rightarrow\; y = 4x + \dfrac{13}{5}$.

70.   The product of the slopes of two perpendicular lines is -1 thus $m = -2$. Then $x_1 = -1$ and $y_1 = 1$.

The slope-intercept equation is $y = -2(x + 1) + 1 \;\Rightarrow\; y = -2x - 2 + 1 \;\Rightarrow\; y = -2x - 1$.

71.   By inspection we see that the slope is $m = -1$. The y-intercept is 2. The equation is $y = -x + 2$.

72.   By inspection we see that the slope is $m = 2$. The y-intercept is -3. The equation is $y = 2x - 3$.

73.   No. By substituting for x in the equation we see that when $x = 2$, $y = 1 \neq -1$.

74. Yes. By substituting for x in the equation we see that when $x = 4$, $y = -\dfrac{5}{2}$.

75. The equation of a vertical line is $x = h$. The equation is $x = -4$.

76. The equation of a horizontal line is $y = k$. The equation is $y = -\dfrac{7}{13}$.

77. The line $x = -3$ is vertical, so a line perpendicular to it is horizontal. The horizontal line through $(-2, 1)$ is

    $y = 1$.

78. The line $y = 5$ is horizontal, so a line parallel to it is also horizontal. The horizontal line through $(4, -8)$ is

    $y = -8$.

79. No. The change in the y-values does not remain constant as the x-values increase.

80. Yes. $m = \dfrac{5-1}{0-(-2)} = \dfrac{4}{2} = 2 \;\Rightarrow\; y = 2(x-0) + 5 \;\Rightarrow\; y = 2x + 5$

Applications

81. (a) $m_1 = 1.3, m_2 = 2.35, m_3 = 2.4, m_4 = 2.7$ $\left(\text{i.e. } m_1 = \dfrac{132-106}{1940-1920} = \dfrac{26}{20} = 1.3\right)$

    (b) m_1: from 1920 to 1940 the population increased on average by 1.3 million per year.

    m_2: from 1940 to 1960 the population increased on average by 2.35 million per year.

    m_3: from 1960 to 1980 the population increased on average by 2.4 million per year.

    m_4: from 1980 to 2000 the population increased on average by 2.7 million per year.

82. For the first 2 minutes the inlet pipe is open. For the next 3 minutes both pipes are open. For the next 2 minutes only the outlet pipe is open. Finally, for the last 3 minutes both pipes are closed.

83. (a) $f(1910) = -0.0492(1910) + 119.1 \approx 25.1$ years

    (b) Graph $Y_1 = -0.0492X + 119.1$ in [1885, 1965, 10] by [22, 26, 1]. See Figure 83.

    The median age at first marriage for males has decreased over this time period.

    (c) The slope is -0.0492. The median age decreased by about 0.0492 year per year.

    [1885, 1965, 10] by [22, 26, 1]

    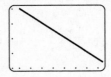

    Figure 83

84. (a) $f(1991) = 2.4$ million

    (b) The number of marriages each year did not change over this time period.

85. (a) $f(x) = 8x$

    (b) The slope of the graph of f is 8.

(c) The total fat changes at a rate of 8 grams per cup of milk.

86.   (a) See Figure 86.

(b) Using the first and last data points to find the linear function results in $f(x) = -0.31x + 633.6$.

(c) $f(2000) = -0.31(2000 - 1990) + 16.7 = -0.31(10) + 16.7 = -3.1 + 16.7 = 13.6$

The birth rate is about 13.6 per 1000 people. *Answers may vary.*

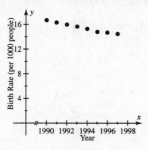

Figure 86

87.   (a) From the table $f(1995) = 113$. In 1995 there were 113 unhealthy days.

(b) $D = \{1995, 1997, 1999, 2000, 2003\}$; $R = \{56, 60, 87, 88, 113\}$

(c) The number of unhealthy days decreased and then increased over this time period.

88.   (a) See Figure 88. There is a linear relationship.

(b) Using the points (0, 32) and (100, 212) the slope is $m = \dfrac{212 - 32}{100 - 0} = \dfrac{180}{100} = \dfrac{9}{5}$. The y-intercept is 32.

The function is given by $f(x) = \dfrac{9}{5}x + 32$. The slope of $\dfrac{9}{5}$ means that

a 1°C change equals a $\dfrac{9}{5}$°F change.

(c) $f(20) = \dfrac{9}{5}(20) + 32 = 36 + 32 = 68°$ F.

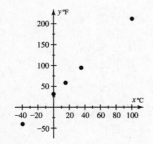

Figure 88

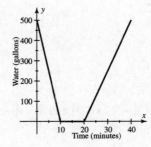

Figure 89

89.   See Figure 89.

90.   (a) Because the rate of change is 25,000 cases per year, $m = 25,000$. Here $x_1 = 2000$ and $y_1 = 875,000$.

The equation is $f(x) = 25,000(x - 2000) + 875,000$.

(b) $f(2007) = 25,000(2007 - 2000) + 875,000 = 1,050,000$ cases. This is the total (cumulative) number of

U.S. AIDS cases reported in 2007.

91.   The graph should be increasing and should initially be a positive amount. b

92.   The graph should be a horizontal line above the *x*-axis since this distance is a positive constant. e

93.   The graph should be increasing and should start at zero. a

94.   The graph should be decreasing since this film format is nearly extinct. d

95.   The graph should be a horizontal line below the *x*-axis since this temperature is a negative constant. f

96.   The graph should start below the *x*-axis and then increase to a value above the *x*-axis. c

Chapter 2 Test

1. $f(4) = 3(4)^2 - \sqrt{4} = 3(16) - 2 = 48 - 2 = 46; \ (4, 46)$

2. $C(x) = 4x; \ \ C(5) = 4(5) = 20;$ 5 pounds of candy costs $20.

3.   (a) See Figure 3a.

       (b) See Figure 3b.

       (c) See Figure 3c.

       (d) See Figure 3d.

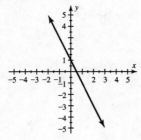

Figure 3a

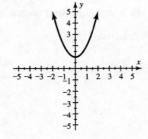

Figure 3b

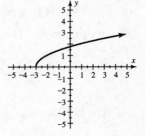

Figure 3c

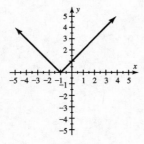

Figure 3d

4. $f(-3) = 0; \ f(0) = -3; \ \ D: \{x \mid -3 \le x \le 3\}$ and $R: \{x \mid -3 \le y \le 0\}$

5. Symbolic: $f(x) = x^2 - 5$

Numerical: The table is shown in Figure 5a.

Graphical: The graph is shown in Figure 5b.

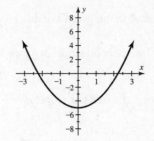

x	−3	−2	−1	0	1	2	3
$y = f(x)$	4	−1	−4	−5	−4	−1	4

Figure 5a   Figure 5b

6. No. It does not pass the vertical line test.

7. (a) $D = \{-2, -1, 0, 5\}, R = \{3, 5, 7\}$

   (b) Any real number is a valid input for this function.

   D: all real numbers

   (c) The radicand must be greater than or equal to zero.

   $D: x \geq -4$

   (d) Any real number is a valid input for this function.

   D: all real numbers

   (e) The denominator of this function cannot equal zero.

   $D: x \neq 5$

8. It is; $f(x) = -8x + 6$

9. $m = \dfrac{-2 - 7}{6 - (-3)} = \dfrac{-9}{9} = -1$

10. $-\dfrac{1}{2}$. The y-value decreases 1 unit for every 2 units of increase in the x-value.

11. (a) The line passes through $(-2, 0)$ and $(0, -4)$.

   $m = \dfrac{-4 - 0}{0 - (-2)} = \dfrac{-4}{2} = -2 \Rightarrow y = -2x - 4$

   (b) The line passing through $(-2, 5)$ and $(-1, 3)$ is $\dfrac{3 - 5}{-1 - (-2)} = \dfrac{-2}{1} = -2$. The product of the slopes of

   perpendicular lines is −1. $m = \dfrac{1}{2} \Rightarrow y = \dfrac{1}{2}(x + 5) + 2 \Rightarrow y = \dfrac{1}{2}x + \dfrac{5}{2} + 2 \Rightarrow y = \dfrac{1}{2}x + \dfrac{9}{2}$

   (c) $m = \dfrac{\frac{3}{2} - (-2)}{-5 - 1} = \dfrac{\frac{3}{2} + \frac{4}{2}}{-6} = \dfrac{\frac{7}{2}}{-6} = \dfrac{7}{2} \cdot \left(-\dfrac{1}{6}\right) = -\dfrac{7}{12}$

$$\Rightarrow y = -\frac{7}{12}(x-1)-2 \Rightarrow y = -\frac{7}{12}x - \frac{17}{12}$$

12.   The slope is $m = \frac{0-4}{2-0} = \frac{-4}{2} = -2$. From the table, the x-intercept is 2 and the y-intercept is 4.

13.   A line parallel to the given line has slope $m = -3$.

   The slope-intercept equation is $y = -3\left(x - \frac{1}{3}\right) + 2 \Rightarrow y = -3x + 1 + 2 \Rightarrow y = -3x + 3$.

14.   By inspection the line has slope $m = \frac{2}{3}$. The y-intercept is -2. The slope-intercept equation is $y = \frac{2}{3}x - 2$.

   A line perpendicular to this line that passes through the origin would have slope $m = -\frac{3}{2}$ and y-intercept

   0. The equation of the perpendicular line is $y = -\frac{3}{2}x$.

15.   A vertical line has equation $x = h$. The equation is $x = \frac{2}{3}$. A horizontal line has equation $y = k$. The

   equation is $y = -\frac{1}{7}$.

16.   (a) $m_1 = \frac{11-7}{1980-1970} = \frac{4}{10} = 0.4$; Similarly, $m_2 = 0$, $m_3 = -0.4$

   (b) m_1: From 1970 to 1980 the number of welfare beneficiaries increased by 0.4 million per year on

   average.

   m_1: From 1980 to 1990 there was no change in the number of welfare beneficiaries.

   m_1: From 1990 to 2000 the number of welfare beneficiaries decreased by 0.4 million per year on

   average.

17.   See Figure 17.

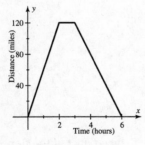

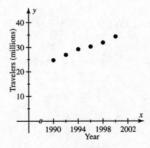

Figure 17   Figure 18

18.   (a) See Figure 18.

   (b) Using the points (1990, 24.8) and (2000, 34.4): $m = \frac{34.4 - 24.8}{2000 - 1990} = \frac{9.6}{10} = 0.96$, $x_1 = 1990$, $y_1 = 24.8$

Answers may vary.

(c) $f(2002) = 0.96(2002 - 1990) + 24.8 = 0.96(12) + 24.8 = 11.52 + 24.8 \approx 36.3$

In 2002 the number of travelers was about 36.3 million. *Answers may vary.*

Chapter 2 Extended and Discovery Exercises

1. (a) The graph for tank A is linear. See Figures 1a. The graph for tank B is nonlinear. See Figure 1b.

   (b) Tank B flows faster at first so it is the first to be half empty.

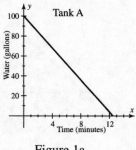
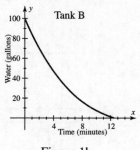
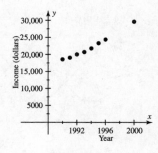

Figure 1a   Figure 1b   Figure 2

2. (a) See Figure 2.

   (b) Since the data appears to be nearly linear with an annual increase of about $1100 per year we may find a point-slope form of a linear equation using the point (1990, 18,666) and slope $m = 1100$.

   The equation is $f(x) = 1100(x - 1990) + 18,666$. *Answers may vary.*

   (c) $f(1998) = 1100(1998 - 1990) + 18,666 = 8800 + 18,666 = \$27,466$

3. The line that passes through (0, 0) and (5, 3) has slope $m = \dfrac{3-0}{5-0} = \dfrac{3}{5}$ and y-intercept 0.

   The equation for this line is $y = \dfrac{3}{5}x$. The line parallel to this line has the same slope and y-intercept 5.

   The equation for this parallel line is $y = \dfrac{3}{5}x + 5$. The perpendicular line which passes through (0, 0) has

   slope $m = -\dfrac{5}{3}$ and y-intercept 0. Its equation is $y = -\dfrac{5}{3}x$. Finally, the fourth side of the rectangle passes

   through (5, 3) and has slope $m = -\dfrac{5}{3}$. Its equation is $y = -\dfrac{5}{3}(x - 5) + 3 \Rightarrow y = -\dfrac{5}{3}x + \dfrac{34}{3}$.

4.   The line that passes through (0, 0) and (2, 2) has slope $m = \dfrac{2-0}{2-0} = \dfrac{2}{2} = 1$ and y-intercept 0.

The equation for this line is $y = x$. The line parallel to this line has the same slope and passes through (1, 3). The equation for this parallel line is $y = 1(x-1) + 3 \Rightarrow y = x + 2$. The perpendicular line which passes through (0, 0) has slope $m = -1$ and y-intercept 0. Its equation is $y = -x$. Finally, the fourth side of the rectangle passes through (1, 3) and has slope $m = -1$. Its equation is $y = -1(x-1) + 3 \Rightarrow y = -x + 4$. See Figure 4.

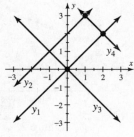

Figure 4

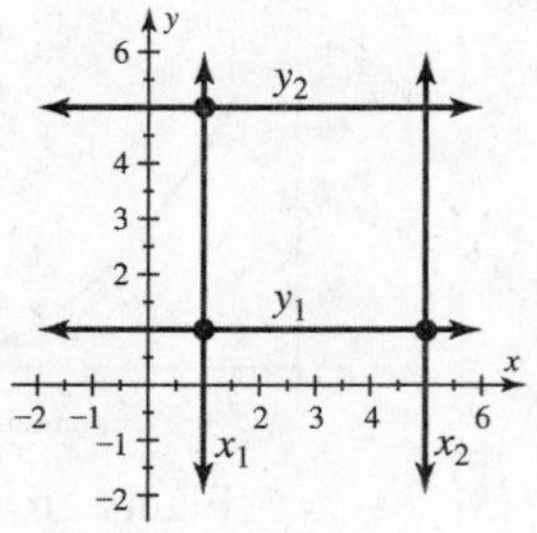

Figure 5

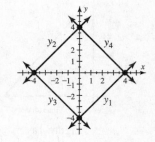

Figure 6

5.   The line that passes through (1, 1) and (5, 1) is horizontal and has equation $y = 1$. The line parallel to this line is also horizontal and passes through (5, 5). Its equation is $y = 5$. The perpendicular line which passes through the point (1, 1) is vertical and has equation $x = 1$. Finally, the other vertical line passes through (5, 1) and has equation $x = 5$. See Figure 5.

6.   The line that passes through (4, 0) and (0, 4) has slope $m = \dfrac{4-0}{0-4} = \dfrac{4}{-4} = -1$ and y-intercept 4.

The equation for this line is $y = -x + 4$. The line parallel to this line has the same slope and y-intercept –4. The equation for this parallel line is $y = -x - 4$. The perpendicular line which passes through (0, 4) has slope $m = 1$ and y-intercept 4. Its equation is $y = x + 4$. Finally, the fourth side of the rectangle has slope $m = 1$ and y-intercept –4. Its equation is $y = x - 4$. See Figure 6.

7. (a) See Figure 7.

    (b) $m_1 = -0.24$, $m_2 = -0.98$, $m_3 = -0.97$, $m_4 = -0.95$, $m_5 = -0.93$, $m_6 = -0.85$, $m_7 = -0.76$, $m_8 = -0.63$ ie. we

    may calculate the slope of the first line segment as follows: $m_1 = \dfrac{69.9 - 72.3}{10 - 0} = \dfrac{-2.4}{10} = -0.24$ m_1: from

    age 0 to age 10 the remaining life expectancy decreases at the rate of 0.24 years per year The remaining

    slopes may be interpreted in a similar way.

    (c) A 20-year-old woman is expected to live 60.1 more years making her total life expectancy 80.1 years.

    A 70-year-old woman is expected to live 15.5 more years making her total life expectancy 85.5 years.

    The 70-year-old woman has already lived through much of the risk that a 20-year-old must still face.

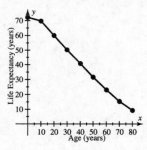

Figure 7

8. (a) When the fish hatches it weighs about 7 mg. It weighs about 105 mg at 6 weeks of age and about 158 mg

    at 12 weeks of age. *Answers may vary.*

    (b) From hatching to 6 weeks: $\dfrac{105 - 7}{6 - 0} = \dfrac{98}{6} \approx 16.3$ mg/week.

    From 6 weeks to 12 weeks: $\dfrac{158 - 105}{12 - 6} = \dfrac{53}{6} \approx 8.8$ mg/week.

    (c) On average the fish gains about 16.3 mg per week during the first 6 weeks of its life and about 8.8 mg per

    week during the second 6 weeks of its life.

    (d) The fish gains weight the fastest during the first 6 weeks of its life.

9. (a) The carbon dioxide levels are increasing.

    (b) The carbon dioxide levels oscillate each year.

    (c) Seasonal changes in plant growth affect the carbon dioxide levels on a small scale. This can be attributed

    to the use of carbon dioxide by many plants during the process of photosynthesis.

10. (a) The graph shows a general increasing trend which is similar to the graph for Hawaii but the yearly

    oscillations are larger.

    (b) The climate in Alaska is more extreme so the seasonal changes in plant growth are more dramatic.

    Both graphs show an overall increasing trend due to the use of fossil fuels and other factors.

Chapters 1–2 Cumulative Review Exercises

1. Natural number: $\sqrt[3]{8}$

    Whole number: $\sqrt[3]{8}$

Integer: $-3, \sqrt[3]{8}$

Rational number: $-3, \frac{3}{4}, -5.8, \sqrt[3]{8}$

Irrational number: $\sqrt{7}$

2.   Natural number: $\frac{50}{10}$

Whole number: $0, \frac{50}{10}$

Integer: $0, -5, -\sqrt{9}, \frac{50}{10}$

Rational number: $0, -5, \frac{1}{2}, -\sqrt{9}, \frac{50}{10}$

Irrational number: $\sqrt{8}$

3.   Distributive property; $6x + 8x = (6+8)x = 14x$

4.   Commutative property for addition

5. $-5^2 + 3 + 4 \cdot 5 = -25 + 3 + 4 \cdot 5 = -25 + 3 + 20 = -2$

6. $(x+1)x + (x+1)5 = (x+1)(x+5)$

7. $-(a-b) = -a - (-b) = -a + b$

8. $\dfrac{c}{a+b}$

9.   Entering $(1/3) \div ((3/4)(7/8)) + (5/6)$ ▶ Frac into a graphing calculator gives $\dfrac{169}{126}$.

10.   (a) $\left(\dfrac{2^{-3}}{3^{-2}}\right)^2 = \left(\dfrac{3^2}{2^3}\right)^2 = \left(\dfrac{9}{8}\right)^2 = \dfrac{81}{64}$

(b) $\dfrac{\left(3x^2 y^{-3}\right)^4}{x^3 \left(y^4\right)^{-2}} = \dfrac{3^4 x^8 y^{-12}}{x^3 y^{-8}} = 81 x^{8-3} y^{-12-(-8)} = 81 x^5 y^{-4}$

$$= \dfrac{81 x^5}{y^4}$$

(c) $\left(\dfrac{ab^{-2}}{a^{-3}b^4}\right)^{-2} = \left(\dfrac{a^{-3}b^4}{ab^{-2}}\right)^2 = \dfrac{\left(a^{-3}b^4\right)^2}{\left(ab^{-2}\right)^2} = \dfrac{a^{-6}b^8}{a^2 b^{-4}}$

$$= a^{-6-2} b^{8-(-4)} = a^{-8} b^{12} = \dfrac{b^{12}}{a^8}$$

11. 9.54×10^3

12. $f(4) = 2(4)^2 + \sqrt{4} = 2(16) + 2 = 32 + 2 = 34$

13. $B = \dfrac{1}{2}(4)(3)^2 = \dfrac{1}{2}(4)(9) = 18$

14.   (a) $D = \{-3, 0, 2\}$

   (b)   The denominator of the function cannot equal zero. $D: x \neq -6$

   (c)   The radicand of the function must be greater than or equal to zero. $D: x \geq -4$

15.   16.

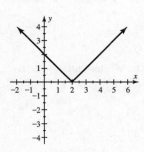

Figure 15

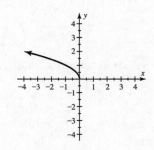

Figure 16

17.   18.

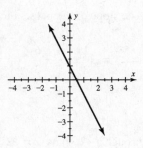

Figure 17

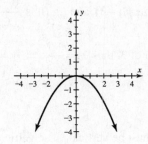

Figure 18

19.   20.

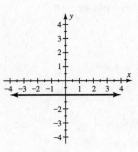

Figure 19

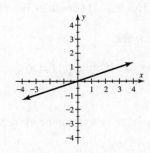

Figure 20

21.   (a)   Yes, the graph passes the vertical line test so it represents a function.

   (b) D: all real numbers; R: all real numbers

   (c)   From the graph, $f(-1) = 1.5$ and $f(0) = 1$.

   (d) x-intercept:$(2, 0)$; y-intercept:$(0, 1)$

   (e)   For each 2-unit increase in x, the values of $f(x)$ decrease by 1 unit, so $m = -\dfrac{1}{2}$.

(f) $f(x) = -\dfrac{1}{2}x + 1$

22. $m = \dfrac{-4-5}{2-(-4)} = \dfrac{-9}{6} = -\dfrac{3}{2} \;\Rightarrow\; y = -\dfrac{3}{2}(x+4) + 5$

$\Rightarrow\; y = -\dfrac{3}{2}x - 6 + 5 \;\Rightarrow\; y = -\dfrac{3}{2}x - 1$

23.   The slope of the line passing through $(1, -3)$ and $(2, 0)$ is $\dfrac{0-(-3)}{2-1} = \dfrac{3}{1} = 3$. The slope of the perpendicular

line must be the negative reciprocal of 3, thus $m = -\dfrac{1}{3} \;\Rightarrow\; y = -\dfrac{1}{3}(x+1) + 2 \;\Rightarrow\; y = -\dfrac{1}{3}x - \dfrac{1}{3} + \dfrac{6}{3}$

$\Rightarrow\; y = -\dfrac{1}{3}x + \dfrac{5}{3}.$

24.   A line perpendicular to the x-axis must be vertical. The vertical line passing through $(-2, 4)$ is $x = -2$.

25. $m = \dfrac{-3-0}{2-1} = \dfrac{-3}{1} = -3$; the x-intercept is given in the table: 1; By substituting $x = 1$ and $y = 0$ into

$y = -3x + b,$ we find that $b = 3$.

26.   For each 2-unit increase in x, the y- values decrease by 3 units, so $m = -\dfrac{3}{2}$. The y-intercept is $\dfrac{3}{2}$, so

$y = -\dfrac{3}{2}x + \dfrac{3}{2}.$

Applications

27.   Commutative property for addition; $120 + 80 = 200 = 80 + 120$

28.   Since Volume $= \text{Length} \times \text{Width} \times \text{Height},$ $2(\text{Length}) \times \text{Width} \times 2(\text{Height}) = 4(\text{Volume})$ so

$4(400 \text{ cubic inches}) = 1600 \text{ cubic inches}.$

29.   (a) We can estimate the 1995 injury rate by finding the average of the 1992 and 1998

rates. $(8.3 + 6.7) \div 2 = 15 \div 2 = 7.5$

We can estimate the 2000 injury rate by finding the average of the 1999 and 2001

rates. $(6.3 + 5.4) \div 2 = (11.7) \div 2 = 5.85$

(b) 5.85; *Answers may vary.*

30.   The person earns \$9 per hour.

31. See Figure 31.

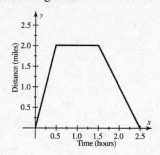

Figure 31

32. (a) See Figure 32a. Median age is increasing.

[1820, 1995, 20] by [0, 40, 10]

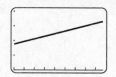

Figure 32a

(b)

Year (x)	1820	1840	1860	1880	1900	1920	1940
Median Age	16.7	18.5	20.3	22.1	23.9	25.7	27.5

$f(1900) = 23.9$; In 1900 the median age was 23.9.

(c) Each year the median age increased by 0.09 year, on average.

33. (a) $f(x) = 10x$

(b) See Figure 33b.

(c) 10

(d) Total fat increases at a rate of 10 g of fat per slice of pizza.

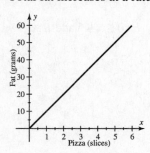

Figure 33b

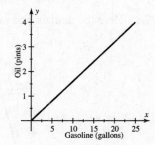

Figure 34b

34. (a) $f(6) = \dfrac{4}{25}(6) = \dfrac{24}{25} = 0.96$ pint

(b) See Figure 34b.

(c) $\dfrac{4}{25} = 0.16$

(d) 0.16 pint of oil should be added per gallon of gasoline.

35.   (a) $f(1960) = 3.4(1960 - 1960) + 87.8 = 3.4(0) + 87.8 = 87.8$. In 1960, there were 87.8 million tons of

waste.

(b) See Figure 35b.

Figure 35b

(c) 3.4; solid waste increased by 3.4 million tons per year.

36.   (a) $m_1 = \dfrac{200 - 300}{2 - 0} = \dfrac{-100}{2} = -50$

$m_2 = 0$

$m_3 = \dfrac{0 - 200}{7 - 3} = \dfrac{-200}{4} = -50$

(b) m_1: The driver is traveling toward home at a rate of 50 miles per hour.

m_2: The car is not moving.

m_3: The driver is traveling toward home at a rate of 50 miles per hour.

(c) The car started 300 miles from home and moved toward home at 50 mph for 2 hours to a location 200

miles from home. The car was then parked for 1 hour. Finally, the car moved toward home at 50 mph

for 4 hours, at which point it was home.

Critical Thinking Solutions for Chapter 2

Section 2.1

•   The graph is shown in Figure 2.1. Here the domain and range are given by $D: 0 \le x \le 5$ and $R: 0 \le y \le 250$.

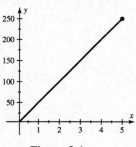

Figure 2.1

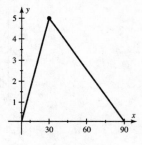

Figure 2.3

Section 2.2

•   A third point will verify that the first two points are plotted correctly.

Section 2.3

•   The graph is shown in Figure 2.3.

Section 2.4

- A graph of a function cannot have more than one y-intercept since it would not pass the vertical line test.

  A graph of a function can have more than one x-intercept since it could still pass the vertical line test.

Chapter 3: Linear Equations and Inequalities

3.1: Linear Equations

Concepts

1. $ax + b = 0$, where $a \neq 0$

2. One

3. Yes, since $4(1) - 1 = 3$ and $3(1) = 3$.

4. Yes, we obtain the equation $x = 2$ if we divide both sides of the equation $3x = 6$ by 3.

5. 4

6. -2

7. An equals sign (=)

8. Three methods for solving linear equations are numerical, graphical and symbolic.

9. The solution is the *x*-coordinate of the intersection point. That is, the solution is 3.

10. The standard form of the equation of a line is $ax + by = c$, where a and b cannot both be 0.

11. $a + c = b + c$

12. $ac = bc$, for $c \neq 0$

13. No, since $5 - 6 = -1 \neq -2$.

14. Yes, since $-\frac{1}{2}(2) + 1 = 0$ and $\frac{1}{3}(2) - \frac{2}{3} = 0$.

15. Yes, since $3\left[2\left(-\frac{2}{3}\right) + 3\right] = 6\left(-\frac{2}{3}\right) + 9 = -4 + 9 = 5$ and $\frac{13}{3} - \left(-\frac{2}{3}\right) = \frac{15}{3} = 5$.

16. No, since $2 - 1 = 1$ but $2 - (2 + 1) = 2 - 3 = -1$.

17. Yes, since $-(2(-2) - 3) + 2(-2) = 3$ and $1 - (-2) = 3$.

18. Yes, since $-\frac{1}{3}(4 - 1) + \frac{2}{5}(1 + 1) = -\frac{1}{3}(3) + \frac{2}{5}(2) = -1 + \frac{4}{5} = -\frac{1}{5}$.

Symbolic Solutions

19. $x - 4 = 10 \Rightarrow x = 14$; Check: $14 - 4 = 10 \Rightarrow 10 = 10$, so $x = 14$

21. $5 + x = 3 \Rightarrow x = -2$; Check: $5 + (-2) = 3 \Rightarrow 3 = 3$, so $x = -2$

23. $\frac{1}{2} - x = 2 \Rightarrow \frac{1}{2} - x = \frac{4}{2} \Rightarrow -x = \frac{3}{2} \Rightarrow x = -\frac{3}{2}$; Check: $\frac{1}{2} - \left(-\frac{3}{2}\right) = 2$

    $\Rightarrow \frac{4}{2} = 2 \Rightarrow 2 = 2$, so $x = -\frac{3}{2}$

25. $-6x = -18 \Rightarrow x = 3$; Check: $-6(3) = -18 \Rightarrow -18 = -18$, so $x = 3$

27. $-\dfrac{1}{3}x = 4 \Rightarrow x = -12$; Check: $-\dfrac{1}{3}(-12) = 4 \Rightarrow 4 = 4$, so $x = -12$

29. $2x + 3 = 13 \Rightarrow 2x = 10 \Rightarrow x = 5$; Check: $2(5) + 3 = 13 \Rightarrow 10 + 3 = 13 \Rightarrow 13 = 13$, so $x = 5$

31. $3x - 7 = 8 \Rightarrow 3x = 15 \Rightarrow x = 5$; Check: $3(5) - 7 = 8 \Rightarrow 15 - 7 = 8 \Rightarrow 8 = 8$, so $x = 5$

33. $2x = 8 - \dfrac{1}{2}x \Rightarrow 2x + \dfrac{1}{2}x = 8 \Rightarrow \dfrac{5}{2}x = 8 \Rightarrow x = 8\left(\dfrac{2}{5}\right) \Rightarrow x = \dfrac{16}{5}$; Check:

$2\left(\dfrac{16}{5}\right) = 8 - \dfrac{1}{2}\left(\dfrac{16}{5}\right) \Rightarrow \dfrac{32}{5} = 8 - \dfrac{8}{5} \Rightarrow \dfrac{32}{5} = \dfrac{40}{5} - \dfrac{8}{5} \Rightarrow \dfrac{32}{5} = \dfrac{32}{5}$, so $x = \dfrac{16}{5}$

35. $3x - 1 = 11(1 - x) \Rightarrow 3x - 1 = 11 - 11x \Rightarrow 14x = 12 \Rightarrow x = \dfrac{12}{14} \Rightarrow x = \dfrac{6}{7}$; Check:

$3\left(\dfrac{6}{7}\right) - 1 = 11\left(1 - \dfrac{6}{7}\right) \Rightarrow \dfrac{18}{7} - \dfrac{7}{7} = 11\left(\dfrac{7}{7} - \dfrac{6}{7}\right) \Rightarrow \dfrac{11}{7} = 11\left(\dfrac{1}{7}\right) \Rightarrow \dfrac{11}{7} = \dfrac{11}{7}$, so $x = \dfrac{6}{7}$

37. $x + 4 = 2 - \dfrac{1}{3}x \Rightarrow x + \dfrac{1}{3}x = -2 \Rightarrow \dfrac{4}{3}x = -2 \Rightarrow x = -2\left(\dfrac{3}{4}\right) \Rightarrow x = -\dfrac{3}{2}$; Check:

$-\dfrac{3}{2} + 4 = 2 - \dfrac{1}{3}\left(-\dfrac{3}{2}\right) \Rightarrow -\dfrac{3}{2} + \dfrac{8}{2} = \dfrac{4}{2} + \dfrac{1}{2} \Rightarrow \dfrac{5}{2} = \dfrac{5}{2}$, so $x = -\dfrac{3}{2}$

39. $2(x - 1) = 5 - 2x \Rightarrow 2x - 2 = 5 - 2x \Rightarrow 4x = 7 \Rightarrow x = \dfrac{7}{4}$; Check:

$2\left(\dfrac{7}{4} - 1\right) = 5 - 2\left(\dfrac{7}{4}\right) \Rightarrow 2\left(\dfrac{7}{4} - \dfrac{4}{4}\right) = \dfrac{20}{4} - \dfrac{14}{4} \Rightarrow 2\left(\dfrac{3}{4}\right) = \dfrac{6}{4} \Rightarrow \dfrac{6}{4} = \dfrac{6}{4}$, so $x = \dfrac{7}{4}$

41. $\dfrac{2x + 1}{3} = \dfrac{2x - 1}{2} \Rightarrow 2(2x + 1) = 3(2x - 1) \Rightarrow 4x + 2 = 6x - 3 \Rightarrow 5 = 2x \Rightarrow x = \dfrac{5}{2}$; Check:

$\dfrac{2\left(\frac{5}{2}\right) + 1}{3} = \dfrac{2\left(\frac{5}{2}\right) - 1}{2} \Rightarrow \dfrac{5 + 1}{3} = \dfrac{5 - 1}{2} \Rightarrow \dfrac{6}{3} = \dfrac{4}{2} \Rightarrow 2 = 2$, so $x = \dfrac{5}{2}$

43. $4.2x - 6.2 = 1 - 1.1x \Rightarrow 5.3x = 7.2 \Rightarrow x = \dfrac{7.2}{5.3} \Rightarrow x = \dfrac{72}{53} \approx 1.36$; Check:

$4.2\left(\dfrac{72}{53}\right) - 6.2 = 1 - 1.1\left(\dfrac{72}{53}\right) \Rightarrow \dfrac{302.4}{53} - \dfrac{328.6}{53} = \dfrac{53}{53} - \dfrac{79.2}{53} \Rightarrow -\dfrac{26.2}{53} = -\dfrac{26.2}{53}$, so $x = \dfrac{72}{53}$

45. $\dfrac{1}{2}x - \dfrac{3}{2} = 4 \Rightarrow \dfrac{1}{2}x = 4 + \dfrac{3}{2} \Rightarrow \dfrac{1}{2}x = \dfrac{11}{2} \Rightarrow x = \dfrac{11}{2}\left(\dfrac{2}{1}\right) \Rightarrow x = 11$; Check:

$\dfrac{1}{2}(11) - \dfrac{3}{2} = 4 \Rightarrow \dfrac{11}{2} - \dfrac{3}{2} = 4 \Rightarrow \dfrac{8}{2} = 4 \Rightarrow 4 = 4$, so $x = 11$

47. $4(x - 1980) + 6 = 18 \Rightarrow 4x - 7920 + 6 = 18 \Rightarrow 4x = 7932 \Rightarrow x = \dfrac{7932}{4} \Rightarrow x = 1983$; Check:

$4(1983 - 1980) + 6 = 18 \Rightarrow 4(3) + 6 = 18 \Rightarrow 12 + 6 = 18 \Rightarrow 18 = 18$, so $x = 1983$

49. $2(y - 3) + 5(1 - 2y) = 4y + 1 \Rightarrow 2y - 6 + 5 - 10y = 4y + 1 \Rightarrow -12y = 2 \Rightarrow y = -\dfrac{1}{6}$; Check:

$$2\left(-\frac{1}{6}-3\right)+5\left(1-2\left(-\frac{1}{6}\right)\right)=4\left(-\frac{1}{6}\right)+1 \Rightarrow 2\left(-\frac{1}{6}-\frac{18}{6}\right)+5\left(\frac{6}{6}+\frac{2}{6}\right)=-\frac{4}{6}+\frac{6}{6}$$

$$\Rightarrow 2\left(-\frac{19}{6}\right)+5\left(\frac{8}{6}\right)=\frac{2}{6} \Rightarrow -\frac{38}{6}+\frac{40}{6}=\frac{2}{6} \Rightarrow \frac{2}{6}=\frac{2}{6}, \text{ so } y=-\frac{1}{6}$$

51. $-3(2-3z)+2z=1-3z \Rightarrow -6+9z+2z=1-3z \Rightarrow 14z=7 \Rightarrow z=\frac{1}{2}$; Check:

$$-3\left(2-3\left(\frac{1}{2}\right)\right)+2\left(\frac{1}{2}\right)=1-3\left(\frac{1}{2}\right) \Rightarrow -3\left(\frac{4}{2}-\frac{3}{2}\right)+\frac{2}{2}=\frac{2}{2}-\frac{3}{2}$$

$$\Rightarrow -3\left(\frac{1}{2}\right)+\frac{2}{2}=-\frac{1}{2} \Rightarrow -\frac{3}{2}+\frac{2}{2}=-\frac{1}{2} \Rightarrow -\frac{1}{2}=-\frac{1}{2}, \text{ so } z=\frac{1}{2}$$

53. Clear fractions by multiplying each side of the equation by 6.

$$\frac{2}{3}(t-3)+\frac{1}{2}t=5 \Rightarrow 4(t-3)+3t=30 \Rightarrow 4t-12+3t=30 \Rightarrow 7t=42 \Rightarrow t=6\text{ ; Check:}$$

$$\frac{2}{3}(6-3)+\frac{1}{2}(6)=5 \Rightarrow \frac{2}{3}(3)+3=5 \Rightarrow 2+3=5 \Rightarrow 5=5, \text{ so } t=6$$

55. Clear fractions by multiplying each side of the equation by 12.

$$\frac{3k}{4}-\frac{2k}{3}=\frac{1}{6} \Rightarrow 9k-8k=2 \Rightarrow k=2\text{ ; Check: }\frac{3(2)}{4}-\frac{2(2)}{3}=\frac{1}{6} \Rightarrow \frac{6}{4}-\frac{4}{3}=\frac{1}{6} \Rightarrow \frac{18}{12}-\frac{16}{12}=\frac{2}{12} \Rightarrow \frac{2}{12}=\frac{2}{12}, \text{ so}$$

$$k=2$$

57. $0.2(n-2)+0.4n=0.05 \Rightarrow 0.2n-0.4+0.4n=0.05 \Rightarrow 0.6n=0.45 \Rightarrow n=\frac{0.45}{0.6} \Rightarrow n=0.75$; Check:

$$0.2(0.75-2)+0.4(0.75)=0.05 \Rightarrow 0.2(-1.25)+0.3=0.05 \Rightarrow -0.25+0.3=0.05 \Rightarrow 0.05=0.05, \text{ so}$$

$$n=0.75$$

59. $0.7y-0.8(y-1)=2 \Rightarrow 0.7y-0.8y+0.8=2 \Rightarrow -0.1y=1.2 \Rightarrow y=\frac{1.2}{-0.1} \Rightarrow y=-12$; Check:

$$0.7(-12)-0.8(-12-1)=2 \Rightarrow -8.4-0.8(-13)=2 \Rightarrow -8.4+10.4=2 \Rightarrow 2=2, \text{ so } y=-12$$

61. $ax+b=0 \Rightarrow ax=-b \Rightarrow x=-\frac{b}{a}, a \neq 0$

Numerical Solutions

63. The completed table is shown in Figure 63. From the table we see that $-4x+8=0$ when $x=2$.

x	1	2	3	4	5
$-4x+8$	4	0	-4	-8	-12

Figure 63

65. The completed table is shown in Figure 65. From the table we see that $4 - 2x = x + 7$ when $x = -1$.

x	−2	−1	0	1	2
$4 - 2x$	8	6	4	2	0
$x + 7$	5	6	7	8	9

Figure 65

67. From a table of values (not shown) we find that $x - 3 = 7$ when $x = 10$.

69. From a table of values (not shown) we find that $2y - \dfrac{1}{2} = \dfrac{3}{2}$ when $y = 1$.

71. From a table of values (not shown) we find that $3(z - 1) + 1.5 = 2z$ when $z = 1.5$.

Graphical Solutions

73. The graphs intersect when $x = -2$.

75. The graphs intersect when $x = -1$.

77. Graph $Y_1 = 5 - 2X$ and $Y_2 = 7$ in [−10, 10, 1] by [−10, 10, 1]. See Figure 77. The solution is $x = -1$.

[−10, 10, 1] by [−10, 10, 1]

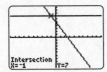

Figure 77

79. Graph $Y_1 = 2X - (X + 2)$ and $Y_2 = -2$ in [−10, 10, 1] by [−10, 10, 1]. See Figure 79. The solution is $x = 0$.

[−10, 10, 1] by [−10, 10, 1]   [−10, 10, 1] by [−10, 10, 1]

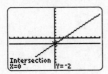

Figure 79

81. Graph $Y_1 = 3(X + 2) + 1$ and $Y_2 = X + 1$ in [−10, 10, 1] by [−10, 10, 1]. See Figure 81. The solution is

$x = -3$.

[−10, 10, 1] by [−10, 10, 1]

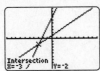

Figure 81

[1980, 2020, 10] by [0, 150, 10]

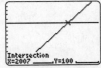

Figure 83

83. Graph $Y_1 = 5(X-1990)+15$ and $Y_2 = 100$ in [1980, 2020, 10] by [0, 150, 10]. See Figure 83. Here $x = 2007$.

85. Graph $Y_1 = \sqrt{2}+4(X-\pi)$ and $Y_2 = 1/7X$ in [-2, 8, 1] by [-5, 5, 1]. The solution is $x \approx 2.89$.

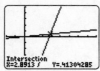
Figure 85

Solving Linear Equations by More Than One Method

87. (a) Table $Y_1 = 2X-1$ and $Y_2 = 13$ with TblStart = 1 and ΔTbl = 1. See Figure 87a. $x = 7$

    (b) Graph $Y_1 = 2X-1$ and $Y_2 = 13$ in [0, 10, 1] by [0, 20, 2]. See Figure 87b. $x = 7$

    (c) $2x-1 = 13 \Rightarrow 2x = 14 \Rightarrow x = \dfrac{14}{2} \Rightarrow x = 7$

[0, 10, 1] by [0, 20, 2]

     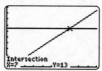

Figure 87a   Figure 87b

89. (a) Table $Y_1 = 3X-5-(X+1)$ and $Y_2 = 0$ with TblStart = 0 and ΔTbl = 1. See Figure 89a. $x = 3$

    (b) Graph $Y_1 = 3X-5-(X+1)$ and $Y_2 = 0$ in [-10, 10, 1] by [-10, 10, 1]. See Figure 89b. $x = 3$

    (c) $3x-5-(x+1) = 0 \Rightarrow 3x-5-x-1 = 0 \Rightarrow 2x = 6 \Rightarrow x = \dfrac{6}{2} \Rightarrow x = 3$

[-10, 10, 1] by [-10, 10, 1]

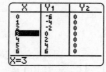   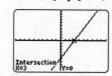

Figure 89a   Figure 89b

91. $2x+3 = 2x \Rightarrow 3 = 0$, false; contradiction

93. $2x+3x = 5x \Rightarrow (2+3)x = 5x \Rightarrow 5x = 5x$; identity

95. $2(x-1) = 2x-2 \Rightarrow 2x-2 = 2x-2 \Rightarrow -2 = -2$; identity

97. $5-7(4-3x) = x-4(2-4x) \Rightarrow 5-28+21x = x-8+16x$

    $\Rightarrow -23+21x = 17x-8 \Rightarrow -23+4x = -8 \Rightarrow 4x = 15$

    $x = \dfrac{15}{4}$; conditional

99. Graph of f: $m = \dfrac{-1-9}{1-(-4)} = \dfrac{-10}{5} = -2$; $y-9 = -2(x-(-4))$

$$\Rightarrow y - 9 = -2x - 8 \Rightarrow y = -2x + 1$$

Graph of g: $m = \dfrac{13-1}{0-(-3)} = \dfrac{12}{3} = 4; \quad y - 13 = 4(x - 0) \Rightarrow$

$$y - 13 = 4x \Rightarrow y = 4x + 13$$

$$-2x + 1 = 4x + 13 \Rightarrow -6x + 1 = 13 \Rightarrow -6x = 12 \Rightarrow x = -2;$$

$$y = -2(-2) + 1 \Rightarrow y = 4 + 1 \Rightarrow y = 5; (-2, 5)$$

Finding Intercepts

101. (a) To find the *x*-intercept, let $y = 0$ and solve for x: $x + 0 = 4 \Rightarrow x = 4$; to find the *y*-intercept, let $x = 0$ and

solve for y: $0 + y = 4 \Rightarrow y = 4$

   (b) See Figure 101b.

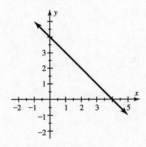

Figure 101b

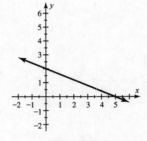

Figure 103b

   (c) $x + y = 4 \Rightarrow y = -x + 4$

103. (a) To find the *x*-intercept, let $y = 0$ and solve for x: $2x + 5(0) = 10$

$\Rightarrow 2x = 10 \Rightarrow x = 5$; to find the *y*-intercept, let $x = 0$ and solve for y: $2(0) + 5y = 10 \Rightarrow 5y = 10 \Rightarrow y = 2$

   (b) See Figure 103b.

   (c) $2x + 5y = 10 \Rightarrow 5y = -2x + 10 \Rightarrow y = -\dfrac{2}{5}x + 2$

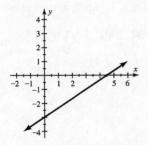

Figure 105b

105. (a) To find the *x*-intercept, let $y = 0$ and solve for x: $2x - 3(0) = 9 \Rightarrow 2x = 9 \Rightarrow x = \dfrac{9}{2}$; to find the *y*-

intercept, let $x = 0$ and solve for y: $2(0) - 3y = 9 \Rightarrow -3y = 9 \Rightarrow y = -3$

   (b) See Figure 105b.

(c) $2x - 3y = 9 \Rightarrow -3y = -2x + 9 \Rightarrow y = \dfrac{2}{3}x - 3$

107. (a) To find the *x*-intercept, let $y = 0$ and solve for $x: -\dfrac{2}{3}x - 0 = 2 \Rightarrow -\dfrac{2}{3}x = 2 \Rightarrow x = -3$; to find the *y*-

intercept, let $x = 0$ and solve for $y: -\dfrac{2}{3}(0) - y = 2 \Rightarrow -y = 2 \Rightarrow y = -2$

(b) See Figure 107b.

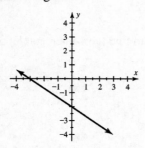

Figure 107b

(c) $-\dfrac{2}{3}x - y = 2 \Rightarrow -y = \dfrac{2}{3}x + 2 \Rightarrow y = -\dfrac{2}{3}x - 2$

Applications

109. (a) $m = \dfrac{16.6 - 10.3}{2004 - 1960} = \dfrac{6.3}{44} = \dfrac{63}{440}$; $S(x) = \dfrac{63}{440}(x - 1960) + 10.3$

(b) $S(2000) = \dfrac{63}{440}(2000 - 1960) + 10.3 = \dfrac{63}{440}(40) + 10.3 \approx 5.7 + 10.3 = 16$; about 16 pounds

(c) $20 = \dfrac{63}{440}(x - 1960) + 10.3 \Rightarrow 9.7 = \dfrac{63}{440}(x - 1960) \Rightarrow 67.7 \approx x - 1960 \Rightarrow 2027.7 \approx x$; about 2028

111. (a) $70x - 138{,}532 = 908 \Rightarrow 70x = 139{,}440 \Rightarrow x = \dfrac{139{,}440}{70} \Rightarrow x = 1992$

(b) Table $Y_1 = 70X - 138532$ and $Y_2 = 908$ with TblStart = 1988 and ΔTbl = 1. See Figure 111b. $x = 1992$

(c) Graph $Y_1 = 70X - 138532$ and $Y_2 = 908$ in [1988, 1995, 1] by [0, 1000, 100]. See Figure

111c. $x = 1992$

[1988, 1995, 1] by [0, 1000, 100]

   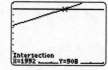

Figure 111b   Figure 111c

113. (a) The slope is $m = \dfrac{1.8 - 1.4}{2002 - 1998} = \dfrac{0.4}{4} = 0.1$.

Then using the point (1998, 1.4) we have $y = 0.1(x - 1998) + 1.4$ or $y = 0.1x - 198.4$.

(b) $0.1x - 198.4 = 1.7 \Rightarrow 0.1x = 200.1 \Rightarrow x = 2001$

(c) $0.1x - 198.4 = 2 \Rightarrow 0.1x = 200.4 \Rightarrow x = 2004$

115. $51.6(x - 1985) + 9.1 = -31.9(x - 1985) + 167.7 \Rightarrow 51.6x - 102,416.9 = -31.9x + 63,489.2 \Rightarrow$

$83.5x = 165,906.1 \Rightarrow x \approx 1986.89.$ Sales were about equal in late 1986 or around 1987.

3.2: Introduction to Problem Solving

Concepts

1.   We divide each side by 2 to obtain $y = \dfrac{x}{2}$.

2. $4x - 7$

3.   The perimeter is $P = 2W + 2L$.

4.   read the problem carefully

5. $43.1\% = 0.431$

6. $0.20 \times 50 = 10$

Solving for a Variable

7. $4x + 3y = 12 \Rightarrow 3y = -4x + 12 \Rightarrow y = \dfrac{-4x + 12}{3} \Rightarrow y = -\dfrac{4}{3}x + 4$

9. $5(2x - 3y) = 2x \Rightarrow 10x - 15y = 2x \Rightarrow 8x = 15y \Rightarrow x = \dfrac{15}{8}y$

11. $S = 6ab \Rightarrow \dfrac{S}{6a} = \dfrac{6ab}{6a} \Rightarrow b = \dfrac{S}{6a}$

13. $\dfrac{r + t}{2} = 7 \Rightarrow r + t = 14 \Rightarrow t = 14 - r$

15. $-3x + y = 8 \Rightarrow y = 3x + 8 \Rightarrow f(x) = 3x + 8$

17. $4x = 2\pi y \Rightarrow \dfrac{4x}{2\pi} = \dfrac{2\pi y}{2\pi} \Rightarrow \dfrac{2x}{\pi} = y \Rightarrow f(x) = \dfrac{2x}{\pi}$

19. $\dfrac{3y}{8} = x \Rightarrow 3y = 8x \Rightarrow y = \dfrac{8x}{3} \Rightarrow f(x) = \dfrac{8x}{3}$

Writing and Solving Equations

21.   (a) $x + 2 = 12$

       (b) $x + 2 = 12 \Rightarrow x = 10$

23.   (a) $\dfrac{x}{5} = x + 1$

       (b) $\dfrac{x}{5} = x + 1 \Rightarrow x = 5x + 5 \Rightarrow -4x = 5 \Rightarrow x = -\dfrac{5}{4}$

25.   (a) $\dfrac{x + 5}{2} = 7$

(b) $\dfrac{x+5}{2} = 7 \Rightarrow x+5 = 14 \Rightarrow x = 9$

27.　(a) $\dfrac{x}{2} = 17$

　　(b) $\dfrac{x}{2} = 17 \Rightarrow x = 34$

29.　(a) $x + (x+1) + (x+2) = 30$

　　(b) $x + (x+1) + (x+2) = 30 \Rightarrow 3x+3 = 30 \Rightarrow 3x = 27 \Rightarrow x = \dfrac{27}{3} \Rightarrow x = 9$

Applications

31.　(a) Solve $P = 2L + 2W$ for L. $P = 2L + 2W \Rightarrow \dfrac{P}{2} = L + W \Rightarrow L = \dfrac{P}{2} - W$

　　(b) $L = \dfrac{86}{2} - 19 = 43 - 19 = 24$ feet

33.　$A = 2\pi rh \Rightarrow \dfrac{A}{2\pi r} = \dfrac{2\pi rh}{2\pi r} \Rightarrow h = \dfrac{A}{2\pi r}$

35.　$F = \dfrac{9}{5}C + 32 \Rightarrow F - 32 = \dfrac{9}{5}C \Rightarrow \dfrac{5}{9}(F-32) = C \Rightarrow C = \dfrac{5}{9}(F-32)$

37.　$A = \dfrac{1}{2}bh \Rightarrow 2A = bh \Rightarrow \dfrac{2A}{b} = \dfrac{bh}{b} \Rightarrow h = \dfrac{2A}{b}$

39.　Let x represent the smallest integer. Then $x+1$ and $x+2$ represent the next consecutive integers.

　　$x + (x+1) + (x+2) = 135 \Rightarrow 3x+3 = 135 \Rightarrow 3x = 132 \Rightarrow x = 44$. The integers are 44, 45, and 46.

41.　Let x represent the smallest odd integer. Then $x+2$ and $x+4$ represent the next consecutive odd integers.

　　$2\left[x + (x+2) + (x+4)\right] = 150 \Rightarrow 3x+6 = 75 \Rightarrow 3x = 69 \Rightarrow x = 23$. The integers are 23, 25, and 27.

43.　Two sides of length x plus two sides of length $x+8$ equals 48.

　　$2x + 2(x+8) = 48 \Rightarrow 2x + 2x + 16 = 48 \Rightarrow 4x = 32 \Rightarrow x = \dfrac{32}{4} \Rightarrow x = 8$ feet

45.　Let x represent the length of a side of the square (and the diameter of each circle). Then $2\pi x + 4x = 41$

　　$\Rightarrow (2\pi + 4)x = 41 \Rightarrow x = \dfrac{41}{2\pi + 4} \Rightarrow x \approx 4$.

　　The wire should be cut at about $4(4) = 16$ inches for the square and $\pi(4) \approx 12.5$ inches for each circle.

47.　Use the formula $d = rt$. Let x represent the speed of one car and $x+6$ the speed of the other

　　car. $355 = x(2.5) + (x+6)(2.5) \Rightarrow 355 = 2.5x + 2.5x + 15 \Rightarrow 340 = 5x \Rightarrow x = 68$. One car is traveling at 68

　　miles per hour, and the other car is traveling at 74 miles per hour.

49.　Mixing 3 liters of 30% acid with x liters of 80% acid results in $x+3$ liters of 60% acid.

$0.30(3) + 0.80x = 0.60(x+3) \Rightarrow 0.9 + 0.8x = 0.6x + 1.8 \Rightarrow 0.2x = 0.9 \Rightarrow x = 4.5$

The chemist used 4.5 liters of 80% sulfuric acid.

51.   Let x represent the loan amount at 6% interest. Then the remaining amount at 4% interest is $5000 - x$.

$0.06x + 0.04(5000 - x) = 244 \Rightarrow 0.06x + 200 - 0.04x = 244 \Rightarrow 0.02x = 44 \Rightarrow x = 2200$

The loans were for $2200 at 6% and $2800 at 4%.

53.   Let x represent the measure of the smallest angle. Then the largest angle has measure $2x$ and the third angle

has measure $2x - 10$. $x + 2x + 2x - 10 = 180 \Rightarrow 5x = 190 \Rightarrow x = 38$. The angle measures are

$38°$, $66°$, and $76°$.

55.   (a) $900(40) = 36,000 \text{ ft}^3/\text{hr}$

(b) There should be no more than 66 people in the room. $900x = 60,000 \Rightarrow x = \dfrac{60,000}{900} \Rightarrow x \approx 66.7$

57.   Let x represent the cost of a gallon of gasoline. $873 = \dfrac{15,000}{50}x \Rightarrow 873 = 300x \Rightarrow x = \$2.91$

59.   Let x represent the area of Nevada. $0.09x = 9900 \Rightarrow x = 110,000$ square miles

61.   Let x represent the number of Wal-Mart employees in 2002. $x + 0.57x = 2.2 \Rightarrow 1.57x = 2.2 \Rightarrow x \approx 1.401$

There were about 1.4 million Wal-Mart employees in 2002.

63.   Let x be the number of people surveyed. Remember to write 32% as 0.32.

$0.32x = 480 \Rightarrow x = \dfrac{480}{0.32} \Rightarrow x = 1500$ people

65.   Since heart disease accounts for 29.1% more deaths than cancer, the number of deaths that can be attributed

to heart disease is $1.291(550,000) = 710,050$. The total number of deaths is

$550,000 + 710,050 = 1,260,050$.

That is, about 1,260,000 deaths can be attributed to heart disease *and* cancer.

Checking Basic Concepts for Sections 3.1 & 3.2

1.   A linear equation has one solution.

2.   Multiply each side of the equation by 6 to clear fractions.

$\dfrac{1}{2}z - (1 - 2z) = \dfrac{2}{3}z \Rightarrow 3z - 6(1 - 2z) = 4z \Rightarrow 3z - 6 + 12z = 4z \Rightarrow 11z = 6 \Rightarrow z = \dfrac{6}{11}$

3. (a) $2(3x+4)+3=-1 \Rightarrow 6x+8+3=-1 \Rightarrow 6x=-12 \Rightarrow x=\dfrac{-12}{6} \Rightarrow x=-2$

   (b) Graph $Y_1 = 2(3X+4)+3$ and $Y_2 = -1$ in $[-10, 10, 1]$ by $[-10, 10, 1]$. See Figure 3b. $x=-2$

   (c) Table $Y_1 = 2(3X+4)+3$ and $Y_2 = -1$ with TblStart $= -5$ and ΔTbl $= 1$. See Figure 3c. $x=-2$

   The answers agree.

   $[-10, 10, 1]$ by $[-10, 10, 1]$

   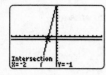   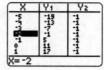

   Figure 3b   Figure 3c

4. Let x represent the time the athlete spends jogging at 10 miles per hour. Then $1.2-x$ represents the time

   spent jogging at 8 miles per hour. $10x+8(1.2-x)=10.2 \Rightarrow 10x+9.6-8x=10.2 \Rightarrow 2x=0.6 \Rightarrow x=0.3$

   The athlete jogged for 0.3 hour at 10 miles per hour and 0.9 hour at 8 miles per hour.

5. To find the x-intercept, Let $y=0$ and solve for x: $5x-4(0)=20 \Rightarrow 5x=20 \Rightarrow x=4$; to find the y-intercept,

   let $x=0$ and solve for y: $5(0)-4y=20 \Rightarrow -4y=20 \Rightarrow y=-5$.

6. (a) $3(x-2)=3x-6 \Rightarrow 3x-6=3x-6 \Rightarrow -6=-6$; identity

   (b) $2x+3x=5x-2 \Rightarrow (2+3)x=5x-2 \Rightarrow 5x=5x-2 \Rightarrow 0=-2$; contradiction

3.3: Linear Inequalities

Concepts

1. An example of a linear inequality is $3x+2<5$. *Answers may vary.*

2. Yes. For example, any positive number will satisfy the inequality $x+2>2$.

3. Yes. Subtract 5 from each side.

4. Yes. Add 7 to each side.

5. Yes. They have the same solution set.

6. No, $-4x<8$ is equivalent to $x>-2$.

7. An equation has an equals sign while an inequality has an inequality symbol.

8. Three methods for solving a linear inequality are numerical, graphical and symbolic.

9. Yes. $2-5 \le 3$

10. No. $\dfrac{3}{2}(-2)-\dfrac{1}{2} \not\ge 1-(-2)$

11. No. $2(5)-3 \not> 5(5)-\left[2(5)+1\right]$

12. Yes. $2(\pi-1)<3(\pi+1)$

13. (a) $x-2=0 \Rightarrow x=2 \Rightarrow \{x \mid x=2\}$

(b) $x - 2 < 0 \Rightarrow x < 2 \Rightarrow \{x \mid x < 2\}$

(c) $x - 2 > 0 \Rightarrow x > 2 \Rightarrow \{x \mid x > 2\}$

14. (a) $x + 5 = 0 \Rightarrow x = -5 \Rightarrow \{x \mid x = -5\}$

(b) $x + 5 < 0 \Rightarrow x < -5 \Rightarrow \{x \mid x < -5\}$

(c) $x + 5 > 0 \Rightarrow x > -5 \Rightarrow \{x \mid x > -5\}$

15. (a) $-2x + 6 = 0 \Rightarrow -2x = -6 \Rightarrow x = 3 \Rightarrow \{x \mid x = 3\}$

(b) $-2x + 6 < 0 \Rightarrow -2x < -6 \Rightarrow x > 3 \Rightarrow \{x \mid x > 3\}$

(c) $-2x + 6 > 0 \Rightarrow -2x > -6 \Rightarrow x < 3 \Rightarrow \{x \mid x < 3\}$

16. (a) $-6 - 6x = 0 \Rightarrow -6x = 6 \Rightarrow x = -1 \Rightarrow \{x \mid x = -1\}$

(b) $-6 - 6x < 0 \Rightarrow -6x < 6 \Rightarrow x > -1 \Rightarrow \{x \mid x > -1\}$

(c) $-6 - 6x > 0 \Rightarrow -6x > 6 \Rightarrow x < -1 \Rightarrow \{x \mid x < -1\}$

Symbolic Solutions

17. $x + 3 \leq 5 \Rightarrow x \leq 2 \Rightarrow \{x \mid x \leq 2\}$

19. $\dfrac{1}{4}x > 9 \Rightarrow x > 36 \Rightarrow \{x \mid x > 36\}$

21. $4 - 3x \leq -\dfrac{2}{3} \Rightarrow -3x \leq -\dfrac{2}{3} - 4 \Rightarrow -3x \leq -\dfrac{14}{3} \Rightarrow x \geq -\dfrac{1}{3}\left(-\dfrac{14}{3}\right) \Rightarrow x \geq \dfrac{14}{9} \Rightarrow \left\{x \mid x \geq \dfrac{14}{9}\right\}$

23. $7 - \dfrac{1}{2}x < x - \dfrac{3}{2} \Rightarrow 14 - x < 2x - 3 \Rightarrow -3x < -17 \Rightarrow x > \dfrac{17}{3} \Rightarrow \left\{x \mid x > \dfrac{17}{3}\right\}$

25. $\dfrac{5}{2}(2x - 3) < 6 - 2x \Rightarrow 5x - \dfrac{15}{2} < 6 - 2x \Rightarrow 7x < \dfrac{27}{2} \Rightarrow x < \dfrac{1}{7}\left(\dfrac{27}{2}\right) \Rightarrow x < \dfrac{27}{14} \Rightarrow \left\{x \mid x < \dfrac{27}{14}\right\}$

27. $\dfrac{3x - 2}{-2} \leq \dfrac{x - 4}{-5} \Rightarrow -15x + 10 \leq -2x + 8 \Rightarrow -13x \leq -2 \Rightarrow x \geq \dfrac{2}{13} \Rightarrow \left\{x \mid x \geq \dfrac{2}{13}\right\}$

29. $3(x - 2000) + 15 < 45 \Rightarrow 3x < 6030 \Rightarrow x < \dfrac{6030}{3} \Rightarrow x < 2010 \Rightarrow \{x \mid x < 2010\}$

31. $0.4x - 0.7 < 1.3 \Rightarrow 0.4x < 2 \Rightarrow x < 5 \Rightarrow \{x \mid x < 5\}$

33. $\dfrac{4}{5}x - \dfrac{1}{5} \geq -5 \Rightarrow 4x - 1 \geq -25 \Rightarrow 4x \geq -24 \Rightarrow x \geq -6 \Rightarrow \{x \mid x \geq -6\}$

35. $-\dfrac{1}{3}(z - 3) - \dfrac{1}{4} \geq \dfrac{1}{4}(5 - z) \Rightarrow -4(z - 3) - 3 \geq 3(5 - z) \Rightarrow -4z + 12 - 3 \geq 15 - 3z \Rightarrow$

$-4z + 9 \geq 15 - 3z \Rightarrow -z \geq 6 \Rightarrow z \leq -6 \Rightarrow \{z \mid z \leq -6\}$

37. $\frac{3}{4}(2t-5) \leq \frac{1}{2}(4t-6)+1 \Rightarrow 3(2t-5) \leq 2(4t-6)+4 \Rightarrow 6t-15 \leq 8t-12+4 \Rightarrow$

$6t-15 \leq 8t-8 \Rightarrow -2t \leq 7 \Rightarrow t \geq -\frac{7}{2} \Rightarrow \left\{t \,\middle|\, t \geq -\frac{7}{2}\right\}$

39. $\frac{1}{2}(4-(x+3))+4 > -\frac{1}{3}(2x-(1-x)) \Rightarrow 3(4-x-3)+24 > -2(2x-1+x) \Rightarrow$

$12-3x-9+24 > -4x+2-2x \Rightarrow -3x+27 > -6x+2 \Rightarrow 3x > -25 \Rightarrow x > -\frac{25}{3} \Rightarrow \left\{x \,\middle|\, x > -\frac{25}{3}\right\}$

41. $0.05+0.08x < 0.01x-0.04(3-4x) \Rightarrow 5+8x < x-4(3-4x)$

$\Rightarrow 5+8x < x-12+16x \Rightarrow -9x < -17 \Rightarrow x > \frac{17}{9} \Rightarrow \left\{x \,\middle|\, x > \frac{17}{9}\right\}$

Numerical Solutions

43. See Figure 43. $\left\{x \,\middle|\, x \geq 3\right\}$

x	1	2	3	4	5
$-2x+6$	4	2	0	-2	-4

Figure 43

45. Table $Y_1 = X-3$ and $Y_2 = 0$ with TblStart = 0 and ΔTbl = 1. See Figure 45. $\left\{x \,\middle|\, x > 3\right\}$

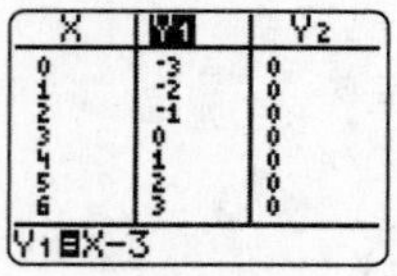

Figure 45 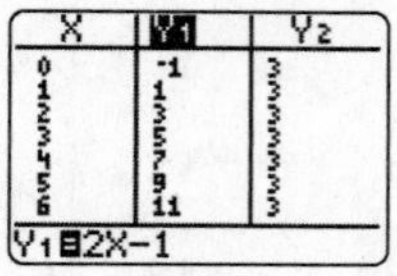   Figure 47

47. Table $Y_1 = 2X-1$ and $Y_2 = 3$ with TblStart = 0 and ΔTbl = 1. See Figure 47. $\left\{x \,\middle|\, x \geq 2\right\}$

Graphical Solutions

49. (a) The graph crosses the x-axis at the point $(-1, 0)$. The solution set is $\left\{x \,\middle|\, x = -1\right\}$.

    (b) The graph is below the x-axis whenever $x < -1$. The solution set is $\left\{x \,\middle|\, x < -1\right\}$.

    (c) The graph is above the x-axis whenever $x > -1$. The solution set is $\left\{x \,\middle|\, x > -1\right\}$.

51. (a) The graph crosses the x-axis at the point $\left(\frac{1}{2}, 0\right)$. The solution set is $\left\{x \,\middle|\, x = \frac{1}{2}\right\}$.

    (b) The graph is below the x-axis whenever $x > \frac{1}{2}$. The solution set is $\left\{x \,\middle|\, x > \frac{1}{2}\right\}$.

    (c) The graph is above the x-axis whenever $x < \frac{1}{2}$. The solution set is $\left\{x \,\middle|\, x < \frac{1}{2}\right\}$.

53. The graph of y_1 is below the graph of y_2 whenever $x \geq 1$. The solution set is $\left\{x \,\middle|\, x \geq 1\right\}$.

55. The graph of y_1 is above the graph of y_2 whenever $x > 2$. The solution set is $\left\{x \,\middle|\, x > 2\right\}$.

57. (a) Car 1 is traveling faster. The slope of the line for Car 1 is greater than the slope of the line for Car 2.

(b) They are the same distance from St. Louis when $x = 5$ hours. The distance from St. Louis is 400 miles.

(c) Car 2 is farther from St. Louis than Car 1 for times in the interval $0 \le x < 5$.

59. Graph $Y_1 = X - 1$ and $Y_2 = 0$ in [–10, 10, 1] by [–10, 10, 1]. See Figure 59. $\{x \mid x < 1\}$

[–10, 10, 1] by [–10, 10, 1]

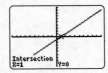

Figure 59

[–10, 10, 1] by [–10, 10, 1]

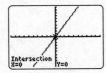

Figure 61

61. Graph $Y_1 = 2X$ and $Y_2 = 0$ in [–10, 10, 1] by [–10, 10, 1]. See Figure 61. $\{x \mid x \ge 0\}$

63. Graph $Y_1 = 4 - 2X$ and $Y_2 = 8$ in [–10, 10, 1] by [–10, 10, 1]. See Figure 63. $\{x \mid x \ge -2\}$

[–10, 10, 1] by [–10, 10, 1]

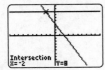

Figure 63

[–10, 10, 1] by [–10, 10, 1]

Figure 65

65. Graph $Y_1 = X - (2X + 4)$ and $Y_2 = 0$ in [–10, 10, 1] by [–10, 10, 1]. See Figure 65. $\{x \mid x < -4\}$

67. Graph $Y_1 = 2(X + 2) + 5$ and $Y_2 = -X$ in [–10, 10, 1] by [–10, 10, 1]. See Figure 67. $\{x \mid x < -3\}$

[–10, 10, 1] by [–10, 10, 1]

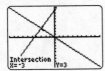

Figure 67

[1985, 2000, 5] by [−50, 50, 10]

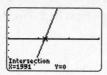

Figure 69

69.   Graph $Y_1 = 25(X - 1995) + 100$ and $Y_2 = 0$ in [1985, 2000, 5] by [−50, 50, 10]. See Figure

69. $\{x \mid x \leq 1991\}$

71.   Graph $Y_1 = \pi\left(\sqrt{(2)}X - 1.2\right) + \sqrt{(3)}X$ and $Y_2 = (\pi + 1)/3$ in [−10, 10, 1] by [−10, 10, 1]. $\{x \mid x > 0.834\}$

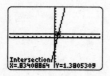

Figure 71

73.   Reverse the inequality symbol: $\{x \mid x < a\}$

Solving Linear Inequalities by More Than One Method

75.   (a) Table $Y_1 = 5X - 2$ and $Y_2 = 8$ with TblStart = 0 and ΔTbl = 1. See Figure 75a. $\{x \mid x < 2\}$

(b) Graph $Y_1 = 5X - 2$ and $Y_2 = 8$ in [−10, 10, 1] by [−10, 10, 1]. See Figure 75b. $\{x \mid x < 2\}$

(c) $5x - 2 < 8 \Rightarrow 5x < 10 \Rightarrow x < \dfrac{10}{5} \Rightarrow x < 2 \Rightarrow \{x \mid x < 2\}$

[−10, 10, 1] by [−10, 10, 1]

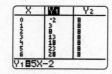   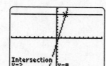

Figure 75a   Figure 75b

77.   (a) Table $Y_1 = 3X - 3$ and $Y_2 = 2X$ with TblStart = 0 and ΔTbl = 1. See Figure 77a. $\{x \mid x \leq 3\}$

(b) Graph $Y_1 = 3X - 3$ and $Y_2 = 2X$ in [−10, 10, 1] by [−10, 10, 1]. See Figure 77b. $\{x \mid x \leq 3\}$

(c) $2x \geq 3x - 3 \Rightarrow x \leq 3 \Rightarrow \{x \mid x \leq 3\}$

[−10, 10, 1] by [−10, 10, 1]

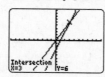

Figure 77a   Figure 77b

Inequalities and Functions

79.   (a) $x-5=7 \Rightarrow x=12 \Rightarrow \{x | x=12\}$

(b) $x-5>7 \Rightarrow x>12 \Rightarrow \{x | x>12\}$

81.   (a) $4-3x \geq 11 \Rightarrow -3x \geq 7 \Rightarrow x \leq -\dfrac{7}{3} \Rightarrow \left\{x \middle| x \leq -\dfrac{7}{3}\right\}$

(b) $4-3x \leq 11 \Rightarrow -3x \leq 7 \Rightarrow x \geq -\dfrac{7}{3} \Rightarrow \left\{x \middle| x \geq -\dfrac{7}{3}\right\}$

83.   (a) $3(x+7)+1 \leq -5 \Rightarrow 3x+22 \leq -5 \Rightarrow 3x \leq -27 \Rightarrow x \leq -9 \Rightarrow \{x | x \leq -9\}$

(b) $3(x+7)+1 \geq -5 \Rightarrow 3x+22 \geq -5 \Rightarrow 3x \geq -27 \Rightarrow x \geq -9 \Rightarrow \{x | x \geq -9\}$

Applications

85.   From the graph the deficit was less than \$1 trillion in 1980 or before.

87.   (a)   The band paid \$200 for the blank CDs because $P=-200$ when $x=0$.

(b)   The band charged \$5 for each recorded CD because the slope of the line is 5.

(c)   Since the slope of the line is 5 and the y-intercept is -200, the equation is $P=5x-200$.

(d)   The break-even point occurs when $P=0$.   From the graph, this is $x=40$ CDs.

(e)   The band will make a profit if it sells more than 40 CDs or when $x>40$.

89.   Two widths plus two lengths must be less than 50.

$$2x+2(x+5)<50 \Rightarrow 2x+2x+10<50 \Rightarrow 4x<40 \Rightarrow x<\frac{40}{4} \Rightarrow x<10 \text{ ft}$$

91.   (a)   The first car is traveling 70 mph and the second car is traveling 60 mph.

(b)   They are the same distance away when $70x=60x+35 \Rightarrow 10x=35 \Rightarrow x=\dfrac{35}{10} \Rightarrow x=3.5$ hr.

(c)   The first car is farther away when time has elapsed past 3.5 hours.

93. $60-19x>0 \Rightarrow 60>19x \Rightarrow \dfrac{60}{19}>x \Rightarrow x<3.16$ (approximate).   Below about 3.16 miles.

95. $19 \leq \dfrac{705W}{(70)^2} \Rightarrow 19 \leq \dfrac{705W}{4900} \Rightarrow 93,100 \leq 705W$

$\Rightarrow 132 \leq W$

$25 \geq \dfrac{705W}{(70)^2} \Rightarrow 25 \geq \dfrac{705W}{4900} \Rightarrow 122,500 \geq 705W$

$\Rightarrow 174 \geq W$

W is between 132 pounds and 174 pounds.

97.   First, find the slope of the line that passes through the points (1994, 1.3) and (2000, 1.8);

$$m=\frac{1.8-1.3}{2000-1994}=\frac{0.5}{6} \approx 0.083.$$ The equation is $y=0.083(x-1994)+1.3$ or $y=0.083x-164.202$.

Solve this as an inequality when $y \geq 1.55$. $0.083x - 164.202 \geq 1.55 \Rightarrow 0.083x \geq 165.752 \Rightarrow x \geq 1997.01$

AIDS funding was greater than or equal to \$1.55 billion in 1997 or after.

99.   (a) $0.96x - 14.4 = 6.7 \Rightarrow 0.96x = 21.1 \Rightarrow x = \dfrac{21.1}{0.96} \Rightarrow x \approx 22$

A bass that weighs 6.7 pounds is about 22 inches long.

(b)   Bass which are less than 22 inches long weigh less than 6.7 pounds.

3.4: Compound Inequalities

Concepts

1.   An example of a compound inequality containing the word *and* is $x > 1$ and $x \leq 7$;   *Answers may vary.*

2.   An example of a compound inequality containing the word *or* is $x \leq 3$ or $x > 5$;   *Answers may vary.*

3.   No, $1 \not> 3$.

4.   Yes, $1 < 3$.

5.   Yes. The inequality can be written in either form.

6.   Three ways for solving a compound inequality are numerically, graphically and symbolically.

7.   Yes, $x = 2$ satisfies both inequalities. No, $x = 6$ does not satisfy $x - 1 < 5$.

8.   No, $x = -2$ does not satisfy $2x + 1 \geq 4$. Yes, $x = 3$ satisfies both inequalities.

9.   No, $x = 0$ does not satisfy either inequality. Yes, $x = 3$ satisfies $2x \geq 3$.

10.   Yes, $x = -5$ satisfies $x + 1 \leq -4$. No, $x = 2$ does not satisfy either inequality.

11.   No, $x = -3$ does not satisfy $2 - x \leq 4$. Yes, $x = 0$ satisfies both inequalities.

12.   Yes, $x = -1$ satisfies $3x \leq 3$. Yes, $x = 1$ satisfies either inequality.

Interval Notation

13. $[2, 10]$

15. $(5, 8]$

17. $(-\infty, 4)$

19. $(-2, \infty)$

21. $[-2, 5)$

23. $(-8, 8]$

25. $(3, \infty)$

27. $(-\infty, -2] \cup [4, \infty)$

29. $(-\infty, 1) \cup [5, \infty)$

31. $(-3, 5]$

33. $(-\infty, -2)$

35. $(-\infty, 4)$

37. $(-\infty, 1) \cup (2, \infty)$

Symbolic Solutions

39. $x \le 3$ and $x \ge -1 \Rightarrow -1 \le x \le 3;\ \{x | -1 \le x \le 3\};$ See Figure 39.

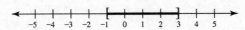

Figure 39

41. $2x < 5$ and $2x > -4 \Rightarrow x < \dfrac{5}{2}$ and $x > -2 \Rightarrow -2 < x < 2.5;\ \{x | -2 < x < 2.5\};$ See Figure 41.

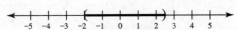

Figure 41

43. $x \le -1$ or $x \ge 2;\ \{x | x \le -1 \text{ or } x \ge 2\};$ See Figure 43.

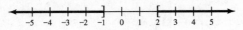

Figure 43

45. $5 - x > 1$ or $x + 3 \ge -1 \Rightarrow -x > -4$ or $x \ge -4 \Rightarrow x < 4$ or $x \ge -4;$ All real numbers; See Figure 45.

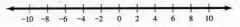

Figure 45

47. $x - 3 \le 4 \Rightarrow x \le 7$ and $x + 5 \ge -1 \Rightarrow x \ge -6$

The solutions must satisfy both of these inequalities. The interval is $[-6, 7]$.

49. $3t - 1 > -1 \Rightarrow 3t > 0 \Rightarrow t > 0$ and $2t - \dfrac{1}{2} > 6 \Rightarrow 2t > \dfrac{13}{2} \Rightarrow t > \dfrac{13}{4}$

The solutions must satisfy both of these inequalities. The interval is $\left(\dfrac{13}{4}, \infty\right)$.

51. $x - 4 \ge -3 \Rightarrow x \ge 1$ or $x - 4 \le 3 \Rightarrow x \le 7$

The solutions may satisfy either one or both of these inequalities. The interval is $(-\infty, \infty)$.

53. $-x < 1 \Rightarrow x > -1$ or $5x + 1 < -10 \Rightarrow 5x < -11 \Rightarrow x < -\dfrac{11}{5}$

The solutions may satisfy either one or both of these inequalities. The interval is $\left(-\infty, -\dfrac{11}{5}\right) \cup (-1, \infty)$.

55. $1 - 7x < -48 \Rightarrow -7x < -49 \Rightarrow x > 7$ and $3x + 1 \le -9 \Rightarrow 3x \le -10 \Rightarrow x \le -\dfrac{10}{3}$

The solutions must satisfy both of these inequalities. This is not possible. No solutions.

57. $-2 \le t + 4 < 5 \Rightarrow -6 \le t < 1;\ [-6, 1)$

59. $-\dfrac{5}{8} \le y - \dfrac{3}{8} < 1 \Rightarrow -\dfrac{1}{4} \le y < \dfrac{11}{8}; \left[-\dfrac{1}{4}, \dfrac{11}{8}\right)$

61. $-27 \le 3x \le 9 \Rightarrow -9 \le x \le 3; \ [-9, \ 3]$

63. $\dfrac{1}{2} < -2y \le 8 \Rightarrow -\dfrac{1}{4} > y \ge -4 \Rightarrow -4 \le y < -\dfrac{1}{4}; \left[-4, -\dfrac{1}{4}\right)$

65. $-4 < 5z + 1 \le 6 \Rightarrow -5 < 5z \le 5 \Rightarrow -1 < z \le 1; (-1, 1]$

67. $3 \le 4 - n \le 6 \Rightarrow -1 \le -n \le 2 \Rightarrow 1 \ge n \ge -2 \Rightarrow -2 \le n \le 1; [-2, 1]$

69. $-1 < 2z - 1 < 3 \Rightarrow 0 < 2z < 4 \Rightarrow 0 < z < 2; (0, 2)$

71. $-2 \le 5 - \dfrac{1}{3}m < 2 \Rightarrow -7 \le -\dfrac{1}{3}m < -3 \Rightarrow 21 \ge m > 9 \Rightarrow 9 < m \le 21; (9, 21]$

73. $100 \le 10(5x - 2) \le 200 \Rightarrow 10 \le 5x - 2 \le 20 \Rightarrow 12 \le 5x \le 22 \Rightarrow \dfrac{12}{5} \le x \le \dfrac{22}{5}; \left[\dfrac{12}{5}, \dfrac{22}{5}\right]$

75. $-3 < \dfrac{3z + 1}{4} < 1 \Rightarrow -12 < 3z + 1 < 4 \Rightarrow -13 < 3z < 3 \Rightarrow -\dfrac{13}{3} < z < 1; \left(-\dfrac{13}{3}, 1\right)$

77. $-\dfrac{5}{2} \le \dfrac{2 - m}{4} \le \dfrac{1}{2} \Rightarrow -10 \le 2 - m \le 2 \Rightarrow -12 \le -m \le 0 \Rightarrow 12 \ge m \ge 0 \Rightarrow 0 \le m \le 12; [0, 12]$

Numerical and Graphical Solutions

79.   The values of $3x$ are between -3 and 6 when $-1 \le x \le 2$. The interval is $[-1, 2]$.

81.   The values of $1 - x$ are between -1 and 2 when $-1 < x < 2$. The interval is $(-1, 2)$.

83.   The values of y_1 are between the lines $y = -2$ and $y = 2$ when $-3 \le x \le 1$. The interval is $[-3, 1]$.

85.   The values of y_1 are below the line $y = -2$ and above the line $y = 2$ when $x < -2$ or $x > 0$.

The interval is $(-\infty, -2) \cup (0, \infty)$.

87.   (a) The car is moving toward Omaha since the distance is decreasing.

(b) The car is 100 miles from Omaha when $x = 4$ hr. The car is 200 miles from Omaha when $x = 2$ hr.

(c) The car is 100 to 200 miles from Omaha when the elapsed time is between 2 and 4 hours.

(d) The car's distance from Omaha is greater than or equal to 200 miles during the first 2 hours.

89.   Numerical: Table $Y_1 = 2X - 4$ with TblStart = 0 and ΔTbl = 1. See Figure 89a.

Graphical: Graph $Y_1 = -2$, $Y_2 = 2X - 4$ and $Y_3 = 4$ in [−5, 5, 1] by [−5, 5, 1]. See Figure 89b.

The solution is $[1, 4]$.

[−5, 5, 1] by [−5, 5, 1]

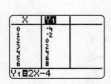
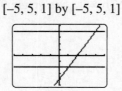

Figure 89a   Figure 89b

91.   Numerical: Table $Y_1 = X + 1$ with TblStart = −4 and ΔTbl = 1. See Figure 91a.

Graphical: Graph $Y_1 = -1$, $Y_2 = X + 1$ and $Y_3 = 1$ in [−5, 5, 1] by [−5, 5, 1]. See Figure 91b.

The solution is $(-\infty, -2) \cup (0, \infty)$.

[−5, 5, 1] by [−5, 5, 1]

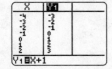
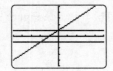

Figure 91a   Figure 91b

93.   Numerical: Table $Y_1 = 25(X - 2000) + 45$ with TblStart = 2000 and ΔTbl = 1.

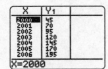

Graphical: Graph $Y_1 = 95$, $Y_2 = 25(X - 2000) + 45$ and $Y_3 = 295$ in [2000, 2020, 1] by [0, 325, 25].

The solution is $[2002, 2010]$.

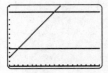

Using More Than One Method

95. $4 \le 5x - 1 \le 14 \Rightarrow 5 \le 5x \le 15 \Rightarrow 1 \le x \le 3; [1, 3]$

Graphical: Graph $Y_1 = 4$, $Y_2 = 5X - 1$ and $Y_3 = 14$ in [0, 10, 1] by [0, 20, 1]. See Figure 95a.

Numerical: Table $Y_1 = 5X - 1$ with TblStart = 0 and ΔTbl = 1. See Figure 95b.

The solution is $[1, 3]$.

[0, 10, 1] by [0, 20, 1]

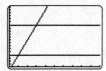   

Figure 95a   Figure 95b

97. $4 - x \ge 1$ or $4 - x < 3 \Rightarrow -x \ge -3$ or $-x < -1 \Rightarrow x \le 3$ or $x > 1; (-\infty, \infty)$

Graphical: Graph $Y_1 = 1$, $Y_2 = 4 - X$ and $Y_3 = 3$ in [−10, 10, 1] by [−10, 10, 1]. See Figure 97a.

Numerical: Table $Y_1 = 4 - X$ with TblStart = −1 and ΔTbl = 1. See Figure 97b.

The solution is $(-\infty, \infty)$.

[−10, 10, 1] by [−10, 10, 1]

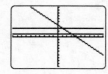   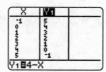

Figure 97a   Figure 97b

99. $2x + 1 < 3$ or $2x + 1 \ge 7 \Rightarrow x < 1$ or $x \ge 3; (-\infty, 1) \cup [3, \infty)$

Graphical: Graph $Y_1 = 3$, $Y_2 = 2X + 1$ and $Y_3 = 7$ in [−10, 10, 1] by [−10, 10, 1].

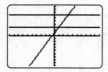

Numerical: Table $Y_1 = 2X + 1$ with TblStart = 0 and ΔTbl = 1.

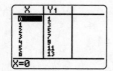

The solution is $(-\infty, 1) \cup [3, \infty)$.

101. $c < x + b \le d \Rightarrow c - b < x \le d - b; (c - b, d - b]$

Applications

103.   (a) $178.5 \le 5.7(x-1990)+150 \le 207$

$\Rightarrow 28.5 \le 5.7(x-1990) \le 57 \Rightarrow 5 \le x-1990 \le 10$

$\Rightarrow 1995 \le x \le 2000;$ The average price ranged from \$178,500 to \$207,000 from 1995 to 2000.

(b) Table $Y_1 = 5.7(X-1990)+150$ with TblStart = 1995 and ΔTbl = 1. The solution is 1995 to 2000.

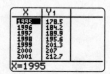

105. $51.3 \le 60-5.8x \le 57.1 \Rightarrow -8.7 \le -5.8x \le -2.9$

$\Rightarrow 1.5 \ge x \ge 0.5;$ The dew point ranges from $57.1°F$ to $51.3°F$ from 0.5 to 1.5 miles high.

107.   The perimeter of the rectangle is given by $2(x+3)+2(2x)$.

$40 \le 2(x+3)+2(2x) \le 60 \Rightarrow 40 \le 6x+6 \le 60 \Rightarrow 34 \le 6x \le 54 \Rightarrow 5.\overline{6} \le x \le 9$

109. $22.5 \le 70-19x \le 41.5 \Rightarrow -47.5 \le -19x \le -28.5 \Rightarrow 2.5 \ge x \ge 1.5 \Rightarrow 1.5 \le x \le 2.5$

The air temperature is from $22.5°F$ to $41.5°F$ at altitudes from 1.5 to 2.5 miles.

111. $-90 \le \dfrac{9}{5}C+32 \le 98 \Rightarrow -122 \le \dfrac{9}{5}C \le 66 \Rightarrow -122\left(\dfrac{5}{9}\right) \le C \le 66\left(\dfrac{5}{9}\right) \Rightarrow -67.\overline{7} \le C \le 36.\overline{6}$

The temperature range in Celsius is $-67.\overline{7}°C$ to $36.\overline{6}°C$.

Checking Basic Concepts for Sections 3.3 & 3.4

1. $4-3x < \dfrac{1}{2}x \Rightarrow 8-6x < x \Rightarrow 8 < 7x \Rightarrow \dfrac{8}{7} < x \Rightarrow \left\{x \middle| x > \dfrac{8}{7}\right\}$

2.   (a) The two lines intersect at (2, 0). The solution is $\{x \mid x = 2\}$.

(b) The line $y = \dfrac{1}{2}x-1$ is below the line $y = 2-x$ when $x < 2$. The solution set is $\{x \mid x < 2\}$.

(c) The line $y = \dfrac{1}{2}x-1$ intersects and is above the line $y = 2-x$ when $x \ge 2$. The solution set is $\{x \mid x \ge 2\}$.

3.   (a) Yes, since $2(3)-1 = 5 \ge 3$.

(b) No, since $(3)+2 = 5 \not< 4$.

4.   (a) $-5 \le 2x+1 \le 3 \Rightarrow -6 \le 2x \le 2 \Rightarrow -3 \le x \le 1; [-3, 1]$

(b) $1-x \le -2 \Rightarrow -x \le -3 \Rightarrow x \ge 3$ or $1-x > 2 \Rightarrow -x > 1 \Rightarrow x < -1; (-\infty, -1) \cup [3, \infty)$

(c) $-2 < \dfrac{4-3x}{2} \le 6 \Rightarrow -4 < 4-3x \le 12 \Rightarrow -8 < -3x \le 8 \Rightarrow \dfrac{8}{3} > x \ge -\dfrac{8}{3}; \left[-\dfrac{8}{3}, \dfrac{8}{3}\right)$

3.5: Absolute Value Equations and Inequalities

Concepts

1.   An example of an absolute value equation is $|3x+2|=6$. *Answers may vary.*

2.   An example of an absolute value inequality is $|2x-1|\leq 17$. *Answers may vary.*

3.   Yes, since $|-3|=3$.

4.   Yes, since $|-4|=4>3$.

5.   Yes

6.   No, it is equivalent to $-3<x<3$.

7.   2 times

8.   0 times

9.   No, $|2(-3)-5|=|-11|=11\neq 1$. Yes, $|2(3)-5|=|1|=1$.

10.   Yes, $|5-6(1)|=|-1|=1$. No, $|5-6(0)|=|5|=5\neq 1$.

11.   No, $|7-4(-2)|=|15|=15\not\leq 5$. Yes, $|7-4(2)|=|-1|=1\leq 1$.

12.   No, $|2+(-4)|=|-2|=2\not< 2$. Yes, $|2+(-1)|=|1|=1<2$.

13.   Yes, $\left|7\left(-\dfrac{4}{7}\right)+4\right|=|-4+4|=|0|=0>-1$. Yes, $|7(2)+4|=|18|=18>-1$.

14.   No, $\left|12\left(-\dfrac{1}{4}\right)+3\right|=|-3+3|=|0|=0\not\geq 3$. Yes, $|12(2)+3|=|27|=27\geq 3$.

15. $x=0$ or $x=4$

16. $x=-2$ or $x=1$

Symbolic Solutions

17. $x=-7$ or $x=7$

19.   An absolute value can not be negative. No solution.

21. $4x=9\Rightarrow x=\dfrac{9}{4}$ or $4x=-9\Rightarrow x=-\dfrac{9}{4}$

23.   Since $|-2x|-6=2\Rightarrow|-2x|=8$: $-2x=8\Rightarrow x=-4$ or $-2x=-8\Rightarrow x=4$

25. $2x+1=11\Rightarrow 2x=10\Rightarrow x=5$ or $2x+1=-11\Rightarrow 2x=-12\Rightarrow x=-6$

27.   Since $|-2x+3|+3=4\Rightarrow|-2x+3|=1$:

       $-2x+3=1\Rightarrow -2x=-2\Rightarrow x=1$ or $-2x+3=-1\Rightarrow -2x=-4\Rightarrow x=2$

29. $\dfrac{1}{2}x-1=5\Rightarrow\dfrac{1}{2}x=6\Rightarrow x=12$ or $\dfrac{1}{2}x-1=-5\Rightarrow\dfrac{1}{2}x=-4\Rightarrow x=-8$

31.   An absolute value can not be negative. No solution.

33.   Since $\left|\dfrac{2}{3}z-1\right|-3=8 \Rightarrow \left|\dfrac{2}{3}z-1\right|=11$:

$\dfrac{2}{3}z-1=11 \Rightarrow \dfrac{2}{3}z=12 \Rightarrow z=18$ or $\dfrac{2}{3}z-1=-11 \Rightarrow \dfrac{2}{3}z=-10 \Rightarrow z=-15$

35. $z-1=2z \Rightarrow -z=1 \Rightarrow z=-1$ or $z-1=-2z \Rightarrow 3z=1 \Rightarrow z=\dfrac{1}{3}$

37. $3t+1=2t-4 \Rightarrow t=-5$ or $3t+1=-2t+4 \Rightarrow 5t=3 \Rightarrow t=\dfrac{3}{5}$

39. $\dfrac{1}{4}x=3+\dfrac{1}{4}x \Rightarrow 0=3$ (no solution)   or $\dfrac{1}{4}x=-3-\dfrac{1}{4}x \Rightarrow \dfrac{1}{2}x=-3 \Rightarrow x=-6$

41.   (a) $2x=8 \Rightarrow x=4$ or $2x=-8 \Rightarrow x=-4$

(b) $\{x|-4<x<4\}$

(c) $\{x|x<-4 \text{ or } x>4\}$

43.   (a) $5-4x=3 \Rightarrow -4x=-2 \Rightarrow x=\dfrac{1}{2}$ or $5-4x=-3 \Rightarrow -4x=-8 \Rightarrow x=2$

(b) $\left\{x\left|\dfrac{1}{2}\le x\le 2\right.\right\}$

(c) $\left\{x\left|x\le\dfrac{1}{2} \text{ or } x\ge 2\right.\right\}$

45.   The solutions to $|x|\le 3$ satisfy $c\le x\le d$ where c and d are the solutions to $|x|=3$.

$|x|=3$ is equivalent to $x=-3$ and $x=3$. The interval is $[-3,\ 3]$.

47.   The solutions to $|k|>4$ satisfy $k<c$ or $k>d$ where c and d are the solutions to $|k|=4$.

$|k|=4$ is equivalent to $k=-4$ and $k=4$. The interval is $(-\infty,\ -4)\cup(4,\ \infty)$.

49.   The inequality $|t|\le -3$ has no solutions because absolute value is never negative.

51.   The inequality $|z|>0$ is true for any value of z except $z=0$. The interval is $(-\infty,\ 0)\cup(0,\ \infty)$.

53.   The solutions to $|2x|>7$ satisfy $x<c$ or $x>d$ where c and d are the solutions to $|2x|=7$.

$|2x|=7$ is equivalent to $2x=-7 \Rightarrow x=-\dfrac{7}{2}$ or $2x=7 \Rightarrow x=\dfrac{7}{2}$.

The interval is $\left(-\infty,\ -\dfrac{7}{2}\right)\cup\left(\dfrac{7}{2},\ \infty\right)$.

55.   The solutions to $|-4x+4|<16$ satisfy $c<x<d$ where c and d are the solutions to $|-4x+4|=16$.

$|-4x+4|=16$ is equivalent to $-4x+4=-16 \Rightarrow x=5$ or $-4x+4=16 \Rightarrow x=-3$.

The interval is $(-3,\ 5)$.

57. First divide each side of $2|x+5| \geq 8$ by 2 to obtain $|x+5| \geq 4$.

    The solutions to $|x+5| \geq 4$ satisfy $x \leq c$ or $x \geq d$ where c and d are the solutions to $|x+5| = 4$.

    $|x+5| = 4$ is equivalent to $x+5 = -4 \Rightarrow x = -9$ or $x+5 = 4 \Rightarrow x = -1$.

    The interval is $(-\infty, -9] \cup [-1, \infty)$.

59. First add 1 to each side of $|8-6x| - 1 \leq 2$ to obtain $|8-6x| \leq 3$.

    The solutions to $|8-6x| \leq 3$ satisfy $c \leq x \leq d$ where c and d are the solutions to $|8-6x| = 3$.

    $|8-6x| = 3$ is equivalent to $8-6x = -3 \Rightarrow x = \dfrac{11}{6}$ or $8-6x = 3 \Rightarrow x = \dfrac{5}{6}$. The interval is $\left[\dfrac{5}{6}, \dfrac{11}{6}\right]$.

61. First subtract 5 from each side of $5 + \left|\dfrac{2-x}{3}\right| \leq 9$ to obtain $\left|\dfrac{2-x}{3}\right| \leq 4$.

    The solutions to $\left|\dfrac{2-x}{3}\right| \leq 4$ satisfy $c \leq x \leq d$ where c and d are the solutions to $\left|\dfrac{2-x}{3}\right| = 4$.

    $\left|\dfrac{2-x}{3}\right| = 4$ is equivalent to $\dfrac{2-x}{3} = -4 \Rightarrow x = 14$ or $\dfrac{2-x}{3} = 4 \Rightarrow x = -10$. The interval is $[-10, 14]$.

63. The inequality $|2x-1| \leq -3$ has no solutions because absolute value is never negative.

65. First add 1 to each side of $|x+1| - 1 > -3$ to obtain $|x+1| > -2$.

    The inequality $|x+1| > -2$ is true for all values of x because absolute value is never negative.

    The interval is $(-\infty, \infty)$.

67. The inequality $|2z-4| \leq -1$ has no solutions because absolute value is never negative.

69. The inequality $|3z-1| > -3$ is true for all values of z because absolute value is never negative.

    The interval is $(-\infty, \infty)$.

71. The solutions to $\left|\dfrac{2-t}{3}\right| \geq 5$ satisfy $t \leq c$ or $t \geq d$ where c and d are the solutions to $\left|\dfrac{2-t}{3}\right| = 5$.

    $\left|\dfrac{2-t}{3}\right| = 5$ is equivalent to $\dfrac{2-t}{3} = -5 \Rightarrow t = 17$ or $\dfrac{2-t}{3} = 5 \Rightarrow x = -13$.

    The interval is $(-\infty, -13] \cup [17, \infty)$.

73. The solutions to $|t-1| \leq 0.1$ satisfy $c \leq t \leq d$ where c and d are the solutions to $|t-1| = 0.1$.

    $|t-1| = 0.1$ is equivalent to $t-1 = -0.1 \Rightarrow t = 0.9$ or $t-1 = 0.1 \Rightarrow t = 1.1$. The interval is $[0.9, 1.1]$.

75. The solutions to $|b-10| > 0.5$ satisfy $b < c$ or $b > d$ where c and d are the solutions to $|b-10| = 0.5$.

    $|b-10| = 0.5$ is equivalent to $b-10 = -0.5 \Rightarrow b = 9.5$ or $b-10 = 0.5 \Rightarrow b = 10.5$.

    The interval is $(-\infty, 9.5) \cup (10.5, \infty)$.

Numerical and Graphical Solutions

77. (a) From the table, $y = 2$ when $x = -1$ or $x = 3$.

    (b) $y < 2$ when $-1 < x < 3$. The interval is $(-1, 3)$.

    (c) $y > 2$ when $x < -1$ or $x > 3$. The interval is $(-\infty, -1) \cup (3, \infty)$.

79. (a) From the graph $y_1 = 1$ when $x = -1$ or $x = 0$.

    (b) $y_1 \leq 1$ when $-1 \leq x \leq 0$. The interval is $[-1, 0]$.

    (c) $y_1 \geq 1$ when $x \leq -1$ or $x \geq 0$. The interval is $(-\infty, -1] \cup [0, \infty)$.

81. Graph $Y_1 = \text{abs}(X)$ and $Y_2 = 1$ in $[-3, 3, 1]$ by $[-3, 3, 1]$. See Figures 81a and 81b. $(-\infty, -1] \cup [1, \infty)$

    $[-3, 3, 1]$ by $[-3, 3, 1]$ $[-3, 3, 1]$ by $[-3, 3, 1]$

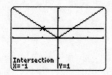   

    Figure 81a   Figure 81b

83. Graph $Y_1 = \text{abs}(X - 1)$ and $Y_2 = 3$ in $[-5, 5, 1]$ by $[-5, 5, 1]$. See Figures 83a and 83b. $[-2, 4]$

    $[-5, 5, 1]$ by $[-5, 5, 1]$ $[-5, 5, 1]$ by $[-5, 5, 1]$

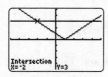   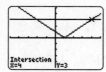

    Figure 83a   Figure 83b

85. Graph $Y_1 = \text{abs}(4 - 2X)$ and $Y_2 = 2$ in $[0, 5, 1]$ by $[0, 5, 1]$. See Figures 85a and 85b. $(-\infty, 1) \cup (3, \infty)$

    $[0, 5, 1]$ by $[0, 5, 1]$ $[0, 5, 1]$ by $[0, 5, 1]$

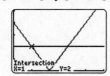   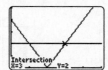

    Figure 85a   Figure 85b

87. Graph $Y_1 = \text{abs}(10 - 3X)$ and $Y_2 = 4$ in $[0, 6, 1]$ by $[0, 6, 1]$. See Figures 87a and 87b. $\left(2, 4.\overline{6}\right)$

    $[0, 6, 1]$ by $[0, 6, 1]$ $[0, 6, 1]$ by $[0, 6, 1]$

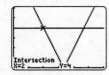   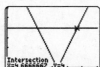

    Figure 87a   Figure 87b

89.   Graph $Y_1 = \text{abs}(8.1 - X)$ and $Y_2 = -2$ in [0, 15, 1] by [−5, 5, 1]. See Figure 89. $(-\infty, \infty)$

[0, 15, 1] by [−5, 5, 1]

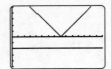

Figure 89

Using More Than One Method

91.   (a) $3x = 9 \Rightarrow x = 3$ or $3x = -9 \Rightarrow x = -3;$ $\{x \mid -3 \le x \le 3\}$

(b)   Graph $Y_1 = \text{abs}(3X)$ and $Y_2 = 9$ in [−5, 5, 1] by [−5, 15, 1]. See Figures 91a and 91b.

(c)   Table $Y_1 = \text{abs}(3X)$ with TblStart = −7 and ΔTbl = 2. See Figure 91c.

The solution set is $\{x \mid -3 \le x \le 3\}$.

[−5, 5, 1] by [−5, 15, 1]   [−5, 5, 1] by [−5, 15, 1]

  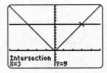 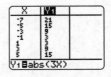

Figure 91a   Figure 91b   Figure 91c

93.   (a) $2x - 5 = 1 \Rightarrow 2x = 6 \Rightarrow x = 3$ or $2x - 5 = -1 \Rightarrow 2x = 4 \Rightarrow x = 2;$ $\{x \mid x < 2 \text{ or } x > 3\}$

(b)   Graph $Y_1 = \text{abs}(2X - 5)$ and $Y_2 = 1$ in [0, 5, 1] by [−1, 2, 1]. See Figures 93a and 93b.

(c)   Table $Y_1 = \text{abs}(2X - 5)$ with TblStart = 1 and ΔTbl = 0.5. See Figure 93c.

The solution set is $\{x \mid x < 2 \text{ or } x > 3\}$.

[0, 5, 1] by [−1, 2, 1]   [0, 5, 1] by [−1, 2, 1]

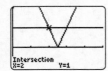 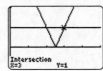 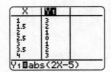

Figure 93a   Figure 93b   Figure 93c

95. $|x| \le 4$

97. $|y| > 2$

99. $|2x + 1| \le 0.3$

101. $|\pi x| \ge 7$

103.   two

Applications

105. (a) $T - 43 = 24 \Rightarrow T = 67$ or $T - 43 = -24 \Rightarrow T = 19$; $\{T | 19 \leq T \leq 67\}$

   (b) The monthly average temperatures in Marquette, Michigan vary from 19°F to 67°F.

107. (a) $T - 10 = 36 \Rightarrow T = 46$ or $T - 10 = -36 \Rightarrow T = -26$; $\{T | -26 \leq T \leq 46\}$

   (b) The monthly average temperatures in Chesterfield, Canada vary from –26°F to 46°F.

109. (a) $A = (29,028 + 22,834 + 20,320 + 19,340 + 18,510 + 16,066 + 7,310) \div 7 \approx 19,058$ feet

   (b) Africa and Europe have elevations within 1000 feet of A.

   (c) South America, North America, Africa, Europe and Antarctica have elevations within 5000 feet of A.

111. $d - 2.5 = 0.002 \Rightarrow d = 2.502$ or $d - 2.5 = -0.002 \Rightarrow d = 2.498$; $\{d | 2.498 \leq d \leq 2.502\}$

   The diameter can vary from 2.498 inches to 2.502 inches.

113. $|d - 3.8| \leq 0.03$

115. The solutions to $\left| \dfrac{x - 20}{20} \right| < 0.05$ satisfy $c < x < d$ where c and d are the solutions to $\left| \dfrac{x - 20}{20} \right| = 0.05$.

   $\left| \dfrac{x - 20}{20} \right| = 0.05$ is equivalent to $\dfrac{x - 20}{20} = -0.05 \Rightarrow x = 19$ or $\dfrac{x - 20}{20} = 0.05 \Rightarrow x = 21$.

   The interval is $(19, 21)$. The values must be between 19 and 21, exclusively.

Checking Basic Concepts for Section 3.5

1. $\left| \dfrac{3}{4} x - 1 \right| - 3 = 5 \Rightarrow \left| \dfrac{3}{4} x - 1 \right| = 8$

   $\dfrac{3}{4} x - 1 = -8 \Rightarrow \dfrac{3}{4} x = -7 \Rightarrow x = -\dfrac{28}{3}$ or $\dfrac{3}{4} x - 1 = 8 \Rightarrow \dfrac{3}{4} x = 9 \Rightarrow x = 12$

2. (a) $3x - 6 = 8 \Rightarrow 3x = 14 \Rightarrow x = \dfrac{14}{3}$ or $3x - 6 = -8 \Rightarrow 3x = -2 \Rightarrow x = -\dfrac{2}{3}$

   (b) The solutions to $|3x - 6| < 8$ satisfy $c < x < d$ where c and d are the solutions to $|3x - 6| = 8$.

   From part (a), the interval is $\left(-\dfrac{2}{3}, \dfrac{14}{3} \right)$.

   (c) The solutions to $|3x - 6| > 8$ satisfy $x < c$ or $x > d$ where c and d are the solutions to $|3x - 6| = 8$.

   From part (a), the interval is $\left(-\infty, -\dfrac{2}{3} \right) \cup \left(\dfrac{14}{3}, \infty \right)$.

3. The solutions to $|-2(3 - x)| < 6$ satisfy $c < x < d$ where c and d are the solutions to $|-2(3 - x)| = 6$.

   $|-2(3 - x)| = 6$ is equivalent to $-2(3 - x) = -6 \Rightarrow x = 0$ or $-2(3 - x) = 6 \Rightarrow x = 6$.

   The interval is $(0, 6)$. Similarly, the solution to $|-2(3 - x)| \geq 6$ is the interval $(-\infty, 0] \cup [6, \infty)$.

4. (a) From the graph $y = 2$ when $x = 1$ or $x = 3$.

(b) $y \le 2$ when $1 \le x \le 3$. The interval is $[1, 3]$.

(c) $y \ge 2$ when $x \le 1$ or $x \ge 3$. The interval is $(-\infty, 1] \cup [3, \infty)$.

Chapter 3 Review Exercises

Section 3.1

1. The completed table is shown in Figure 1. From the table we see that $3x - 6 = 0$ when $x = 2$.

x	0	1	2	3	4
$3x - 6$	-6	-3	0	3	6

x	-1	0	1	2	3
$5 - 2x$	7	5	3	1	-1

Figure 1   Figure 2

2. The completed table is shown in Figure 2. From the table we see that $5 - 2x = 3$ when $x = 1$.

3. Yes it is a solution since $\dfrac{1}{2}\left(-\dfrac{2}{3}\right) - 2\left(2\left(-\dfrac{2}{3}\right) + 3\right) = -\dfrac{11}{3}$ and $4\left(-\dfrac{2}{3}\right) - 1 = -\dfrac{11}{3}$.

4. Since the graphs intersect at the point $(1, 3)$, the solution is $x = 1$.

5. Graph $Y_1 = 5 - 2X$ and $Y_2 = -1 + X$ in $[-10, 10, 1]$ by $[-10, 10, 1]$. See Figure 5. $x = 2$

$[-10, 10, 1]$ by $[-10, 10, 1]$

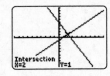

Figure 5   Figure 6

6. Table $Y_1 = 4X - 3$ and $Y_2 = 5 + 2X$ with TblStart = 0 and ΔTbl = 1. See Figure 6. $x = 4$

7. $2x - 7 = 21 \;\Rightarrow\; 2x = 28 \;\Rightarrow\; x = \dfrac{28}{2} \;\Rightarrow\; x = 14$

8. $1 - 7x = -\dfrac{5}{2} \;\Rightarrow\; -7x = -\dfrac{5}{2} - 1 \;\Rightarrow\; -7x = -\dfrac{7}{2} \;\Rightarrow\; x = -\dfrac{1}{7}\left(-\dfrac{7}{2}\right) \;\Rightarrow\; x = \dfrac{1}{2}$

9. $-2(4x - 1) = 1 - x \;\Rightarrow\; -8x + 2 = 1 - x \;\Rightarrow\; 1 = 7x \;\Rightarrow\; \dfrac{1}{7} = x \;\Rightarrow\; x = \dfrac{1}{7}$

10. $-\dfrac{3}{4}(x - 1) + 5 = 6 \;\Rightarrow\; -\dfrac{3}{4}x + \dfrac{3}{4} + 5 = 6 \;\Rightarrow\; -\dfrac{3}{4}x = \dfrac{1}{4} \;\Rightarrow\; x = -\dfrac{4}{3}\left(\dfrac{1}{4}\right) \;\Rightarrow\; x = -\dfrac{1}{3}$

11. $\dfrac{x - 4}{3} = 2 \;\Rightarrow\; x - 4 = 6 \;\Rightarrow\; x = 10$

12. $\dfrac{2x - 3}{2} = \dfrac{x + 3}{5} \;\Rightarrow\; 5(2x - 3) = 2(x + 3) \;\Rightarrow\; 10x - 15 = 2x + 6 \;\Rightarrow\; 8x = 21 \;\Rightarrow\; x = \dfrac{21}{8}$

13. $-2(z - 1960) + 32 = 8 \;\Rightarrow\; -2(z - 1960) = -24 \;\Rightarrow\; z - 1960 = 12 \;\Rightarrow\; z = 1972$

14. $5 - (3 - 2z) + 4 = 5(2z - 3) - (3z - 2) \;\Rightarrow\; 2z + 6 = 7z - 13 \;\Rightarrow\; -5z = -19 \;\Rightarrow\; z = \dfrac{19}{5}$

15. $\dfrac{2}{3}\left(\dfrac{t - 3}{2}\right) + 4 = \dfrac{1}{3}t - (1 - t) \;\Rightarrow\; \dfrac{t - 3}{3} + 4 = \dfrac{4}{3}t - 1 \;\Rightarrow\; t - 3 + 12 = 4t - 3 \;\Rightarrow\;$

$t + 9 = 4t - 3 \implies -3t = -12 \implies t = 4$

16. $0.3r - 0.12(r - 1) = 0.4r + 2.1 \implies 0.18r + 0.12 = 0.4r + 2.1 \implies -0.22r = 1.98 \implies r = -9$

17. $-4(5 - 3x) = 12x - 20 \implies -20 + 12x = 12x - 20 \implies -20 = -20;$ identity

18. $4x - (x - 3) = 3x - 3 \implies 4x - x + 3 = 3x - 3 \implies 3x + 3 = 3x - 3 \implies 3 = -3;$ contradiction

19. To find the x-intercept, let $y = 0$ and solve for $x : 4x - 5(0) = 20 \implies 4x = 20 \implies x = 5;$ to find the y-intercept,

   let $x = 0$ and solve for $y : 4(0) - 5y = 20 \implies -5y = 20 \implies y = -4;$

   $$4x - 5y = 20 \implies -5y = -4x + 20 \implies y = \frac{4}{5}x - 4$$

20. To find the x-intercept, let $y = 0$ and solve for $x : 2x + \frac{1}{2}(0) = -6 \implies 2x = -6 \implies x = -3;$ to find the y-intercept,

   let $x = 0$ and solve for $y : 2(0) + \frac{1}{2}y = -6 \implies \frac{1}{2}y = -6 \implies y = -12;$

   $$2x + \frac{1}{2}y = -6 \implies \frac{1}{2}y = -2x - 6 \implies y = -4x - 12$$

Section 3.2

21. $5x - 4y = 20 \implies 5x - 20 = 4y \implies \dfrac{5x - 20}{4} = y \implies y = \dfrac{5}{4}x - 5$

22. $-\dfrac{1}{3}x + \dfrac{1}{2}y = 1 \implies \dfrac{1}{2}y = \dfrac{1}{3}x + 1 \implies y = \dfrac{2}{3}x + 2$

23. $2a + 3b = a \implies a = -3b$

24. $4m - 5n = 6m + 2n \implies -7n = 2m \implies n = -\dfrac{2}{7}m$

25. $A = \dfrac{1}{2}h(a + b) \implies 2A = h(a + b) \implies \dfrac{2A}{h} = a + b \implies \dfrac{2A}{h} - a = b \implies b = \dfrac{2A}{h} - a$

26. $V = \dfrac{1}{3}\pi r^2 h \implies 3V = \pi r^2 h \implies \dfrac{3V}{\pi r^2} = h \implies h = \dfrac{3V}{\pi r^2}$

27. $\dfrac{1}{2}x - \dfrac{3}{4}y = 2 \implies 2x - 3y = 8 \implies -3y = -2x + 8 \implies y = \dfrac{2}{3}x - \dfrac{8}{3} \implies f(x) = \dfrac{2}{3}x - \dfrac{8}{3}$

28. $-7(x - 4y) = y + 1 \implies -7x + 28y = y + 1 \implies 27y = 7x + 1 \implies y = \dfrac{7}{27}x + \dfrac{1}{27} \implies f(x) = \dfrac{7}{27}x + \dfrac{1}{27}$

29. $3x + 4y = 10 - y \implies 5y = -3x + 10 \implies y = -\dfrac{3}{5}x + 2 \implies f(x) = -\dfrac{3}{5}x + 2$

30. $\dfrac{y}{x} = 3 \implies y = 3x \implies f(x) = 3x$

31. $2x + 25 = 19; \ 2x + 25 = 19 \implies 2x = -6 \implies x = \dfrac{-6}{2} \implies x = -3$

32. $2x - 5 = x + 1; \ 2x - 5 = x + 1 \implies x = 6$

Section 3.3

33.   From the table $f(x) < 5$ when $x > -1; \{x | x > -1\}$.

34. $2(3-x)+4 < 0 \Rightarrow 6-2x+4 < 0 \Rightarrow -2x+10 < 0 \Rightarrow -2x < -10 \Rightarrow x > 5; \{x | x > 5\}$

35.   From the graph $y_1 \geq y_2$ when $x \geq 2; \{x | x \geq 2\}$.

36.   (a) $5-4x = -2 \Rightarrow -4x = -7 \Rightarrow x = \dfrac{7}{4}$

      (b) $5-4x < -2 \Rightarrow -4x < -7 \Rightarrow x > \dfrac{7}{4} \Rightarrow \left\{ x \Big| x > \dfrac{7}{4} \right\}$

      (c) $5-4x > -2 \Rightarrow -4x > -7 \Rightarrow x < \dfrac{7}{4} \Rightarrow \left\{ x \Big| x < \dfrac{7}{4} \right\}$

37. $-2x+1 \leq 3 \Rightarrow -2x \leq 2 \Rightarrow x \geq -1; \{x | x \geq -1\}$

38. $x-5 \geq 2x+3 \Rightarrow -x \geq 8 \Rightarrow x \leq -8; \{x | x \leq -8\}$

39. $\dfrac{3x-1}{4} > \dfrac{1}{2} \Rightarrow 3x-1 > 2 \Rightarrow 3x > 3 \Rightarrow x > 1; \{x | x > 1\}$

40. $-3.2(x-2) < 1.6x \Rightarrow -3.2x+6.4 < 1.6x \Rightarrow 6.4 < 4.8x \Rightarrow \dfrac{6.4}{4.8} < x \Rightarrow x > \dfrac{4}{3}; \left\{ x \Big| x > \dfrac{4}{3} \right\}$

41. $\dfrac{2}{5}t-(5-t)+3 > 2\left(\dfrac{3-t}{5}\right) \Rightarrow 2t-25+5t+15 > 6-2t \Rightarrow 7t-10 > 6-2t \Rightarrow$

      $9t > 16 \Rightarrow t > \dfrac{16}{9}; \left\{ t \Big| t > \dfrac{16}{9} \right\}$

42. $\dfrac{2(3x-5)}{5} - 3 \geq \dfrac{2-x}{3} + 2 \Rightarrow 18x-30-45 \geq 10-5x+30 \Rightarrow 18x-75 \geq -5x+40 \Rightarrow$

      $23x \geq 115 \Rightarrow x \geq 5; \{x | x \geq 5\}$

43. $0.05t-0.15 \leq 0.03-0.75(t-2) \Rightarrow 0.05t-0.15 \leq -0.75t+1.53 \Rightarrow$

      $0.8t \leq 1.68 \Rightarrow t \leq 2.1; \{t | t \leq 2.1\}$

44. $0.6z-1.55(z+5) < 1-0.35(2z-1) \Rightarrow 0.6z-1.55z-7.75 < 1-0.7z+0.35 \Rightarrow$

      $-0.95z-7.75 < -0.7z+1.35 \Rightarrow -0.25z < 9.1 \Rightarrow z > -36.4; \{z | z > -36.4\}$

45.   Graph $Y_1 = 0.72(\pi-1.3x)$ and $Y_2 = 0.54$ in $[-10,10,1]$ by $[-10,10,1]$. The solution is $\{x | x \leq 1.840\}$.

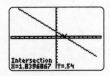

      Figure 45

46.   Graph $Y_1 = \sqrt{(2)} - (4 - \pi x)$ and $Y_2 = \sqrt{(7)}$ in $[-10, 10, 1]$ by $[-10, 10, 1]$. The solution is $\{x | x > 1.665\}$.

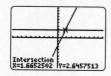

Figure 46

Section 3.4

47. $x + 1 \le 3 \Rightarrow x \le 2$ and $x + 1 \ge -1 \Rightarrow x \ge -2$; $[-2, 2]$.   See Figure 47.

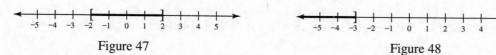

Figure 47   Figure 48

48. $2x + 7 < 5 \Rightarrow 2x < -2 \Rightarrow x < -1$ and $-2x \ge 6 \Rightarrow x \le -3$; $(-\infty, -3]$.   See Figure 48.

49. $5x - 1 \le 3 \Rightarrow 5x \le 4 \Rightarrow x \le \dfrac{4}{5}$ or $1 - x < -1 \Rightarrow 2 < x$; $\left(-\infty, \dfrac{4}{5}\right] \cup (2, \infty)$.   See Figure 49.

Figure 49   Figure 50

50. $3x + 1 > -1 \Rightarrow 3x > -2 \Rightarrow x > -\dfrac{2}{3}$ or $3x + 1 < 10 \Rightarrow 3x < 9 \Rightarrow x < 3$; $(-\infty, \infty)$.   See Figure 50.

51. $2x + 2$ is between -2 and 4 when $-2 \le x \le 1$; $[-2, 1]$

52.   (a) The intersection point of y_1 and y_2 is $(-4, -100)$. The solution is $x = -4$.

       (b) The intersection point of y_2 and y_3 is $(2, 50)$. The solution is $x = 2$.

       (c) y_2 is between y_1 and y_3 when $-4 \le x \le 2$; $[-4, 2]$

       (d) y_2 is below y_3 when $x < 2$; $(-\infty, 2)$

53.   (a) The intersection point of y_1 and y_2 is $(2, 2)$. The solution is $x = 2$.

       (b) y_1 is below y_2 when $x > 2$; $(2, \infty)$

       (c) y_1 is above y_2 when $x < 2$; $(-\infty, 2)$

54.   (a) The intersection point of $f(x)$ and $g(x)$ is $(4, 20)$. The solution is $x = 4$.

       (b) The intersection point of $g(x)$ and $h(x)$ is $(2, 40)$. The solution is $x = 2$.

       (c) $g(x)$ is between $f(x)$ and $h(x)$ when $2 < x < 4$; $(2, 4)$

55. $\left[-3, \dfrac{2}{3}\right]$

56. $(-6, 45]$

57. $\left(-\infty, \dfrac{7}{2}\right)$

58. $[1.8, \infty)$

59. $(-3, 4)$

60. $(-\infty, 4) \cup (10, \infty)$

61. $-4 < x+1 < 6 \Rightarrow -5 < x < 5$; The solution set is $(-5, 5)$.

62. $20 \le 2x+4 \le 60 \Rightarrow 16 \le 2x \le 56 \Rightarrow 8 \le x \le 28$; The solution set is $[8, 28]$.

63. $-3 < 4 - \dfrac{1}{3}x < 7 \Rightarrow -7 < -\dfrac{1}{3}x < 3 \Rightarrow 21 > x > -9 \Rightarrow -9 < x < 21$; The solution set is $(-9, 21)$.

64. $2 \le \dfrac{1}{2}x - 2 \le 12 \Rightarrow 4 \le \dfrac{1}{2}x \le 14 \Rightarrow 8 \le x \le 28$; The solution set is $[8, 28]$.

65. $-3 \le \dfrac{4-5x}{3} - 2 < 3 \Rightarrow -9 \le 4 - 5x - 6 < 9 \Rightarrow -9 \le -5x - 2 < 9 \Rightarrow -7 \le -5x < 11 \Rightarrow$

    $\dfrac{7}{5} \ge x > -\dfrac{11}{5} \Rightarrow -\dfrac{11}{5} < x \le \dfrac{7}{5}; \left(-\dfrac{11}{5}, \dfrac{7}{5}\right]$

66. $30 \le \dfrac{2x-6}{5} - 4 < 50 \Rightarrow 150 \le 2x - 6 - 20 < 250 \Rightarrow 150 \le 2x - 26 < 250 \Rightarrow$

    $176 \le 2x < 276 \Rightarrow 88 \le x < 138; [88, 138)$

Section 3.5

67. No, $|12(-3) - 24| = |-60| = 60 \ne 24$. No, $|12(2) - 24| = |0| = 0 \ne 24$.

68. No, $\left|5 - 3\left(\dfrac{4}{3}\right)\right| = |1| = 1 \not> 3$. Yes, $|5 - 3(0)| = |5| = 5 > 3$.

69. No, $|3(-3) - 6| = |-15| = 15 \not\le 6$. Yes, $|3(4) - 6| = |6| = 6 \le 6$.

70. No, $|2 + 3(-3)| + 4 = |-7| + 4 = 7 + 4 = 11 \not< 11$. Yes, $\left|2 + 3\left(\dfrac{2}{3}\right)\right| + 4 = |4| + 4 = 4 + 4 = 8 < 11$.

71. (a) From the table, $y_1 = 2$ when $x = 0$ or $x = 4$.

    (b) From the table, $y_1 < 2$ when $0 < x < 4$; $(0, 4)$.

    (c) From the table, $y_1 > 2$ when $x < 0$ or $x > 4$; $(-\infty, 0) \cup (4, \infty)$.

72. (a) From the graph, $|2x + 2| = 4$ when $x = -3$ or $x = 1$.

    (b) From the graph, $|2x + 2| \le 4$ when $-3 \le x \le 1$; $[-3, 1]$.

    (c) From the graph, $|2x + 2| \ge 4$ when $x \le -3$ or $x \ge 1$; $(-\infty, -3] \cup [1, \infty)$.

73. $x = -22$ or $x = 22$

74. $2x - 9 = 7 \Rightarrow 2x = 16 \Rightarrow x = 8$ or $2x - 9 = -7 \Rightarrow 2x = 2 \Rightarrow x = 1$

75. $4 - \frac{1}{2}x = 17 \Rightarrow -\frac{1}{2}x = 13 \Rightarrow x = -26$ or $4 - \frac{1}{2}x = -17 \Rightarrow -\frac{1}{2}x = -21 \Rightarrow x = 42$

76.   First note that $\frac{1}{3}|3x-1| + 1 = 9 \Rightarrow \frac{1}{3}|3x-1| = 8 \Rightarrow |3x-1| = 24$.

$3x - 1 = 24 \Rightarrow 3x = 25 \Rightarrow x = \frac{25}{3}$ or $3x - 1 = -24 \Rightarrow 3x = -23 \Rightarrow x = -\frac{23}{3}$

77. $2x - 5 = 5 - 3x \Rightarrow 5x = 10 \Rightarrow x = 2$ or $2x - 5 = -5 + 3x \Rightarrow -x = 0 \Rightarrow x = 0$

78. $-3 + 3x = -2x + 6 \Rightarrow 5x = 9 \Rightarrow x = \frac{9}{5}$ or $-3 + 3x = 2x - 6 \Rightarrow x = -3$

79.   (a) $x + 1 = 7 \Rightarrow x = 6$ or $x + 1 = -7 \Rightarrow x = -8$

(b) The solutions to $|x+1| \le 7$ satisfy $c \le x \le d$ where c and d are the solutions to $|x+1| = 7$.

From part (a), the interval is $[-8, 6]$.

(c) The solutions to $|x+1| \ge 7$ satisfy $x \le c$ or $x \ge d$ where c and d are the solutions to $|x+1| = 7$.

From part (a), the interval is $(-\infty, -8] \cup [6, \infty)$.

80.   (a) $1 - 2x = 6 \Rightarrow -2x = 5 \Rightarrow x = -\frac{5}{2}$ or $1 - 2x = -6 \Rightarrow -2x = -7 \Rightarrow x = \frac{7}{2}$

(b) The solutions to $|1-2x| \le 6$ satisfy $c \le x \le d$ where c and d are the solutions to $|1-2x| = 6$.

From part (a), the interval is $\left[-\frac{5}{2}, \frac{7}{2} \right]$.

(c) The solutions to $|1-2x| \ge 6$ satisfy $x \le c$ or $x \ge d$ where c and d are the solutions to $|1-2x| = 6$.

From part (a), the interval is $\left(-\infty, -\frac{5}{2} \right] \cup \left[\frac{7}{2}, \infty \right)$.

81.   The solutions to $|x| > 3$ satisfy $x < c$ or $x > d$ where c and d are the solutions to $|x| = 3$.

$|x| = 3$ is equivalent to $x = -3$ or $x = 3$. The interval is $(-\infty, -3) \cup (3, \infty)$.

82.   The solutions to $|-5x| < 20$ satisfy $c < x < d$ where c and d are the solutions to $|-5x| = 20$.

$|-5x| = 20$ is equivalent to $-5x = -20 \Rightarrow x = 4$ or $-5x = 20 \Rightarrow x = -4$. The interval is $(-4, 4)$.

83.   The solutions to $|4x-2| \le 14$ satisfy $c \le x \le d$ where c and d are the solutions to $|4x-2| = 14$.

$|4x-2| = 14$ is equivalent to $4x - 2 = -14 \Rightarrow x = -3$ or $4x - 2 = 14 \Rightarrow x = 4$. The interval is $[-3, 4]$.

84.   The solutions to $\left| 1 - \frac{4}{5}x \right| \ge 3$ satisfy $x \le c$ or $x \ge d$ where c and d are the solutions to $\left| 1 - \frac{4}{5}x \right| = 3$.

$\left| 1 - \frac{4}{5}x \right| = 3$ is equivalent to $1 - \frac{4}{5}x = -3 \Rightarrow x = 5$ or $1 - \frac{4}{5}x = 3 \Rightarrow x = -\frac{5}{2}$.

The interval is $\left(-\infty, -\frac{5}{2} \right] \cup [5, \infty)$.

85.   The solutions to $|t-4.5| \le 0.1$ satisfy $c \le t \le d$ where c and d are the solutions to $|t-4.5| = 0.1$.

$|t-4.5| = 0.1$ is equivalent to $t-4.5 = -0.1 \Rightarrow t = 4.4$ or $t-4.5 = 0.1 \Rightarrow t = 4.6$.

The interval is $[4.4, \ 4.6]$.

86.   First divide each side of $-2|13t-5| \ge -4$ by -2 to obtain $|13t-5| \le 2$.

The solutions to $|13t-5| \le 2$ satisfy $c \le t \le d$ where c and d are the solutions to $|13t-5| = 2$.

$|13t-5| = 2$ is equivalent to $13t-5 = -2 \Rightarrow t = \dfrac{3}{13}$ or $13t-5 = 2 \Rightarrow t = \dfrac{7}{13}$.

The interval is $\left[\dfrac{3}{13}, \ \dfrac{7}{13}\right]$.

87.   The inequality $|5-4x| > -5$ is true for all values of x because absolute value is never negative.

The interval is $(-\infty, \ \infty)$.

88.   Since absolute value can never be negative, $|2t-3| \le 0$ is equivalent to $|2t-3| = 0$.

$|2t-3| = 0$ is equivalent to $2t-3 = 0 \Rightarrow t = \dfrac{3}{2}$. The only solution is $\dfrac{3}{2}$.

89.   Graph $Y_1 = \text{abs}(2X)$ and $Y_2 = 3$ in $[-3, 3, 1]$ by $[0, 5, 1]$. See Figures 89a and 89b.

From the graph, $|2x| \ge 3$ in the interval $(-\infty, \ -1.5] \cup [1.5, \ \infty)$.

$[-3, 3, 1]$ by $[0, 5, 1]$ $[-3, 3, 1]$ by $[0, 5, 1]$

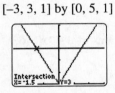   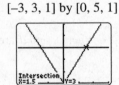

Figure 89a   Figure 89b

$[-7, 7, 1]$ by $[0, 4, 1]$ $[-7, 7, 1]$ by $[0, 4, 1]$

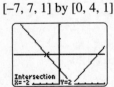   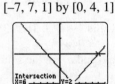

Figure 90a   Figure 90b

90.   Graph $Y_1 = \text{abs}((1/2)X - 1)$ and $Y_2 = 2$ in $[-7, 7, 1]$ by $[0, 4, 1]$. See Figures 90a and 90b.

From the graph, $\left|\dfrac{1}{2}x - 1\right| \le 2$ in the interval $[-2, \ 6]$.

91. $|x| \le 0.05$

92. $|5x-1| > 4$

Applications

93. Let x represent the loan amount at 5%. Then $7700 - x$ represents the loan amount at 7%.

$$0.05x + 0.07(7700 - x) = 469 \Rightarrow 0.05x + 539 - 0.07x = 469 \Rightarrow -0.02x = -70 \Rightarrow x = 3500$$

The loan amount at 5% was $3500 and the loan amount at 7% was $7700 - 3500 = \$4200$.

94. Let x represent the amount of time spent running at 8 mph. Then $1.4 - x$ represents the time running at 10 mph.

$$8x + 10(1.4 - x) = 12.8 \Rightarrow 8x + 14 - 10x = 12.8 \Rightarrow -2x = -1.2 \Rightarrow x = 0.6$$

The athlete ran for 0.6 hour at 8 mph and for $1.4 - 0.6 = 0.8$ hour at 10 mph.

95. Let x represent the cost of the house if it was located in San Francisco.

$$0.54x = 415,000 \Rightarrow x = \frac{415,000}{0.54} \approx 768,518.5$$

The house would cost about $769,000 in San Francisco.

96. (a) $m = \dfrac{498 - 1369}{1999 - 1980} = \dfrac{-871}{19} \approx -45.84$

$$y = -45.84(x - 1980) + 1369 \Rightarrow y = -45.84x + 92,136; \ f(x) = -45.84x + 92,136 \ (\text{approximate})$$

(b) $f(1988) = -45.84(1988) + 92,136 \approx 1006$. *Answers may vary slightly.*

97. (a) The distance between the riders is zero (they meet) when $x = 3$ hours.

(b) The distance between the riders is 20 when $x = 2$ hours and $x = 4$ hours.

(c) The riders are less than 20 miles apart between 2 and 4 hours exclusively.

(d) The sum of the speeds of the two bicyclists is 20 mph.

98. (a) The annual interest will be $300 for a loan of $2000.

(b) The annual interest will be more than $300 for loans of more than $2000.

(c) The annual interest will be less than $300 for loans of less than $2000.

99. (a) Car 1 is traveling faster since its graph has the steeper slope.

(b) The cars are both 200 miles from Austin at $x = 3$ hours.

(c) Car 1 is closer to Austin than Car 2 at times before $x = 3$ hours.

100. There were more than 10 deaths in 1989, 1990, 1991, 1993 and 1995.

101. (a) The median age of the U.S. population is increasing at a rate of 0.09 year per year.

(b) $0.09x - 147.1 = 23 \Rightarrow 0.09x = 170.1 \Rightarrow x = 1890$. Median age was 23 in 1890.

102. $490 \le 17.7x - 34,636 \le 670 \Rightarrow 35,126 \le 17.7x \le 35,306 \Rightarrow 1984.5 \le x \le 1994.7$ (approximate)

The number of female officers was from 490 to 670 from about 1985 to 1995.

103. (a) $1.11x - 23.3 = 12 \Rightarrow 1.11x = 35.3 \Rightarrow x \approx 31.8$ inches

(b) Walleyes with lengths less than 31.8 inches would weigh less than 12 pounds.

104. (a) When $x = 0$ (ground level), $T(x) = 60$ so the T-intercept is $b = 60$. Since the temperature decreases by

    $19°$ for each one-mile increase in altitude, the slope is $a = -19$. The formula is $T(x) = -19x + 60$.

    (b) $20 \le -19x + 60 \le 40 \Rightarrow -40 \le -19x \le -20 \Rightarrow 2.11 \ge x \ge 1.05 \Rightarrow 1.05 \le x \le 2.11$ (approximate)

    The temperature is from $40°F$ to $20°F$ for altitudes from about 1.05 to 2.11 miles.

105. Note that the car must travel a total of $200 + 395 = 595$ miles. Use the formula $d = rt$.

    $$595 = 70t \Rightarrow t = \frac{595}{70} \Rightarrow t = 8.5 \text{ hours}$$

106. Let x be the width of the rectangle. Then its length is $2x + 5$ and its perimeter is $2x + 2(2x + 5)$.

    $$2x + 2(2x + 5) = 88 \Rightarrow 2x + 4x + 10 = 88 \Rightarrow 6x = 78 \Rightarrow x = 13$$

    The width is 13 feet and the length is $2(13) + 5 = 31$ feet. The rectangle is 13 ft by 31 ft.

107. $-48 \le \frac{9}{5}C + 32 \le 107 \Rightarrow -80 \le \frac{9}{5}C \le 75 \Rightarrow -80\left(\frac{5}{9}\right) \le C \le 75\left(\frac{5}{9}\right) \Rightarrow -44.\overline{4} \le C \le 41.\overline{6}$

    The temperature range in Celsius is $-44.\overline{4}°C$ to $41.\overline{6}°C$.

108. $|L - 160| \le 1$; $L - 160 = 1 \Rightarrow L = 161$ or $L - 160 = -1 \Rightarrow L = 159$; $\{L | 159 \le L \le 161\}$

109. (a) $|A - 3.9| \le 1.7$

    (b) $A - 3.9 = 1.7 \Rightarrow A = 5.6$ or $A - 3.9 = -1.7 \Rightarrow A = 2.2$; $\{A | 2.2 \le A \le 5.6\}$

110. The solutions to $\left|\frac{T - 35}{35}\right| < 0.08$ satisfy $c < T < d$ where c and d are the solutions to $\left|\frac{T - 35}{35}\right| = 0.08$.

    $\left|\frac{T - 35}{35}\right| = 0.08$ is equivalent to $\frac{T - 35}{35} = -0.08 \Rightarrow x = 32.2$ and $\frac{T - 35}{35} = 0.08 \Rightarrow x = 37.8$.

    The interval is $(32.2, 37.8)$. The values must be between 32.2 and 37.8, exclusively.

Chapter 3 Test

1. Yes; $3 - 4\left(1 - 2 \cdot \frac{1}{6}\right) = 3 - 4\left(1 - \frac{2}{6}\right) = 3 - 4\left(\frac{4}{6}\right)$

    $= 3 - \frac{16}{6} = \frac{18}{6} - \frac{16}{6} = \frac{2}{6} = \frac{1}{3}$ and $2 \cdot \frac{1}{6} = \frac{2}{6} = \frac{1}{3}$

2.

x	-2	-1	0	1	2
$2 - 3x$	8	5	2	-1	-4

    The solution set for $2 - 3x \le -1$ is $\{x | x \ge 1\}$.

3. $3 - 5x = 18 \Rightarrow -5x = 15 \Rightarrow x = -3$. Check: $3 - 5(-3) = 3 - (-15) = 3 + 15 = 18$

4. (a) The intersection point of y_1 and y_2 is $(2, 1)$. The solution is $x = 2$.

    (b) y_1 is above y_2 when $x \le 2$; $(-\infty, 2]$

    (c) y_1 is below y_2 when $x \ge 2$; $[2, \infty)$

5. The graphs of $y_1 = 4 - 2x$ and $y_2 = 1 + x$ (not shown) intersect at the point (1, 2). The solution is $x = 1$.

6. (a) $-\dfrac{2}{3}(3x - 2) + 1 = x \Rightarrow -2x + \dfrac{4}{3} + 1 = x$

    $\Rightarrow \dfrac{7}{3} = 3x \Rightarrow \dfrac{1}{3}\left(\dfrac{7}{3}\right) = x \Rightarrow x = \dfrac{7}{9}$

    (b) $\dfrac{2z + 1}{3} = \dfrac{3(1 - z)}{2} \Rightarrow 2(2z + 1) = 3[3(1 - z)]$

    $\Rightarrow 4z + 2 = 9 - 9z \Rightarrow 13z = 7 \Rightarrow z = \dfrac{7}{13}$

    (c) $0.4(2 - 3x) = 0.5x - 0.2 \Rightarrow 4(2 - 3x) = 5x - 2$

    $\Rightarrow 8 - 12x = 5x - 2 \Rightarrow 10 = 17x \Rightarrow x = \dfrac{10}{17}$

7. x-intercept: $-6x + 3(0) = 9 \Rightarrow -6x = 9 \Rightarrow x = -\dfrac{3}{2}$

    y-intercept: $-6(0) + 3y = 9 \Rightarrow 3y = 9 \Rightarrow y = 3$

    $-6x + 3y = 9 \Rightarrow 3y = 6x + 9 \Rightarrow y = 2x + 3$

8. $2 + 5x = x - 4;\ 2 + 5x = x - 4 \Rightarrow 4x = -6 \Rightarrow x = \dfrac{-6}{4} \Rightarrow x = -\dfrac{3}{2}$

9. $3x - 2y = 6 \Rightarrow -2y = -3x + 6 \Rightarrow y = \dfrac{3}{2}x - 3;$

    $f(x) = \dfrac{3}{2}x - 3$

10. $-\dfrac{5}{2}x + \dfrac{1}{2} \le 2 \Rightarrow -5x + 1 \le 4 \Rightarrow -5x \le 3 \Rightarrow x \ge -\dfrac{3}{5};\ \left[-\dfrac{3}{5},\ \infty\right)$

11. $3.1(3 - x) < 2.9x \Rightarrow 9.3 - 3.1x < 2.9x \Rightarrow -6x < -9.3 \Rightarrow x > 1.55;\ (1.55,\ \infty)$

12. $2x + 6 < 2$ and $-3x \ge 3 \Rightarrow 2x < -4$ and $x \le -1 \Rightarrow x < -2$ and $x \le -1;$ See Figure 12.

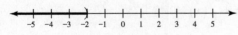

Figure 12

13. From the table, $-3x < -3$ when $x > 1$ and $-3x > 6$ when $x < -2$. The interval is $(-\infty,\ -2) \cup (1,\ \infty)$.

14. (a) The intersection point of y_1 and y_2 is $(-5,\ -300)$. The solution is $x = -5$.

    (b) The intersection point of y_2 and y_3 is $(5,\ 100)$. The solution is $x = 5$.

    (c) y_2 is between y_1 and y_3 when $-5 \le x \le 5$. The interval is $[-5,\ 5]$.

    (d) y_2 is below y_3 when $x < 5$. The interval is $(-\infty,\ 5)$.

15. $-2 < 2 + \dfrac{1}{2}x < 2 \Rightarrow -4 < \dfrac{1}{2}x < 0 \Rightarrow -8 < x < 0; \ (-8, 0)$

16. $2 - \dfrac{1}{3}x = 6 \Rightarrow -\dfrac{1}{3}x = 4 \Rightarrow x = -12 \ $ or $ \ 2 - \dfrac{1}{3}x = -6 \Rightarrow -\dfrac{1}{3}x = -8 \Rightarrow x = 24$

17. (a) $-5 \le x \le 5; [-5, 5]$

(b) The solution set is all real numbers except 0, or $(-\infty, 0) \cup (0, \infty)$.

(c) There are no solutions for this inequality. There are no values of x for which $|2x + 7|$ is negative.

(d) $5x = 10 \Rightarrow x = 2 \ $ or $ \ 5x = -10 \Rightarrow x = -2$

The solution set is $x < -2$ or $x > 2$, or $(-\infty, -2) \cup (2, \infty)$.

(e) $-2|2 - 5x| + 1 \le 5 \Rightarrow |2 - 5x| \ge -2$

This is true for all values of x, or $(-\infty, \infty)$.

(f) $-5 < 3 - x < 5 \Rightarrow -8 < -x < 2 \Rightarrow -2 < x < 8; (-2, 8)$

18. (a) $1 - 5x = 3 \Rightarrow -5x = 2 \Rightarrow x = -\dfrac{2}{5} \ $ or $ \ 1 - 5x = -3 \Rightarrow -5x = -4 \Rightarrow x = \dfrac{4}{5}$

(b) The solutions to $|1 - 5x| \le 3$ satisfy $c \le x \le d$ where c and d are the solutions to $|1 - 5x| = 3$.

From part (a), the interval is $\left[-\dfrac{2}{5}, \ \dfrac{4}{5} \right]$.

(c) The solutions to $|1 - 5x| \ge 3$ satisfy $x \le c$ or $x \ge d$ where c and d are the solutions to $|1 - 5x| = 3$.

From part (a), the interval is $\left(-\infty, \ -\dfrac{2}{5} \right] \cup \left[\dfrac{4}{5}, \ \infty \right)$.

19. (a) A sports drink with 16 grams of carbohydrates will have 64 calories.

(b) A sports drink with more than 16 grams of carbohydrates will have more than 64 calories.

(c) A sports drink with less than 16 grams of carbohydrates will have less than 64 calories.

(d) Each gram of carbohydrates corresponds to 4 calories.

20. (a) First divide the weight 150 by 2 to get 75. The function is $f(x) = 0.4x + 75$.

(b) $0.4x + 75 = 89 \Rightarrow 0.4x = 14 \Rightarrow x = \dfrac{14}{0.4} \Rightarrow x = 35$ minutes

21. $d = \dfrac{1}{2}gt^2 \Rightarrow 2d = gt^2 \Rightarrow \dfrac{2d}{t^2} = g \Rightarrow g = \dfrac{2d}{t^2}$

22. Let x represent the loan amount at 4%. Then $5000 - x$ represents the loan amount at 6%.

$0.04x + 0.06(5000 - x) = 230 \Rightarrow 0.04x + 300 - 0.06x = 230 \Rightarrow -0.02x = -70 \Rightarrow x = 3500$

The loan amount at 4% was $3500 and the loan amount at 6% was $5000 - 3500 = \$1500$.

23. x: measure of smallest angle

$2x$: measure of largest angle

$x+16$: measure of third angle

$x+2x+x+16=180 \Rightarrow 4x+16=180 \Rightarrow 4x=164$

$\Rightarrow x=41, \ 2x=82, \ x+16=57$

The measures of the angles are $41°$, $57°$, and $82°$.

24. x : retail price of jacket

$x-0.25x$: sale price of jacket

$x-0.25x=81 \Rightarrow 0.75x=81 \Rightarrow x=108$

The retail price was \$108.

Chapter 3 Extended and Discovery Exercises

1.   (a) The temperature at ground level is $T(0)=90-19(0)=90\,°\mathrm{F}$.

The dew point at ground level is $D(0)=70-5.8(0)=70\,°\mathrm{F}$.

(b) $90-19x=70-5.8x \Rightarrow 20=13.2x \Rightarrow \dfrac{20}{13.2}=x \Rightarrow x \approx 1.52$

The temperature and dew point are equal at an altitude of about 1.52 miles.

(c) Clouds will not form at altitudes below about 1.52 miles.

(d) Clouds will form at altitudes above about 1.52 miles.

2.   (a) At some altitudes it is likely that the air temperature will reach the dew point.

(b) Clouds will form at much lower altitudes than usual. In this case, it may become foggy.

3.   (a) The scatterplot is shown in Figure 3a.

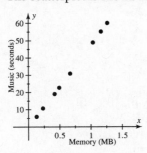

Figure 3a

(b) The relationship appears to be linear. This seems reasonable if one considers that when the number of megabytes doubles, the number of seconds should also double.

(c) Using the points (0.129, 6.010) and (1.260, 60.18) $m=\dfrac{60.18-6.010}{1.260-0.129} \approx 47.9$, $x_1=0.129$ and $y_1=6.010$.

The equation is $y \approx 47.9(x-0.129)+6.010 \Rightarrow y \approx 47.9x-0.1691$.

Each additional megabyte of memory can record approximately 47.9 seconds of music.

(d) The scatterplot and the line are shown in Figure 3d.

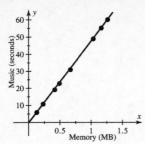

Figure 3d

(e) $47.9x - 0.1691 = 120$

(f) Symbolic: $47.9x - 0.1691 = 120 \Rightarrow 47.9x = 120.1691 \Rightarrow x \approx 2.5$ MB

4.   (a) Using (25,000, 12) and (100,000, 96), $m = \dfrac{96 - 12}{100,000 - 25,000} \approx 0.00112$, $x_1 = 25,000$ and $y_1 = 12$.

(b) A graph of the line and the data points are shown in Figure 4b.

(c) $28.8 \leq 0.00112(x - 25,000) + 12 \leq 51.2$

(d) Graph $Y_1 = 0.00112(X - 25,000) + 12$, $Y_2 = 28.8$ and $Y_3 = 51.2$ in [0, 110,000, 10,000] by [0, 100, 10].

See Figures 4d and 4e. $\{x | 40,000 \leq x \leq 60,000\}$

Symbolic: $28.8 \leq 0.00112(x - 25,000) + 12 \leq 51.2 \Rightarrow 40,000 \leq x \leq 60,000$; $\{x | 40,000 \leq x \leq 60,000\}$

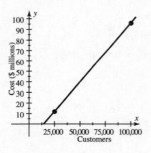

Figure 4b

[0, 110,000, 10,000] by [0, 100, 10]   [0, 110,000, 10,000] by [0, 100, 10]

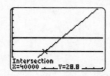   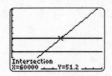

Figure 4d   Figure 4e

Chapters 1-3 Cumulative Review Exercises

1.   Natural: $\dfrac{8}{2} = 4$; Whole: 0, $\dfrac{8}{2} = 4$; Integer: -7, 0, $\dfrac{8}{2} = 4$; Rational: -7, $-\dfrac{3}{5}$, 0, $\dfrac{8}{2} = 4$, $5.\overline{12}$;

Irrational: $\sqrt{5}$

2.   Natural: $\sqrt{9} = 3$; Whole: $\dfrac{0}{9}$, $\sqrt{9} = 3$; Integer: $-\dfrac{6}{3}$, $\dfrac{0}{9}$, $\sqrt{9} = 3$;

Rational: $-\dfrac{6}{3}, \dfrac{0}{9}, \sqrt{9} = 3, 4.\overline{6}$; Irrational: π

3. Distributive

4. Associative

5. $11 + 26 + (-1) + 14 = 11 + (-1) + 26 + 14 = 10 + 40 = 50$

6. $7 \cdot 98 = 7(100 - 2) = 700 - 14 = 686$

7.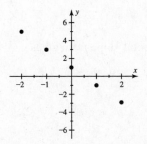

8. $-(-5y - 4) = 5y + 4$

9. $\dfrac{24x^{-4}y^2}{8xy^{-5}} = \dfrac{24}{8} \cdot x^{-4-1}y^{2-(-5)} = 3x^{-5}y^7 = \dfrac{3y^7}{x^5}$

10. $\left(\dfrac{3a^2}{4b^3}\right)^{-3} = \left(\dfrac{4b^3}{3a^2}\right)^3 = \dfrac{4^3 b^9}{3^3 a^6} = \dfrac{64b^9}{27a^6}$

11. $15 - 2^3 \div 4 = 15 - 8 \div 4 = 15 - 2 = 13$

12. Move the decimal point 5 places to the right: $0.000059 = 5.9 \times 10^{-5}$.

13. $J = \sqrt{38 - (13)} = \sqrt{25} = 5$

14. $r = 3 - (-3)^3 = 3 - (-27) = 3 + 27 = 30$

15. By testing data points in each formula, we find that the best model is (ii).

16. The domain corresponds to the x-coordinates and the range corresponds to the y-coordinates.

    $D = \{-2,\ 0,\ 1,\ 3\};\ R = \{0,\ 2,\ 4\}$

17. When $x = -2$, $y = -2(-2) + 1 = 5$. The other ordered pairs can be found similarly.

    The ordered pairs are $(-2,\ 5)$, $(-1,\ 3)$, $(0,\ 1)$, $(1,\ -1)$, $(2,\ -3)$. See Figure 17.

18. When $x = -2$, $y = \dfrac{(-2)^3 + 4}{2} = \dfrac{-8 + 4}{2} = \dfrac{-4}{2} = -2$. The other ordered pairs can be found similarly.

    The ordered pairs are $(-2,\ -2)$, $\left(-1,\ \dfrac{3}{2}\right)$, $(0,\ 2)$, $\left(1,\ \dfrac{5}{2}\right)$, $(2,\ 6)$. See Figure 18.

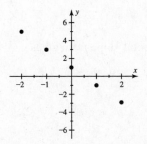

Figure 17

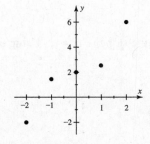

Figure 18

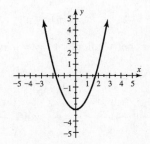

Figure 19

19. See Figure 19.

20.   The function is defined for all values of the variable except 0. The domain is $\{x \mid x \neq 0\}$.

21.   From the graph, when $x = 0$, $y = 2$. Therefore $f(0) = 2$. Similarly, $f(-2) = -2$.

22.   From the table, when $x = 0$, $f(x) = 2$. Therefore $f(0) = 2$. Similarly, $f(-2) = -4$.

23.   Yes; $a = -3$, $b = 11$

24.   No. A linear function cannot contain a square root.

25.   The slope is $m = \dfrac{11 - 7}{2 - 1} = 4$. Since $f(x) = 3$ when $x = 0$ the y-intercept is 3. Here $f(x) = 4x + 3$.

26.   See Figure 26.

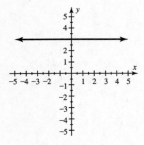   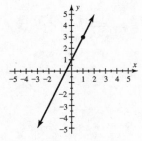

Figure 26   Figure 29

27.   The equation is in the form $f(x) = mx + b$. The slope is 5 and the y-intercept is -4.

28. $m = \dfrac{6 - (-2)}{-1 - 3} = \dfrac{8}{-4} = -2$

29.   See Figure 29.

30.   The graph falls 1 unit for each 2 units of run. The slope is $-\dfrac{1}{2}$. The y-intercept is -1. So $y = -\dfrac{1}{2}x - 1$.

31.   Since $f(x)$ increases by 3 for each unit increase in x, the slope is 3. From the point $(-1, 0)$, the x-intercept

is -1.

From the point $(0, 3)$, the y-intercept is 3.

32.   Vertical lines have equations of the form $x = k$. The equation of the vertical line passing through $(5, 7)$ is

$x = 5$.

33.   Since the line is parallel to $y = 3x - 4$, the slope is $m = 3$. Using the point-slope form gives

$y = 3(x - 2) + 1 \Rightarrow y = 3x - 6 + 1 \Rightarrow y = 3x - 5$

34.   Since the line is perpendicular to $y = -x - 5$, the slope is $m = 1$. Using the point-slope form gives

$y = 1(x - (-3)) + 0 \Rightarrow y = x + 3$

35.   Yes since $\dfrac{1}{2}\left(\dfrac{8}{5}\right) - 4\left(\dfrac{8}{5} - 1\right) = -\dfrac{8}{5}$ and $\dfrac{1}{4}\left(\dfrac{8}{5}\right) - 2 = -\dfrac{8}{5}$.

36.   The graphs intersect at the point $(1, -2)$, so $y_1 = y_2$ when $x = 1$.

37. $\dfrac{3}{4}(x-2)+4=2 \Rightarrow \dfrac{3}{4}(x-2)=-2 \Rightarrow x-2=-\dfrac{8}{3} \Rightarrow x=-\dfrac{2}{3}$

38. $\dfrac{2}{3}\left(\dfrac{t-7}{4}\right)-2=\dfrac{1}{3}t-(2t+3) \Rightarrow \dfrac{t-7}{6}-2=\dfrac{1}{3}t-2t-3 \Rightarrow t-7-12=2t-12t-18 \Rightarrow$

   $t-19=-10t-18 \Rightarrow 11t=1 \Rightarrow t=\dfrac{1}{11}$

39. To find the x-intercept, let $y=0$ and solve for x: $x-\dfrac{1}{3}(0)=2 \Rightarrow x=2$; to find the y-intercept, let $x=0$

   and solve for y: $0-\dfrac{1}{3}y=2 \Rightarrow -\dfrac{1}{3}y=2 \Rightarrow y=-6$;

   $x-\dfrac{1}{3}y=2 \Rightarrow -\dfrac{1}{3}y=-x+2 \Rightarrow y=3x-6$

40. To find the x-intercept, let $y=0$ and solve for x: $5x-4(0)=-10 \Rightarrow 5x=-10 \Rightarrow x=-2$; to find the

   y-intercept, let $x=0$ and solve for y: $5(0)-4y=-10 \Rightarrow -4y=-10 \Rightarrow y=\dfrac{5}{2}$;

   $5x-4y=-10 \Rightarrow -4y=-5x-10 \Rightarrow y=\dfrac{5}{4}x+\dfrac{5}{2}$

41. Here $y<2$ when $x>0$. $\{x|x>0\}$

42. Here y_1 is below y_2 when $x \le -3$. The interval is $(-\infty, -3]$.

43. $\dfrac{4x-9}{6} > \dfrac{1}{2} \Rightarrow 4x-9>3 \Rightarrow 4x>12 \Rightarrow x>3$. The interval is $(3, \infty)$.

44. $\dfrac{2}{3}z-2 \le \dfrac{1}{4}z-(2z+2) \Rightarrow 8z-24 \le 3z-24z-24 \Rightarrow 8z-24 \le -21z-24 \Rightarrow$

   $29z \le 0 \Rightarrow z \le 0$. The interval is $(-\infty, 0]$.

45. $x+2>1 \Rightarrow x>-1$ and $2x-1 \le 9 \Rightarrow 2x \le 10 \Rightarrow x \le 5$

   The solutions must satisfy both of these inequalities. The interval is $(-1, 5]$. See Figure 45.

   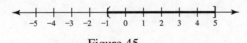

   Figure 45

46. $4x+7<1 \Rightarrow 4x<-6 \Rightarrow x<-\dfrac{3}{2}$ or $3x+2 \ge 11 \Rightarrow 3x \ge 9 \Rightarrow x \ge 3$. The solutions may satisfy only one (or

   both) of these inequalities. The interval is $\left(-\infty, -\dfrac{3}{2}\right) \cup [3, \infty)$. See Figure 46.

   Figure 46

47. $-7 \le 2x-3 \le 5 \Rightarrow -4 \le 2x \le 8 \Rightarrow -2 \le x \le 4$. The interval is $[-2, 4]$.

48. $-8 \le -\dfrac{1}{2}x - 3 \le 5 \Rightarrow -5 \le -\dfrac{1}{2}x \le 8 \Rightarrow 10 \ge x \ge -16 \Rightarrow -16 \le x \le 10$. The interval is $[-16, \ 10]$.

49. The solutions to $|3x+5| > 13$ satisfy $x < c$ or $x > d$ where c and d are the solutions to $|3x+5| = 13$.

    $|3x+5| = 13$ is equivalent to $3x+5 = -13 \Rightarrow x = -6$ or $3x+5 = 13 \Rightarrow x = \dfrac{8}{3}$.

    The interval is $\left(-\infty, \ -6\right) \cup \left(\dfrac{8}{3}, \ \infty\right)$.

50. First divide each side of $-3|2t-11| \ge -9$ by -3 to obtain $|2t-11| \le 3$.

    The solutions to $|2t-11| \le 3$ satisfy $c \le t \le d$ where c and d are the solutions to $|2t-11| = 3$.

    $|2t-11| = 3$ is equivalent to $2t-11 = -3 \Rightarrow t = 4$ or $2t-11 = 3 \Rightarrow t = 7$. The interval is $[4, \ 7]$.

51. (a) $y_1 = 2$ when $x = -3$ or when $x = 1$.

    (b) $y_1 \le 2$ when $-3 \le x \le 1$. $[-3, \ 1]$

    (c) $y_1 \ge 2$ when $x \le -3$ or $x \ge 1$. $(-\infty, \ -3] \cup [1, \ \infty)$

52. $\dfrac{2}{3}x - 4 = -8 \Rightarrow \dfrac{2}{3}x = -4 \Rightarrow x = -6$ or $\dfrac{2}{3}x - 4 = 8 \Rightarrow \dfrac{2}{3}x = 12 \Rightarrow x = 18$

53. (a) $A = 4300(1.05)^{10} \approx \$7004.25$

    (b) $A = 11{,}000(1.05)^{6} \approx \$14{,}741.05$

54. For each 2-hour increase in time, the car travels an additional 130 miles. The car's speed is 65 mph. $d = 65t$

55. (a) There are 120 milligrams of sodium in a 12-ounce can so there are 10 milligrams of sodium per ounce.

    The formula is $f(x) = 10x$.

    (b) The slope of this line is 10.

    (c) The soda contains 10 milligrams of sodium per ounce.

56. See Figure 56.

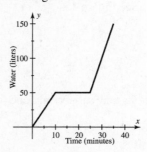

    Figure 56

57. (a) The bicyclist is traveling away from Euclid because the slope of the line for the bicyclist is positive.

    (b) The lines intersect at the point (3, 200). When 3 hours has elapsed, they are both 200 miles from Euclid.

    (c) The motorcyclist is farther from Euclid from 0 to 3 hours.

58. $410 \le 17.7(x-1960)+56 \le 605 \Rightarrow 354 \le 17.7(x-1960) \le 549 \Rightarrow$

$20 \le x-1960 \le 31.02 \Rightarrow 1980 \le x \le 1991$ (approximate). From 1980 to 1991.

Critical Thinking Solutions for Chapter 3

Section 3.1

- The answer must satisfy the given equation. A different equation could have an error in it.

Section 3.3

- We may substitute *x*-values from the solution set into the inequality and check them.
  If they do not check, then our solution set is incorrect.

- To get the desired solution set, we need only to reverse the inequality symbol. $\{x|x \le 2\}$

Section 3.4

- 1.   See Figure 1a. There are no numbers that satisfy *both* inequalities.

  2.   See Figure 1b. Every real number is *either* greater than 2 *or* less than 5.

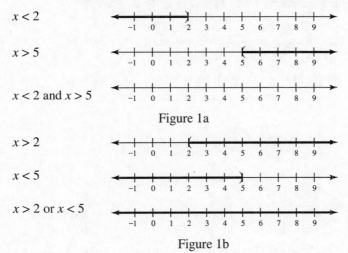

Figure 1a

Figure 1b

- Students exhale more carbon dioxide during an exam than they do during a lecture. This increase may be due to anxiety experienced during examinations.

Chapter 4: Systems of Linear Equations

4.1: Systems of Linear Equations in Two Variables

Concepts

1.   No, two straight lines cannot have exactly two points of intersection.

2. $3x + 2y = 5$

  $2x - y = 0$ *Answers may vary.*

3.   Numerically and graphically.

4.   There is one solution.

5.   There are no solutions.

6.   There are infinitely many solutions. The two lines are identical.

7. $(1, -2)$ since it satisfies both equations. The ordered pair $(4, 4)$ does not satisfy $3x + y = 1$.

8. $(3, 1)$ since it satisfies both equations. The ordered pair $(-3, -1)$ does not satisfy $3x + y = 10$.

9. $\left(-1, \dfrac{13}{3}\right)$ since it satisfies both equations. The ordered pair $(4, 6)$ does not satisfy $4x + 3y = 9$.

10. $(4, 0)$ since it satisfies both equations. The ordered pair $(3, 5)$ does not satisfy $5x - 4y = 20$.

Graphical Solutions

11.   See Figure 11. The solution is $(1, 1)$.

13.   See Figure 13. The solution is $(0, 1)$.

15.   See Figure 15. The solution is $(3, -2)$.

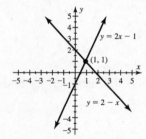   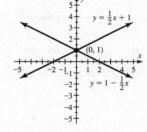   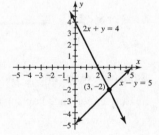

Figure 11   Figure 13   Figure 15

17.   See Figure 17. The solution is $(4, -1)$.

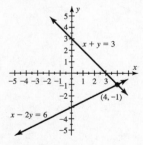

Figure 17

19.   The solution is the intersection point $(3, 1)$.

21. The two lines do not intersect. There are no solutions.

23. The system can be solved graphically by solving each equation for y and then graphing.

    $-x + y = 1 \Rightarrow y = x + 1$ and $x + y = 3 \Rightarrow y = 3 - x$

    The graphs (not shown) intersect at the point $(1, 2)$. The system is consistent. The equations are independent.

25. The system can be solved graphically by solving each equation for y and then graphing.

    $2x + y = 5 \Rightarrow y = 5 - 2x$ and $-2x + y = -3 \Rightarrow y = 2x - 3$

    The graphs (not shown) intersect at the point $(2, 1)$. The system is consistent. The equations are independent.

27. The system can be solved graphically by solving each equation for y and then graphing.

    $x + y = 3 \Rightarrow y = 3 - x$ and $2x + 2y = 6 \Rightarrow y = 3 - x$

    The graphs (not shown) are identical, $\{(x, y) | x + y = 3\}$. The system is consistent. The equations are dependent.

29. The system can be solved graphically by solving each equation for y and then graphing.

    $3x - y = 0 \Rightarrow y = 3x$ and $2x + y = 5 \Rightarrow y = 5 - 2x$

    The graphs (not shown) intersect at the point $(1, 3)$. The system is consistent. The equations are independent.

31. The system can be solved graphically by solving each equation for y and then graphing.

    $-2x + y = 3 \Rightarrow y = 2x + 3$ and $4x - 2y = 2 \Rightarrow y = 2x - 1$

    The graphs (not shown) are parallel. There are no solutions. The system is inconsistent.

33. The system can be solved graphically by solving each equation for y and then graphing.

    $x + y = 6 \Rightarrow y = 6 - x$ and $x - y = 2 \Rightarrow y = x - 2$

    The graphs (not shown) intersect at the point $(4, 2)$. The system is consistent. The equations are independent.

35. The system can be solved graphically by solving each equation for y and then graphing.

    $x - y = 4 \Rightarrow y = x - 4$ and $2x - 2y = 4 \Rightarrow y = x - 2$

    The graphs (not shown) are parallel. There are no solutions. The system is inconsistent.

37. The system can be solved graphically by solving each equation for y and then graphing.

    $6x - 4y = -2 \Rightarrow y = \dfrac{3x + 1}{2}$ and $-3x + 2y = 1 \Rightarrow y = \dfrac{3x + 1}{2}$

    The graphs (not shown) are identical, $\{(x, y) | -3x + 2y = 1\}$. The system is consistent. The equations are dependent.

39. The system can be solved graphically by solving each equation for y and then graphing.

    $4x + 3y = 2 \Rightarrow y = \dfrac{2 - 4x}{3}$ and $5x + 2y = 6 \Rightarrow y = \dfrac{6 - 5x}{2}$

    The graphs (not shown) intersect at the point $(2, -2)$. The system is consistent. The equations are independent.

41. The system can be solved graphically by solving each equation for y and then graphing.

$$2x + 2y = 4 \Rightarrow y = 2 - x \text{ and } x - 3y = -2 \Rightarrow y = \frac{x+2}{3}$$

The graphs (not shown) intersect at the point (1, 1).

43.   The system can be solved graphically by solving each equation for y and then graphing.

$$-\frac{1}{2}x - \frac{1}{2}y = \frac{3}{2} \Rightarrow y = -x - 3 \text{ and } x - \frac{1}{2}y = 3 \Rightarrow y = 2x - 6$$

The graphs (not shown) intersect at the point $(1, -4)$.

45.   (a) $x + y = 4$: To find the x-intercept, let $y = 0$ and solve for x: $x + 0 = 4 \Rightarrow x = 4$; to find the y-intercept, let

$x = 0$ and solve for y: $0 + y = 4 \Rightarrow y = 4$.

$x - y = -2$: To find the x-intercept, let $y = 0$ and solve for x: $x - 0 = -2 \Rightarrow x = -2$; to find the y-intercept,

let $x = 0$ and solve for y: $0 - y = -2 \Rightarrow -y = -2 \Rightarrow y = 2$.

(b)   See Figure 45b. The solution is (1, 3).

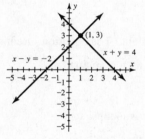

Figure 45b   Figure 47b

47.   (a) $-2x + 3y = 12$: To find the x-intercept, let $y = 0$ and solve for x: $-2x + 3(0) = 12 \Rightarrow -2x = 12 \Rightarrow x = -6$;

to find the y-intercept, let $x = 0$ and solve for y: $-2(0) + 3y = 12 \Rightarrow 3y = 12 \Rightarrow y = 4$.

$-2x + y = 8$: To find the x-intercept, let $y = 0$ and solve for x: $-2x + 0 = 8 \Rightarrow -2x = 8 \Rightarrow x = -4$; to find

the y-intercept, let $x = 0$ and solve for y: $-2(0) + y = 8 \Rightarrow y = 8$.

(b)   See Figure 47b. The solution is (–3, 2).

49.   Note that $\frac{1}{4}x + \frac{1}{2}y = \frac{3}{20} \Rightarrow y = -\frac{1}{2}x + \frac{3}{10}$ and $\frac{1}{8}x - y = -\frac{3}{10} \Rightarrow y = \frac{1}{8}x + \frac{3}{10}$.

Graph $Y_1 = (-1/2)X + (3/10)$ and $Y_2 = (1/8)X + (3/10)$ in [–1, 1, 0.1] by [–1, 1, 0.1]. See Figure 49.

The unique solution is the intersection point (0, 0.3).

[–1, 1, 0.1] by [–1, 1, 0.1]

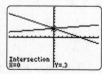

Figure 49

51.   Note that $0.1x + 0.2y = 0.25 \Rightarrow y = -\frac{1}{2}x + \frac{5}{4}$ and $0.7x - 0.3y = 0.9 \Rightarrow y = \frac{7}{3}x - 3$.

Graph $Y_1 = (-1/2)X + (5/4)$ and $Y_2 = (7/3)X - 3$ in $[-4, 4, 1]$ by $[-4, 4, 1]$. See Figure 51.

The unique solution is the intersection point (1.5, 0.5).

[–4, 4, 1] by [–4, 4, 1]

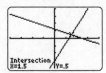

Figure 51

53.   Note that $0.1x + 0.2y = 50 \Rightarrow y = -0.5x + 250$ and $0.3x - 0.1y = 10 \Rightarrow y = 3x - 100$.

Graph $Y_1 = -0.5X + 250$ and $Y_2 = 3X - 100$ in $[0, 300, 50]$ by $[0, 300, 50]$. See Figure 53.

The unique solution is the intersection point (100, 200).

[0, 300, 50] by [0, 300, 50]   [–10, 10, 1] by [–10, 10, 1]

   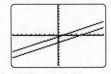

Figure 53   Figure 55

55.   Note that $x - 2y = 5 \Rightarrow y = \dfrac{1}{2}x - \dfrac{5}{2}$ and $-2x + 4y = -2 \Rightarrow y = \dfrac{1}{2}x - \dfrac{1}{2}$.

Graph $Y_1 = (1/2)X - (5/2)$ and $Y_2 = (1/2)X - (1/2)$ in $[-10, 10, 1]$ by $[-10, 10, 1]$. See Figure 55.

Since the lines are parallel, there is no solution.

Numerical Solutions

57.   Since $y_1 = y_2 = 3$ when $x = 2$, the solution is (2, 3).

59.   Since $y_1 = y_2 = 1$ when $x = 1$, the solution is (1, 1).

61.   Note that $x + y = 3 \Rightarrow y = 3 - x$ and $x - y = 7 \Rightarrow y = x - 7$.

Table $Y_1 = 3 - X$ and $Y_2 = X - 7$ with TblStart = 0 and ΔTbl = 1. See Figure 61.

Since $Y_1 = Y_2 = -2$ when $X = 5$, the solution is (5, –2).

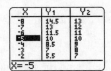

Figure 61   Figure 63

63.   Note that $3x + 2y = 5 \Rightarrow y = -\dfrac{3}{2}x + \dfrac{5}{2}$ and $-x - y = -5 \Rightarrow y = -x + 5$.

Table $Y_1 = (-3/2)X + (5/2)$ and $Y_2 = -X + 5$ with TblStart = –8 and ΔTbl = 1. See Figure 63.

Since $Y_1 = Y_2 = 10$ when $X = -5$, the solution is (–5, 10).

65. Note that $0.5x - 0.1y = 0.1 \Rightarrow y = 5x - 1$ and $0.1x - 0.3y = -0.4 \Rightarrow y = \frac{1}{3}x + \frac{4}{3}$.

    Table $Y_1 = 5X - 1$ and $Y_2 = (1/3)X + (4/3)$ with TblStart $= -1$ and ΔTbl $= 0.5$. See Figure 65.

    Since $Y_1 = Y_2 = 1.5$ when $X = 0.5$, the solution is $(0.5, 1.5)$.

X	Y₁	Y₂
-1	-6	1.1667
.5	-3.5	1.3333
0	-1	1.5
.5	1.5	1.5
1	4	1.6667
1.5	6.5	1.8333
2	9	2

    X=.5

    Figure 65

67. $x + ay = 1 \Rightarrow ay = -x + 1 \Rightarrow y = -\frac{1}{a}x + \frac{1}{a}$; $2x + 2ay = 4 \Rightarrow 2ay = -2x + 4 \Rightarrow y = -\frac{1}{a}x + \frac{2}{a}$. So the graphs of

    the two lines have the same slope $\left(-\dfrac{1}{a}\right)$ but different *y*-intercepts. Therefore, the lines are parallel and do not

    intersect, so there are no solutions.

Writing and Solving Equations

69. (a) Let x and y represent the two numbers. Then the system needed is $x + y = 18$ and $x - y = 6$.

    (b) Note that $x + y = 18 \Rightarrow y = -x + 18$ and $x - y = 6 \Rightarrow y = x - 6$.

    The graphs of these equations (not shown) intersect at the point $(12, 6)$. The numbers are 12 and 6.

71. (a) Let x and y represent the time spent running at 6 mph and 8 mph respectively.

    Then the system needed is $x + y = 1$ and $6x + 8y = 7$.

    (b) Note that $x + y = 1 \Rightarrow y = 1 - x$ and $6x + 8y = 7 \Rightarrow y = \dfrac{7 - 6x}{8}$.

    The graphs of these equations (not shown) intersect at the point $(0.5, 0.5)$.

    The athlete ran for 0.5 hour at 6 mph and for 0.5 hour at 8 mph.

73. (a) Let x and y represent the length and width of the rectangle respectively.

    Then the system needed is $2x + 2y = 76$ and $x - y = 4$.

    (b) Note that $2x + 2y = 76 \Rightarrow y = -x + 38$ and $x - y = 4 \Rightarrow y = x - 4$.

    The graphs of these equations (not shown) intersect at the point $(21, 17)$.

    The length of the rectangle is 21 inches and the width is 17 inches.

75. (a) Let x and y represent the measure of the largest angle and the measure of one of the equal angles respectively.

    Then the system needed is $x + 2y = 180$ and $x - y = 60$.

    (b) Note that $x + 2y = 180 \Rightarrow y = -\dfrac{1}{2}x + 90$ and $x - y = 60 \Rightarrow y = x - 60$.

    The graphs of these equations (not shown) intersect at the point $(100, 40)$.

    The measure of the largest angle is $100°$ and the measure of each of the smaller angles is $40°$.

Applications

77.   Let x and y represent the number of times Kentucky and UCLA have appeared respectively. Then the system needed is $x + y = 80$ and $x - y = 8$. By solving each equation for y and graphing each (figure not shown) we find the intersection point is (44, 36). Kentucky appeared 44 times and UCLA appeared 36 times.

79.   Let x and y represent the number of home runs hit by McGwire and Sosa respectively. Then the system needed is $x + y = 136$ and $x - y = 4$. By solving each equation for y and graphing each (figure not shown) we find the intersection point is (70, 66). McGwire hit 70 home runs and Sosa hit 66.

81.   Let x represent the amount invested at 4% interest, and y represent the amount invested at 5% interest. Then the system needed is $x + y = 600$ and $y = x + 100$. By solving the first equation for y and graphing both equations (figure not shown) we find the intersection point is (250, 350). $250 is invested at 4% and $350 is invested at 5%.

4.2: The Substitution and Elimination Methods

Concepts

1.   Substitution and elimination.

2.   No, all methods will produce the same solution. However, calculator solutions are sometimes rounded.

3.   Elimination yields a contradiction.

4.   Elimination yields an identity.

5.   Solve the first equation for y.

6.   Multiply the second equation by –2.

Substitution Method

7.   Substituting $y = 2x$ into the second equation yields the following:

$3x + (2x) = 5 \Rightarrow 5x = 5 \Rightarrow x = 1$ and so $y = 2(1) \Rightarrow y = 2$. The solution is (1, 2).

9.   Substituting $x = 2y - 1$ into the second equation yields the following:

$(2y - 1) + 5y = 20 \Rightarrow 7y = 21 \Rightarrow y = 3$ and so $x = 2(3) - 1 \Rightarrow x = 5$. The solution is (5, 3).

11.   Note that $x - 2y = 0 \Rightarrow x = 2y$. Substituting $x = 2y$ into the second equation yields the following:

$3(2y) + y = 7 \Rightarrow 7y = 7 \Rightarrow y = 1$ and so $x = 2(1) \Rightarrow x = 2$. The solution is (2, 1).

This result is supported by the graph's intersection point of (2, 1).

13.   Substituting $y = 3x$ into the second equation yields the following:

$x + (3x) = 4 \Rightarrow 4x = 4 \Rightarrow x = 1$ and so $y = 3(1) = 3$. The solution is (1, 3).

By solving each equation for y and graphing each (figure not shown), the intersection point is (1, 3).

15.   Note that $x + y = 2 \Rightarrow y = 2 - x$. Substituting $y = 2 - x$ into the second equation yields:

$2x - (2 - x) = 1 \Rightarrow 3x - 2 = 1 \Rightarrow 3x = 3 \Rightarrow x = 1$ and so $y = 2 - (1) = 1$. The solution is (1, 1).

By solving each equation for y and graphing each (figure not shown), the intersection point is (1, 1).

17.   Note that $x - 2y = -4 \Rightarrow x = 2y - 4$. Substituting $x = 2y - 4$ into the first equation yields:

$2(2y - 4) + 3y = 6 \Rightarrow 7y = 14 \Rightarrow y = 2$ and so $x = 2(2) - 4 = 0$. The solution is $(0, 2)$.

By solving each equation for y and graphing each (figure not shown), the intersection point is $(0, 2)$.

19.   Note that $4x + y = 3 \Rightarrow y = -4x + 3$. Substituting $y = -4x + 3$ into the second equation yields:

$2x - 3(-4x + 3) = -2 \Rightarrow 2x + 12x - 9 = -2 \Rightarrow 14x = 7 \Rightarrow x = \dfrac{1}{2}$ and so $y = -4\left(\dfrac{1}{2}\right) + 3 \Rightarrow y = 1$. The solution is

$\left(\dfrac{1}{2}, 1\right)$.

21.   Note that $x - 2y = 0 \Rightarrow x = 2y$. Substituting $x = 2y$ into the second equation yields:

$-3(2y) + 2y = -1 \Rightarrow -6y + 2y = -1 \Rightarrow -4y = -1 \Rightarrow y = \dfrac{1}{4}$ and so $x = 2\left(\dfrac{1}{4}\right) = \dfrac{1}{2}$. The solution is $\left(\dfrac{1}{2}, \dfrac{1}{4}\right)$.

23.   Note that $-x + 3y = 0 \Rightarrow x = 3y$. Substituting $x = 3y$ into the first equation yields:

$2(3y) - 5y = -1 \Rightarrow y = -1$ and so $x = 3(-1) \Rightarrow x = -3$. The solution is $(-3, -1)$.

25.   Note that $x - y = 1 \Rightarrow x = y + 1$. Substituting $x = y + 1$ into the second equation yields:

$2(y + 1) + 6y = -2 \Rightarrow 8y = -4 \Rightarrow y = -\dfrac{1}{2}$ and so $x = \left(-\dfrac{1}{2}\right) + 1 \Rightarrow x = \dfrac{1}{2}$. The solution is $\left(\dfrac{1}{2}, -\dfrac{1}{2}\right)$.

27.   Note that $2x - y = 6 \Rightarrow y = 2x - 6$. Substituting $y = 2x - 6$ into the first equation yields:

$\dfrac{1}{2}x - \dfrac{1}{2}(2x - 6) = 1 \Rightarrow -\dfrac{1}{2}x = -2 \Rightarrow x = 4$ and so $y = 2(4) - 6 = 2$. The solution is $(4, 2)$.

29.   Note that $\dfrac{1}{6}x - \dfrac{1}{3}y = -1 \Rightarrow x = 2y - 6$. Substituting $x = 2y - 6$ into the second equation yields:

$\dfrac{1}{3}(2y - 6) + \dfrac{5}{6}y = 7 \Rightarrow \dfrac{3}{2}y = 9 \Rightarrow y = 6$ and so $x = 2(6) - 6 = 6$. The solution is $(6, 6)$.

31.   Note that $\dfrac{1}{4}x - \dfrac{1}{3}y = 3 \Rightarrow x = \dfrac{4}{3}y + 12$. Substituting $x = \dfrac{4}{3}y + 12$ into the first equation yields:

$\dfrac{1}{2}\left(\dfrac{4}{3}y + 12\right) + \dfrac{2}{3}y = -2 \Rightarrow \dfrac{4}{3}y = -8 \Rightarrow y = -6$ and so $x = \dfrac{4}{3}(-6) + 12 \Rightarrow x = 4$. The solution is $(4, -6)$.

33.   Note that $0.1x + 0.4y = 1.3 \Rightarrow x = -4y + 13$. Substituting $x = -4y + 13$ into the second equation yields:

$0.3(-4y + 13) - 0.2y = 1.1 \Rightarrow -1.4y = -2.8 \Rightarrow y = 2$ and so $x = -4(2) + 13 \Rightarrow x = 5$.

The solution is $(5, 2)$.

35.   Note that $x - y = 5 \Rightarrow x = y + 5$. Substituting $x = y + 5$ into the second equation yields:

$2(y + 5) - 2y = 10 \Rightarrow 10 = 10$ and this result is an identity. The system is dependent.

37.   Note that $-\dfrac{5}{3}x - y = 2 \Rightarrow y = -\dfrac{5}{3}x - 2$. Substituting $y = -\dfrac{5}{3}x - 2$ into the first equation yields:

$5x + 3\left(-\dfrac{5}{3}x - 2\right) = 6 \Rightarrow 5x - 5x - 6 = 6 \Rightarrow -6 = 6,$ a contradiction. Therefore the system is inconsistent and there are no solutions.

39.   Note that $x + 3y = -2 \Rightarrow x = -3y - 2$. Substituting $x = -3y - 2$ into the second equation yields:

$-\dfrac{1}{2}(-3y - 2) - \dfrac{3}{2}y = 1 \Rightarrow \dfrac{3}{2}y + 1 - \dfrac{3}{2}y = 1 \Rightarrow 1 = 1,$ an identity. Therefore the system is dependent and the

solution set is $\{(x, y) \mid x + 3y = -2\}$.

Elimination Method

41.   Adding the two equations will eliminate the variable y.

$\begin{array}{l} x - y = 5 \\ \underline{x + y = 9} \\ \quad 2x = 14 \end{array}$ Thus, $x = 7$. And so $(7) - y = 5 \Rightarrow y = 2$. The solution is $(7, 2)$.

This result is supported by the graph's intersection point of $(7, 2)$.

43.   Adding the two equations will eliminate the variable y.

$\begin{array}{l} x + y = 3 \\ \underline{x - y = 1} \\ \quad 2x = 4 \end{array}$ Thus, $x = 2$. And so $(2) + y = 3 \Rightarrow y = 1$. The solution is $(2, 1)$.

By solving each equation for y and graphing each (figure not shown), the intersection point is $(2, 1)$.

45.   Multiplying the first equation by 2 and adding the two equations will eliminate the variable y.

$\begin{array}{l} 8x - 2y = 8 \\ \underline{x + 2y = 1} \\ \quad 9x = 9 \end{array}$ Thus, $x = 1$. And so $(1) + 2y = 1 \Rightarrow y = 0$. The solution is $(1, 0)$.

By solving each equation for y and graphing each (figure not shown), the intersection point is $(1, 0)$.

47.   Adding the two equations will eliminate the variable y.

$\begin{array}{l} x + y = 4 \\ \underline{x - y = 2} \\ \quad 2x = 6 \end{array}$ Thus, $x = 3$. And so $3 + y = 4 \Rightarrow y = 1$. The solution is $(3, 1)$.

49.   Multiplying the first equation by -1 and adding the two equations will eliminate the variable x.

$\begin{array}{l} 3x + y = -5 \\ \underline{-3x + 2y = -1} \\ \quad 3y = -6 \end{array}$ Thus $y = -2$. And so $-3x - (-2) = 5 \Rightarrow -3x = 3 \Rightarrow x = -1$. The solution is $(-1, -2)$.

51.   Adding the two equations will eliminate the variable y.

$\begin{array}{l} 2x + y = 4 \\ \underline{2x - y = -2} \\ \quad 4x = 2 \end{array}$ Thus, $x = \dfrac{1}{2}$. And so $2\left(\dfrac{1}{2}\right) + y = 4 \Rightarrow y = 3$. The solution is $\left(\dfrac{1}{2}, 3\right)$.

53.   Multiplying the second equation by -2 and adding the two equations will eliminate the variable x.

$6x - 4y = 12$

$\dfrac{-6x - 10y = 12}{-14y = 24}$ Thus, $y = -\dfrac{12}{7}$. And so $6x - 4\left(-\dfrac{12}{7}\right) = 12 \Rightarrow 6x = \dfrac{36}{7} \Rightarrow x = \dfrac{6}{7}$.

The solution is $\left(\dfrac{6}{7}, -\dfrac{12}{7}\right)$.

55.   Multiplying the second equation by 2 and adding the two equations will eliminate both variables.

$2x - 4y = 5$

$\dfrac{-2x + 4y = 18}{0 = 23}$ This is always false and the system is inconsistent. No solutions.

57.   Multiplying the first equation by –2 and adding the two equations will eliminate both variables.

$-4x - 2y = -4$

$\dfrac{4x + 2y = 4}{0 = 0}$ This is always true and the system is dependent with solutions: $\{(x,\, y) | 2x + y = 2\}$.

59.   Multiply the first equation by $-\dfrac{1}{2}$, the second equation by $\dfrac{2}{15}$ and add the equations to eliminate the variable y.

$-x - 2y = 11$

$\dfrac{10x + 2y = -16}{9x = -5}$ Thus, $x = -\dfrac{5}{9}$. And so $2\left(-\dfrac{5}{9}\right) + 4y = -22 \Rightarrow 4y = -\dfrac{188}{9} \Rightarrow y = -\dfrac{47}{9}$.

The solution is $\left(-\dfrac{5}{9}, -\dfrac{47}{9}\right)$.

61.   Multiply the first equation by 30, the second equation by –20 and add the equations to eliminate the variable y.

$9x + 6y = 24$

$\dfrac{-8x - 6y = -22}{x = 2}$ Thus, $x = 2$. And so $0.3(2) + 0.2y = 0.8 \Rightarrow 0.2y = 0.2 \Rightarrow y = 1$. The solution is $(2, 1)$.

63.   Note that $2x - y = -13 \Rightarrow y = 2x + 13$. Substituting $y = 2x + 13$ into the first equation yields:

$2x + 3(2x + 13) = 7 \Rightarrow 8x = -32 \Rightarrow x = -4$ and so $x = 2(-4) + 13 = 5$. The solution is $(-4, 5)$.

65.   Note that $5u + v = 2 \Rightarrow v = 2 - 5u$. Substituting $v = 2 - 5u$ into the first equation yields:

$3u - 5(2 - 5u) = 4 \Rightarrow 28u = 14 \Rightarrow u = \dfrac{1}{2}$ and so $v = 2 - 5\left(\dfrac{1}{2}\right) = -\dfrac{1}{2}$. The solution is $\left(\dfrac{1}{2}, -\dfrac{1}{2}\right)$.

67.   Multiplying the first equation by 2 and adding the two equations will eliminate both variables.

$4r - 6t = 14$

$\dfrac{-4r + 6t = -14}{0 = 0}$ This is an identity. The system is dependent with solutions of the form $\{(r,\, t) | 2r - 3t = 7\}$.

69.   Multiplying the first equation by –1 and adding the two equations will eliminate both variables.

$$-m + n = -5$$
$$\underline{m - n = 7}$$
$$0 = 2 \qquad \text{This is a contradiction. The system is inconsistent and has no solutions.}$$

71. Note that $2x - 3y = 2 \Rightarrow x = \frac{3}{2}y + 1$. Substituting $x = \frac{3}{2}y + 1$ into the second equation yields:

$$3\left(\frac{3}{2}y + 1\right) - 5y = 4 \Rightarrow -\frac{1}{2}y = 1 \Rightarrow y = -2 \text{ and so } x = \frac{3}{2}(-2) + 1 = -2. \text{ The solution is } (-2, -2).$$

73. Note that $0.1x - 0.3y = -5 \Rightarrow x = 3y - 50$. Substituting $x = 3y - 50$ into the second equation yields:

$$0.5(3y - 50) + 1.1y = 27 \Rightarrow 2.6y = 52 \Rightarrow y = 20 \text{ and so } x = 3(20) - 50 = 10. \text{ The solution is } (10, \ 20).$$

75. Multiplying the first equation by -4 and the second equation by 4 will eliminate the variable z when adding.

$$-2y + 2z = 4$$
$$\underline{3y - 2z = 4}$$
$$y = 8 \qquad \text{Thus, } y = 8. \text{ And so } -2(8) + 2z = 4 \Rightarrow z = 10. \text{ The solution is } (8, \ 10).$$

77. Multiplying the first equation by -5 and adding the two equations will eliminate the variable x.

$$-x + 2y = 3$$
$$\underline{x \ - y = 1}$$
$$y = 4 \qquad \text{Thus, } y = 4. \text{ And so } x - (4) = 1 \Rightarrow x = 5. \text{ The solution is } (5, \ 4).$$

Using More Than One Method

79. Note that $x - y = 4 \Rightarrow y = x - 4$ and $x + y = 6 \Rightarrow y = -x + 6$.

Graphical: Graph $Y_1 = X - 4$ and $Y_2 = -X + 6$ in $[0, 10, 1]$ by $[0, 2, 1]$. See Figure 79a.

The unique solution is the intersection point $(5, 1)$.

Numerical: Table $Y_1 = X - 4$ and $Y_2 = -X + 6$ with TblStart $= 0$ and ΔTbl $= 1$. See Figure 79b.

Since $Y_1 = Y_2 = 1$ when $X = 5$, the solution is $(5, 1)$.

Symbolic: Adding the two equations will eliminate the variable y.

$$x - y = 4$$
$$\underline{x + y = 6}$$
$$2x = 10 \qquad \text{Thus, } x = 5. \text{ And so } (5) - y = 4 \Rightarrow -y = -1 \Rightarrow y = 1. \text{ The solution is } (5, 1).$$

$[0, 10, 1]$ by $[0, 2, 1]$

 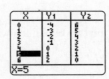

Figure 79a   Figure 79b

81. Note that $5x + 2y = 9 \Rightarrow y = -\frac{5}{2}x + \frac{9}{2}$ and $3x - y = 1 \Rightarrow y = 3x - 1$.

Graphical: Graph $Y_1 = (-5/2)X + (9/2)$ and $Y_2 = 3X - 1$ in [0, 3, 1] by [0, 3, 1]. See Figure 81a.

The unique solution is the intersection point (1, 2).

Numerical: Table $Y_1 = (-5/2)X + (9/2)$ and $Y_2 = 3X - 1$ with TblStart = 0 and ΔTbl = 1. See Figure 81b.

Since $Y_1 = Y_2 = 2$ when $X = 1$, the solution is (1, 2).

Symbolic: Multiplying the second equation by 2 and adding the two equations will eliminate the variable y.

$$5x + 2y = 9$$
$$\underline{6x - 2y = 2}$$
$$11x = 11$$

Thus, $x = 1$. And so $5(1) + 2y = 9 \Rightarrow 2y = 4 \Rightarrow y = 2$. The solution is (1, 2).

[0, 3, 1] by [0, 3, 1]

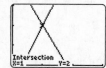
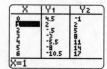

Figure 81a   Figure 81b

83.   Note that $3x - 2y = 8.5 \Rightarrow y = \dfrac{3x - 8.5}{2}$ and $2x + 4y = 3 \Rightarrow y = \dfrac{-2x + 3}{4}$.

Graphical: Graph $Y_1 = (3X - 8.5)/2$ and $Y_2 = (-2X + 3)/4$ in [-5, 5, 1] by [-5, 5, 1]. See Figure 83a.

The unique solution is the intersection point $(2.5, -0.5)$.

Numerical: Table $Y_1 = (3X - 8.5)/2$ and $Y_2 = (-2X + 3)/4$ with TblStart = 0 and ΔTbl = 0.5. See Figure 83b.

Since $Y_1 = Y_2 = -0.5$ when $X = 2.5$, the solution is $(2.5, -0.5)$.

Symbolic: Multiplying the first equation by 2 and adding the two equations will eliminate the variable y.

$$6x - 4y = 17$$
$$\underline{2x + 4y = 3}$$
$$8x = 20$$

Thus, $x = 2.5$. And so $2(2.5) + 4y = 3 \Rightarrow y = -0.5$. The solution is $(2.5, -0.5)$.

[-5, 5, 1] by [-5, 5, 1]

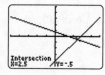
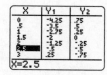

Figure 83a   Figure 83b

85.   Multiplying the second equation by -1 and adding the two equations will eliminate the variable y:

$$ax + y = 4$$
$$\underline{-x - y = -4}$$
$$(a - 1)x = 0$$

Thus $x = 0$ since $a \neq 1$. And so $0 + y = 4 \Rightarrow y = 4$. The solution is (0, 4).

Applications

87. Let x and y represent the number fat calories burned by the heavier and lighter athlete respectively. Then the system needed is $x - y = 58$ and $x + y = 290$. Adding the two equations will eliminate the variable y.

$$x - y = 58$$
$$\underline{x + y = 290}$$
$$2x = 348$$

Thus, $x = 174$. And so $(174) - y = 58 \Rightarrow y = -116 \Rightarrow y = 116$.

The heavier athlete burned 174 fat calories or $174 \div 9 = 19.\overline{3}$ fat grams. The lighter athlete burned 116 fat calories or $116 \div 9 = 12.\overline{8}$ fat grams.

89. Let x and y represent the amount of 10% and 80% solution respectively. Then the system needed is $x + y = 4$ and $0.10x + 0.80y = 0.50(4)$. Multiplying the second equation by -10 and adding the two equations will eliminate the variable x.

$$x + y = 4$$
$$\underline{-x - 8y = -20}$$
$$-7y = -16$$

Thus, $y = \dfrac{16}{7}$. And so $x + \left(\dfrac{16}{7}\right) = 4 \Rightarrow x = \dfrac{12}{7}$. Drain and replace $\dfrac{16}{7}$ gallons.

91. Let x and y represent the number of premium and regular rooms respectively. Then the system needed is $x + y = 50$ and $115x + 80y = 4945$. Multiplying the first equation by -80 and adding will eliminate y.

$$-80x - 80y = -4000$$
$$\underline{115x + 80y = 4945}$$
$$35x = 945$$

Thus, $x = 27$. And so $(27) + y = 50 \Rightarrow y = 23$.

The hotel rented 27 premium rooms and 23 regular rooms.

93. Let x and y represent the larger and smaller angles respectively. Then the system needed is $x + y = 180$ and $x - 2y = 30$. Multiplying the first equation by 2 and adding will eliminate y.

$$2x + 2y = 360$$
$$\underline{x - 2y = 30}$$
$$3x = 390$$

Thus, $x = 130$. And so $(130) + y = 180 \Rightarrow y = 50$. The angles are $50°$ and $130°$.

95. Let x and y represent the numbers of males and females (in millions) in 2003, respectively. Then the system needed is $x + y = 291$ and $y = x + 5$. Substituting $y = x + 5$ into the first equation yields:

$x + x + 5 = 291 \Rightarrow 2x = 286 \Rightarrow x = 143$ and so $y = 143 + 5 \Rightarrow y = 148$. In 2003, there were 143 million males and 148 million females in the United States.

97. Let x and y represent the average speed of the plane and the jet stream respectively. Using the formula $d = rt$ and converting the times to minutes, the system is $2400 = (x - y)(250)$ and $2400 = (x + y)(225)$.

These equations may be written as follows: $5x - 5y = 48$ and $3x + 3y = 32$.

Multiplying the first equation by 3, the second by 5 and adding the equations will eliminate the variable y.

$$15x - 15y = 144$$
$$\underline{15x + 15y = 160}$$
$$30x = 304$$

Thus, $x = \dfrac{152}{15}$. And so $3\left(\dfrac{152}{15}\right) + 3y = 32 \Rightarrow 3y = \dfrac{8}{5} \Rightarrow y = \dfrac{8}{15}$.

Plane: $\dfrac{152}{15}$ miles per minute or 608 mph; jet stream: $\dfrac{8}{15}$ miles per minute or 32 mph.

99. Let x and y represent the speed of the boat and the speed of the current respectively. Then the system needed is $x + y = 30$ and $x - y = 20$. Adding the two equations will eliminate y.

$$x + y = 30$$
$$\underline{x - y = 20}$$
$$2x = 50$$

Thus, $x = 25$. And so $(25) + y = 30 \Rightarrow y = 5$.

The speed of the boat is 25 mph and the speed of the current is 5 mph.

101. Let x and y represent the amount borrowed at 8% and at 9% respectively. The system needed is $x + y = 3500$ and $0.08x + 0.09y = 294$. Multiplying the first equation by –8, the second by 100 and adding the equations will eliminate the variable x.

$$-8x - 8y = -28,000$$
$$\underline{8x + 9y = 29,400}$$
$$y = 1400$$

Thus, $y = 1400$. And so $x + 1400 = 3500 \Rightarrow x = 2100$.

There was $2100 borrowed at 8% and $1400 borrowed at 9%.

103. Let x and y represent the cost of each day credit and each night credit respectively. The system needed is $12x + 6y = 1800$ and $x - y = 30$. Multiplying the second by 6 and adding the equations will eliminate y.

$$12x + 6y = 1800$$
$$\underline{6x - 6y = 180}$$
$$18x = 1980$$

Thus, $x = 110$. And so $(110) - y = 30 \Rightarrow y = 80$.

Day credits cost $110 each and night credits cost $80 each.

105. Multiplying the second equation by $\dfrac{2}{\sqrt{3}}$ and adding the equations will eliminate W_2.

$$W_1 - W_2 = 0$$
$$\underline{W_1 + W_2 \approx 231}$$
$$2W_1 \approx 231$$

Thus, $W_1 \approx 115.5$. And so $(115.5) - W_2 = 0 \Rightarrow W_2 = 115.5$.

The weight exerted on each rafter is about 115.5 pounds.

107. Let x and y represent the length and width of the court respectively. The system needed is $2x + 2y = 296$ and is $x - y = 44$. Multiplying the first equation by 0.5 and adding the equations will eliminate the variable y.

(a)
$$x + y = 148$$
$$\underline{x - y = 44}$$
$$2x = 192$$

Thus, $x = 96$. And so $2(96) + 2y = 296 \Rightarrow 2y = 104 \Rightarrow y = 52$.

The court has a length 96 feet and a width 52 feet.

Note that $2x + 2y = 296 \Rightarrow y = -x + 148$ and $x - y = 44 \Rightarrow y = x - 44$.

(b) Graph $Y_1 = -X + 148$ and $Y_2 = X - 44$ in [0, 150, 10] by [0, 100, 10]. See Figure 107b.

The unique solution is the intersection point (96, 52).

(c) Table $Y_1 = -X + 148$ and $Y_2 = X - 44$ with TblStart = 92 and ΔTbl = 1. See Figure 107c.

Since $Y_1 = Y_2 = 52$ when X = 96, the solution is (96, 52).

[0, 150, 10] by [0, 100, 10]

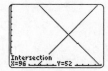   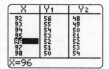

Figure 107b   Figure 107c

Checking Basic Concepts for Sections 4.1 & 4.2

1. Note that $2x - y = 5 \Rightarrow y = 2x - 5$ and $-x + 3y = 0 \Rightarrow y = \dfrac{1}{3}x.$

(a) Graph $Y_1 = 2X - 5$ and $Y_2 = (1/3)X$ in [0, 5, 1] by [0, 2, 1]. See Figure 1a.

The unique solution is the intersection point (3, 1).

(b) Table $Y_1 = 2X - 5$ and $Y_2 = (1/3)X$ with TblStart = 0 and ΔTbl = 1. See Figure 1b.

Since $Y_1 = Y_2 = 1$ when X = 3, the solution is (3, 1).

Yes, the answers agree.

[0, 5, 1] by [0, 2, 1]

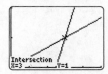   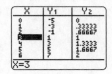

Figure 1a   Figure 1b

2. Note that $x + 4y = 14 \Rightarrow x = 14 - 4y$. Substituting $x = 14 - 4y$ into the first equation yields the following:

$4(14 - 4y) - 3y = -1 \Rightarrow -19y = -57 \Rightarrow y = 3$ and so $x = 14 - 4(3) \Rightarrow x = 2$. The solution is (2, 3).

3. Multiply the first equation by 2, the second equation by 3 and add the equations to eliminate the variable y.

$\begin{array}{l} 8x - 6y = -34 \\ \underline{-18x + 6y = 69} \\ -10x = 35 \end{array}$ Thus, $x = -\dfrac{7}{2}.$ And so $4\left(-\dfrac{7}{2}\right) - 3y = -17 \Rightarrow -3y = -3 \Rightarrow y = 1.$

The solution is $\left(-\dfrac{7}{2}, 1\right).$ The system is not dependent. The system is not inconsistent.

4. (a) Let x and y represent the larger and smaller angles respectively. Then the system needed is $x + y = 90$

and $x - y = 40.$

(b) Adding the two equations will eliminate y.

$$x + y = 90$$

$$\frac{x - y = 40}{2x = 130}$$ Thus, $x = 65$. And so $(65) + y = 90 \Rightarrow y = 25$. The angles are $65°$ and $25°$.

4.3: Systems of Linear Inequalities

Concepts

1.   Two

2.   Yes, many points may satisfy a system of inequalities.

3.   Yes, since $5(3) - 2(1) = 13 > 8$. No, since $5(2) - 2(1) = 8 \not< 8$.

4.   No, since $2(3) - 4 = 2 \not< 1$. The second inequality is not satisfied. $(-2, 0)$ is a solution because

$$-2 - 2(0) = -2 \geq -8 \text{ and } 2(-2) - 0 = -4 < 1.$$

5.   solid

6.   dashed

Linear Inequalities

7.   See Figure 7.

9.   See Figure 9.

11.   See Figure 11.

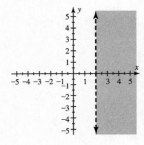   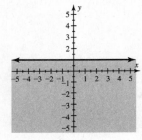   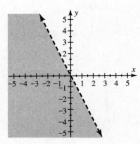

Figure 7   Figure 9   Figure 11

13.   See Figure 13.

15.   See Figure 15.

17.   See Figure 17.

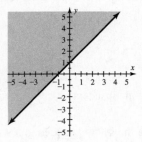   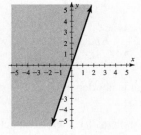   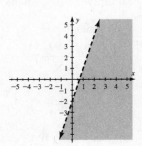

Figure 13   Figure 15   Figure 17

19.   See Figure 19.

21.   See Figure 21.

23.   See Figure 23.

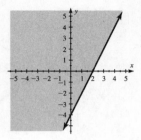

Figure 19

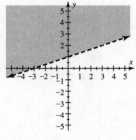

Figure 21

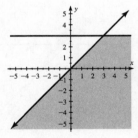

Figure 23

25.   See Figure 25.

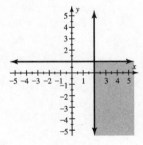

Figure 25

Systems of Inequalities

27.   See Figure 27.

29.   See Figure 29.

31.   See Figure 31.

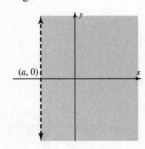

Figure 27

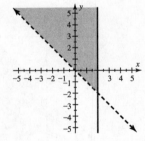

Figure 29

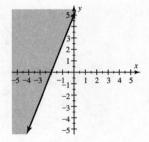

Figure 31

33.   See Figure 33.

35.   See Figure 35.

37.   See Figure 37.

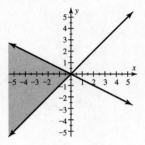

Figure 33

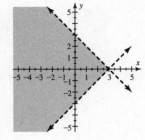

Figure 35

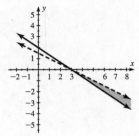

Figure 37

39.   See Figure 39.

41.   See Figure 41.

43.   See Figure 43.

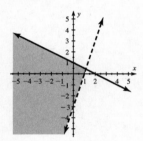

Figure 39

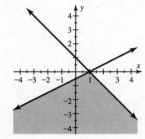

Figure 41

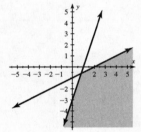

Figure 43

45.   See Figure 45.

47.   See Figure 47.

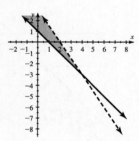

Figure 45

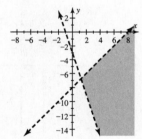

Figure 47

49. This region can be shaded using the *Shade* feature of the TI-83, found under the DRAW menu.

Shade$(-3, 5)$ in $[-10, 10, 1]$ by $[-10, 10, 1]$. See Figure 49.

[–10, 10, 1] by [–10, 10, 1]   [–10, 10, 1] by [–10, 10, 1]

   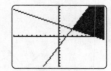

Figure 49   Figure 51

51. This region can be shaded using the *Shade* feature of the TI-83, found under the DRAW menu.

Note that $x + 2y \geq 8 \Rightarrow y \geq -\dfrac{1}{2}x + 4$ and $6x - 3y \geq 10 \Rightarrow y \leq 2x - \dfrac{10}{3}$.

Shade$\big((-1/2)X + 4, 2X - (10/3)\big)$ in $[-10, 10, 1]$ by $[-10, 10, 1]$. See Figure 51.

53. This region can be shaded using the *Shade* feature of the TI-83, found under the DRAW menu.

Note that $0.9x + 1.7y \leq 3.2 \Rightarrow y \leq \dfrac{-0.9}{1.7}x + \dfrac{3.2}{1.7}$ and $1.9x - 0.7y \leq 1.3 \Rightarrow y \geq \dfrac{1.9}{0.7}x - \dfrac{1.3}{0.7}$.

Shade$\big((1.9/0.7)X - (1.3/0.7), (-0.9/1.7)X + (3.2/1.7)\big)$ in $[-10, 10, 1]$ by $[-10, 10, 1]$. See Figure 53.

[–10, 10, 1] by [–10, 10, 1]

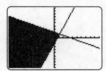

Figure 53

55. *b*; The point $(1, 8)$ satisfies this inequality. Only graph *b* contains this point.

57. *a*; The point $(1, -7)$ satisfies this inequality. Only graph *a* contains this point.

59. The equation of a horizontal line through the point $(0, 2)$ is $y = 2$. The inequality is $y \geq 2$.

61. The equation of a vertical line through the point $(-2, 0)$ is $x = -2$. The equation of a line through the points $(0, 0)$ and $(1, 1)$ is $y = x$. The system of inequalities is $y \geq x$ and $x \geq -2$.

Applications

63.   First note that the amount of candy cannot be negative so $x \geq 0$ and $y \geq 0$. Also, if the cost must be less than

$15, the inequality $3x + 5y < 15$ must be satisfied. See Figure 63.

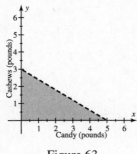

Figure 63   Figure 65

65.   First note that the number of CD players and radios cannot be negative so $x \geq 0$ and $y \geq 0$. Then, if the

business must manufacture at least as many radios as CD players, the inequality $y \geq x$ must be

satisfied. Also, the total number made each day cannot exceed 40 so the inequality $x + y \leq 40$ must be

satisfied. See Figure 65.

67.   (a) The range is approximately 105 to 134 bpm.

(b) The inequalities are $y \leq -0.6(x - 20) + 140$ and $y \geq -0.5(x - 20) + 110$.

69.   The person weighs less than recommended.

71.   The inequalities are $25h - 7w \leq 800$ and $5h - w \geq 170$.

73.   First note that the amounts of candy and peanuts cannot be negative so $x \geq 0$ and $y \geq 0$. Also,

$3x \leq 6 \Rightarrow x \leq 2$ and $2y \leq 8 \Rightarrow y \leq 4$. See Figure 73.

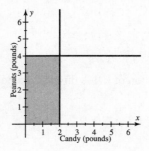

Figure 73

4.4: Introduction to Linear Programming

Concepts

1.   linear programming

2.   objective

3.   feasible solutions

4.   linear inequalities

5.   vertex

6.   objective

Regions of Feasible Solutions

7.   See Figure 7.

9.   See Figure 9.

11.   See Figure 11.

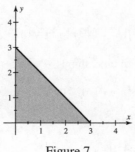

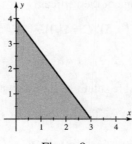

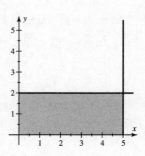

Figure 7   Figure 9   Figure 11

13.   See Figure 13.

15.   See Figure 15.

17.   See Figure 17.

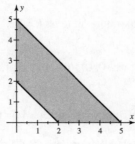

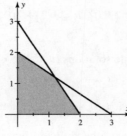

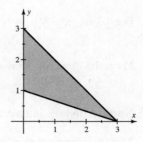

Figure 13   Figure 15   Figure 17

19.   See Figure 19.

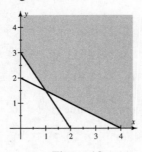

Figure 19

Linear Programming

21.   The maximum value of R occurs at one of the vertices.

For $(1, 1)$, $R = 4(1) + 5(1) = 9$. For $(1, 3)$, $R = 4(1) + 5(3) = 19$.   For $(4, 1)$, $R = 4(4) + 5(1) = 21$.

For $(4, 3)$, $R = 4(4) + 5(3) = 31$. The maximum value is $R = 31$.

23.   The maximum value of R occurs at one of the vertices.

For $(0, 2)$, $R = (0) + 3(2) = 6$. For $(0, 5)$, $R = (0) + 3(5) = 15$.   For $(3, 3)$, $R = (3) + 3(3) = 12$.

For $(5,0)$, $R = (5)+3(0) = 5$. For $(2,0)$, $R = (2)+3(0) = 2$. The maximum value is $R = 15$.

25.   The minimum value of C occurs at one of the vertices.

For $(1,1)$, $C = 2(1)+3(1) = 5$. For $(1,3)$, $C = 2(1)+3(3) = 11$.   For $(4,1)$, $C = 2(4)+3(1) = 11$.

For $(4,3)$, $C = 2(4)+3(3) = 17$. The minimum value is $C = 5$.

27.   The minimum value of C occurs at one of the vertices.

For $(0,2)$, $C = 5(0)+(2) = 2$. For $(0,5)$, $C = 5(0)+(5) = 5$.   For $(3,3)$, $C = 5(3)+(3) = 18$.

For $(5,0)$, $C = 5(5)+(0) = 25$. For $(2,0)$, $C = 5(2)+(0) = 10$. The minimum value is $C = 2$.

29.   From the graph of the region of feasible solutions (not shown), the vertices are (0, 0), (0, 150), and (150, 0).

The maximum value of R occurs at one of the vertices. For $(0,0)$, $R = 3(0)+5(0) = 0$.

For $(0,150)$, $R = 3(0)+5(150) = 750$. For $(150,0)$, $R = 3(150)+5(0) = 450$.

The maximum value is $R = 750$.

31.   From the graph of the region of feasible solutions (not shown), the vertices are (0, 0), (0, 6), and (3, 0).

The maximum value of R occurs at one of the vertices. For $(0,0)$, $R = 3(0)+2(0) = 0$.

For $(0,6)$, $R = 3(0)+2(6) = 12$. For $(3,0)$, $R = 3(3)+2(0) = 9$. The maximum value is $R = 12$.

33.   From the graph of the region of feasible solutions (not shown), the vertices are (0, 0), (0, 2), (1.5, 1.5) and (2, 0).

Note: to find the intersection point (1.5, 1.5), solve the system of equations $3x + y = 6$ and $x + 3y = 6$.

The maximum value of R occurs at one of the vertices. For $(0,0)$, $R = 12(0)+9(0) = 0$.

For $(0,2)$, $R = 12(0)+9(2) = 18$. For $(1.5,1.5)$, $R = 12(1.5)+9(1.5) = 31.5$.

For $(2,0)$, $R = 12(2)+9(0) = 24$. The maximum value is $R = 31.5$.

35.   From the graph of the region of feasible solutions (not shown), the vertices are (0, 2), (4, 0), and (2, 0).

The maximum value of R occurs at one of the vertices. For $(0,2)$, $R = 4(0)+5(2) = 10$.

For $(4,0)$, $R = 4(4)+5(0) = 16$. For $(2,0)$, $R = 4(2)+5(0) = 8$. The maximum value is $R = 16$.

37.   From the graph of the region of feasible solutions (not shown), the vertices are (0, 0), (0, 2), (3, 2) and (3, 0).

The minimum value of C occurs at one of the vertices. For $(0,0)$, $C = (0)+2(0) = 0$.

For $(0,2)$, $C = (0)+2(2) = 4$. For $(3,2)$, $C = (3)+2(2) = 7$. For $(3,0)$, $C = (3)+2(0) = 3$.

The minimum value is $C = 0$.

39.   From the graph of the region of feasible solutions (not shown), the vertices are (0, 4), and (4, 0).

Note that this region is unbounded. The minimum value of C occurs at one of the vertices.

For $(0,4)$, $C = 8(0)+15(4) = 60$. For $(4,0)$, $C = 8(4)+15(0) = 32$. The minimum value is $C = 32$.

41.   From the graph of the region of feasible solutions (not shown), the vertices are (0, 2), (0, 6), (3, 0) and (2, 0).

The minimum value of C occurs at one of the vertices. For $(0, 2)$, $C = 30(0) + 40(2) = 80$.

For $(0, 6)$, $C = 30(0) + 40(6) = 240$. For $(3, 0)$, $C = 30(3) + 40(0) = 90$.

For $(2, 0)$, $C = 30(2) + 40(0) = 60$. The minimum value is $C = 60$.

43.   Let x and y represent the amount of candy and coffee respectively. The business can sell no more than a total of 100 pounds so $x + y \leq 100$. Also, because at least 20 pounds of candy must be sold each day $x \geq 20$. Finally, the amount of coffee cannot be negative so $y \geq 0$. Here the revenue function is $R = 4x + 6y$. From the graph of the region of feasible solutions (not shown), the vertices are (20, 0), (20, 80), and (100, 0). The maximum value of R occurs at one of the vertices. For $(20, 0)$, $R = 4(20) + 6(0) = 80$.

For $(20, 80)$, $R = 4(20) + 6(80) = 560$. For $(100, 0)$, $R = 4(100) + 6(0) = 400$.

To maximize revenue, the business should sell 20 pounds of candy and 80 pounds of coffee.

45.   Let x and y represent the amount of Brand X and Brand Y respectively. Since each ounce of Brand X contains 20 units of vitamin A, each ounce of Brand Y contains 10 units of vitamin A and the total amount of vitamin A must be at least 40 units, $20x + 10y \geq 40$. Since each ounce of Brand X contains 10 units of vitamin C, each ounce of Brand Y contains 10 units of vitamin C and the total amount of vitamin C must be at least 30 units, $10x + 10y \geq 30$. Finally the quantities cannot be negative so $x \geq 0$ and $y \geq 0$. Here the cost function is $C = 0.90x + 0.60y$. From the graph of the region of feasible solutions (not shown), the vertices are (0, 4), (1, 2) and (3, 0). Note that this region is unbounded. To find the intersection point (1, 2) solve the system of equations $20x + 10y = 40$ and $10x + 10y = 30$. The minimum value of C occurs at one of the vertices.

For $(0, 4)$, $C = 0.90(0) + 0.60(4) = 2.40$. For $(1, 2)$, $C = 0.90(1) + 0.60(2) = 2.10$.

For $(3, 0)$, $C = 0.90(3) + 0.60(0) = 2.70$.

To minimize cost, 1 ounce of Brand X and 2 ounces of Brand Y should be mixed.

47.   Let x and y represent the number of hamsters and mice respectively. Since the total number of animals cannot exceed 50, $x + y \leq 50$. Because no more than 20 hamsters can be raised, $x \leq 20$. Here the revenue function is $R = 15x + 10y$. From the graph of the region of feasible solutions (not shown), the vertices are (0, 50), (20, 30) and (20, 0). To find (20, 30) solve the equations $x + y = 50$ and $x = 20$. The maximum value of R occurs at one of the vertices. For $(0, 50)$, $R = 15(0) + 10(50) = 500$. For $(20, 30)$, $R = 15(20) + 10(30) = 600$.

For $(20, 0)$, $R = 15(20) + 10(0) = 300$. The maximum revenue is \$600.

Checking Basic Concepts for Sections 4.3 & 4.4

1.   The equation of a line through the points (0, 3) and (1, 1) is $y = -2x + 3$. The inequality is $y \leq -2x + 3$.

2.   See Figure 2.

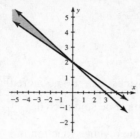

Figure 2

3.   From the graph of the region of feasible solutions (not shown), the vertices are (0, 0), (0, 5), (2, 3) and (4, 0).

To find the intersection point (2, 3) solve the system of equations $3x + 2y = 12$ and $4x + 4y = 20$.

The maximum value of R occurs at one of the vertices. For $(0, 0)$, $R = 3(0) + 2(0) = 0$.

For $(0, 5)$, $R = 3(0) + 2(5) = 10$. For $(2, 3)$, $R = 3(2) + 2(3) = 12$. For $(4, 0)$, $R = 3(4) + 2(0) = 12$.

The maximum value is $R = 12$.

4.5: Systems of Linear Equations in Three Variables

Concepts

1.   No, three planes cannot intersect at exactly 2 points.

2.

$$x + y + z = 5 \qquad \text{\textit{Answers may vary.}}$$
$$2x - 3y + z = 7$$
$$x + 2y - 4z = 2$$

3.   Yes, since $1 + 2 + 3 = 6$.

4.   No, a solution must be an ordered triple.

5.   Two

6.   Three

7.   No

8.   Infinitely many

Solving Linear Systems

9.   (1, 2, 3) satisfies all three inequalities.

11.   (–1, 1, 2) satisfies all three inequalities.

13.   Substitute $z = 1$ into the second equation: $2y + (1) = -1 \Rightarrow 2y = -2 \Rightarrow y = -1$

Substitute $z = 1$ and $y = -1$ into the first equation: $x + (-1) - (1) = 1 \Rightarrow x - 2 = 1 \Rightarrow x = 3$

The solution is (3, –1, 1).

15.   Substitute $z = 2$ into the second equation: $2y + 3(2) = 3 \Rightarrow 2y = -3 \Rightarrow y = -\dfrac{3}{2}$

Substitute $z = 2$ and $y = -\dfrac{3}{2}$ into the first equation: $-x - 3\left(-\dfrac{3}{2}\right) + (2) = -2 \Rightarrow -x = -\dfrac{17}{2} \Rightarrow x = \dfrac{17}{2}$

The solution is $\left(\dfrac{17}{2}, -\dfrac{3}{2}, 2\right)$.

17.   Substitute $c = -2$ into the second equation: $-3b + (-2) = 4 \Rightarrow -3b = 6 \Rightarrow b = -2$

Substitute $c = -2$ and $b = -2$ into the first equation: $a - (-2) + 2(-2) = 3 \Rightarrow a - 2 = 3 \Rightarrow a = 5$

The solution is $(5, -2, -2)$.

19.   Add the first two equations together to eliminate the variable x.

$$\begin{array}{r} x + y - z = 11 \\ -x + 2y + 3z = -1 \\ \hline 3y + 2z = 10 \end{array}$$

From the third equation, $2z = 4 \Rightarrow z = 2$. And so $3y + 2(2) = 10 \Rightarrow 3y = 6 \Rightarrow y = 2$.

Substitute $z = 2$ and $y = 2$ into the first equation: $x + (2) - (2) = 11 \Rightarrow x = 11$

The solution is $(11, 2, 2)$.

21.   Add the first two equations together to eliminate both of the variables x and z.

$$\begin{array}{r} x + y - z = -2 \\ -x \quad\;\; + z = 1 \\ \hline y = -1 \end{array}$$

Substitute $y = -1$ into the third equation, $(-1) + 2z = 3 \Rightarrow 2z = 4 \Rightarrow z = 2$.

Substitute $z = 2$ and $y = -1$ into the first equation: $x + (-1) - (2) = -2 \Rightarrow x = 1$

The solution is $(1, -1, 2)$.

23.   Add the second and third equations together to eliminate the variable y.

$$\begin{array}{r} y + z = -1 \\ -y + 3z = 9 \\ \hline 4z = 8 \end{array}$$

And so $z = 2$. Substitute $z = 2$ into the second equation, $y + (2) = -1 \Rightarrow y = -3$.

Substitute $z = 2$ and $y = -3$ into the first equation: $x + (-3) - 2(2) = -7 \Rightarrow x = 0$

The solution is $(0, -3, 2)$.

25.   Multiply the second equation by -2 and add the first and second equations to eliminate the variables y and z.

$$\begin{array}{r} x + 2y + 2z = 1 \\ -2x - 2y - 2z = 0 \\ \hline -x = 1 \end{array}$$

And so $x = -1$. Add the first and third equations together to eliminate the variables x and y.

$$\begin{array}{r} x + 2y + 2z = 1 \\ -x - 2y + 3z = -11 \\ \hline 5z = -10 \end{array}$$

And so $z = -2$. Substitute $x = -1$ and $z = -2$ into the second equation: $(-1) + y + (-2) = 0 \Rightarrow y = 3$

The solution is (–1, 3, –2).

27.   Multiply the second equation by -1 and add the first and second equations to eliminate the variables y and z.

$$x+y+z=5$$
$$\underline{-y-z=-6}$$
$$x=-1$$

And so $x=-1$. Substitute $x=-1$ into the third equation: $(-1)+z=3 \Rightarrow z=4$

Substitute $x=-1$ and $z=4$ into the first equation: $(-1)+y+(4)=5 \Rightarrow y=2$

The solution is $(-1, 2, 4)$.

29.   Add the second and third equations to eliminate the variables x and z.

$$-x+y+2z=1$$
$$\underline{x+y-2z=9}$$
$$2y=10$$

And so, $y=5$. Now add the first and second equations to eliminate the variable x.

$$x+2y+3z=24$$
$$\underline{-x+y+2z=1}$$
$$3y+5z=25$$

Now substitute $y=5$ in this new equation: $3(5)+5z=25 \Rightarrow 5z=10 \Rightarrow z=2$

Finally substitute $y=5$ and $z=2$ in the first equation: $x+2(5)+3(2)=24 \Rightarrow x=8$

The solution is $(8,5,2)$.

31.   Add the first and second equations to eliminate the variable y.

$$x+y+z=2$$
$$\underline{x-y+z=1}$$
$$2x+2z=3$$

Multiply the third equation by -2 and add to this new equation to eliminate the variables x and z.

$$2x+2z=3$$
$$\underline{-2x+-2z=-6}$$
$$0=-3 \qquad \text{This is a contradiction, so there are no solutions.}$$

33.   Add the first two equations to eliminate the variable y.

$$x+y+z=6$$
$$\underline{x-y+z=2}$$
$$2x+2z=8 \qquad \Rightarrow x+z=4 \Rightarrow x=4-z$$

Add the second and third equations to eliminate the variables x and z.

$$x-y+z=2$$
$$\underline{-x+5y-z=6}$$
$$4y=8 \qquad \Rightarrow y=2.$$

The system is dependent, and the solutions are all ordered triples of the form $(4-z, 2, z)$.

35. Add the first and second equations to eliminate the variables y and z.

$$2x+y+z=3$$
$$2x-y-z=9$$
$$\overline{4x=12}$$

And so, $x=3$. Add the second and third equations together to eliminate the variable y.

$$2x-y-z=9$$
$$x+y-z=0$$
$$\overline{3x-2z=9}$$

Now substitute $x=3$ in this new equation: $3(3)-2z=9 \Rightarrow -2z=0 \Rightarrow z=0$

Finally substitute $x=3$ and $z=0$ in the third equation: $(3)+y-(0)=0 \Rightarrow y=-3$

The solution is $(3,-3,0)$.

37. Multiply the first equation by -1 and add the first and second equations to eliminate the variable x.

$$-2x-6y+2z=-47$$
$$2x+y+3z=-28$$
$$\overline{-5y+5z=-75}$$

Multiply the third equation by 2 and add the first and third equations to eliminate the variables x and z.

$$2x+6y-2z=47$$
$$-2x+2y+2z=-7$$
$$\overline{8y=40}$$

And so $y=5$. Substitute $y=5$ into the first *new* equation: $-5(5)+5z=-75 \Rightarrow 5z=-50 \Rightarrow z=-10$

Substitute $y=5$ and $z=-10$ into the *original* third equation: $-x+(5)+(-10)=-\frac{7}{2} \Rightarrow x=-\frac{3}{2}$

The solution is $\left(-\frac{3}{2}, 5, -10\right)$.

39. Multiply the second equation by -1 and add the first and second equations to eliminate the variable y.

$$x+3y-4z=\frac{13}{2}$$
$$2x-3y+z=-\frac{1}{2}$$
$$\overline{3x-3z=6}$$

Multiply this *new* equation by -1 and add it to the third equation to eliminate the variable x.

$$-3x+3z=-6$$
$$3x+z=4$$
$$\overline{4z=-2}$$

And so $z=-\frac{1}{2}$. Substitute $z=-\frac{1}{2}$ into the first *new* equation: $3x-3\left(-\frac{1}{2}\right)=6 \Rightarrow 3x=\frac{9}{2} \Rightarrow x=\frac{3}{2}$

Substitute $x = \dfrac{3}{2}$ and $z = -\dfrac{1}{2}$ into the *original* first equation: $\left(\dfrac{3}{2}\right) + 3y - 4\left(-\dfrac{1}{2}\right) = \dfrac{13}{2} \Rightarrow y = 1$

The solution is $\left(\dfrac{3}{2}, 1, -\dfrac{1}{2}\right)$.

41. Multiply the second equation by -1 and add to the third equation to eliminate the variables x, y, and z.

$$-x + y - z = -1$$
$$\underline{x - y + z = 3}$$
$$0 = 2 \qquad \text{This is a contradiction. There are no solutions.}$$

43. Add the first two equations to eliminate the variable y.

$$x + y + z = 5$$
$$\underline{x - y + z = 3}$$
$$2x + 2z = 8 \qquad \Rightarrow x + z = 4 \Rightarrow x = 4 - z$$

Multiply the first equation by -1 and add the second equation to eliminate the variables x and z.

$$-x - y - z = -5$$
$$\underline{x - y + z = 3}$$
$$-2y = -2 \qquad \Rightarrow y = 1$$

The system is dependent, and the solutions are all ordered pairs of the form $(4 - z, 1, z)$.

45. Multiply the first equation by -1 and add the second equation to eliminate the variables x, y, and z.

$$-x - y - z = -a$$
$$\underline{x + y + z = 2a}$$
$$0 = a \qquad \text{But } a \neq 0, \text{ so this is a contradiction.}$$

There are no solutions.

Applications

47. (a) $\quad x + 2y + 4z = 10$
$$\qquad\quad x + 4y + 6z = 15$$
$$\qquad\qquad\quad 3y + 2z = 6$$

(b) Using techniques similar to those used in exercises 19-44, the solution is $(2, 1, 1.5)$.

A hamburger costs $2.00, fries cost $1.00 and a soft drink costs $1.50.

49. (a) $\quad x + y + z = 180$
$$\qquad\quad x - z = 55$$
$$\qquad\quad x - y - z = -10$$

(b) Using techniques similar to those used in exercises 19-44, the solution is $(85, 65, 30)$.

The angles are $x = 85°$, $y = 65°$, and $z = 30°$.

(c) These values check.

51. Add the first two equations to eliminate the variables y and z.

$$x + y + z = 180$$
$$\underline{x - y - z = 40}$$
$$2x \qquad = 220 \quad \Rightarrow x = 110$$

$$x - z = 90 \Rightarrow z = x - 90 \Rightarrow z = 110 - 90 \Rightarrow z = 20$$

Substitute the values for x and z into the first equation and solve for y.

$$110 + y + 20 = 180 \Rightarrow y = 50$$

The angle measures are 110°, 50°, and 20°.

53.   (a) $a + 600b + 4c = 525$
              $a + 400b + 2c = 365$
              $a + 900b + 5c = 805$

       (b)   Using techniques similar to those used in exercises 19-44, the solution is (5, 1, –20).

              That is, $a = 5$, $b = 1$, and $c = -20$ and so the equation is $F = 5 + A - 20W$.

       (c)   When $A = 500$ and $W = 3$, $F = 5 + (500) - 20(3) = 445$ fawns.

55.   (a) $N + P + K = 80$
              $N + P - K = \ 8$
                $9P - K = \ 0$

       (b)   Using techniques similar to those used in exercises 19-44, the solution is (40, 4, 36).

              The sample contains 40 pounds of nitrogen, 4 pounds of phosphorus and 36 pounds of potassium.

57.   Let x, y and z represent the amounts invested at 8%, 10% and 15% respectively. The system needed is:

$$x \qquad + y \qquad + z = 30,000$$
$$0.08x + 0.10y + 0.15z = 3550$$
$$x \qquad + y \qquad - z = 2000$$

Using techniques similar to those used in exercises 19-44, the solution is (7500, 8500, 14,000).

There was $7500 invested at 8%, $8500 invested at 10% and $14,000 invested at 15%.

4.6: Matrix Solutions of Linear Systems

Concepts

1.   A rectangular array of numbers

2. $\begin{bmatrix} 2 & 1 & 3 \\ 0 & -4 & 2 \end{bmatrix}$; 2×3; *Answers may vary.*

3. $\left[\begin{array}{cc|c} 1 & 3 & 10 \\ 2 & -6 & 4 \end{array}\right]$; 2×3; *Answers may vary.*

4.   3×4

5. $\left[\begin{array}{cc|c} 1 & 0 & -3 \\ 0 & 1 & 5 \end{array}\right]$; *Answers may vary.*

6.   4, 2, –3

Dimensions of Matrices and Augmented Matrices

7. 3×3

9. 3×2

11. $\begin{bmatrix} 1 & -3 & | & 1 \\ -1 & 3 & | & -1 \end{bmatrix}$

13. $\begin{bmatrix} 2 & -1 & 2 & | & -4 \\ 1 & -2 & 0 & | & 2 \\ -1 & 1 & -2 & | & -6 \end{bmatrix}$

15. $\begin{aligned} x+2y &= -6 \\ 5x\ -y &= 4 \end{aligned}$

17. $\begin{aligned} x-y+2z &= 6 \\ 2x\ +y-2z &= 1 \\ -x+2y\ -z &= 3 \end{aligned}$

19. $\begin{aligned} x &= 4 \\ y &= -2 \\ z &= 7 \end{aligned}$

Gaussian Elimination

21. $\begin{bmatrix} 1 & 1 & | & 4 \\ 1 & 3 & | & 10 \end{bmatrix} R_2 - R_1 \to \begin{bmatrix} 1 & 1 & | & 4 \\ 0 & 2 & | & 6 \end{bmatrix} (1/2)R_2 \to \begin{bmatrix} 1 & 1 & | & 4 \\ 0 & 1 & | & 3 \end{bmatrix} R_1 - R_2 \to \begin{bmatrix} 1 & 0 & | & 1 \\ 0 & 1 & | & 3 \end{bmatrix}$

The solution is $(1, 3)$.

23. $\begin{bmatrix} 2 & 3 & | & 3 \\ -2 & 2 & | & 7 \end{bmatrix} R_2 + R_1 \to \begin{bmatrix} 2 & 3 & | & 3 \\ 0 & 5 & | & 10 \end{bmatrix} (1/5)R_2 \to \begin{bmatrix} 2 & 3 & | & 3 \\ 0 & 1 & | & 2 \end{bmatrix} R_1 - 3R_2 \to \begin{bmatrix} 2 & 0 & | & -3 \\ 0 & 1 & | & 2 \end{bmatrix} (1/2)R_2 \to \begin{bmatrix} 1 & 0 & | & -\frac{3}{2} \\ 0 & 1 & | & 2 \end{bmatrix}$

The solution is $\left(-\dfrac{3}{2}, 2\right)$.

25. $\begin{bmatrix} 1 & -1 & | & 5 \\ 1 & 3 & | & -1 \end{bmatrix} R_2 - R_1 \to \begin{bmatrix} 1 & -1 & | & 5 \\ 0 & 4 & | & -6 \end{bmatrix} (1/4)R_2 \to \begin{bmatrix} 1 & -1 & | & 5 \\ 0 & 1 & | & -\frac{3}{2} \end{bmatrix} R_1 + R_2 \to \begin{bmatrix} 1 & 0 & | & \frac{7}{2} \\ 0 & 1 & | & -\frac{3}{2} \end{bmatrix}$

The solution is $\left(\dfrac{7}{2}, -\dfrac{3}{2}\right)$.

27. $\begin{bmatrix} 4 & -8 & | & -10 \\ 1 & 1 & | & 2 \end{bmatrix} \begin{matrix} Exchange \\ R_2 \leftrightarrow R_1 \end{matrix} \begin{bmatrix} 1 & 1 & | & 2 \\ 4 & -8 & | & -10 \end{bmatrix} (-1/2)R_2 \to \begin{bmatrix} 1 & 1 & | & 2 \\ -2 & 4 & | & 5 \end{bmatrix} R_2 + 2R_1 \to \begin{bmatrix} 1 & 1 & | & 2 \\ 0 & 6 & | & 9 \end{bmatrix}$

$(1/6)R_2 \to \begin{bmatrix} 1 & 1 & | & 2 \\ 0 & 1 & | & \frac{3}{2} \end{bmatrix} R_1 - R_2 \to \begin{bmatrix} 1 & 0 & | & \frac{1}{2} \\ 0 & 1 & | & \frac{3}{2} \end{bmatrix}$

The solution is $\left(\dfrac{1}{2}, \dfrac{3}{2}\right)$.

29. $\begin{bmatrix} 1 & 1 & 1 & | & 6 \\ 0 & 2 & -1 & | & 1 \\ 0 & 1 & 1 & | & 5 \end{bmatrix} \begin{matrix} R_1 - R_3 \to \\ R_2 - 2R_3 \to \end{matrix} \begin{bmatrix} 1 & 0 & 0 & | & 1 \\ 0 & 0 & -3 & | & -9 \\ 0 & 1 & 1 & | & 5 \end{bmatrix} \begin{matrix} Exchange \\ R_2 \leftrightarrow R_3 \end{matrix} \begin{bmatrix} 1 & 0 & 0 & | & 1 \\ 0 & 1 & 1 & | & 5 \\ 0 & 0 & -3 & | & -9 \end{bmatrix} (-1/3)R_3 \to \begin{bmatrix} 1 & 0 & 0 & | & 1 \\ 0 & 1 & 1 & | & 5 \\ 0 & 0 & 1 & | & 3 \end{bmatrix}$

$$R_2 - R_3 \rightarrow \begin{bmatrix} 1 & 0 & 0 & | & 1 \\ 0 & 1 & 0 & | & 2 \\ 0 & 0 & 1 & | & 3 \end{bmatrix}$$

The solution is $(1, 2, 3)$.

31. $\begin{bmatrix} 1 & 2 & 3 & | & 6 \\ -1 & 3 & 4 & | & 0 \\ 1 & 1 & -2 & | & -6 \end{bmatrix} \begin{matrix} \\ R_2 + R_1 \rightarrow \\ R_3 - R_1 \rightarrow \end{matrix} \begin{bmatrix} 1 & 2 & 3 & | & 6 \\ 0 & 5 & 7 & | & 6 \\ 0 & -1 & -5 & | & -12 \end{bmatrix} (-1)R_3 \rightarrow \begin{bmatrix} 1 & 2 & 3 & | & 6 \\ 0 & 5 & 7 & | & 6 \\ 0 & 1 & 5 & | & 12 \end{bmatrix}$

$\begin{matrix} R_1 - 2R_3 \rightarrow \\ R_2 - 5R_3 \rightarrow \\ {} \end{matrix} \begin{bmatrix} 1 & 0 & -7 & | & -18 \\ 0 & 0 & -18 & | & -54 \\ 0 & 1 & 5 & | & 12 \end{bmatrix} \begin{matrix} \\ Exchange \\ R_2 \leftrightarrow R_3 \end{matrix} \begin{bmatrix} 1 & 0 & -7 & | & -18 \\ 0 & 1 & 5 & | & 12 \\ 0 & 0 & -18 & | & -54 \end{bmatrix} (-1/18)R_3 \rightarrow \begin{bmatrix} 1 & 0 & -7 & | & -18 \\ 0 & 1 & 5 & | & 12 \\ 0 & 0 & 1 & | & 3 \end{bmatrix}$

$\begin{matrix} R_1 + 7R_3 \rightarrow \\ R_2 - 5R_3 \rightarrow \\ {} \end{matrix} \begin{bmatrix} 1 & 0 & 0 & | & 3 \\ 0 & 1 & 0 & | & -3 \\ 0 & 0 & 1 & | & 3 \end{bmatrix}$

The solution is $(3, -3, 3)$.

33. $\begin{bmatrix} 1 & 1 & 1 & | & 0 \\ 2 & 1 & 2 & | & -1 \\ 1 & 1 & 0 & | & 0 \end{bmatrix} \begin{matrix} \\ R_2 - 2R_1 \rightarrow \\ R_3 - R_1 \rightarrow \end{matrix} \begin{bmatrix} 1 & 1 & 1 & | & 0 \\ 0 & -1 & 0 & | & -1 \\ 0 & 0 & -1 & | & 0 \end{bmatrix} \begin{matrix} \\ (-1)R_2 \rightarrow \\ (-1)R_3 \rightarrow \end{matrix} \begin{bmatrix} 1 & 1 & 1 & | & 0 \\ 0 & 1 & 0 & | & 1 \\ 0 & 0 & 1 & | & 0 \end{bmatrix} R_1 - R_2 \rightarrow \begin{bmatrix} 1 & 0 & 1 & | & -1 \\ 0 & 1 & 0 & | & 1 \\ 0 & 0 & 1 & | & 0 \end{bmatrix}$

$R_1 - R_3 \rightarrow \begin{bmatrix} 1 & 0 & 0 & | & -1 \\ 0 & 1 & 0 & | & 1 \\ 0 & 0 & 1 & | & 0 \end{bmatrix}$

The solution is $(-1, 1, 0)$.

The solution is $(4, -1, -1)$.

35. $\begin{bmatrix} 1 & 1 & 1 & | & 3 \\ -1 & 0 & -1 & | & -2 \\ 1 & 1 & 2 & | & 4 \end{bmatrix} \begin{matrix} \\ R_2 + R_1 \rightarrow \\ R_3 - R_1 \rightarrow \end{matrix} \begin{bmatrix} 1 & 1 & 1 & | & 3 \\ 0 & 1 & 0 & | & 1 \\ 0 & 0 & 1 & | & 1 \end{bmatrix} R_1 - R_2 \rightarrow \begin{bmatrix} 1 & 0 & 1 & | & 2 \\ 0 & 1 & 0 & | & 1 \\ 0 & 0 & 1 & | & 1 \end{bmatrix} R_1 - R_3 \rightarrow \begin{bmatrix} 1 & 0 & 0 & | & 1 \\ 0 & 1 & 0 & | & 1 \\ 0 & 0 & 1 & | & 1 \end{bmatrix}$

The solution is $(1, 1, 1)$.

37. $\begin{bmatrix} 1 & 2 & 1 & | & 3 \\ 2 & 1 & -1 & | & -6 \\ -1 & -1 & 2 & | & 5 \end{bmatrix} \begin{matrix} \\ R_2 - 2R_1 \rightarrow \\ R_3 + R_1 \rightarrow \end{matrix} \begin{bmatrix} 1 & 2 & 1 & | & 3 \\ 0 & -3 & -3 & | & -12 \\ 0 & 1 & 3 & | & 8 \end{bmatrix} (-1/3)R_2 \rightarrow \begin{bmatrix} 1 & 2 & 1 & | & 3 \\ 0 & 1 & 1 & | & 4 \\ 0 & 1 & 3 & | & 8 \end{bmatrix}$

$\begin{matrix} {} \\ {} \\ R_3 - R_2 \rightarrow \end{matrix} \begin{bmatrix} 1 & 2 & 1 & | & 3 \\ 0 & 1 & 1 & | & 4 \\ 0 & 0 & 2 & | & 4 \end{bmatrix} \begin{matrix} R_1 - 2R_2 \rightarrow \\ {} \\ (1/2)R_3 \rightarrow \end{matrix} \begin{bmatrix} 1 & 0 & -1 & | & -5 \\ 0 & 1 & 1 & | & 4 \\ 0 & 0 & 1 & | & 2 \end{bmatrix} \begin{matrix} R_1 + R_3 \rightarrow \\ R_2 - R_3 \rightarrow \\ {} \end{matrix} \begin{bmatrix} 1 & 0 & 0 & | & -3 \\ 0 & 1 & 0 & | & 2 \\ 0 & 0 & 1 & | & 2 \end{bmatrix}$

The solution is $(-3, 2, 2)$.

39. See example 6 in the text for graphing calculator instructions.

$[A] = \begin{bmatrix} 1 & 4 & | & 13 \\ 5 & -3 & | & -50 \end{bmatrix}$; $\text{rref}([A]) = \begin{bmatrix} 1 & 0 & | & -7 \\ 0 & 1 & | & 5 \end{bmatrix}$;   The solution is $(-7, 5)$.

41. See example 6 in the text for graphing calculator instructions.

$[A] = \begin{bmatrix} 2 & -1 & 3 & | & 9 \\ -4 & 5 & 2 & | & 12 \\ 2 & 0 & 7 & | & 23 \end{bmatrix}$; $\text{rref}([A]) = \begin{bmatrix} 1 & 0 & 0 & | & 1 \\ 0 & 1 & 0 & | & 2 \\ 0 & 0 & 1 & | & 3 \end{bmatrix}$;   The solution is $(1, 2, 3)$.

43. See example 6 in the text for graphing calculator instructions.

$$[A] = \begin{bmatrix} 6 & 2 & 1 & | & 4 \\ -2 & 4 & 1 & | & -3 \\ 2 & -8 & 0 & | & -2 \end{bmatrix}; \ \text{rref}([A]) = \begin{bmatrix} 1 & 0 & 0 & | & 1 \\ 0 & 1 & 0 & | & 0.5 \\ 0 & 0 & 1 & | & -3 \end{bmatrix}; \ \text{The solution is } (1, 0.5, -3).$$

45.   See example 6 in the text for graphing calculator instructions.

$$[A] = \begin{bmatrix} 4 & 3 & 12 & | & -9.25 \\ -1 & 15 & 8 & | & -4.75 \\ 0 & 6 & 7 & | & -5.5 \end{bmatrix}; \ \text{rref}([A]) = \begin{bmatrix} 1 & 0 & 0 & | & 0.5 \\ 0 & 1 & 0 & | & 0.25 \\ 0 & 0 & 1 & | & -1 \end{bmatrix}; \ \text{The solution is } (0.5, 0.25, -1).$$

47.   See example 6 in the text for graphing calculator instructions.

$$[A] = \begin{bmatrix} 1.2 & -0.9 & 2.7 & | & 5.37 \\ 3.1 & -5.1 & 7.2 & | & 14.81 \\ 0.2 & 1.8 & -3.6 & | & -6.38 \end{bmatrix}; \ \text{rref}([A]) = \begin{bmatrix} 1 & 0 & 0 & | & 0.5 \\ 0 & 1 & 0 & | & -0.2 \\ 0 & 0 & 1 & | & 1.7 \end{bmatrix}; \ \text{The solution is } (0.5, -0.2, 1.7).$$

49. $$\begin{bmatrix} 1 & 2 & | & 4 \\ -2 & -4 & | & -8 \end{bmatrix} R_2 + 2R_1 \rightarrow \begin{bmatrix} 1 & 2 & | & 4 \\ 0 & 0 & | & 0 \end{bmatrix}$$

Row 2 represents the equation $0 = 0$. The system is dependent.

51. $$\begin{bmatrix} 1 & 1 & 1 & | & 3 \\ 1 & 1 & -1 & | & 1 \\ 1 & 1 & 0 & | & 3 \end{bmatrix} \begin{matrix} \\ R_2 - R_1 \rightarrow \\ R_3 - R_1 \rightarrow \end{matrix} \begin{bmatrix} 1 & 1 & 1 & | & 3 \\ 0 & 0 & -2 & | & -2 \\ 0 & 0 & -1 & | & 0 \end{bmatrix} R_2 - 2R_3 \rightarrow \begin{bmatrix} 1 & 1 & 1 & | & 3 \\ 0 & 0 & 0 & | & -2 \\ 0 & 0 & -1 & | & 0 \end{bmatrix}$$

Row 2 represents the equation $0 = -2$. The system is inconsistent.

53. $$\begin{bmatrix} 1 & 2 & 3 & | & 14 \\ 2 & -3 & -2 & | & -10 \\ 3 & -1 & 1 & | & 4 \end{bmatrix} \begin{matrix} \\ R_2 - 2R_1 \rightarrow \\ R_3 - 3R_1 \rightarrow \end{matrix} \begin{bmatrix} 1 & 2 & 3 & | & 14 \\ 0 & -7 & -8 & | & -38 \\ 0 & -7 & -8 & | & -38 \end{bmatrix} R_3 - R_2 \rightarrow \begin{bmatrix} 1 & 2 & 3 & | & 14 \\ 0 & -7 & -8 & | & -38 \\ 0 & 0 & 0 & | & 0 \end{bmatrix}$$

Row 3 represents the equation $0 = 0$. The system is dependent.

55. $$\begin{bmatrix} a & 0 & 0 & | & 1 \\ 0 & b & 0 & | & 1 \\ 0 & 0 & ab & | & 2 \end{bmatrix} \begin{matrix} (1/a)R_1 \rightarrow \\ (1/b)R_2 \rightarrow \\ (1/(ab))R_3 \rightarrow \end{matrix} \begin{bmatrix} 1 & 0 & 0 & | & \frac{1}{a} \\ 0 & 1 & 0 & | & \frac{1}{b} \\ 0 & 0 & 1 & | & \frac{2}{ab} \end{bmatrix}$$

The solution is $\left(\frac{1}{a}, \frac{1}{b}, \frac{2}{ab} \right)$.

Applications

57.   The equation found in example 7 is $W = -374 + 19H + 6L$. When $H = 12$ and $L = 60$,

$$W = -374 + 19(12) + 6(60) = 214 \text{ lb.}$$

59.   (a) $a + 2b + 1.4c = 3$

$a + 1.5b + 0.65c = 2$

$a + 4b + 3.4c = 6$

(b) Using the graphing calculator to solve the system, the solution is $a = 0.6$, $b = 0.5$, $c = 1$.

So the equation is $H = 0.6 + 0.5M + P$.

(c) When $M = 3$ and $P = 2$, $H \approx 0.6 + 0.5(3) + (2) = 4.1 \approx 4$ people.

61.   Let x, y and z represent the time spent running at 5, 6 and 8 mph respectively. The system needed is:

$$\begin{array}{l} x + y + z = 2 \\ 5x + 6y + 8z = 12.5 \\ x - z = 0 \end{array} \quad \text{and so } [A] = \begin{bmatrix} 1 & 1 & 1 & | & 2 \\ 5 & 6 & 8 & | & 12.5 \\ 1 & 0 & -1 & | & 0 \end{bmatrix}; \ \text{rref}([A]) = \begin{bmatrix} 1 & 0 & 0 & | & 0.5 \\ 0 & 1 & 0 & | & 1 \\ 0 & 0 & 1 & | & 0.5 \end{bmatrix}$$

The solution is (0.5, 1, 0.5). The runner ran 0.5 hr at 5 mph, 1 hr at 6 mph and 0.5 hr at 8 mph.

63. Let x, y and z represent the amount invested at 5%, 8% and 12% respectively. The system needed is:

$$\begin{array}{rcrcrcl} x & + & y & + & z & = & 3000 \\ 0.05x & + & 0.08y & + & 0.12z & = & 285 \\ 3x & & & - & z & = & 0 \end{array} \quad \text{and so } [A] = \begin{bmatrix} 1 & 1 & 1 & | & 3000 \\ 0.05 & 0.08 & 0.12 & | & 285 \\ 3 & 0 & -1 & | & 0 \end{bmatrix}; \text{rref}([A]) = \begin{bmatrix} 1 & 0 & 0 & | & 500 \\ 0 & 1 & 0 & | & 1000 \\ 0 & 0 & 1 & | & 1500 \end{bmatrix}$$

The solution is (500, 1000, 1500). There was $500 invested at 5%, $1000 at 8% and $1500 at 12%.

Checking Basic Concepts for Sections 4.5 & 4.6

1. (1, 3, –1) satisfies all three equations.

2. Multiply the first equation by –2 and add the first and second equations to eliminate the variable x.

$$\begin{array}{rcl} -2x + 2y - 2z & = & -4 \\ 2x - 3y + z & = & -1 \\ \hline -y - z & = & -5 \end{array}$$

Add the first and third equations to eliminate the variables x and y.

$$\begin{array}{rcl} x - y + z & = & 2 \\ -x + y + z & = & 4 \\ \hline 2z & = & 6 \end{array}$$

And so $z = 3$. Substitute $z = 3$ into the first *new* equation: $-y - (3) = -5 \Rightarrow -y = -2 \Rightarrow y = 2$

Substitute $y = 2$ and $z = 3$ into the *original* first equation: $x - (2) + (3) = 2 \Rightarrow x = 1$

The solution is (1, 2, 3).

3. (a)
$$\begin{bmatrix} 1 & 2 & 1 & | & 1 \\ 1 & 1 & 1 & | & -1 \\ 0 & 1 & 1 & | & 1 \end{bmatrix} R_2 - R_1 \rightarrow \begin{bmatrix} 1 & 2 & 1 & | & 1 \\ 0 & -1 & 0 & | & -2 \\ 1 & 1 & 1 & | & 1 \end{bmatrix} \begin{array}{l} R_1 + 2R_2 \rightarrow \\ -1R_2 \rightarrow \\ R_3 + R_2 \rightarrow \end{array} \begin{bmatrix} 1 & 0 & 1 & | & -3 \\ 0 & 1 & 0 & | & 2 \\ 0 & 0 & 1 & | & -1 \end{bmatrix}$$

$$R_1 - R_3 \rightarrow \begin{bmatrix} 1 & 0 & 0 & | & -2 \\ 0 & 1 & 0 & | & 2 \\ 0 & 0 & 1 & | & -1 \end{bmatrix}; \text{ The solution is } (-2, 2, -1).$$

(b) $[A] = \begin{bmatrix} 1 & 2 & 1 & | & 1 \\ 1 & 1 & 1 & | & -1 \\ 0 & 1 & 1 & | & 1 \end{bmatrix}$; rref $([A]) = \begin{bmatrix} 1 & 0 & 0 & | & -2 \\ 0 & 1 & 0 & | & 2 \\ 0 & 0 & 1 & | & -1 \end{bmatrix}$; The solution is $(-2, 2, -1)$.

4.7: Determinants

concepts

1. square

2. number

3. system of linear equations

4. 0

Calculating Determinants

5. $\det A = 1(-8) - 3(-2) = -8 + 6 = -2$

7. $\det A = -3(-1) - 8(7) = 3 - 56 = -53$

9. $\det A = 23(-13) - 6(4) = -299 - 24 = -323$

11. $\det A = 1\big[(1)(7)-(-4)(-3)\big]-0\big[(-1)(7)-(-4)(2)\big]+0\big[(-1)(-3)-(1)(2)\big] = -5-0+0 = -5$

13. $\det A = 2\big[(-2)(8)-(1)(6)\big]-1\big[(-1)(8)-(1)(0)\big]+0\big[(-1)(6)-(-2)(0)\big] = -44+8+0 = -36$

15. $\det A = (-1)\big[(-3)(7)-(-3)(5)\big]-3\big[(3)(7)-(-3)(5)\big]+2\big[(3)(5)-(-3)(5)\big] = 6-108+60 = -42$

17. $\det A = 5\big[(-2)(5)-(0)(0)\big]-0\big[(0)(5)-(0)(0)\big]+0\big[(0)(0)-(-2)(0)\big] = -50-0+0 = -50$

19. $\det A = 0\big[(3)(9)-(5)(-9)\big]-0\big[(2)(9)-(5)(-3)\big]+0\big[(2)(-9)-(3)(-3)\big] = 0-0+0 = 0$

21.   Using the calculator we find $\det([A]) = -3555$.

23.   Using the calculator we find $\det([A]) = -7466.5$.

25. $\det A = a\,(bc-0) - 0\,(0\cdot c-0) + 0\,(0-0\cdot b) = abc$

Calculating Area

27.   The triangle has vertices $(3, 2)$, $(5, 8)$ and $(9, 5)$. The matrix needed is $A = \begin{bmatrix} 3 & 5 & 9 \\ 2 & 8 & 5 \\ 1 & 1 & 1 \end{bmatrix}$.

The area is $D = \left| \dfrac{1}{2} \det([A]) \right| = 15 \text{ ft}^2$.

29.   The triangle has vertices $(-6, -4)$, $(2, 6)$ and $(6, -2)$. The matrix needed is $A = \begin{bmatrix} -6 & 2 & 6 \\ -4 & 6 & -2 \\ 1 & 1 & 1 \end{bmatrix}$.

The area is $D = \left| \dfrac{1}{2} \det([A]) \right| = 52 \text{ ft}^2$.

31.   Split the figure into two triangles with vertices $(2, 1)$, $(3, 6)$, $(9, 3)$ and vertices $(3, 6)$, $(7, 7)$, $(9, 3)$.

The matrices needed are $A = \begin{bmatrix} 2 & 3 & 9 \\ 1 & 6 & 3 \\ 1 & 1 & 1 \end{bmatrix}$ and $B = \begin{bmatrix} 3 & 7 & 9 \\ 6 & 7 & 3 \\ 1 & 1 & 1 \end{bmatrix}$.

The area is $D = \left| \dfrac{1}{2} \det([A]) \right| + \left| \dfrac{1}{2} \det([B]) \right| = 16.5 + 9 = 25.5 \text{ ft}^2$.

Cramer's Rule

33. $E = \det \begin{bmatrix} 4 & 3 \\ 20 & -4 \end{bmatrix} = -16-60 = -76; \quad F = \det \begin{bmatrix} 5 & 4 \\ 6 & 20 \end{bmatrix} = 100-24 = 76$

$D = \det \begin{bmatrix} 5 & 3 \\ 6 & -4 \end{bmatrix} = -20-18 = -38;$ The solution is $x = \dfrac{E}{D} = \dfrac{-76}{-38} = 2$ and $y = \dfrac{F}{D} = \dfrac{76}{-38} = -2.$

35. $E = \det \begin{bmatrix} -3 & -5 \\ -8 & 6 \end{bmatrix} = -18-40 = -58; \quad F = \det \begin{bmatrix} 7 & -3 \\ -4 & -8 \end{bmatrix} = -56-12 = -68$

$D = \det \begin{bmatrix} 7 & -5 \\ -4 & 6 \end{bmatrix} = 42-20 = 22;$ The solution is $x = \dfrac{E}{D} = \dfrac{-58}{22} = -\dfrac{29}{11}$ and $y = \dfrac{F}{D} = \dfrac{-68}{22} = -\dfrac{34}{11}.$

37. $E = \det\begin{bmatrix} -61 & -3 \\ -23 & -4 \end{bmatrix} = 244 - 69 = 175;$ $F = \det\begin{bmatrix} 8 & -61 \\ -1 & -23 \end{bmatrix} = -184 - 61 = -245$

   $D = \det\begin{bmatrix} 8 & -3 \\ -1 & -4 \end{bmatrix} = -32 - 3 = -35;$ The solution is $x = \dfrac{E}{D} = \dfrac{175}{-35} = -5$ and $y = \dfrac{F}{D} = \dfrac{-245}{-35} = 7.$

Checking Basic Concepts for Section 4.7

1.   (a) $\det A = -3(3) - (-2)(4) = -9 + 8 = -1$

   (b) $\det A = 1\big[(1)(-1) - (2)(1)\big] - 5\big[(-2)(-1) - (2)(3)\big] + 0\big[(-2)(1) - (1)(3)\big] = -3 + 20 + 0 = 17$

2. $E = \det\begin{bmatrix} -14 & -1 \\ -36 & -4 \end{bmatrix} = 56 - 36 = 20;$ $F = \det\begin{bmatrix} 2 & -14 \\ 3 & -36 \end{bmatrix} = -72 - (-42) = -30$

   $D = \det\begin{bmatrix} 2 & -1 \\ 3 & -4 \end{bmatrix} = -8 - (-3) = -5;$ The solution is $x = \dfrac{E}{D} = \dfrac{20}{-5} = -4$ and $y = \dfrac{F}{D} = \dfrac{-30}{-5} = 6.$

3.   The triangle has vertices $(-1, 2)$, $(5, 6)$ and $(2, -3)$. The matrix needed is $A = \begin{bmatrix} -1 & 5 & 2 \\ 2 & 6 & -3 \\ 1 & 1 & 1 \end{bmatrix}.$

   The area is $D = \left| \dfrac{1}{2}\det([A]) \right| = 21$ square units.

Chapter 4 Review Exercises

Section 4.1

1. $(3, 2)$ is a solution since it satisfies both equations.

2. $(4, -3)$ is a solution since it satisfies both equations.

3.   The unique solution is the intersection point $(-1, -3)$.

4.   Since $Y_1 = Y_2 = 8.5$ when $x = 3.5$, the solution is $(3.5, 8.5)$.

5.   Note that $x + y = 6 \Rightarrow y = -x + 6$ and $x - y = -4 \Rightarrow y = x + 4$.

   Graph $Y_1 = -X + 6$ and $Y_2 = X + 4$ in $[0, 5, 1]$ by $[0, 8, 1]$. See Figure 5. The intersection point is $(1, 5)$.

   Since there is one unique intersection point, the system is consistent. The equations are independent.

   $[0, 5, 1]$ by $[0, 8, 1]$ $[-10, 10, 1]$ by $[-10, 10, 1]$

          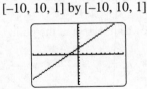

   Figure 5   Figure 6

6.   Note that $x - y = -2 \Rightarrow y = x + 2$ and $-2x + 2y = 4 \Rightarrow y = x + 2$.

   Graph $Y_1 = X + 2$ and $Y_2 = X + 2$ in $[-10, 10, 1]$ by $[-10, 10, 1]$. See Figure 6. The graphs coincide. The

   solution is $\{(x, y) \mid x - y = -2\}$. The system is consistent, and the equations are dependent.

7.   Note that $4x + 2y = 1 \Rightarrow y = -2x + \dfrac{1}{2}$ and $2x + y = 5 \Rightarrow y = -2x + 5$.

Graph $Y_1 = -2X + (1/2)$ and $Y_2 = -2X + 5$ in $[-10, 10, 1]$ by $[-10, 10, 1]$. See Figure 7.

Since the two lines are parallel, the system has no solutions. The system is inconsistent.

8. Note that $x - 3y = 5 \Rightarrow y = \dfrac{1}{3}x - \dfrac{5}{3}$ and $x + 5y = -3 \Rightarrow y = -\dfrac{1}{5}x - \dfrac{3}{5}$.

Graph $Y_1 = (1/3)X - (5/3)$ and $Y_2 = (-1/5)X - (3/5)$ in $[0, 3, 1]$ by $[-3, 0, 1]$. See Figure 8. The intersection point is $(2, -1)$.

Since there is one unique intersection point, the system is consistent. The equations are independent.

$[-10, 10, 1]$ by $[-10, 10, 1]$ $[0, 3, 1]$ by $[-3, 0, 1]$

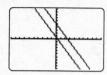
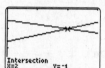
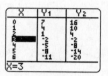
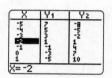

Figure 7   Figure 8   Figure 9   Figure 10

9. Note that $3x + y = 7 \Rightarrow y = -3x + 7$ and $6x + y = 16 \Rightarrow y = -6x + 16$.

Table $Y_1 = -3X + 7$ and $Y_2 = -6X + 16$ with TblStart $= 0$ and ΔTbl $= 1$. See Figure 9.

Since $Y_1 = Y_2 = -2$ when $X = 3$, the solution is $(3, -2)$.

10. Note that $4x + 2y = -6 \Rightarrow y = -2x - 3$ and $3x - y = -7 \Rightarrow y = 3x + 7$.

Table $Y_1 = -2X - 3$ and $Y_2 = 3X + 7$ with TblStart $= -5$ and ΔTbl $= 1$. See Figure 10.

Since $Y_1 = Y_2 = 1$ when $X = -2$, the solution is $(-2, 1)$.

11. (a) Let x and y represent the two numbers. Then the system needed is $x + y = 25$ and $x - y = 10$.

(b) Note that $x + y = 25 \Rightarrow y = -x + 25$ and $x - y = 10 \Rightarrow y = x - 10$.

The graphs of these equations (not shown) intersect at the point $(17.5, 7.5)$. The numbers are 17.5 and 7.5.

12. (a) Let x and y represent the two numbers. Then the system needed is $3x - 2y = 19$ and $x + y = 18$.

(b) Note that $3x - 2y = 19 \Rightarrow y = \dfrac{3}{2}x - \dfrac{19}{2}$ and $x + y = 18 \Rightarrow y = -x + 18$.

The graphs of these equations (not shown) intersect at the point $(11, 7)$. The numbers are 11 and 7.

13. Note that $\pi x - 2.1y = \sqrt{2} \Rightarrow -2.1y = -\pi x + \sqrt{2} \Rightarrow y = -\dfrac{1}{2.1}(-\pi x + \sqrt{2})$ and $\sqrt{3}x + y = \dfrac{5}{6} \Rightarrow y = -\sqrt{3}x + \dfrac{5}{6}$.

Graph $Y_1 = -1/2.1(-\pi X + \sqrt{(2)})$ and $Y_2 = -\sqrt{(3)}X + 5/6$ in $[-10, 10, 1]$ by $[-10, 10, 1]$. The solution is approximately $(0.467, 0.025)$.

14. Note that $\sqrt{5}x - \pi y = \dfrac{2}{7} \Rightarrow -\pi y = -\sqrt{5}x + \dfrac{2}{7} \Rightarrow y = -\dfrac{1}{\pi}\left(-\sqrt{5}x + \dfrac{2}{7}\right)$ and

$x + 0.3y = \pi \Rightarrow 0.3y = -x + \pi \Rightarrow y = \dfrac{1}{0.3}(-x + \pi)$. Graph $Y_1 = -1/\pi\,(-\sqrt{(5)}X + 2/7)$

and $Y_2 = 1/0.3\,(-X + \pi)$ in $[-10, 10, 1]$ by $[-10, 10, 1]$. The solution is approximately $(2.611, 1.768)$.

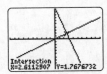

Section 4.2

15. Note that $x + 2y = -1 \Rightarrow x = -2y - 1$. Substituting $x = -2y - 1$ into the first equation yields the following:

$2(-2y - 1) + 5y = -1 \Rightarrow y = 1$ and so $x = -2(1) - 1 \Rightarrow x = -3$. The solution is $(-3, 1)$.

16. Note that $3x + y = 6 \Rightarrow y = -3x + 6$. Substituting $y = -3x + 6$ into the second equation yields the following:

$4x + 5(-3x + 6) = 8 \Rightarrow -11x = -22 \Rightarrow x = 2$ and so $y = -3(2) + 6 \Rightarrow y = 0$. The solution is $(2, 0)$.

17. Note that $4x + 2y = 0 \Rightarrow y = -2x$. Substituting $y = -2x$ into the first equation yields the following:

$2x - 3(-2x) = -8 \Rightarrow 8x = -8 \Rightarrow x = -1$ and so $y = -2(-1) = 2$. The solution is $(-1, 2)$.

18. Note that $5x + 3y = -1 \Rightarrow y = \dfrac{-5x - 1}{3}$. Substituting $y = \dfrac{-5x - 1}{3}$ in the second equation yields the following:

$3x - 5\left(\dfrac{-5x - 1}{3}\right) = -21 \Rightarrow 9x + 25x + 5 = -63 \Rightarrow 34x = -68 \Rightarrow x = -2$ and so $y = \dfrac{-5(-2) - 1}{3} = 3$.

The solution is $(-2, 3)$.

19. Adding the two equations will eliminate the variable y.

$\begin{aligned} 3x + y &= 4 \\ 2x - y &= -2 \\ \hline 5x &= 2 \end{aligned}$ Thus, $x = \dfrac{2}{5}$. And so $3\left(\dfrac{2}{5}\right) + y = 4 \Rightarrow y = 4 - \dfrac{6}{5} \Rightarrow y = \dfrac{14}{5}$. The solution is $\left(\dfrac{2}{5}, \dfrac{14}{5}\right)$.

20. Multiply the first equation by 2, the second equation by 3 and add the equations to eliminate the variable y.

$\begin{aligned} 4x + 6y &= -26 \\ 9x - 6y &= 0 \\ \hline 13x &= -26 \end{aligned}$

Thus, $x = -2$. And so $2(-2) + 3y = -13 \Rightarrow 3y = -9 \Rightarrow y = -3$. The solution is $(-2, -3)$.

21. Multiplying the first equation by 2 and adding the two equations will eliminate both variables.

$\begin{aligned} 6x - 2y &= 10 \\ -6x + 2y &= -10 \\ \hline 0 &= 0 \end{aligned}$ This is always true and the system is dependent with solutions: $\{(x, y)\,|\,3x - y = 5\}$.

22. Multiplying the second equation by 2 and adding the two equations will eliminate both variables.

$$8x - 6y = 7$$
$$\underline{-8x + 6y = 22}$$
$$0 = 29$$ This is always false and the system is inconsistent. No solution.

Section 4.3

23.   See Figure 23.

24.   See Figure 24.

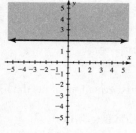

Figure 23

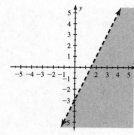

Figure 24

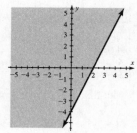

Figure 25

25.   See Figure 25.

26.   See Figure 26.

27.   See Figure 27.

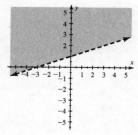

Figure 26

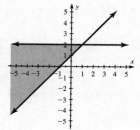

Figure 27

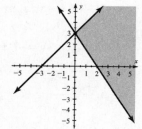

Figure 28

28.   See Figure 28.

29.   See Figure 29.

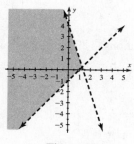

Figure 29

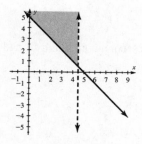

Figure 30

30.   See Figure 30.

31.   The equation of a vertical line through the point $(1, 0)$ is $x = 1$. The equation of a horizontal line through the point $(0, -1)$ is $y = -1$. The system of inequalities is $y < -1$ and $x > 1$.

32.   The equation of a line through the points $(-1, -1)$ and $(0, 1)$ is $y = 2x + 1$. The equation of a line through the points $(0, 4)$ and $(4, 0)$ is $y = -x + 4$. The system of inequalities is $y \le -x + 4$ and $y \ge 2x + 1$.

Section 4.4

33.   The maximum value of *R* occurs at one of the vertices.

For $(1, 1)$, $R = 7(1) + 8(1) = 15$. For $(1, 3)$, $R = 7(1) + 8(3) = 31$. For $(3, 1)$, $R = 7(3) + 8(1) = 29$.

The maximum value is $R = 31$.

34.   The minimum value of *R* occurs at one of the vertices.

For $(1, 1)$, $C = (1) + 2(1) = 3$. For $(1, 3)$, $C = (1) + 2(3) = 7$. For $(3, 1)$, $C = (3) + 2(1) = 5$.

The minimum value is $C = 3$.

35.   From the graph of the region of feasible solutions (not shown), the vertices are (0, 1), (0, 3), (3, 0), and (1, 0).

The maximum value of *R* occurs at one of the vertices. For $(0, 1)$, $R = 2(0) + (1) = 1$.

For $(0, 3)$, $R = 2(0) + (3) = 3$. For $(3, 0)$, $R = 2(3) + (0) = 6$. For $(1, 0)$, $R = 2(1) + (0) = 2$.

The maximum value is $R = 6$.

36.   From the graph of the region of feasible solutions (not shown), the vertices are (0, 0), (0, 4), (3, 3), and (4, 0).

Note: to find the intersection point (3, 3), solve the system of equations $3x + y = 12$ and $x + 3y = 12$.

The maximum value of *R* occurs at one of the vertices. For $(0, 0)$, $R = 6(0) + 9(0) = 0$.

For $(0, 4)$, $R = 6(0) + 9(4) = 36$. For $(3, 3)$, $R = 6(3) + 9(3) = 45$. For $(4, 0)$, $R = 6(4) + 9(0) = 24$.

The maximum value is $R = 45$.

Section 4.5

37.   Yes, since $3 + (-4) + 5 = 4$.

38.   (1, –1, 2) is a solution since it satisfies all three of the equations.

39.   Add the first two equations together to eliminate the variable *x*.

$$\begin{array}{r} x - y - 2z = -11 \\ -x + 2y + 3z = 16 \\ \hline y + z = 5 \end{array}$$

From the third equation, $3z = 6 \Rightarrow z = 2$. And so $y + (2) = 5 \Rightarrow y = 3$.

Substitute $z = 2$ and $y = 3$ into the first equation: $x - (3) - 2(2) = -11 \Rightarrow x = -4$

The solution is (–4, 3, 2).

40.   Multiply the second equation by –5 and the third equation by 3. Add these equations to eliminate the variable *z*.

$$\begin{array}{r} 10x - 5y - 15z = 10 \\ 3x - 6y + 15z = -78 \\ \hline 13x - 11y = -68 \end{array}$$

Multiply the first equation by 11 and add it to this *new* equation to eliminate the variable *y*.

$$11x + 11y = 44$$
$$\underline{13x - 11y = -68}$$
$$24x = -24$$

And so $x = -1$. Substitute $x = -1$ into the first equation: $(-1) + y = 4 \Rightarrow y = 5$

Substitute $x = -1$ and $y = 5$ into the second equation: $-2(-1) + (5) + 3z = -2 \Rightarrow 3z = -9 \Rightarrow z = -3$

The solution is $(-1, 5, -3)$.

41. Multiply the second equation by -1 and add it to the third equation to eliminate the variable z.

$$-x - 2y - z = -7$$
$$\underline{-2x + y + z = 7}$$
$$-3x - y = 0$$

Multiply the first equation by -1 and add it to this *new* equation to eliminate the variable y.

$$-2x + y = 5$$
$$\underline{-3x - y = 0}$$
$$-5x = 5$$

And so $x = -1$. Substitute $x = -1$ into the first equation: $2(-1) - y = -5 \Rightarrow y = 3$

Substitute $x = -1$ and $y = 3$ into the second equation: $-2(-1) + (3) + z = 7 \Rightarrow z = 2$

The solution is $(-1, 3, 2)$.

42. Multiply the second equation by 2 and add the first and second equations to eliminate the variable x.

$$2x + 3y + z = 6$$
$$\underline{-2x + 4y + 4z = 6}$$
$$7y + 5z = 12$$

Add the second and third equations together to eliminate the variable x.

$$-x + 2y + 2z = 3$$
$$\underline{x + y + 2z = 4}$$
$$3y + 4z = 7$$

Multiply the first *new* equation by 4 and the second *new* equation by -5. Add these to eliminate the variable z.

$$28y + 20z = 48$$
$$\underline{-15y - 20z = -35}$$
$$13y = 13$$

And so $y = 1$. Substitute $y = 1$ into the second *new* equation: $3(1) + 4z = 7 \Rightarrow 4z = 4 \Rightarrow z = 1$

Substitute $y = 1$ and $z = 1$ into the *original* third equation: $x + (1) + 2(1) = 4 \Rightarrow x = 1$

The solution is $(1, 1, 1)$.

43. Add the first two equations to eliminate the variable y.

$$x - y + 3z = 2$$
$$2x + y + 4z = 3$$
$$3x \quad + 7z = 5$$

Multiply the second equation by –2 and add to the third equation to eliminate the variable y.

$$-4x - 2y - 8z = -6$$
$$\underline{x + 2y + \ z = 5}$$
$$-3x \quad -7z = -1$$

Add this new equation to the result of the first sum to eliminate the variables x and z.

$$3x + 7z = 5$$
$$\underline{-3x - 7z = -1}$$
$$0 = 4 \qquad \text{This is a contradiction. There are no solutions.}$$

44.   Solve the third equation for x: $x + z = 2 \Rightarrow x = 2 - z$. Substitute $x = 2 - z$ into the second equation:

$2 - z + y - z = 1 \Rightarrow y - 2z = -1 \Rightarrow y = 2z - 1$. The system is dependent, and all solutions are of the form of the

ordered triple $(2 - z, 2z - 1, z)$.

Section 4.6

45. $\begin{bmatrix} 1 & 1 & 1 & | & -6 \\ 1 & 2 & 1 & | & -8 \\ 0 & 1 & 1 & | & -5 \end{bmatrix} R_2 - R_1 \rightarrow \begin{bmatrix} 1 & 1 & 1 & | & -6 \\ 0 & 1 & 0 & | & -2 \\ 0 & 1 & 1 & | & -5 \end{bmatrix} \begin{matrix} R_1 - R_3 \rightarrow \\ \\ R_3 - R_2 \rightarrow \end{matrix} \begin{bmatrix} 1 & 0 & 0 & | & -1 \\ 0 & 1 & 0 & | & -2 \\ 0 & 0 & 1 & | & -3 \end{bmatrix};$ The solution is $(-1, -2, -3)$.

46. $\begin{bmatrix} 1 & 1 & 1 & | & -3 \\ -1 & 1 & 0 & | & 5 \\ 0 & 1 & 1 & | & -1 \end{bmatrix} \begin{matrix} R_1 - R_3 \rightarrow \\ R_2 + R_1 \rightarrow \end{matrix} \begin{bmatrix} 1 & 0 & 0 & | & -2 \\ 0 & 2 & 1 & | & 2 \\ 0 & 1 & 1 & | & -1 \end{bmatrix} R_2 - R_3 \rightarrow \begin{bmatrix} 1 & 0 & 0 & | & -2 \\ 0 & 1 & 0 & | & 3 \\ 0 & 1 & 1 & | & -1 \end{bmatrix} R_3 - R_2 \rightarrow \begin{bmatrix} 1 & 0 & 0 & | & -2 \\ 0 & 1 & 0 & | & 3 \\ 0 & 0 & 1 & | & -4 \end{bmatrix}$

The solution is $(-2, 3, -4)$.

47. $\begin{bmatrix} 1 & 2 & -1 & | & 1 \\ -1 & 1 & -2 & | & 5 \\ 0 & 2 & 1 & | & 10 \end{bmatrix} \begin{matrix} R_1 - R_3 \rightarrow \\ R_2 + R_1 \rightarrow \end{matrix} \begin{bmatrix} 1 & 0 & -2 & | & -9 \\ 0 & 3 & -3 & | & 6 \\ 0 & 2 & 1 & | & 10 \end{bmatrix} (1/3)R_2 \rightarrow \begin{bmatrix} 1 & 0 & -2 & | & -9 \\ 0 & 1 & -1 & | & 2 \\ 0 & 2 & 1 & | & 10 \end{bmatrix}$

$R_3 - 2R_2 \rightarrow \begin{bmatrix} 1 & 0 & -2 & | & -9 \\ 0 & 1 & -1 & | & 2 \\ 0 & 0 & 3 & | & 6 \end{bmatrix} (1/3)R_3 \rightarrow \begin{bmatrix} 1 & 0 & -2 & | & -9 \\ 0 & 1 & -1 & | & 2 \\ 0 & 0 & 1 & | & 2 \end{bmatrix} \begin{matrix} R_1 + 2R_3 \rightarrow \\ R_2 + R_3 \rightarrow \end{matrix} \begin{bmatrix} 1 & 0 & 0 & | & -5 \\ 0 & 1 & 0 & | & 4 \\ 0 & 0 & 1 & | & 2 \end{bmatrix}$

The solution is $(-5, 4, 2)$.

48. $\begin{bmatrix} 2 & 2 & -2 & | & -14 \\ -2 & -3 & 2 & | & 12 \\ 1 & 1 & -4 & | & -22 \end{bmatrix} \begin{matrix} (1/2)R_1 \rightarrow \\ R_2 + R_1 \rightarrow \\ R_3 - (1/2)R_1 \rightarrow \end{matrix} \begin{bmatrix} 1 & 1 & -1 & | & -7 \\ 0 & -1 & 0 & | & -2 \\ 0 & 0 & -3 & | & -15 \end{bmatrix} \begin{matrix} R_1 + R_2 \rightarrow \\ -1R_2 \rightarrow \\ (-1/3)R_3 \rightarrow \end{matrix} \begin{bmatrix} 1 & 0 & -1 & | & -9 \\ 0 & 1 & 0 & | & 2 \\ 0 & 0 & 1 & | & 5 \end{bmatrix}$

$R_1 + R_3 \rightarrow \begin{bmatrix} 1 & 0 & 0 & | & -4 \\ 0 & 1 & 0 & | & 2 \\ 0 & 0 & 1 & | & 5 \end{bmatrix};$ The solution is $(-4, 2, 5)$.

49.   See example 6 in section 4.6 in the text for graphing calculator instructions.

$[A] = \begin{bmatrix} 3 & -2 & 6 & | & -17 \\ -2 & -1 & 5 & | & 20 \\ 0 & 4 & 7 & | & 30 \end{bmatrix}; \text{rref}([A]) = \begin{bmatrix} 1 & 0 & 0 & | & -7 \\ 0 & 1 & 0 & | & 4 \\ 0 & 0 & 1 & | & 2 \end{bmatrix};$ The solution is $(-7, 4, 2)$.

50.   See example 6 in section 4.6 in the text for graphing calculator instructions.

$$[A] = \begin{bmatrix} 19 & -13 & -7 & 7.4 \\ 22 & 33 & -8 & 110.5 \\ 10 & -56 & 9 & 23.7 \end{bmatrix}; \text{rref}([A]) = \begin{bmatrix} 1 & 0 & 0 & 5.4 \\ 0 & 1 & 0 & 2.1 \\ 0 & 0 & 1 & 9.7 \end{bmatrix}; \quad \text{The solution is } (5.4, 2.1, 9.7)$$

Section 4.7

51. $\det A = 6(2) - (-4)(-5) = 12 - 20 = -8$

52. $\det A = 0(9) - 5(-6) = 0 + 30 = 30$

53. $\det A = 3\big[(4)(1) - (-3)(7)\big] - 1\big[(-5)(1) - (-3)(-3)\big] + 0\big[(-5)(7) - (4)(-3)\big] = 75 - (-14) + 0 = 89$

54. $\det A = -2\big[(1)(8) - (-5)(-3)\big] - 2\big[(-1)(8) - (-5)(-7)\big] + 3\big[(-1)(-3) - (1)(-7)\big] = 14 - (-86) + 30 = 130$

55. Using the calculator we find $\det([A]) = 181,845$

56. Using the calculator we find $\det([A]) = 67.688$

57. The triangle has vertices $(-4, 6)$, $(-2, -4)$ and $(6, 2)$. The matrix needed is $A = \begin{bmatrix} -4 & -2 & 6 \\ 6 & -4 & 2 \\ 1 & 1 & 1 \end{bmatrix}$.

     The area is $D = \left| \dfrac{1}{2} \det([A]) \right| = 46 \text{ ft}^2$.

58. The triangle has vertices $(-12, -8)$, $(4, 8)$ and $(8, -4)$. The matrix needed is $A = \begin{bmatrix} -12 & 4 & 8 \\ -8 & 8 & -4 \\ 1 & 1 & 1 \end{bmatrix}$.

     The area is $D = \left| \dfrac{1}{2} \det([A]) \right| = 128 \text{ ft}^2$.

59. $E = \det \begin{bmatrix} 8 & 6 \\ 18 & -8 \end{bmatrix} = -64 - 108 = -172; \quad F = \det \begin{bmatrix} 7 & 8 \\ 5 & 18 \end{bmatrix} = 126 - 40 = 86$

     $D = \det \begin{bmatrix} 7 & 6 \\ 5 & -8 \end{bmatrix} = -56 - 30 = -86;$ The solution is $x = \dfrac{E}{D} = \dfrac{-172}{-86} = 2$ and $y = \dfrac{F}{D} = \dfrac{86}{-86} = -1.$

60. $E = \det \begin{bmatrix} 25 & 5 \\ -3 & 4 \end{bmatrix} = 100 + 15 = 115; \quad F = \det \begin{bmatrix} -2 & 25 \\ 3 & -3 \end{bmatrix} = 6 - 75 = -69$

     $D = \det \begin{bmatrix} -2 & 5 \\ 3 & 4 \end{bmatrix} = -8 - 15 = -23;$ The solution is $x = \dfrac{E}{D} = \dfrac{115}{-23} = -5$ and $y = \dfrac{F}{D} = \dfrac{-69}{-23} = 3.$

61. $E = \det \begin{bmatrix} 1.5 & -6 \\ 8 & -5 \end{bmatrix} = -7.5 + 48 = 40.5; \quad F = \det \begin{bmatrix} 3 & 1.5 \\ 7 & 8 \end{bmatrix} = 24 - 10.5 = 13.5$

     $D = \det \begin{bmatrix} 3 & -6 \\ 7 & -5 \end{bmatrix} = -15 + 42 = 27;$ The solution is $x = \dfrac{E}{D} = \dfrac{40.5}{27} = \dfrac{3}{2}$ and $y = \dfrac{F}{D} = \dfrac{13.5}{27} = \dfrac{1}{2}.$

62. $E = \det \begin{bmatrix} -47 & 4 \\ 63 & -7 \end{bmatrix} = 329 - 252 = 77; \quad F = \det \begin{bmatrix} -5 & -47 \\ 6 & 63 \end{bmatrix} = -315 + 282 = -33$

     $D = \det \begin{bmatrix} -5 & 4 \\ 6 & -7 \end{bmatrix} = 35 - 24 = 11;$ The solution is $x = \dfrac{E}{D} = \dfrac{77}{11} = 7$ and $y = \dfrac{F}{D} = \dfrac{-33}{11} = -3.$

Applications

63. Let x and y represent pedestrian fatalities for 1994 and 2004 respectively. Then the system needed is

$x + y = 10,130$ and $x - y = 848$. Adding the two equations will eliminate the variable y.

$x + y = 10,130$

$\underline{x - y = 848}$

$2x = 10,978$ Thus, $x = 5489$. And so $(5489) + y = 10,130 \Rightarrow y = 4641$.

There were 5489 pedestrian fatalities in 1994 and 4641 in 2004.

64. Let x and y represent the time spent on the stair climber and the bicycle respectively. Then the system needed

is $x + y = 30$ and $11.5x + 9y = 290$. Multiplying the first equation by -9 and adding the two equations will

eliminate the variable y.

$-9x - 9y = -270$

$\underline{11.5x + 9y = 290}$

$2.5x = 20$ Thus, $x = 8$. And so $(8) + y = 30 \Rightarrow y = 22$.

The athlete spent 8 minutes on the stair climber and 22 minutes on the stationary bicycle.

65. First note that the amounts of candy and cashews cannot be negative, so $x \geq 0$ and $y \geq 0$. Also, if the total

cost must be less than or equal to $20 then the inequality $4x + 5y \leq 20$ must be satisfied. See Figure 65.

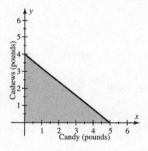

Figure 65

66. Let x and y represent the amounts of the unsubsidized and subsidized loans, respectively. Then the system of

equations is $x + y = 4000$ and $x = y - 500$. Substitute $x = y - 500$ into the first equation:

$y - 500 + y = 4000 \Rightarrow 2y = 4500 \Rightarrow y = 2250$, and so $x = 2250 - 500 \Rightarrow x = 1750$. The unsubsidized loan is

for $1750 and the subsidized loan is for $2250.

67. Let x and y represent the number of shirts and pants respectively. Since a shirt requires 20 minutes of cutting,

pants require 10 minutes of cutting and the cutting machine is only available for 360 minutes each day,

$20x + 10y \leq 360$. Since a shirt requires 10 minutes of sewing, pants require 20 minutes of sewing and the

sewing machine is only available for 480 minutes each day, $10x + 20y \leq 480$. Since the number of shirts and

pants cannot be negative $x \geq 0$ and $y \geq 0$. Here the profit function is $P = 20x + 25y$. From the graph of the

region of feasible solutions (not shown), the vertices are (0, 0), (0, 24), (8, 20) and (18, 0). To find (8, 20)

solve the system of equations $20x + 10y = 360$ and $10x + 20y = 480$.

The maximum value of P occurs at one of the vertices.

For $(0,\ 0)$, $P = 20(0) + 25(0) = 0$. For $(0,\ 24)$, $P = 20(0) + 25(24) = 600$.

For $(8,\ 20)$, $P = 20(8) + 25(20) = 660$. For $(18,\ 0)$, $P = 20(18) + 25(0) = 360$.

The maximum profit of \$660 is attained when 8 shirts and 20 pants are sold.

68. Let x and y represent the number of gallons of 30% and 55% solution respectively. Then the system needed is $x + y = 4$ and $0.30x + 0.55y = 1.6$. Multiplying the first equation by –3, the second equation by 10 and adding the two equations will eliminate the variable x.

$$-3x -\ 3y = -12$$
$$\underline{3x + 5.5y =\ 16}$$
$$2.5y = 4 \qquad \text{Thus, } y = 1.6.\text{ And so } x + (1.6) = 4 \Rightarrow x = 2.4.$$

The mechanic should add 2.4 gallons of 30% solution and 1.6 gallons of 55% solution.

69. Let x and y represent the speed of the boat and the current respectively. Using the formula $d = rt$ the system is $18 = (x + y)(1)$ and $18 = (x - y)(1.5)$. These equations may be written: $x + y = 18$ and $1.5x - 1.5y = 18$.

Multiplying the first equation by 1.5 and adding the equations together will eliminate the variable y.

$$1.5x + 1.5y = 27$$
$$\underline{1.5x - 1.5y = 18}$$
$$3x = 45 \qquad \text{Thus, } x = 15.\text{ And so } (15) + y = 18 \Rightarrow y = 3.$$

The boat travels at 15 mph and the river flows at 3 mph.

70. Let x and y represent the number of \$8 and \$12 tickets respectively. Then the system needed is $x + y = 480$ and $8x + 12y = 4620$. Note that $x + y = 480 \Rightarrow y = -x + 480$ and $8x + 12y = 4620 \Rightarrow y = -\dfrac{2}{3}x + 385$.

Multiplying the first equation by –8 and adding the two equations will eliminate the variable x.

$$-8x -8y = -3840$$
$$\underline{8x + 12y =\ 4620}$$
$$4y = 780 \qquad \text{Thus, } y = 195.\text{ And so } x + (195) = 480 \Rightarrow x = 285.\text{ The solution is } (285,\ 195).$$

There were 285 tickets sold costing \$8 each and 195 tickets sold costing \$12 each.

71. (a) $m + 3c + 5b = 14$
$$m + 2c + 4b = 11$$
$$c + 3b = 5$$

(b) Using a graphing calculator to solve the system, the solution is (3, 2, 1).

A malt costs \$3.00, cones cost \$2.00 and an ice cream bar costs \$1.00.

72. Let x, y and z represent the measure of the largest, middle and smallest angle respectively. The system needed is:

$$x + y + z = 180$$
$$x - y - z = 20$$
$$x \quad\ - z = 85$$

Using a graphing calculator to solve the system, the solution is (100, 65, 15).

The measures of the three angles are 100°, 65° and 15°.

73. Let x, y and z represent the amount of \$1.50, \$2.00 and \$2.50 candy respectively. The system needed is:

$$
\begin{aligned}
x + y + z &= 12 \\
1.50x + 2.00y + 2.50z &= 26 \\
-y + z &= 2
\end{aligned}
$$

Using a graphing calculator to solve the system, the solution is (2, 4, 6).

There should be 2 lb of \$1.50 candy, 4 lb of \$2.00 candy and 6 lb of \$2.50 candy.

74. (a) $a + 202b + 63c = 40$

$a + 365b + 70c = 50$

$a + 446b + 77c = 55$

(b) Using the graphing calculator to solve the system, the solution is $a \approx 27.134$, $b \approx 0.061$, $c \approx 0.009$.

So the equation is $C \approx 27.134 + 0.061W + 0.009L$.

(c) When $W = 300$ and $L = 68$, $C \approx 27.134 + 0.061(300) + 0.009(68) = 46.046 \approx 46$ inches.

Chapter 4 Test

1. Note that $2x + y = 7 \Rightarrow y = 7 - 2x$ and $3x - 2y = 7 \Rightarrow y = \dfrac{3x - 7}{2}$.

Graph $Y_1 = 7 - 2X$ and $Y_2 = (3X - 7)/2$ in [−5, 5, 1] by [−5, 5, 1]. See Figure 1.

The graphs intersect at (3, 1). The system is consistent. The equations are independent.

2. Note that $8x - 4y = 3 \Rightarrow y = 2x - \dfrac{3}{4}$ and $-4x + 2y = 6 \Rightarrow y = 2x + 3$.

Graph $Y_1 = 2X - (3/4)$ and $Y_2 = 2X + 3$ in [−5, 5, 1] by [−5, 5, 1]. See Figure 2.

The system is inconsistent since the lines are parallel. No solutions.

[−5, 5, 1] by [−5, 5, 1]   [−5, 5, 1] by [−5, 5, 1]

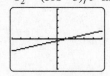   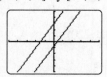

Figure 1   Figure 2

3. Note that $2x - 6y = 2 \Rightarrow y = \dfrac{-2x + 2}{-6}$ and $-3x + 9y = -3 \Rightarrow y = \dfrac{3x - 3}{9}$. Graph $Y_1 = (-2X + 2)/-6$ and

$Y_2 = (3X - 3)/9$ in [−5, 5, 1] by [−5, 5, 1].

The lines are identical, so the system is consistent and the equations are dependent. The solution is

$\{(x, y) \mid x - 3y = 1\}$.

4.   Note that $2x - 5y = 36 \Rightarrow y = \dfrac{-2x + 36}{-5}$ and $-4x + 3y = -23 \Rightarrow y = \dfrac{4x - 23}{3}$. Graph $Y_1 = (-2X + 36)/-5$ and

$Y_2 = (4X - 23)/3$ in $[-10, 10, 1]$ by $[-10, 10, 1]$.

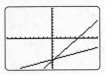

The graphs intersect at $\left(\dfrac{1}{2}, -7\right)$. The system is consistent. The equations are independent.

5.   Note that $2x + 5y = -1 \Rightarrow y = -\dfrac{2}{5}x - \dfrac{1}{5}$. Substituting $y = -\dfrac{2}{5}x - \dfrac{1}{5}$ into the second equation yields:

$3x + 2\left(-\dfrac{2}{5}x - \dfrac{1}{5}\right) = -7 \Rightarrow \dfrac{11}{5}x = -\dfrac{33}{5} \Rightarrow x = -3$ and so $y = -\dfrac{2}{5}(-3) - \dfrac{1}{5} \Rightarrow y = 1$.

The solution is $(-3, 1)$.

6.   (a) Let x and y represent the first and second number respectively. The system is $x - y = 34$ and $x - 2y = 0$.

(b) Multiplying the first equation by -1 and adding the two equations will eliminate the variable x.

$$\begin{array}{r} -x + y = -34 \\ x - 2y = 0 \\ \hline -y = -34 \end{array}$$

Thus, $y = 34$. And so $x - (34) = 34 \Rightarrow x = 68$. The solution is $(68, 34)$.

7.   Note that $-\pi x + \sqrt{3}y = 3.3 \Rightarrow y = \dfrac{\pi x + 3.3}{\sqrt{3}}$ and $\sqrt{5}x + (1 + \sqrt{2})y = 2.1 \Rightarrow y = \dfrac{-\sqrt{5}x + 2.1}{1 + \sqrt{2}}$. Graph

$Y_1 = (\pi X + 3.3)/\sqrt{3}$ and $Y_2 = (-\sqrt{5}X + 2.1)/(1 + \sqrt{2})$ in $[-5, 5, 1]$ by $[-5, 5, 1]$.

The graphs intersect at approximately $(-0.378, 1.220)$.

8.   Note that $3x + y = 17 \Rightarrow y = -3x + 17$ and $2x - 3y = 37 \Rightarrow y = \dfrac{-2x + 37}{-3}$.

X	Y₁	Y₂
5	2	-9
6	-1	-8.333
7	-4	-7.667
8	-7	-7
9	-10	-6.333
10	-13	-5.667
11	-16	-5

X=8

The solution is $(8, -7)$.

9. See Figure 9.

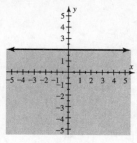

Figure 9

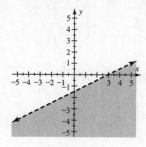

Figure 10

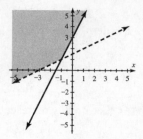

Figure 11

10. See Figure 10.

11. See Figure 11.

12. Multiply the third equation by –1 and add the second and third equations to eliminate the variables y and z.

$$-2x+y+z=5$$
$$\underline{-y-z=3}$$
$$-2x=8$$

And so $x=-4$. Substitute $x=-4$ into the first equation: $(-4)+3y=2 \Rightarrow 3y=6 \Rightarrow y=2$

Substitute $x=-4$ and $y=2$ into the second equation: $-2(-4)+(2)+z=5 \Rightarrow z=-5$

The solution is (–4, 2, –5).

13. Add the first and second equations to eliminate the variable z.

$$x+y-z=1$$
$$\underline{2x-3y+z=0}$$
$$3x-2y=1$$

Multiply the first equation by 2 and add the first and third equations to eliminate the variable z.

$$2x+2y-2z=2$$
$$\underline{x-4y+2z=2}$$
$$3x-2y=4$$

So the two new equations are $3x-2y=1$ and $3x-2y=4$. This is a contradiction, so there are no solutions.

14. The equation of a line through the points $(-2,\ 0)$ and $(0, 2)$ is $y=x+2$. The equation of a line through the points $(0, -1)$ and $(1, 2)$ is $y=3x-1$. The system of inequalities is $y \le x+2$ and $y \ge 3x-1$.

15. (a) $\begin{bmatrix} 1 & 1 & 1 & 2 \\ 1 & -1 & -1 & 3 \\ 2 & 2 & 1 & 6 \end{bmatrix}$

(b) $\begin{bmatrix} 1 & 1 & 1 & 2 \\ 1 & -1 & -1 & 3 \\ 2 & 2 & 1 & 6 \end{bmatrix} \begin{matrix} \\ R_2-R_1 \rightarrow \\ R_3-2R_1 \rightarrow \end{matrix} \begin{bmatrix} 1 & 1 & 1 & 2 \\ 0 & -2 & -2 & 1 \\ 0 & 0 & -1 & 2 \end{bmatrix} \begin{matrix} R_1+(1/2)R_2 \rightarrow \\ (-1/2)R_2 \rightarrow \\ -1R_3 \rightarrow \end{matrix} \begin{bmatrix} 1 & 0 & 0 & \frac{5}{2} \\ 0 & 1 & 1 & -\frac{1}{2} \\ 0 & 0 & 1 & -2 \end{bmatrix}$

$R_2-R_3 \rightarrow \begin{bmatrix} 1 & 0 & 0 & \frac{5}{2} \\ 0 & 1 & 0 & \frac{3}{2} \\ 0 & 0 & 1 & -2 \end{bmatrix}$ The solution is $\left(\dfrac{5}{2}, \dfrac{3}{2}, -2\right)$.

16. $\det A = 3\big[(2)(-3)-(8)(-6)\big]-6\big[(2)(-3)-(8)(-1)\big]+0\big[(2)(-6)-(2)(-1)\big]=126-12+0=114$

17. $E=\det\begin{bmatrix}7 & -3\\ 11 & 2\end{bmatrix}=14-(-33)=47;\quad F=\det\begin{bmatrix}5 & 7\\ -4 & 11\end{bmatrix}=55-(-28)=83$

$D=\det\begin{bmatrix}5 & -3\\ -4 & 2\end{bmatrix}=10-12=-2;$ The solution is $x=\dfrac{E}{D}=\dfrac{47}{-2}=-\dfrac{47}{2}$ and $y=\dfrac{F}{D}=\dfrac{83}{-2}=-\dfrac{83}{2}.$

18.   (a) Let x and y represent private and public tuition respectively. The system needed is

$$x\quad -y=12{,}636$$
$$x-4.6y=0$$

(b) Note that $x-y=12{,}636 \Rightarrow y=x-12{,}636.$ Substituting $y=x-12{,}636$ into the second equation gives:

$$x-4.6\big(x-12{,}636\big)=0 \Rightarrow -3.6x=-58{,}125.6 \Rightarrow x=16{,}146 \text{ and so } y=16{,}146-12{,}636=3510.$$

Private tuition was \$16,146 and public tuition was \$3510.

19.   Let x and y represent the time spent running and on the rowing machine respectively.

The system needed is $x+y=60$ and $12x+9y=669.$

Multiplying the first equation by -9 and adding the two equations will eliminate the variable y.

$$\begin{aligned}-9x-9y&=-540\\ 12x+9y&=669\\ \hline 3x&=129\end{aligned}$$ Thus, $x=43.$ And so $(43)+y=60 \Rightarrow y=17.$ The solution is $(43,\,17).$

The athlete spent 43 minutes running and 17 minutes on the rowing machine.

20.   Let x and y represent the average speed of the plane and the wind respectively. Using the formula $d=rt,$ the

system is $600=(x-y)(2.5)$ and $600=(x+y)(2).$

These equations may be written as follows: $x-y=240$ and $x+y=300.$

Adding the two equations together will eliminate the variable y.

$$\begin{aligned}x-y&=240\\ x+y&=300\\ \hline 2x&=540\end{aligned}$$ Thus, $x=270.$ And so $(270)-y=240 \Rightarrow -y=-30 \Rightarrow y=30.$ The solution is (270, 30).

The speed of the plane is 270 mph and the speed of the wind is 30 mph.

21.   Let x, y and z represent the measure of the largest, middle and smallest angle respectively. The system needed is:

$$\begin{aligned}x+y+z&=180\\ x\quad\ \ -z&=50\\ -x+y+z&=10\end{aligned}\quad \text{and so } [A]=\begin{bmatrix}1 & 1 & 1 & | & 180\\ 1 & 0 & -1 & | & 50\\ -1 & 1 & 1 & | & 10\end{bmatrix};\ \text{rref}([A])=\begin{bmatrix}1 & 0 & 0 & | & 85\\ 0 & 1 & 0 & | & 60\\ 0 & 0 & 1 & | & 35\end{bmatrix}$$

The solution is $(85,\,60,\,35).$ The angles are $85°$, $60°$, and $35°$.

22.   From the graph of the region of feasible solutions (not shown), the vertices are $(0,\,0)$, $(0,\,2)$, and $(3,\,0)$.

The maximum value of R occurs at one of the vertices. For $(0,\,0)$, $R=(0)+2(0)=0.$

For $(0,\,2)$, $R=(0)+2(2)=4.$ For $(3,\,0)$, $R=(3)+2(0)=3.$ The maximum value is $R=4.$

23.   The constraint is $3x + 5y < 30$. See Figure 23.

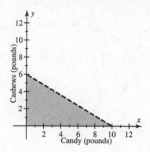

Figure 23

24. x: amount of subsidized loan

$x - 700$: amount of unsubsidized loan

$x + x - 700 = 4400 \Rightarrow 2x - 700 = 4400 \Rightarrow x = 2550, \; x - 700 = 1850$

The unsubsidized loan is for \$2550 and the subsidized loan is for \$1850.

Chapter 4 Extended and Discovery Exercises

1. $D = 1\big[(1)(3) - (1)(2)\big] - 2\big[(1)(3) - (1)(1)\big] + 0\big[(1)(2) - (1)(1)\big] = 1 - 4 + 0 = -3$

$E = 6\big[(1)(3) - (1)(2)\big] - 9\big[(1)(3) - (1)(1)\big] + 9\big[(1)(2) - (1)(1)\big] = 6 - 18 + 9 = -3$

$F = 1\big[(9)(3) - (9)(2)\big] - 2\big[(6)(3) - (9)(1)\big] + 0\big[(6)(1) - (9)(1)\big] = 9 - 18 + 0 = -9$

$G = 1\big[(1)(9) - (1)(9)\big] - 2\big[(1)(9) - (1)(6)\big] + 0\big[(1)(9) - (1)(6)\big] = 0 - 6 + 0 = -6$

$x = \dfrac{E}{D} = \dfrac{-3}{-3} = 1, \quad y = \dfrac{F}{D} = \dfrac{-9}{-3} = 3, \quad z = \dfrac{G}{D} = \dfrac{-6}{-3} = 2$. The solution is $(1, 3, 2)$.

2. $D = 0\big[(-1)(-1) - (1)(-1)\big] - 2\big[(1)(-1) - (1)(1)\big] + 1\big[(1)(-1) - (-1)(1)\big] = 0 + 4 + 0 = 4$

$E = 1\big[(-1)(-1) - (1)(-1)\big] - (-1)\big[(1)(-1) - (1)(1)\big] + 3\big[(1)(-1) - (-1)(1)\big] = 2 - 2 + 0 = 0$

$F = 0\big[(-1)(-1) - (3)(-1)\big] - 2\big[(1)(-1) - (3)(1)\big] + 1\big[(1)(-1) - (-1)(1)\big] = 0 + 8 + 0 = 8$

$G = 0[(-1)(3) - (1)(-1)] - 2[(1)(3) - (1)(1)] + 1[(1)(-1) - (-1)(1)] = 0 - 4 + 0 = -4$

$x = \dfrac{E}{D} = \dfrac{0}{4} = 0, \quad y = \dfrac{F}{D} = \dfrac{8}{4} = 2, \quad z = \dfrac{G}{D} = \dfrac{-4}{4} = -1$. The solution is $(0, 2, -1)$.

3. $D = 1\big[(1)(2) - (1)(0)\big] - 1\big[(0)(2) - (1)(1)\big] + 0\big[(0)(0) - (1)(1)\big] = 2 + 1 + 0 = 3$

$E = 2\big[(1)(2) - (1)(0)\big] - 0\big[(0)(2) - (1)(1)\big] + 1\big[(0)(0) - (1)(1)\big] = 4 + 0 - 1 = 3$

$F = 1\big[(0)(2) - (1)(0)\big] - 1\big[(2)(2) - (1)(1)\big] + 0\big[(0)(2) - (1)(0)\big] = 0 - 3 + 0 = -3$

$G = 1\big[(1)(1) - (1)(0)\big] - 1\big[(0)(1) - (1)(2)\big] + 0\big[(0)(0) - (1)(2)\big] = 1 + 2 + 0 = 3$

$x = \dfrac{E}{D} = \dfrac{3}{3} = 1, \quad y = \dfrac{F}{D} = \dfrac{-3}{3} = -1, \quad z = \dfrac{G}{D} = \dfrac{3}{3} = 1$. The solution is $(1, -1, 1)$.

4. $D = 1\left[(-2)(-3)-(1)(-3)\right]-(-1)\left[(1)(-3)-(1)(2)\right]+0\left[(1)(-3)-(-2)(2)\right] = 9-5+0 = 4$

   $E = 1\left[(-2)(-3)-(1)(-3)\right]-(-2)\left[(1)(-3)-(1)(2)\right]+5\left[(1)(-3)-(-2)(2)\right] = 9-10+5 = 4$

   $F = 1\left[(-2)(-3)-(5)(-3)\right]-(-1)\left[(1)(-3)-(5)(2)\right]+0\left[(1)(-3)-(-2)(2)\right] = 21-13+0 = 8$

   $G = 1[(-2)(5)-(1)(-2)]-(-1)[(1)(5)-(1)(1)]+0[(1)(-2)-(-2)(1)] = -8+4+0 = -4$

   $x = \dfrac{E}{D} = \dfrac{4}{4} = 1, \quad y = \dfrac{F}{D} = \dfrac{8}{4} = 2, \quad z = \dfrac{G}{D} = \dfrac{-4}{4} = -1.$ The solution is $(1, 2, -1)$.

5. $D = 1\left[(1)(2)-(-1)(1)\right]-(-1)\left[(0)(2)-(-1)(2)\right]+2\left[(0)(1)-(1)(2)\right] = 3+2-4 = 1$

   $E = 7\left[(1)(2)-(-1)(1)\right]-5\left[(0)(2)-(-1)(2)\right]+6\left[(0)(1)-(1)(2)\right] = 21-10-12 = -1$

   $F = 1\left[(5)(2)-(6)(1)\right]-(-1)\left[(7)(2)-(6)(2)\right]+2\left[(7)(1)-(5)(2)\right] = 4+2-6 = 0$

   $G = 1\left[(1)(6)-(-1)(5)\right]-(-1)\left[(0)(6)-(-1)(7)\right]+2\left[(0)(5)-(1)(7)\right] = 11+7-14 = 4$

   $x = \dfrac{E}{D} = \dfrac{-1}{1} = -1, \quad y = \dfrac{F}{D} = \dfrac{0}{1} = 0, \quad z = \dfrac{G}{D} = \dfrac{4}{1} = 4.$ The solution is $(-1, 0, 4)$.

6. $D = 1\left[(-3)(-2)-(4)(-1)\right]-2\left[(2)(-2)-(4)(3)\right]+1\left[(2)(-1)-(-3)(3)\right] = 10+32+7 = 49$

   $E = -1\left[(-3)(-2)-(4)(-1)\right]-12\left[(2)(-2)-(4)(3)\right]+(-12)\left[(2)(-1)-(-3)(3)\right] = -10+192-84 = 98$

   $F = 1\left[(12)(-2)-(-12)(-1)\right]-2\left[(-1)(-2)-(-12)(3)\right]+1\left[(-1)(-1)-(12)(3)\right] = -36-76-35 = -147$

   $G = 1\left[(-3)(-12)-(4)(12)\right]-2\left[(2)(-12)-(4)(-1)\right]+1\left[(2)(12)-(-3)(-1)\right] = -12+40+21 = 49$

   $x = \dfrac{E}{D} = \dfrac{98}{49} = 2, \quad y = \dfrac{F}{D} = \dfrac{-147}{49} = -3, \quad z = \dfrac{G}{D} = \dfrac{49}{49} = 1.$ The solution is $(2, -3, 1)$.

7. Since Denver is city 1 and Las Vegas is city 4, we look at either entry a_{14} or a_{41}. The distance is 760 miles.

8. Add entry a_{12} to entry a_{23}. The distance is 360 miles.

9. The dimension would be 20×20 and the matrix would contain 400 elements.

10. The elements on the main diagonal represent the distance from a city to itself, which is always zero.

11. $a_{14} + a_{41} = 760 + 760 = 1520$ miles

12. $a_{11} + a_{44} = 0 + 0 = 0$ miles

13. See Figure 13.

$$\begin{bmatrix} 0 & 130 & 95 & 75 \\ 130 & 0 & 186 & \star \\ 95 & 186 & 0 & 57 \\ 75 & \star & 57 & 0 \end{bmatrix} \qquad \begin{bmatrix} 0 & 97 & \star & \star & 59 \\ 97 & 0 & 113 & 118 & \star \\ \star & 113 & 0 & 94 & \star \\ \star & 118 & 94 & 0 & 177 \\ 59 & \star & \star & 177 & 0 \end{bmatrix}$$

Figure 13   Figure 14

14. See Figure 14.

15. All maps for adjacency matrix A must have the same distances between cities, but the location of each city may vary. The solution is not unique. One possible solution is shown in Figure 15.

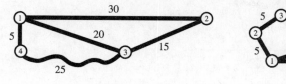

| Figure 15 | Figure 16 |

16. All maps for adjacency matrix A must have the same distances between cities, but the location of each city may vary. The solution is not unique. One possible solution is shown in Figure 16.

17. (a)
$$\begin{bmatrix} 1 & 19 & 57.5 & 32 & \vline & 125 \\ 1 & 26 & 65 & 42 & \vline & 316 \\ 1 & 30 & 72 & 48 & \vline & 436 \\ 1 & 30.5 & 75 & 54 & \vline & 514 \end{bmatrix}$$

(b) $a \approx -552.272,\ b \approx 8.733,\ c \approx 2.859,\ d \approx 10.843$

(c) $N = 24,\ L = 63$ and $C = 39,\ W \approx -552.272 + 8.733(24) + 2.859(63) + 10.843(39) \approx 260$

A bear with a 24-inch neck, 63-inch length and 39-inch chest weighs approximately 260 pounds.

Chapters 1–4 Cumulative Review Exercises

1. Identity

2. Associative

3. $\dfrac{1+2}{3-1} - 2^2 \cdot 4 = \dfrac{3}{2} - 4 \cdot 4 = 1.5 - 16 = -14.5$

4. $-1 + 8 - 19 + 12 = 0$

5. $-\dfrac{a}{b}$

6. (a) $\left(\dfrac{4^2}{3}\right)^{-2} = \left(\dfrac{3}{4^2}\right)^2 = \dfrac{3^2}{4^4} = \dfrac{9}{256}$

(b) $\dfrac{(2ab^{-4})^2}{a^7(b^2)^{-3}} = \dfrac{4a^2 b^{-8}}{a^7 b^{-6}} = 4a^{2-7}b^{-8-(-6)} = 4a^{-5}b^{-2} = \dfrac{4}{a^5 b^2}$

(c) $(x^2 y^3)^{-2}(xy^2)^4 = x^{-4} y^{-6} x^4 y^8 = x^0 y^2 = y^2$

7. $0.0056 = 5.6 \times 10^{-3}$

8. $f(-2) = 5 - 4(-2) = 5 + 8 = 13$

9. $V = \dfrac{1}{3}\pi(3)(2)^2 = \dfrac{1}{3}\pi(3)(4) = 4\pi$

10. The denominator cannot equal zero, so $x \neq -\dfrac{1}{2}$.

11. See Figure 11.

12.   See Figure 12.

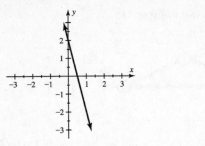
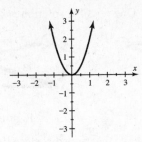

Figure 11   Figure 12

13.   (a) Yes. The graph passes the vertical line test.

(b) D: all real numbers; R: all real numbers

(c) $f(-1) = -1.5$; $f(2) = 0$

(d) x-intercept: 2; y-intercept: -1

(e) Two points on the line are $(2, 0)$ and $(0, -1)$.

$$m = \frac{y_2 - y_1}{x_2 - x_1} = \frac{-1-0}{0-2} = \frac{-1}{-2} = \frac{1}{2}$$

(f) $f(x) = \dfrac{1}{2}x - 1$

14.   The slope of a line perpendicular to $y = -\dfrac{3}{4}x$ is $m = \dfrac{4}{3}$.

$$y - (3) = \frac{4}{3}(x - (-6)) \Rightarrow y = \frac{4}{3}(x + 6) + 3 \Rightarrow y = \frac{4}{3}x + 11$$

15.   Two points on the graph are $(0, 2)$ and $(2, 10)$. $m = \dfrac{y_2 - y_1}{x_2 - x_1} = \dfrac{10-2}{2-0} = \dfrac{8}{2} = 4$; From the table, the y-intercept

is 2, so an equation of the line is $y = 4x + 2$. To find the x-intercept, let $y = 0$ and solve for

$$x: 0 = 4x + 2 \Rightarrow -2 = 4x \Rightarrow x = -\frac{1}{2}.$$

16. $\dfrac{1}{2}(1 - x) + 2x = 3x + 1 \Rightarrow \dfrac{1}{2} - \dfrac{1}{2}x + 2x = 3x + 1$

$$\Rightarrow -\frac{1}{2}x + 2x - 3x = -\frac{1}{2} + 1 \Rightarrow -\frac{3}{2}x = \frac{1}{2} \Rightarrow x = -\frac{2}{3} \cdot \frac{1}{2} \Rightarrow x = -\frac{1}{3}$$

17. $2(1 - x) > 4x - (2 + x) \Rightarrow 2 - 2x > 4x - 2 - x \Rightarrow -2x - 4x + x > -2 - 2 \Rightarrow -5x > -4 \Rightarrow x < \dfrac{4}{5}$; the solutions are

$$\left(-\infty, \frac{4}{5}\right).$$

18. $x + 1 < 2 \Rightarrow x < 1$ or $x + 1 > 4 \Rightarrow x > 3 \Rightarrow (-\infty, 1) \cup (3, \infty)$

19. $-3 \le 2 - 4x < 5 \Rightarrow -5 \le -4x < 3 \Rightarrow \dfrac{5}{4} \ge x > -\dfrac{3}{4} \Rightarrow -\dfrac{3}{4} < x \le \dfrac{5}{4} \Rightarrow \left(-\dfrac{3}{4}, \dfrac{5}{4}\right]$

20. $|3x-2| = 4 \Rightarrow 3x-2 = 4$ or $3x-2 = -4 \Rightarrow 3x = 6$ or $3x = -2 \Rightarrow x = 2$ or $x = -\dfrac{2}{3}$

    (b) $|3x-2| < 4 \Rightarrow -4 < 3x-2 < 4 \Rightarrow -2 < 3x < 6 \Rightarrow -\dfrac{2}{3} < x < 2 \Rightarrow \left(-\dfrac{2}{3}, 2\right)$

    (c) $|3x-2| > 4 \Rightarrow 3x-2 < -4$ or $3x-2 > 4 \Rightarrow 3x < -2$ or $3x > 6 \Rightarrow x < -\dfrac{2}{3}$ or $x > 2 \Rightarrow \left(-\infty, -\dfrac{2}{3}\right) \cup (2, \infty)$

21.   (a) Note that $2x+y = 3 \Rightarrow y = -2x+3$ and $x-2y = -1 \Rightarrow -2y = -x-1 \Rightarrow y = \dfrac{1}{2}x + \dfrac{1}{2}$. Graph

       $Y_1 = -2X+3$ and $Y_2 = 1/2X + 1/2$ in [−5, 5, 1] by [−5, 5, 1]. The solution is (1, 1).

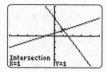

    (b) Note that $4x+3y = 8 \Rightarrow 3y = -4x+8 \Rightarrow y = -\dfrac{4}{3}x + \dfrac{8}{3}$ and $2x+7y = 4 \Rightarrow 7y = -2x+4 \Rightarrow y = -\dfrac{2}{7}x + \dfrac{4}{7}$.

       Table $Y_1 = -4/3X + 8/3$ and $Y_2 = -2/7X + 4/7$ with TblStart = 0 and ΔTbl = 1. The solution is (2, 0).

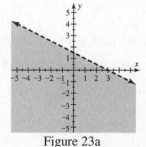

22.   (a) Add the two equations to eliminate the variable y.

$$\begin{array}{r} x+2y = 4 \\ \underline{3x-2y = -2} \\ 4x\qquad = 2 \end{array}$$

       $\Rightarrow x = \dfrac{1}{2}$. Substitute $x = \dfrac{1}{2}$ in the first equation and solve for

       $y: \dfrac{1}{2} + 2y = 4 \Rightarrow 2y = \dfrac{7}{2} \Rightarrow y = \dfrac{7}{4}$. The solution is $\left(\dfrac{1}{2}, \dfrac{7}{4}\right)$.

    (b) Multiply the first equation by 2 and add the equations to eliminate the variables x and y.

$$\begin{array}{r} -2x+2y = 4 \\ \underline{2x-2y = 5} \\ 0 = 9 \end{array}$$

       This is a contradiction. There are no solutions.

23.   (a) Note that $x+2y < 3 \Rightarrow y < -\dfrac{1}{2}x + \dfrac{3}{2}$ See Figure 23a.

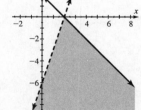

             Figure 23a   Figure 23b

(b) Note that $x + y \le 2 \Rightarrow y \le -x + 2$ and $3x - y > 6 \Rightarrow y < 3x - 6$ See Figure 23b.

24.   Add the second and third equations to eliminate the variable y.

$$\begin{array}{l} 3x + y + z = 12 \\ \underline{x - y - 2z = -1} \\ 4x - z = 11 \end{array}$$

Multiply the second equation by -2 and add to the first equation to eliminate the variable y.

$$\begin{array}{l} -x + 2y + z = 2 \\ \underline{-6x - 2y - 2z = -24} \\ -7x - z = -22 \end{array}$$

Multiply the result of the first sum by -1 and add to this new equation to eliminate the variable z.

$$\begin{array}{l} -4x + z = -11 \\ \underline{-7x - z = -22} \\ -11x = -33 \end{array}$$

$\Rightarrow x = 3$. Then $4(3) - z = 11 \Rightarrow 12 - z = 11 \Rightarrow z = 1$. Substitute the values $x = 3$ and $z = 1$ into the second equation and solve for y.

$$3(3) + y + (1) = 12 \Rightarrow 9 + y + 1 = 12 \Rightarrow y = 2.$$

The solution is $(3, 2, 1)$.

25.

$$\begin{bmatrix} 1 & 1 & 1 & | & -3 \\ 2 & -1 & 1 & | & -11 \\ -1 & -1 & 1 & | & -5 \end{bmatrix} \begin{matrix} -2R_1 + R_2 \rightarrow \\ R_1 + R_3 \rightarrow \end{matrix} \begin{bmatrix} 1 & 1 & 1 & | & -3 \\ 0 & -3 & -1 & | & -5 \\ 0 & 0 & 2 & | & -8 \end{bmatrix} \frac{1}{2}R_3 \rightarrow \begin{bmatrix} 1 & 1 & 1 & | & -3 \\ 0 & -3 & -1 & | & -5 \\ 0 & 0 & 1 & | & -4 \end{bmatrix} \begin{matrix} -R_3 + R_1 \rightarrow \\ R_3 + R_2 \rightarrow \end{matrix}$$

$$\begin{bmatrix} 1 & 1 & 0 & | & 1 \\ 0 & -3 & 0 & | & -9 \\ 0 & 0 & 1 & | & -4 \end{bmatrix} -\frac{1}{3}R_2 \rightarrow \begin{bmatrix} 1 & 1 & 0 & | & 1 \\ 0 & 1 & 0 & | & 3 \\ 0 & 0 & 1 & | & -4 \end{bmatrix} -R_2 + R_1 \rightarrow \begin{bmatrix} 1 & 0 & 0 & | & -2 \\ 0 & 1 & 0 & | & 3 \\ 0 & 0 & 1 & | & -4 \end{bmatrix};$$

The solution is $(-2, 3, -4)$.

26. $1(0 \cdot 3 - 4 \cdot 4) - 2(-1 \cdot 3 - 2 \cdot 4) + 0(-1 \cdot 4 - 2 \cdot 0)$

$= 1(0 - 16) - 2(-3 - 8) + 0(-4 - 0)$

$= -16 - 2(-11) + 0 = -16 + 22 = 6$

27.   (a) $I(x) = 0.07x$

(b) $I(500) = 0.07(500) = \$35$; the interest after one year for $500 at 7% is $35.

28.   (a) $T(1990) = 572.3(1990 - 1980) + 3617 = 572.3(10) + 3617 = 5723 + 3617 = 9340$; average tuition and fees in 1990 were \$9340.

(b) slope = 572.3; tuition and fees increased, on average, by \$572.30 per year from 1980 to 2000.

29.   Let x represent the width. Then $2(x) + 2(x + 7) = 74 \Rightarrow 2x + 2x + 14 = 74 \Rightarrow 4x = 60 \Rightarrow x = 15$; the dimensions are 15 inches by 22 inches.

30. $|P - 12.4| \le 0.2$

31. Let x, y, and z represent the amounts invested at 5%, 6%, and 8%, respectively. Then the system of equations is

$$\begin{array}{ll} x+y+z=5000 & \quad x+y+z=5000 \\ z=y+500 & \Rightarrow \quad -y+z=500 \\ y=x+z-1000 & \quad -x+y-z=-1000 \end{array}$$

Add the first and third equations to eliminate the variables x and z.

$$\begin{array}{l} x+y+z=5000 \\ \underline{-x+y-z=-1000} \\ \quad\quad 2y\quad\ =4000 \end{array}$$

$\Rightarrow y=2000$. Then $z=2000+500=2500$.

Substitute the values $y=2000$ and $z=2500$ into the first equation and solve for

x: $x+2000+2500=5000 \Rightarrow x=500$. So $500 is invested at 5%, $2000 is invested at 6% and $2500 is

invested at 8%.

32. Let x and y represent the populations of Texas and Florida in 2002, respectively. Then the system needed is

$$\begin{array}{l} x+y=38.5 \\ \quad x=y+5.1 \end{array}$$

Substitute $x=y+5.1$ into the first equation and solve for y: $y+5.1+y=38.5 \Rightarrow 2y=33.4 \Rightarrow y=16.7$ and

so $x=16.7+5.1=21.8$. In 2002, the population of Texas was 21.8 million and the population of Florida was

16.7 million.

Critical Thinking Solutions for Chapter 4

Section 4.1

- The tables for y_1 and y_2 would be identical. If the system of two equations is dependent, there are an infinite

  number of solutions.

- The two y-values for a given x-value would never be equal.

Section 4.3

- Yes. The point (2, 2) satisfies *both* inequalities. No. The point (2, 2) does *not* satisfy $2x-y<2$.

- There are no solutions to this system. See Figure 1.

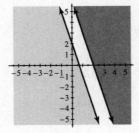

Figure 1

Section 4.5

- (a) There are no solutions. The planes are parallel.

  (b) There are infinitely many solutions; the planes intersect in a line.

- As the number of adults increases, so does the number of fawns. However, as the severity of the winter increases, the number of fawns decreases.

Section 4.6

- The first matrix represents an inconsistent system and the second matrix represents a system of dependent equations.

Section 4.7

- The points are collinear; they all lie on the same line.

Chapter 5: Polynomial Expressions and Functions

5.1: Polynomial Functions

Concepts

1. $3x^2$; *Answers may vary.*

2. The degree is 3. The leading coefficient is -1.

3. No, the powers must match for each variable.

4. $x^4 - 3x + 5$; *Answers may vary.*

5. No, the opposite is $-x^2 - 1$.

6. $4(2)^3(3) = 96$

7. No; a function has only one output for each input.

8. Yes; if the polynomial is linear.

9. $f(2) = (2)^2 + 2 = 4 + 2 = 6$

10. $(f + g)(x) = f(x) + g(x) = x^2 + 2x^2 = 3x^2$

Monomials and Polynomials

11. Yes

13. No, there are two terms.

15. Yes

17. No, the variable is in a denominator.

19. No, the exponent is negative.

21. x^2

23. πr^2

25. xy

27. The degree is 7. The coefficient is 3.

29. The degree is 7. The coefficient is -3.

31. The degree is 6. The coefficient is -1.

33. The degree is 2. The leading coefficient is 5.

35. The degree is 3. The leading coefficient is $-\dfrac{2}{5}$.

37. The degree is 5. The leading coefficient is 1.

39. $x^2 + 4x^2 = 5x^2$

41. $6y^4 - 3y^4 = 3y^4$

43. Not possible

45. $9x^2 - x + 4x - 6x^2 = 3x^2 + 3x$

47. $x^2 + 9xy - y^2 + 4x^2 + y^2 = 5x^2 + 9xy$

49. $4x + 7x^3 y^7 - \dfrac{1}{2}x^3 y^7 + 9x - \dfrac{3}{2}x^3 y^7 = 5x^3 y^7 + 13x$

51. $(3x + 1) + (-x + 1) = 2x + 2$

53. $\left(x^2 - 2x + 15\right) + \left(-3x^2 + 5x - 7\right) = -2x^2 + 3x + 8$

55. $\left(3x^3 - 4x + 3\right) + \left(5x^2 + 4x + 12\right) = 3x^3 + 5x^2 + 15$

57. $\left(x^4 - 3x^2 - 4\right) + \left(-8x^4 + x - \dfrac{1}{2}\right) = -7x^4 - 3x^2 + x - \dfrac{9}{2}$

59. $\left(4r^4 - r + 2\right) + \left(r^3 - 5r\right) = 4r^4 + r^3 - 6r + 2$

61. $\left(4xy - x^2 + y^2\right) + \left(4y^2 - 8xy - x^2\right) = 5y^2 - 4xy - 2x^2$

63. $-6x^5$

65. $-19x^5 + 5x^3 - 3x$

67. $7z^4 - z^2 + 8$

69. $(5x - 3) - (2x + 4) = 5x - 3 - 2x - 4 = 3x - 7$

71. $\left(x^2 - 3x + 1\right) - \left(-5x^2 + 2x - 4\right) = x^2 - 3x + 1 + 5x^2 - 2x + 4 = 6x^2 - 5x + 5$

73. $3\left(4x^4 + 2x^2 - 9\right) - 4\left(x^4 - 2x^2 - 5\right) = 12x^4 + 6x^2 - 27 - 4x^4 + 8x^2 + 20 = 8x^4 + 14x^2 - 7$

75. $4\left(x^4 - 1\right) - \left(4x^4 + 3x + 7\right) = 4x^4 - 4 - 4x^4 - 3x - 7 = -3x - 11$

77. Yes. This is a fourth degree polynomial.

79. No, the exponent on the variable is not positive.

81. No, the variable is located in the denominator.

83. Yes. The degree of the polynomial is 2. It is a quadratic polynomial.

Evaluating Polynomials

85. $-4(2)^2 = -4(4) = -16$

87. $2(2)^2 (3) = 2(4)(3) = 24$

89. $(-3)^2 (4) - (-3)(4)^2 = 9(4) + 3(16) = 36 + 48 = 84$

91. When $x = 1$, $y = -1$ and so $f(1) = -1$. When $x = -2$, $y = 2$ and so $f(-2) = 2$.

93. When $x = -1$, $y = -4$ and so $f(-1) = -4$. When $x = 2$, $y = 2$ and so $f(2) = 2$.

95. $f(-2) = 3(-2)^2 = 3(4) = 12$

97. $f\left(-\dfrac{1}{2}\right)=5-4\left(-\dfrac{1}{2}\right)=5+2=7$

99. $f(-1)=0.5(-1)^4-0.3(-1)^3+5=0.5+0.3+5=5.8$

101. $f(-1)=-(-1)^3=-(-1)=1$

103. $f(-3)=-(-3)^2-3(-3)=-9+9=0$

105. $f(2.4)=1-2(2.4)+(2.4)^2=1-4.8+5.76=1.96$

107. $f(-1)=(-1)^5-5=-1-5=-6$

109. $f(a)=a^2-2a$

111. (a) $(f+g)(2)=(3(2)-1)+(5-(2))=6-1+5-2=8$

    (b) $(f-g)(-1)=(3(-1)-1)-(5-(-1))=-3-1-5-1=-10$

    (c) $(f+g)(x)=(3x-1)+(5-x)=2x+4$

    (d) $(f-g)(x)=(3x-1)-(5-x)=3x-1-5+x=4x-6$

113. (a) $(f+g)(2)=\left(-3(2)^2\right)+\left((2)^2+1\right)=-3(4)+4+1=-7$

    (b) $(f-g)(-1)=\left(-3(-1)^2\right)-\left((-1)^2+1\right)=-3-1-1=-5$

    (c) $(f+g)(x)=\left(-3x^2\right)+\left(x^2+1\right)=-2x^2+1$

    (d) $(f-g)(x)=\left(-3x^2\right)-\left(x^2+1\right)=-4x^2-1$

115. (a) $(f+g)(2)=\left(-(2)^3\right)+\left(3(2)^3\right)=-8+3(8)=16$

    (b) $(f-g)(-1)=\left(-(-1)^3\right)-\left(3(-1)^3\right)=1-(-3)=4$

    (c) $(f+g)(x)=\left(-x^3\right)+\left(3x^3\right)=2x^3$

    (d) $(f-g)(x)=\left(-x^3\right)-\left(3x^3\right)=-4x^3$

117. (a) $(f+g)(2)=\left(2^2-2(2)+1\right)+\left(4(2)^2+3(2)\right)=4-4+1+16+6=23$

    (b) $(f-g)(-1)=\left((-1)^2-2(-1)+1\right)-\left(4(-1)^2+3(-1)\right)=1+2+1-4+3=3$

    (c) $(f+g)(x)=\left(x^2-2x+1\right)+\left(4x^2+3x\right)=5x^2+x+1$

    (d) $(f-g)(x)=\left(x^2-2x+1\right)-\left(4x^2+3x\right)=-3x^2-5x+1$

Applications

119. (a) When $x = 1980$ the y-value is about 1250. When $x = 1995$ the y-value is about 7250.

    In 1980 there were about 1250 women running. In 1995 there were about 7250.

    (b) In 1980 $x = 2$, and so $8.87(2)^2 + 232(2) + 769 = 1268.48 \approx 1268.$

    In 1995 $x = 17$, and so $8.87(17)^2 + 232(17) + 769 = 7276.43 \approx 7276.$ The answers are similar.

    (c) In 1978 $x = 0$, in 1998 $x = 20$, and so

    $$\left(8.87(20)^2 + 232(20) + 769\right) - \left(8.87(0)^2 + 232(0) + 769\right) = 8188$$

121. Three rooms are 8 by x by y and two rooms are 8 by x by z. The polynomial is

    $3(8xy) + 2(8xz) = 24xy + 16xz.$

    When $x = 10$, $y = 12$ and $z = 7$, we have $24(10 \cdot 12) + 16(10 \cdot 7) = 4000 \text{ ft}^3$

123. The volume of one cube is s^3, so the volume of five identical cubes is $5s^3$.

125. A square with side x has area x^2 and a circle with radius x has area πx^2. The polynomial is $x^2 + \pi x^2$.

    When $x = 10$ we have $10^2 + \pi \cdot 10^2 = 100 + 100\pi \approx 414.2 \text{ in}^2$.

127. The table is shown in Figure 127. The athlete's heart rate is between 80 and 110 beats per minute from 4 to 8 minutes after exercise had stopped.

t	3	4	5	6	7	8	9
$y = f(t)$	126.875	110	96.875	87.5	81.875	80	81.875

Figure 127

129. The best model is $f(x)$. This can be verified numerically using a table.

131. (a) $R(x) = 16x$

    (b) $P(x) = R(x) - C(x) = 16x - (4x + 2000) = 12x - 2000$

    (c) $P(3000) = 12(3000) - 2000 = 36,000 - 2000 = 34,000$; the profit is $34,000 for selling 3000 CDs.

5.2: Multiplication of Polynomials

Concepts

1. Distributive

2. $3x$; $2x - 7$; $x^2 - 3x + 1$; *Answers may vary.*

3. $x^{3+5} = x^8$

4. $(2x)^3 = 2^3 \cdot x^3 = 8x^3$ and $\left(x^2\right)^3 = x^{2 \cdot 3} = x^6$

5. $a^2 - b^2$

6. $a^2 + 2ab + b^2$

7. $x^2 - x$

8.   No, it equals $x^2 + 2x + 1$.

Multiplying Polynomials

9. $x^4 \cdot x^8 = x^{4+8} = x^{12}$

11. $-5y^7 \cdot 4y = -20y^{7+1} = -20y^8$

13. $(-xy)\left(4x^3 y^5\right) = -4x^{1+3} \cdot y^{1+5} = -4x^4 y^6$

15. $\left(5y^2 z\right)\left(4x^2 yz^5\right) = 20x^2 \cdot y^{2+1} \cdot z^{1+5} = 20x^2 y^3 z^6$

17. $5(y+2) = 5y + 10$

19. $-2(5x+9) = -10x - 18$

21. $-6y(y-3) = -6y^2 + 18y$

23. $(9-4x)3x = 27x - 12x^2$

25. $-ab\left(a^2 - b^2\right) = -a^3 b + ab^3$

27. $-5m\left(n^3 + m\right) = -5mn^3 - 5m^2$

29. $x^2 + x + 2x + 2 = x^2 + 3x + 2;$ When $x = 5, (5)^2 + 3(5) + 2 = 25 + 15 + 2 = 42$ in^2.

31. $2x^2 + 2x + x + 1 = 2x^2 + 3x + 1;$ When $x = 5, 2(5)^2 + 3(5) + 1 = 50 + 15 + 1 = 66$ in^2.

33. $(x+5)(x+6) = x^2 + 6x + 5x + 30 = x^2 + 11x + 30$

35. $(2x+1)(2x+1) = 4x^2 + 2x + 2x + 1 = 4x^2 + 4x + 1$

37.   No, a square of a sum does not equal the sum of the squares.

39.   No, $(x-2)^2 = x^2 - 4x + 4 \neq x^2 - 4$.

41.   Yes, $(x+3)(x-2) = x^2 + x - 6$.

43.   No, $2z(3z+1) = 6z^2 + 2z \neq 6z^2 + 1$.

45. $(x+5)(x+10) = x^2 + 10x + 5x + 50 = x^2 + 15x + 50$

47. $(x-3)(x-4) = x^2 - 4x - 3x + 12 = x^2 - 7x + 12$

49. $(2z-1)(z+2) = 2z^2 + 4z - z - 2 = 2z^2 + 3z - 2$

51. $(y+3)(y-4) = y^2 - 4y + 3y - 12 = y^2 - y - 12$

53. $(4x-3)(4-9x) = 16x - 36x^2 - 12 + 27x = -36x^2 + 43x - 12$

55. $(-2z+3)(z-2) = -2z^2 + 4z + 3z - 6 = -2z^2 + 7z - 6$

57. $\left(z - \dfrac{1}{2}\right)\left(z + \dfrac{1}{4}\right) = z^2 + \dfrac{1}{4}z - \dfrac{1}{2}z - \dfrac{1}{8} = z^2 - \dfrac{1}{4}z - \dfrac{1}{8}$

59. $\left(x^2+1\right)\left(2x^2-1\right)=2x^4-x^2+2x^2-1=2x^4+x^2-1$

61. $(x+y)(x-2y)=x^2-2xy+xy-2y^2=x^2-xy-2y^2$

63. $4x\left(x^2-2x-3\right)=4x^3-8x^2-12x$

65. $-x\left(x^4-3x^2+1\right)=-x^5+3x^3-x$

67. $\left(2n^2-4n+1\right)\left(3n^2\right)=6n^4-12n^3+3n^2$

69. $(x+1)\left(x^2+2x-3\right)=x^3+2x^2-3x+x^2+2x-3=x^3+3x^2-x-3$

71. $z(2+z)\left(1-z-z^2\right)=\left(2z+z^2\right)\left(1-z-z^2\right)=2z-2z^2-2z^3+z^2-z^3-z^4=-z^4-3z^3-z^2+2z$

73. $2ab^2\left(2a^2-ab+3b^2\right)=4a^3b^2-2a^2b^3+6ab^4$

75. $(2r-4t)\left(3r^2+rt-t^2\right)=6r^3+2r^2t-2rt^2-12r^2t-4rt^2+4t^3=6r^3-10r^2t-6rt^2+4t^3$

77. $\left(2^3\right)^2=2^{3\cdot2}=2^6=64$

79. $2\left(z^3\right)^6=2z^{3\cdot6}=2z^{18}$

81. $(-5x)^2=(-5)^2\cdot x^2=25x^2$

83. $\left(-2xy^2\right)^3=(-2)^3\cdot x^3\cdot y^{2\cdot3}=-8x^3y^6$

85. $\left(-4a^2b^3\right)^2=(-4)^2\cdot a^{2\cdot2}\cdot b^{3\cdot2}=16a^4b^6$

87. $(x-3)(x+3)=x^2-9$

89. $(3-2x)(3+2x)=3^2-(2x)^2=9-4x^2$

91. $(x-y)(x+y)=x^2-y^2$

93. $(x+2)^2=x^2+2(2x)+4=x^2+4x+4$

95. $(2x+1)^2=4x^2+2(2x)+1=4x^2+4x+1$

97. $(x-1)^2=x^2+2(-x)+1=x^2-2x+1$

99. $(3x-2)^2=9x^2+2(-6x)+4=9x^2-12x+4$

101. $3x(x+1)(x-1)=3x\left(x^2-1\right)=3x^3-3x$

103. $3rt(t-2r)(t+2r)=3rt\left(t^2-4r^2\right)=3rt^3-12r^3t$

105. $\left(a^2+2b^2\right)\left(a^2-2b^2\right)=\left(a^2\right)^2-\left(2b^2\right)^2=a^4-4b^4$

107. $\left(3m^3+5n^2\right)^2 = \left(3m^3\right)^2 + 2\left(3m^3\right)\left(5n^2\right) + \left(5n^2\right)^2 = 9m^6 + 30m^3n^2 + 25n^4$

109. $\left(x^3-2y^3\right)^2 = \left(x^3\right)^2 - 2\left(x^3\right)\left(2y^3\right) + \left(2y^3\right)^2 = x^6 - 4x^3y^3 + 4y^6$

111. $\left((x-3)+y\right)\left((x-3)-y\right) = (x-3)^2 - y^2 = x^2 - 6x + 9 - y^2 = x^2 - 6x - y^2 + 9$

113. $\left(r-(t+2)\right)\left(r+(t+2)\right) = r^2 - (t+2)^2 = r^2 - \left(t^2+4t+4\right) = r^2 - t^2 - 4t - 4$

115. $(102)(98) = (100+2)(100-2) = 100^2 - 2^2 = 10{,}000 - 4 = 9996$

117. $(fg)(2) = (2+1)(2-2) = (3)(0) = 0;$

   $(fg)(x) = (x+1)(x-2) = x^2 - 2x + x - 2 = x^2 - x - 2$

119. $(fg)(2) = \left((2)^2\right)\left(3-5(2)\right) = (4)(-7) = -28;$

   $(fg)(x) = \left(x^2\right)(3-5x) = 3x^2 - 5x^3 = -5x^3 + 3x^2$

121. $(fg)(2) = \left(2(2)^2 + 1\right)\left((2)^2 + 2 - 3\right) = (9)(3) = 27;$

   $(fg)(x) = \left(2x^2 + 1\right)\left(x^2 + x - 3\right) = 2x^4 + 2x^3 - 6x^2 + x^2 + x - 3$

   $= 2x^4 + 2x^3 - 5x^2 + x - 3$

Area

123. $x(x+4);$ When $x=20, (20)(20+4) = 20(24) = 480$ ft^2.

125. $(x+3)^2;$ When $x=20, (20+3)^2 = 23^2 = 529$ ft^2.

127. (a) $D = 40 - \dfrac{1}{2}(20) = 40 - 10 = 30;$ demand is 30 thousand units.

   (b) $R = Dp = \left(40 - \dfrac{1}{2}p\right)(p) = 40p - \dfrac{1}{2}p^2$

   (c) $R = 40(30) - \dfrac{1}{2}(30)^2 = 1200 = \dfrac{1}{2}(900) = 1200 - 450 = 750;$ revenue is \$750 thousand.

Applications

129. (a) $N(1+r)^2 = N\left(1+2r+r^2\right) = N + 2Nr + Nr^2 = Nr^2 + 2Nr + N$

   (b) $200(1+0.10)^2 = 242$ and $200(0.10)^2 + 2(200)(0.10) + 200 = 2 + 40 + 200 = 242;$ The answers agree.

131. $x(x+1) = x^2 + x$

133. Let l represent the length of the rectangle. Then since the perimeter is 100 feet, $2l + 2x = 100 \Rightarrow l = 50 - x$.

   Since the area is given by $A = xl$, we may write the area as follows: $x(50-x) = 50x - x^2$.

Checking Basic Concepts for Sections 5.1 & 5.2

1. (a) $8x^2 + 4x - 5x^2 + 3x = 3x^2 + 7x$

    (b) $(5x^2 - 3x + 2) - (3x^2 - 5x^3 + 1) = 5x^2 - 3x + 2 - 3x^2 + 5x^3 - 1 = 5x^3 + 2x^2 - 3x + 1$

2. The polynomial is $x(x + 120)$. When tickets cost \$10 each, the revenue will be $10(10 + 120) = \$1300$.

3. (a) When exercise is stopped, $x = 0$, and so $f(0) = 2(0)^2 - 25(0) + 160 = 0 - 0 + 160 = 160$ bpm.

    (b) After 4 minutes, $x = 4$, and so $f(4) = 2(4)^2 - 25(4) + 160 = 32 - 100 + 160 = 92$ bpm.

4. (a) $-5(x - 6) = -5x + 30$

    (b) $4x^3(3x^2 - 5x) = 12x^5 - 20x^4$

    (c) $(2x - 1)(x + 3) = 2x^2 + 6x - x - 3 = 2x^2 + 5x - 3$

5. (a) $(5x - 6)(5x + 6) = 25x^2 - 36$

    (b) $(3x - 4)^2 = 9x^2 - 24x + 16$

6. See Figure 6.

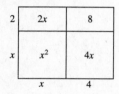

Figure 6

5.3: Factoring Polynomials

Concepts

1. To solve equations

2. Either $x = 0$ or $x - 3 = 0$; the zero-product property

3. Yes, $2x$ is a common factor of each term.

4. No, $4x^2$ is the greatest common factor.

5. No, since $\dfrac{1}{2}(4) = 2$, but $\dfrac{1}{2} \neq 2$ and $4 \neq 2$.

6. Yes, by the zero-product property.

Factoring and Equations

7. $10x - 15 = 5(2x - 3)$

9. $4x + 6y = 2(2x + 3y)$

11. $9r - 15t = 3(3r - 5t)$

13. $2x^3 - 5x = x(2x^2 - 5)$

15. $8a^3 + 10a = 2a\left(4a^2 + 5\right)$

17. $6r^3 - 18r^5 = 6r^3\left(1 - 3r^2\right)$

19. $8x^3 - 4x^2 + 16x = 4x\left(2x^2 - x + 4\right)$

21. $9n^4 - 6n^2 + 3n = 3n\left(3n^3 - 2n + 1\right)$

23. $6t^6 - 4t^4 + 2t^2 = 2t^2\left(3t^4 - 2t^2 + 1\right)$

25. $5x^2y^2 - 15x^2y^3 = 5x^2y^2\left(1 - 3y\right)$

27. $6a^3b^2 - 15a^2b^3 = 3a^2b^2\left(2a - 5b\right)$

29. $18mn^2 + 12m^2n^3 = 6mn^2\left(3 + 2mn\right)$

31. $15x^2y + 10xy - 25x^2y^2 = 5xy\left(3x + 2 - 5xy\right)$

33. $4a^2 - 2ab + 6ab^2 = 2a\left(2a - b + 3b^2\right)$

35. $-2x^2 + 4x - 6 = -2\left(x^2 - 2x + 3\right)$

37. $-8z^4 - 16z^3 = -8z^3\left(z + 2\right)$

39. $-4m^2n^3 - 6mn^2 - 8mn = -2mn\left(2mn^2 + 3n + 4\right)$

41. $m = 0$ or $n = 0$

43. $3z = 0 \Rightarrow z = 0$ or $z + 4 = 0 \Rightarrow z = -4$

45. $r - 1 = 0 \Rightarrow r = 1$ or $r + 3 = 0 \Rightarrow r = -3$

47. $x + 2 = 0 \Rightarrow x = -2$ or $3x - 1 = 0 \Rightarrow 3x = 1 \Rightarrow x = \dfrac{1}{3}$

49. $3x = 0 \Rightarrow x = 0$ or $y - 6 = 0 \Rightarrow y = 6$

51. (a) The graph crossed the x-axis at -3 and 0.

    (b) $x^2 + 3x = 0 \Rightarrow x(x + 3) = 0 \Rightarrow x = 0$ or $x + 3 = 0$, that is $x = 0$ or $x = -3$.

    (c) The zeros of $P(x)$ are -3 and 0.

53. (a) The graph crossed the x-axis at 0 and 2.

    (b) $6x - 3x^2 = 0 \Rightarrow 3x(2 - x) = 0 \Rightarrow 3x = 0$ or $2 - x = 0$, that is $x = 0$ or $x = 2$.

    (c) The zeros of $P(x)$ are 0 and 2.

55. (a) The graph crossed the x-axis at -1 and 1.

    (b) $x^2 - 1 = 0 \Rightarrow (x + 1)(x - 1) = 0 \Rightarrow x + 1 = 0$ or $x - 1 = 0$, that is $x = -1$ or $x = 1$.

    (c) The zeros of $P(x)$ are -1 and 1.

57.   The graph of $y = x^2 - 2x$ (not shown) crosses the x-axis at 0 and 2. The solutions are 0 and 2.

59.   The graph of $y = x - x^2$ (not shown) crosses the x-axis at 0 and 1. The solutions are 0 and 1.

61.   The graph of $y = x^2 + 4x$ (not shown) crosses the x-axis at -4 and 0. The solutions are -4 and 0.

63. $x^2 - x = 0 \Rightarrow x(x-1) = 0 \Rightarrow$ Either $x = 0$ or $x - 1 = 0 \Rightarrow x = 1$.

65. $5x^2 - x = 0 \Rightarrow x(5x-1) \Rightarrow$ Either $x = 0$ or $5x - 1 = 0 \Rightarrow x = \dfrac{1}{5}$.

67. $10x^2 + 5x = 0 \Rightarrow 5x(2x+1) \Rightarrow$ Either $5x = 0 \Rightarrow x = 0$ or $2x + 1 = 0 \Rightarrow x = -\dfrac{1}{2}$.

69. $15x^2 = 10x \Rightarrow 15x^2 - 10x = 0 \Rightarrow 5x(3x-2) \Rightarrow$ Either $5x = 0 \Rightarrow x = 0$ or $3x - 2 = 0 \Rightarrow x = \dfrac{2}{3}$.

71. $25x = 10x^2 \Rightarrow 25x - 10x^2 = 0 \Rightarrow 5x(5-2x) \Rightarrow$ Either $5x = 0 \Rightarrow x = 0$ or $5 - 2x = 0 \Rightarrow x = \dfrac{5}{2}$.

73. $32x^4 - 16x^3 = 0 \Rightarrow 16x^3(2x-1) = 0 \Rightarrow$ Either $16x^3 = 0 \Rightarrow x = 0$ or $2x - 1 = 0 \Rightarrow x = \dfrac{1}{2}$.

75.   Numerically: Table $Y_1 = X^2 + 2X$ with Tblstart $= -5$ and ΔTbl $= 1$.

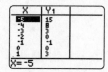

Figure 75a

Graphically: Graph $Y_1 = X^2 + 2X$ in $[-5, 5, 1]$ by $[-5, 5, 1]$.

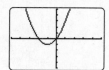

Figure 75b

$x^2 + 2x = 0 \Rightarrow x(x+2) = 0 \Rightarrow x = 0$ or $x + 2 = 0 \Rightarrow x = -2$. The solutions are -2, 0.

77.   Numerically: Table $Y_1 = 2X^2 - 3X$ with Tblstart $= -1$ and ΔTbl $= 0.5$.

Figure 77a

Graphically: Graph $Y_1 = 2X^2 - 3X$ in $[-5, 5, 1]$ by $[-5, 5, 1]$.

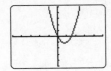

Figure 77b

$2x^2 - 3x = 0 \Rightarrow x(2x-3) = 0 \Rightarrow x = 0$ or $2x - 3 = 0 \Rightarrow x = \dfrac{3}{2}$. The solutions are $0, \dfrac{3}{2}$.

Factoring by Grouping

79. $2x(x+2) + 3(x+2) = (2x+3)(x+2)$

81. $(x-5)x^2 - (x-5)2 = (x-5)(x^2-2)$

83. $x^3 + 3x^2 + 2x + 6 = x^2(x+3) + 2(x+3) = (x+3)(x^2+2)$

85. $6x^3 - 4x^2 + 9x - 6 = 2x^2(3x-2) + 3(3x-2) = (3x-2)(2x^2+3)$

87. $2x^3 - 3x^2 + 2x - 3 = x^2(2x-3) + 1(2x-3) = (x^2+1)(2x-3)$

89. $x^3 - 7x^2 - 3x + 21 = x^2(x-7) - 3(x-7) = (x^2-3)(x-7)$

91. $3x^3 - 15x^2 + 5x - 25 = 3x^2(x-5) + 5(x-5) = (x-5)(3x^2+5)$

93. $xy + x + 3y + 3 = x(y+1) + 3(y+1) = (y+1)(x+3)$

95. $ab - 3a + 2b - 6 = a(b-3) + 2(b-3) = (a+2)(b-3)$

97.   Golf ball: graph b, because the height starts at 0, then increases until it reaches its highest point and then decreases until the height is 0 again.

      Yo – yo: graph a, because the height decreases until its lowest point, then it increases and returns to its starting position.

Applications

99.   (a) $-16t^2 + 128t = 0 \Rightarrow -16t(t-8) = 0 \Rightarrow$ Either $-16t = 0 \Rightarrow t = 0$ or $t - 8 = 0 \Rightarrow t = 8$.

      The golf ball hits the ground after 8 seconds.

      (b) The graph crosses the t-axis at about 8 seconds.

      (c) See Figure 99. Yes, the answer is the same.

      (d) The maximum altitude given by either the graph or the table is 256 feet which is attained after 4 seconds.

t	3	4	5	6	7	8
$y = h(t)$	240	256	240	192	112	0

Figure 99

101. $f(47) = 0.0148(47)^2 + 0.686(47) + 315 \approx 380$; CO_2 concentration in 2005 was 380 ppm.

103. $8y = \pi y^2 \Rightarrow 8y - \pi y^2 = 0 \Rightarrow y(8 - \pi y) = 0 \Rightarrow y = 0$ or $8 - \pi y = 0$, that is $y = 0$ or $y = \dfrac{8}{\pi}$.

The positive value is $\dfrac{8}{\pi}$.

105.   (a) See Figure 105. It models the data quite well.

(b) From the table, the height is about 54,000 feet.

(c) $11t^2 + 6t = 0 \Rightarrow t(11t + 6) = 0 \Rightarrow$ Either $t = 0$ or $11t + 6 = 0 \Rightarrow t = -\dfrac{6}{11}$.

When $t = 0$ the shuttle has not yet left the ground. The value $t = -\dfrac{6}{11}$ has no physical meaning.

t	20	30	40	50	60	70	80
$y = f(t)$	4520	10,080	17,840	27,800	39,960	54,320	70,880

Figure 105

107.   No, there are many possibilities such as 9×24, 8×27 or 13.5×16.

5.4: Factoring Trinomials

Concepts

1.   A polynomial with 3 terms: $x^2 - x + 1$; *Answers may vary.*

2.   The methods for factoring the given expression are symbolic and graphical, grouping and FOIL. *Answers may vary.*

3. $a = 3$, $b = -1$, $c = -3$

4.   Yes, it checks using FOIL.

5. $(x + 3)(x - 1)$

6. $(6x - 3)(x - 2)$ or $3(2x - 1)(x - 2)$ or $6\left(x - \dfrac{1}{2}\right)(x - 2)$

Factoring Trinomials

7.   Yes, it checks using FOIL.

9.   No, $(x + 5)(x - 4) = x^2 + x - 20$.

11.   Yes, it checks using FOIL.

13.   No, $(5m + 2)(2m - 5) = 10m^2 - 21m - 10$.

15. $x^2 + 7x + 10 = (x + 2)(x + 5)$

17. $x^2 + 8x + 12 = (x + 2)(x + 6)$

19. $x^2 - 13x + 36 = (x - 9)(x - 4)$

21. $x^2 - 7x - 8 = (x - 8)(x + 1)$

23. $z^2 + z - 72 = (z - 8)(z + 9)$

25. $t^2 - 15t + 56 = (t-8)(t-7)$

27. $y^2 - 18y + 72 = (y-12)(y-6)$

29. $m^2 - 18m - 40 = (m-20)(m+2)$

31. $n^2 - 20n - 300 = (n-30)(n+10)$

33. $2x^2 + 7x + 3 = (x+3)(2x+1)$

35. $6x^2 - x - 2 = (2x+1)(3x-2)$

37. $4z^2 + 19z + 12 = (z+4)(4z+3)$

39. $6t^2 - 17t + 12 = (2t-3)(3t-4)$

41. $10y^2 + 13y - 3 = (2y+3)(5y-1)$

43. $6m^2 - m - 12 = (2m-3)(3m+4)$

45. $42n^2 + 5n - 25 = (6n+5)(7n-5)$

47. $1 + x - 2x^2 = (1-x)(1+2x)$

49. $20 + 7x - 6x^2 = (5-2x)(4+3x)$

51. $5y^2 + 5y - 30 = 5(y^2 + y - 6) = 5(y-2)(y+3)$

53. $2z^2 + 12z + 16 = 2(z^2 + 6z + 8) = 2(z+2)(z+4)$

55. $z^3 + 9z^2 + 14z = z(z^2 + 9z + 14) = z(z+2)(z+7)$

57. $t^3 - 10t^2 + 21t = t(t^2 - 10t + 21) = t(t-7)(t-3)$

59. $m^4 + 6m^3 + 5m^2 = m^2(m^2 + 6m + 5) = m^2(m+1)(m+5)$

61. $5x^3 + x^2 - 6x = x(5x^2 + x - 6) = x(x-1)(5x+6)$

63. $6x^3 + 21x^2 + 9x = 3x(2x^2 + 7x + 3) = 3x(x+3)(2x+1)$

65. $2x^3 - 14x^2 + 20x = 2x(x^2 - 7x + 10) = 2x(x-5)(x-2)$

67. $60z^3 + 230z^2 - 40z = 10z(6z^2 + 23z - 4) = 10z(z+4)(6z-1)$

69. $4x^4 + 10x^3 - 6x^2 = 2x^2(2x^2 + 5x - 3) = 2x^2(x+3)(2x-1)$

71. $x^2 + (2+3)x + 2 \cdot 3 = x^2 + 2x + 3x + 2 \cdot 3 = x(x+2) + 3(x+2) = (x+2)(x+3)$

73. $x^2 + (a+b)x + ab = x^2 + ax + bx + ab = x(x+a) + b(x+a) = (x+a)(x+b)$

75. $x = 1,\ x = 2,\ x^2 + bx + c = 0 \Rightarrow (x-1)(x-2) = 0 \Rightarrow x^2 - 2x - x + 2 = 0$

$\Rightarrow x^2 - 3x + 2 = 0 \Rightarrow b = -3,\ c = 2$

77. $h(x) = x^2 + 3x + 2 = (x+1)(x+2) \Rightarrow f(x) = x+1,\ g(x) = x+2$

79. $h(x) = x^2 - 2x - 8 = (x+2)(x-4) \Rightarrow f(x) = x+2,\ g(x) = x-4$

Factoring with Graphs and Tables

81.   Since $x = 2$ and $x = 4$ are zeros of the polynomial, the factors are $(x-2)(x-4)$. This checks using FOIL.

83.   Since $x = -1$ and $x = 2$ are zeros of the polynomial, the factors are $2(x+1)(x-2)$. This checks using FOIL.

85.   Since $x = -1$ and $x = 2$ are zeros of the polynomial, the factors are $-(x-2)(x+1)$. This checks using FOIL.

87.   Since $Y_1 = 0$ when $x = -1$ and $x = 4$, the factors are $(x+1)(x-4)$. This checks using FOIL.

89.   Since $Y_1 = 0$ when $x = -1$ and $x = 2$, the factors are $2(x+1)(x-2)$. This checks using FOIL.

91.   Table $Y_1 = X^2 + 3X - 10$ with TblStart $= -6.4$ and ΔTbl $= 1.4$. See Figure 91.

Since $Y_1 = 0$ when $x = -5$ and $x = 2$, the factors are $(x-2)(x+5)$.

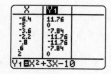   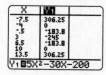

Figure 91   Figure 93

93.   Table $Y_1 = X^2 - 3X - 28$ with TblStart $= -6.2$ and ΔTbl $= 2.2$. See Figure 93.

Since $Y_1 = 0$ when $x = -4$ and $x = 7$, the factors are $(x-7)(x+4)$.

95.   Table $Y_1 = 2X^2 - 14X + 20$ with TblStart $= 0$ and ΔTbl $= 1$. See Figure 95.

Since $Y_1 = 0$ when $x = 2$ and $x = 5$, the factors are $2(x-5)(x-2)$.

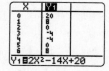

Figure 95   Figure 97

97.   Table $Y_1 = 5X^2 - 30X - 200$ with TblStart $= -7.5$ and ΔTbl $= 3.5$. See Figure 97.

Since $Y_1 = 0$ when $x = -4$ and $x = 10$, the factors are $5(x-10)(x+4)$.

99.   Table $Y_1 = 8X^2 - 44X + 20$ with TblStart = –1 and ΔTbl = 1.5. See Figure 99.

     Since $Y_1 = 0$ when $x = 0.5$ and $x = 5$, the factors are $8(x-5)(x-0.5) = 4(x-5)(2x-1)$.

      Figure 99

Applications

101.   (a) $(35-x)(1200+100x)$

      (b) $(35-x)(1200+100x) = 54,000 \Rightarrow 42,000 + 2300x - 100x^2 = 54,000 \Rightarrow x^2 - 23x + 120 = 0 \Rightarrow$

         $(x-15)(x-8) = 0 \Rightarrow$ Either $x - 15 = 0 \Rightarrow x = 15$ or $x - 8 = 0 \Rightarrow x = 8$

         When $x = 15$ the ticket price is $35 - 15 = \$20$. When $x = 8$ the ticket price is $35 - 8 = \$27$.

103. $x(x+6) = 91 \Rightarrow x^2 + 6x - 91 = 0 \Rightarrow (x+13)(x-7) = 0 \Rightarrow$ Either $x = -13$ or $x = 7$.

     Since the width cannot be negative, the width is 7 feet and the length is $7 + 6 = 13$ feet.

105.   (a) $2t^2 + 88t = 600 \Rightarrow t^2 + 44t - 300 = 0 \Rightarrow (t+50)(t-6) = 0 \Rightarrow$ Either $t = -50$ or $t = 6$.

         Since – 50 seconds has no meaning, the car travels 600 feet in 6 seconds.

      (b)   Table $Y_1 = 2X^2 + 88X - 600$ with TblStart = 2 and ΔTbl = 1. See Figure 105.

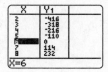

       Figure 105

Checking Basic Concepts for Sections 5.3 & 5.4

1.   (a) $3x^2 - 6x = 3x(x-2)$

      (b) $16x^3 - 8x^2 + 4x = 4x(4x^2 - 2x + 1)$

2.   (a) $x^2 - 2x = 0 \Rightarrow x(x-2) = 0 \Rightarrow$ Either $x = 0$ or $x - 2 = 0 \Rightarrow x = 2$.

      (b) $9x^2 = 81x \Rightarrow 9x(x-9) = 0 \Rightarrow$ Either $9x = 0 \Rightarrow x = 0$ or $x - 9 = 0 \Rightarrow x = 9$.

3.   (a) $x^2 + 3x - 10 = (x-2)(x+5)$

      (b) $x^2 - 3x - 10 = (x-5)(x+2)$

      (c) $8x^2 + 14x + 3 = (2x+3)(4x+1)$

4. $x^2 + 3x + 2 = 0 \Rightarrow (x+1)(x+2) = 0 \Rightarrow$ Either $x + 1 = 0 \Rightarrow x = -1$ or $x + 2 = 0 \Rightarrow x = -2$

5. $-16t^2 + 64t + 8 = 56 \Rightarrow -16t^2 + 64t - 48 = 0 \Rightarrow -16(t-1)(t-3) = 0 \Rightarrow t = 1$ or $t = 3$

After 1 and 3 seconds.

5.5: Special Types of Factoring

Concepts

1. $x^2 - 9$;   *Answers may vary.*

2. $x^2 - 8x + 16$;   *Answers may vary.*

3. $x^3 + 8$;   *Answers may vary.*

4. $a^2 - b^2 = (a-b)(a+b)$

5. $a^2 + 2ab + b^2 = (a+b)^2$

6. $a^3 - b^3 = (a-b)(a^2 + ab + b^2)$

Difference of Two Squares

7.   Yes; $x^2 - 25 = (x-5)(x+5)$

9.   No; $x^3 + y^3 = (x+y)(x^2 - xy + y^2)$

11. $x^2 - 36 = (x-6)(x+6)$

13. $25 - z^2 = (5-z)(5+z)$

15.   The sum of squares cannot be factored.

17. $36x^2 - 100 = 4(9x^2 - 25) = 4(3x-5)(3x+5)$

19. $49a^2 - 64b^2 = (7a-8b)(7a+8b)$

21. $64z^2 - 25z^4 = z^2(64 - 25z^2) = z^2(8-5z)(8+5z)$

23. $5x^3 - 125x = 5x(x^2 - 25) = 5x(x-5)(x+5)$

25.   The sum of squares cannot be factored.

27. $16t^4 - r^2 = (4t^2 - r)(4t^2 + r)$

29. $(x+1)^2 - 25 = ((x+1)-5)((x+1)+5) = (x-4)(x+6)$

31. $100 - (n-4)^2 = (10-(n-4))(10+(n-4)) = (14-n)(6+n)$

33. $y^4 - 16 = (y^2 - 4)(y^2 + 4) = (y-2)(y+2)(y^2 + 4)$

35. $16x^4 - y^4 = (4x^2 - y^2)(4x^2 + y^2) = (2x-y)(2x+y)(4x^2 + y^2)$

37. $x^3 + x^2 - x - 1 = x^2(x+1) - 1(x+1) = (x^2 - 1)(x+1) = (x-1)(x+1)(x+1) = (x-1)(x+1)^2$

39. $4x^3 - 8x^2 - x + 2 = 4x^2(x-2) - 1(x-2) = (4x^2 - 1)(x-2) = (2x-1)(2x+1)(x-2)$

Perfect Square Trinomials

41.   No, the middle term would need to be $8x$.

43.   Yes, $x^2 + 8x + 16 = (x+4)^2$

45.   Yes, $4z^2 - 4z + 1 = (2z-1)^2$

47.   No, the middle term would need to be $-24t$.

49. $x^2 + 2x + 1 = (x+1)^2$

51. $4x^2 + 20x + 25 = (2x+5)^2$

53. $x^2 - 12x + 36 = (x-6)^2$

55. $36z^2 + 12z + 1 = (6z+1)^2$

57. $4y^4 + 4y^3 + y^2 = y^2(4y^2 + 4y + 1) = y^2(2y+1)^2$

59. $9z^3 - 6z^2 + z = z(9z^2 - 6z + 1) = z(3z-1)^2$

61. $9x^2 + 6xy + y^2 = (3x+y)^2$

63. $49a^2 - 28ab + 4b^2 = (7a - 2b)^2$

65. $4x^4 - 4x^3 y + x^2 y^2 = x^2(4x^2 - 4xy + y^2) = x^2(2x-y)^2$

Sum and Difference of Two Cubes

67. $x^3 - 8 = (x-2)(x^2 + 2x + 4)$

69. $y^3 + z^3 = (y+z)(y^2 - yz + z^2)$

71. $27x^3 - 8 = (3x-2)(9x^2 + 6x + 4)$

73. $64z^3 + 27t^3 = (4z + 3t)(16z^2 - 12zt + 9t^2)$

75. $8x^4 + 125x = x(8x^3 + 125) = x(2x+5)(4x^2 - 10x + 25)$

77. $27y - 8x^3 y = y(27 - 8x^3) = y(3 - 2x)(9 + 6x + 4x^2)$

79. $z^6 - 27y^3 = (z^2 - 3y)(z^4 + 3z^2 y + 9y^2)$

81. $125z^6 + 8y^9 = (5z^2 + 2y^3)(25z^4 - 10z^2 y^3 + 4y^6)$

83. $5m^6 + 40n^3 = 5(m^6 + 8n^3) = 5(m^2 + 2n)(m^4 - 2m^2 n + 4n^2)$

General Factoring

85. $25x^2 - 64 = (5x - 8)(5x + 8)$

87. $x^3 + 27 = (x + 3)(x^2 - 3x + 9)$

89. $64x^2 + 16x + 1 = (8x + 1)^2$

91. $3x^2 + 14x + 8 = (x + 4)(3x + 2)$

93. $x^4 + 8x = x(x^3 + 8) = x(x + 2)(x^2 - 2x + 4)$

95. $64x^3 + 8y^3 = 8(8x^3 + y^3) = 8(2x + y)(4x^2 - 2xy + y^2)$

97. $2r^2 - 8t^2 = 2(r^2 - 4t^2) = 2(r - 2t)(r + 2t)$

99. $a^3 + 4a^2b + 4ab^2 = a(a^2 + 4ab + 4b^2) = a(a + 2b)^2$

101. $x^2 - 3x + 2 = (x - 2)(x - 1)$

103. $4z^2 - 25 = (2z - 5)(2z + 5)$

105. $x^4 + 16x^3 + 64x^2 = x^2(x^2 + 16x + 64) = x^2(x + 8)^2$

107. $z^3 - 1 = (z - 1)(z^2 + z + 1)$

109. $3t^2 - 5t - 8 = (t + 1)(3t - 8)$

111. $7a^3 + 20a^2 - 3a = a(7a^2 + 20a - 3) = a(a + 3)(7a - 1)$

113. $x^6 - y^6 = (x^3 - y^3)(x^3 + y^3) = (x - y)(x^2 + xy + y^2)(x + y)(x^2 - xy + y^2)$

115. $100x^2 - 1 = (10x - 1)(10x + 1)$

117. $p^3q^3 - 27 = (pq - 3)(p^2q^2 + 3pq + 9)$

5.6: Summary of Factoring

Concepts

1. Factor out the GCF.

2. $2x$

3. No; the sum of two squares cannot be factored.

4. Grouping

Warm Up

5. $a^2 - a = a(a - 1)$

7. $a^2 - 9 = (a - 3)(a + 3)$

9. $\quad x^2 - 2x + 1 = (x-1)^2$

11. $\quad x^3 - a^3 = (x-a)(x^2 + ax + a^2)$

13. $\quad a^2 + 4$ cannot be factored because it is a sum of squares.

15. $\quad x(x+2) - 3(x+2) = (x-3)(x+2)$

17. $\quad x^3 + 2x^2 + x + 2 = x^2(x+2) + 1(x+2) = (x^2+1)(x+2)$

Factoring

19. $\quad 6x^2 - 14x = 2x(3x-7)$

21. $\quad 2x^3 - 18x = 2x(x^2-9) = 2x(x-3)(x+3)$

23. $\quad 4a^4 - 64 = 4(a^4-16) = 4(a^2-4)(a^2+4) = 4(a-2)(a+2)(a^2+4)$

25. $\quad 6x^3 - 13x^2 - 15x = x(6x^2-13x-15) = x(x-3)(6x+5)$

27. $\quad 2x^4 - 5x^3 - 25x^2 = x^2(2x^2-5x-25) = x^2(x-5)(2x+5)$

29. $\quad 2x^4 + 5x^2 + 3 = (x^2+1)(2x^2+3)$

31. $\quad x^3 + 3x^2 + x + 3 = x^2(x+3) + 1(x+3) = (x^2+1)(x+3)$

33. $\quad 5x^3 - 5x^2 + 10x - 10 = 5(x^3-x^2+2x-2) = 5(x^2(x-1)+2(x-1))$

$\qquad = 5(x^2+2)(x-1)$

35. $\quad ax + bx - ay - by = x(a+b) - y(a+b) = (x-y)(a+b)$

37. $\quad 18x^2 + 12x + 2 = 2(9x^2+6x+1) = 2(3x+1)^2$

39. $\quad -4x^3 + 24x^2 - 36x = -4x(x^2-6x+9) = -4x(x-3)^2$

41. $\quad 8x^3 - 27 = (2x-3)(4x^2+6x+9)$

43. $\quad -x^4 - 8x = -x(x^3+8) = -x(x+2)(x^2-2x+4)$

45. $\quad x^4 - 2x^3 - x + 2 = x^3(x-2) - 1(x-2) = (x^3-1)(x-2)$

$\qquad = (x-1)(x^2+x+1)(x-2)$

47. $\quad r^4 - 16 = (r^2-4)(r^2+4) = (r-2)(r+2)(r^2+4)$

49. $\quad 25x^2 - 4a^2 = (5x-2a)(5x+2a)$

51. $\quad 2x^4 - 2y^4 = 2(x^4-y^4) = 2(x^2-y^2)(x^2+y^2) = 2(x-y)(x+y)(x^2+y^2)$

53. $9x^3 + 6x^2 - 3x = 3x(3x^2 + 2x - 1) = 3x(3x - 1)(x + 1)$

55. $(z - 2)^2 - 9 = ((z - 2) - 3)((z - 2) + 3) = (z - 5)(z + 1)$

57. $3x^5 - 27x^3 + 3x^2 - 27 = 3(x^5 - 9x^3 + x^2 - 9)$

    $= 3(x^3(x^2 - 9) + 1(x^2 - 9)) = 3(x^3 + 1)(x^2 - 9)$

    $= 3(x + 1)(x^2 - x + 1)(x - 3)(x + 3)$

Applications

59. (a) $x^3 - 380x^2 + 2x - 760 = x^2(x - 380) + 2(x - 380)$

    $= (x^2 + 2)(x - 380)$

    (b) $(x^2 + 2)(x - 380) = 0 \Rightarrow x^2 = -2$ (impossible) or $x = 380$; the record high amount was 380 ppm.

61. (a) Volume $= x(x - 1)(x - 1 - 1) = x(x - 1)(x - 2)$

    $= x(x^2 - 3x + 2) = x^3 - 3x^2 + 2x$

    (b) $x^3 - 3x^2 + 2x = 6$

    (c) $x^3 - 3x^2 + 2x = 6 \Rightarrow x^3 - 3x^2 + 2x - 6 = 0$

    $\Rightarrow x^2(x - 3) + 2(x - 3) = 0 \Rightarrow (x^2 + 2)(x - 3) = 0$

    $\Rightarrow x^2 = -2$ (impossible) or $x = 3$; the height is 3 feet.

Checking Basic Concepts for Section 5.5 and 5.6

1. (a) $25x^2 - 16 = (5x - 4)(5x + 4)$

   (b) $x^2 + 12x + 36 = (x + 6)^2$

   (c) $9x^2 - 30x + 25 = (3x - 5)^2$

   (d) $x^3 - 27 = (x - 3)(x^2 + 3x + 9)$

   (e) $81x^4 - 16 = (9x^2 - 4)(9x^2 + 4) = (3x - 2)(3x + 2)(9x^2 + 4)$

2. (a) $x^3 - 2x^2 + 3x - 6 = x^2(x - 2) + 3(x - 2) = (x^2 + 3)(x - 2)$

   (b) $x^2 - 4x - 21 = (x + 3)(x - 7)$

   (c) $12x^3 + 2x^2 - 4x = 2x(6x^2 + x - 2) = 2x(2x - 1)(3x + 2)$

5.7: Polynomial Equations

Concepts

1. Factoring is important so that we can solve polynomial equations. *Answers may vary.*

2. multiplying

3. Subtract 16 from each side.

4. No. The left side of the equation should not be factored because the right side is not zero.

5. Since $x-1$ must equal zero, the solution is 1.

6. No. This equation implies that $x^2 = -4$.

Solving Quadratic Equations

7. The graph of $y = x^2 - 1$ (not shown) crosses the x-axis at $x = -1$ and $x = 1$. The solutions are -1 and 1.

9. The graph of $y = \frac{1}{4}x^2 - 1$ (not shown) crosses the x-axis at $x = -2$ and $x = 2$. The solutions are -2 and 2.

11. The graph of $y = x^2 - x - 2$ (not shown) crosses the x-axis at $x = -1$ and $x = 2$. The solutions are -1 and 2.

13. $z^2 - 64 = 0 \Rightarrow (z-8)(z+8) = 0 \Rightarrow z - 8 = 0$ or $z + 8 = 0 \Rightarrow z = 8$ or $z = -8$. The solutions are -8 and 8.

15. $4y^2 - 1 = 0 \Rightarrow (2y-1)(2y+1) = 0 \Rightarrow 2y - 1 = 0$ or $2y + 1 = 0 \Rightarrow y = \frac{1}{2}$ or $y = -\frac{1}{2}$. The solutions are

    $-\frac{1}{2}$ and $\frac{1}{2}$.

17. $x^2 - 3x - 4 = 0 \Rightarrow (x-4)(x+1) = 0 \Rightarrow x - 4 = 0$ or $x + 1 = 0 \Rightarrow x = 4$ or $x = -1$. The solutions are

    -1 and 4.

19. $x^2 + 4x - 12 = 0 \Rightarrow (x+6)(x-2) = 0 \Rightarrow x + 6 = 0$ or $x - 2 = 0 \Rightarrow x = -6$ or $x = 2$. The solutions are

    -6 and 2.

21. $2x^2 + 5x - 3 = 0 \Rightarrow (2x-1)(x+3) = 0 \Rightarrow 2x - 1 = 0$ or $x + 3 = 0 \Rightarrow x = \frac{1}{2}$ or $x = -3$. The solutions are

    -3 and $\frac{1}{2}$.

23. $2x^2 = 32 \Rightarrow 2x^2 - 32 = 0 \Rightarrow x^2 - 16 = 0 \Rightarrow (x-4)(x+4) = 0 \Rightarrow x - 4 = 0$ or $x + 4 = 0 \Rightarrow$

    $x = 4$ or $x = -4$. The solutions are -4 and 4.

25. $z^2 + 14z + 49 = 0 \Rightarrow (z+7)^2 = 0 \Rightarrow z + 7 = 0 \Rightarrow z = -7$. The solution is -7.

27. $9t^2 + 1 = 6t \Rightarrow 9t^2 - 6t + 1 = 0 \Rightarrow (3t-1)^2 = 0 \Rightarrow 3t - 1 = 0 \Rightarrow t = \frac{1}{3}$. The solution is $\frac{1}{3}$.

29. $15n^2 = 7n + 2 \Rightarrow 15n^2 - 7n - 2 = 0 \Rightarrow (5n+1)(3n-2) = 0 \Rightarrow 5n + 1 = 0$ or $3n - 2 = 0 \Rightarrow$

    $n = -\frac{1}{5}$ or $n = \frac{2}{3}$. The solutions are $-\frac{1}{5}$ and $\frac{2}{3}$.

31. $24m^2 + 23m = 12 \Rightarrow 24m^2 + 23m - 12 = 0 \Rightarrow (3m+4)(8m-3) = 0 \Rightarrow$

    $3m+4 = 0$ or $8m-3 = 0 \Rightarrow m = -\dfrac{4}{3}$ or $m = \dfrac{3}{8}$. The solutions are $-\dfrac{4}{3}$ and $\dfrac{3}{8}$.

33. $x^2 + 12 = 0 \Rightarrow x^2 = -12$. There are no solutions.

35. From the graph $x = -3$ or $x = 2$. $x^2 + x - 6 = 0 \Rightarrow (x+3)(x-2) = 0 \Rightarrow$ Either $x = -3$ or $x = 2$

37. From the graph $x = -2$ or $x = 4$. $2x^2 - 4x - 16 = 0 \Rightarrow 2(x+2)(x-4) = 0 \Rightarrow$ Either $x = -2$ or $x = 4$

39. The parabola should not cross the x-axis. See Figure 39.

41. The parabola should cross the x-axis two times. See Figure 41.

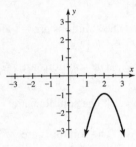

Figure 39

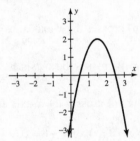

Figure 41

Higher Degree Equations

43. $z^3 = 9z \Rightarrow z^3 - 9z = 0 \Rightarrow z(z^2 - 9) = 0 \Rightarrow z(z-3)(z+3) = 0 \Rightarrow z = 0$ or $z = 3$ or $z = -3$

    The solutions are -3, 0, and 3.

45. $x^3 + x = 0 \Rightarrow x(x^2 + 1) = 0 \Rightarrow x = 0$, $(x^2 = -1$ is not possible$)$

    The solution is 0.

47. $2x^3 - 6x^2 = 20x \Rightarrow 2x^3 - 6x^2 - 20x = 0 \Rightarrow 2x(x^2 - 3x - 10) = 0 \Rightarrow 2x(x-5)(x+2) = 0 \Rightarrow$

    $x = 0$ or $x = 5$ or $x = -2$. The solutions are -2, 0, and 5.

49. $t^4 - 4t^3 - 5t^2 = 0 \Rightarrow t^2(t^2 - 4t - 5) = 0 \Rightarrow t^2(t-5)(t+1) = 0 \Rightarrow t = 0$ or $t = 5$ or $t = -1$

    The solutions are -1, 0, and 5.

51. $x^3 + 7x^2 - 4x - 28 = 0 \Rightarrow x^2(x+7) - 4(x+7) = 0 \Rightarrow (x^2 - 4)(x+7) = 0 \Rightarrow$

    $(x-2)(x+2)(x+7) = 0 \Rightarrow x = 2$ or $x = -2$ or $x = -7$. The solutions are -7, -2, and 2.

53. $n^3 - 2n^2 - 36n + 72 = 0 \Rightarrow n^2(n-2) - 36(n-2) = 0 \Rightarrow (n^2 - 36)(n-2) = 0 \Rightarrow$

    $(n-6)(n+6)(n-2) = 0 \Rightarrow n = 6$ or $n = -6$ or $n = 2$. The solutions are -6, 2, and 6.

55.　(a)　$x^2 - 2x - 15 = 0 \Rightarrow (x+3)(x-5) = 0 \Rightarrow$ Either $x = -3$ or $x = 5$

　　(b)　Graph $Y_1 = X^2 - 2X - 15$ in [−5, 7, 1] by [−20, 10, 2]. See Figures 55a & 55b. Either $x = -3$ or $x = 5$.

　　(c)　Table $Y_1 = X^2 - 2X - 15$ with TblStart = −5 and ΔTbl = 2. See Figure 55c. Either $x = -3$ or $x = 5$.

[−5, 7, 1] by [−20, 10, 2]　　　　　[−5, 7, 1] by [−20, 10, 2]

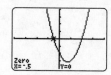　　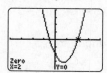　　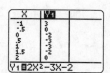

Figure 55a　　　　　　Figure 55b　　　　　　Figure 55c

57.　(a)　$2x^2 - 3x = 2 \Rightarrow 2x^2 - 3x - 2 = 0 \Rightarrow (2x+1)(x-2) = 0 \Rightarrow$ Either $x = -\dfrac{1}{2}$ or $x = 2$

　　(b)　Graph $Y_1 = 2X^2 - 3X - 2$ in [−4, 4, 1] by [−4, 4, 1]. See Figures 57a & 57b. Either $x = -\dfrac{1}{2}$ or $x = 2$.

　　(c)　Table $Y_1 = 2X^2 - 3X - 2$ with TblStart = −1 and ΔTbl = 0.5. See Figure 57c. Either $x = -\dfrac{1}{2}$ or $x = 2$.

[−4, 4, 1] by [−4, 4, 1]　　　　　[−4, 4, 1] by [−4, 4, 1]

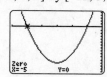　　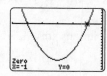　　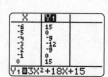

Figure 57a　　　　　　Figure 57b　　　　　　Figure 57c

59.　(a)　$3t^2 + 18t + 15 = 0 \Rightarrow 3(t+5)(t+1) = 0 \Rightarrow$ Either $t = -5$ or $t = -1$

　　(b)　Graph $Y_1 = 3X^2 + 18X + 15$ in [−6, 0, 1] by [−15, 5, 1]. See Figures 59a & 59b. Either $t = -5$ or $t = -1$.

　　(c)　Table $Y_1 = 3X^2 + 18X + 15$ with TblStart = −6 and ΔTbl = 1. See Figure 59c. Either $t = -5$ or $t = -1$.

[−6, 0, 1] by [−15, 5, 1]　　　　　[−6, 0, 1] by [−15, 5, 1]

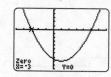　　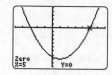　　

Figure 59a　　　　　　Figure 59b　　　　　　Figure 59c

61. (a) $4x^4 + 16x^2 = 16x^3 \Rightarrow 4x^4 - 16x^3 + 16x^2 = 0 \Rightarrow 4x^2(x-2)^2 = 0 \Rightarrow x = 0$ or $x = 2$

(b) Graph $Y_1 = 4X^4 - 16X^3 + 16x^2$ in $[-2, 4, 1]$ by $[-2, 5, 1]$. See Figures 61a & 61b. Here

$x = 0$ or $x = 2$.

(c) Table $Y_1 = 4X^4 - 16X^3 + 16x^2$ with TblStart $= -2$ and ΔTbl $= 1$. See Figure 61c. Here

$x = 0$ or $x = 2$.

$[-2, 4, 1]$ by $[-2, 5, 1]$ $[-2, 4, 1]$ by $[-2, 5, 1]$

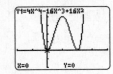   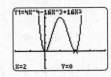   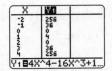

Figure 61a   Figure 61b   Figure 61c

Equations in Quadratic Form

63. $x^4 - 2x^2 - 8 = 0 \Rightarrow (x^2 - 4)(x^2 + 2) = 0 \Rightarrow (x-2)(x+2)(x^2 + 2) = 0 \Rightarrow$

$x = 2$ or $x = -2$, $(x^2 = -2$ is not possible$)$. The solutions are -2 and 2.

65. $x^4 - 26x^2 + 25 = 0 \Rightarrow (x^2 - 25)(x^2 - 1) = 0 \Rightarrow (x-5)(x+5)(x-1)(x+1) = 0 \Rightarrow$

$x = 5$ or $x = -5$ or $x = 1$ or $x = -1$. The solutions are -5, -1, 1, and 5.

67. $x^4 - 13x^2 + 36 = 0 \Rightarrow (x^2 - 9)(x^2 - 4) = 0 \Rightarrow (x-3)(x+3)(x-2)(x+2) = 0 \Rightarrow$

$x = 3$ or $x = -3$ or $x = 2$ or $x = -2$. The solutions are -3, -2, 2, and 3.

69. $x^4 - 16 = 0 \Rightarrow (x^2 - 4)(x^2 + 4) = 0 \Rightarrow (x-2)(x+2)(x^2 + 4) = 0$

$\Rightarrow x = 2$ or $x = -2$, $(x^2 = -4$ is not possible$)$. The solutions are -2 and 2.

71. $9x^4 - 13x^2 + 4 = 0 \Rightarrow (x^2 - 1)(9x^2 - 4) = 0 \Rightarrow (x-1)(x+1)(3x-2)(3x+2) = 0 \Rightarrow$

$x = 1$ or $x = -1$ or $x = \dfrac{2}{3}$ or $x = -\dfrac{2}{3}$. The solutions are -1, $-\dfrac{2}{3}$, $\dfrac{2}{3}$, and 1.

Applications

73. Let x be the height of the picture and $x + 4$ be its width. The overall area is given by $(x+4)(x+8)$.

$(x+4)(x+8) = 525 \Rightarrow x^2 + 12x + 32 = 525 \Rightarrow x^2 + 12x - 493 = 0 \Rightarrow (x-17)(x+29) = 0 \Rightarrow$

$x = 17$ or $x = -29$. The only valid solution is 17. The dimensions are 17 inches by 21 inches.

75. $-0.0001x^2 + 500 = 400 \Rightarrow 0.0001x^2 - 100 = 0 \Rightarrow (0.01x - 10)(0.01x + 10) = 0 \Rightarrow x = 1000$ or $x = -1000$. The

values are -1000 and 1000.

77. Let x be the width of the pool. The area of the pool is given by $x(x+20)$ whereas the total area of the pool

and the sidewalk is $(x+10)(x+30)$. Since the area of just the sidewalk portion is 900 square feet, we have

$$(x+10)(x+30)-x(x+20)=900 \Rightarrow x^2+40x+300-x^2-20x=900 \Rightarrow 20x=600 \Rightarrow x=30$$

The pool is 30 feet by 50 feet.

79. $-16t^2+66t+2=70 \Rightarrow 16t^2-66t+68=0 \Rightarrow 8t^2-33t+34=0 \Rightarrow (8t-17)(t-2)=0 \Rightarrow$

That is $t=\dfrac{17}{8}$ or $t=2$. The baseball is 70 feet in the air at 2 seconds and $\dfrac{17}{8}=2.125$ seconds.

81. $\dfrac{1}{11}x^2+\dfrac{11}{3}x=220 \Rightarrow 3x^2+121x=7260 \Rightarrow 3x^2+121x-7260=0 \Rightarrow (x-33)(3x+220)=0 \Rightarrow$

$x=33$ or $x=-\dfrac{220}{3}$. The only valid solution is 33 mph.

83.   Let x represent the thickness of the iPod. Then $5.904=x(x+1.8)(x+3.5)$.

Graph $Y_1=5.904$ and $Y_2=X(X+1.8)(X+3.5)$ in $[0,\,2.5,\,0.5]$ by $[0,\,10,\,1]$.

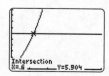

Figure 83

The solution is $x=0.6$, so the dimensions of the iPod are 0.6 inches by 2.4 inches by 4.1 inches.

85.   (a)   The table is shown in Figure 85.

      The elevation begins at 500 feet, decreases, and then increases back to 500 feet.

    (b) $0.0002x^2-0.3x+500=400 \Rightarrow 0.0002x^2-0.3x+100=0 \Rightarrow x^2-1500x+500,000=0 \Rightarrow$

       $(x-500)(x-1000)=0 \Rightarrow x=500$ or $x=1000$. The elevation is 400 feet at 500 feet and 1000 feet.

x	0	300	600	900	1200	1500
$y=E(x)$	500	428	392	392	428	500

Figure 85

Checking Basic Concepts for Section 5.7

1.   The graph crosses the x-axis at $x=-3$ and $x=-1$. The solutions are -3 and -1.

$x^2+4x+3=0 \Rightarrow (x+3)(x+1)=0 \Rightarrow x=-3$ or $x=-1$. The solutions are -3 and -1.

2.   The graph of $y=x^2-9$ (not shown) crosses the x-axis at $x=-3$ and $x=3$. The solutions are -3 and 3.

$x^2-9=0 \Rightarrow (x-3)(x+3)=0 \Rightarrow x=3$ or $x=-3$. The solutions are -3 and 3.

3.   (a) $x^2+9=6x \Rightarrow x^2-6x+9=0 \Rightarrow (x-3)^2=0 \Rightarrow x-3=0 \Rightarrow x=3$

    (b) $x^2+9=0 \Rightarrow x^2=-9$. There is no solution.

    (c) $12x^2+7x-10=0 \Rightarrow (4x+5)(3x-2)=0 \Rightarrow x=-\dfrac{5}{4}$ or $x=\dfrac{2}{3}$

4.   (a) $x^3+5x=0 \Rightarrow x(x^2+5)=0 \Rightarrow x=0,\ \left(x^2=-5 \text{ is not possible}\right)$. The only solution is 0.

(b) $x^4 - 81 = 0 \Rightarrow \left(x^2 - 9\right)\left(x^2 + 9\right) = 0 \Rightarrow (x - 3)(x + 3)\left(x^2 + 9\right) = 0 \Rightarrow x = 3 \text{ or } x = -3$

$\left(x^2 = -9 \text{ is not possible}\right)$. The only solutions are -3 and 3.

(c) $z^4 + z^2 = 20 \Rightarrow z^4 + z^2 - 20 = 0 \Rightarrow \left(z^2 - 4\right)\left(z^2 + 5\right) = 0 \Rightarrow (z - 2)(z + 2)\left(z^2 + 5\right) = 0 \Rightarrow$

$z = 2 \text{ or } z = -2, \ \left(z^2 = -5 \text{ is not possible}\right)$. The only solutions are -2 and 2.

(d) $x^3 + 2x^2 - 49x = 98 \Rightarrow x^3 + 2x^2 - 49x - 98 = 0 \Rightarrow x^2(x + 2) - 49(x + 2) = 0$

$\Rightarrow \left(x^2 - 49\right)(x + 2) = 0 \Rightarrow (x - 7)(x + 7)(x + 2) = 0 \Rightarrow x = 7, -7, \text{ or } -2$. The solutions are

$-7, -2,$ and 7.

Chapter 5 Review Exercises

Section 5.1

1. $x + 5; \ x^2 - 3x + 1;$ *Answers may vary.*

2. $10xy^2;$ When $x = 5$ and $y = 8$, $10(5)(8)^2 = 3200 \text{ in}^3$.

3.   The degree is 5. The coefficient is -4.

4.   The degree is 3. The coefficient is 1.

5.   The degree is 7. The coefficient is 5.

6.   The degree is 10. The coefficient is -9.

7. $5x - 4x + 10x = 11x$

8. $9x^3 - 5x^3 + x^2 = 4x^3 + x^2$

9. $6x^3 y - 4x^3 + 8x^3 y + 5x^3 = 14x^3 y + x^3$

10. $2(2)^3 (-1)^2 - 3(-1) = 16 - (-3) = 16 + 3 = 19$

11.   The degree is 2. The leading coefficient is 5.

12.   The degree is 4. The leading coefficient is 2.

13. $\left(3x^2 - x + 7\right) + \left(5x^2 + 4x - 8\right) = 8x^2 + 3x - 1$

14. $\left(6z^3 + z\right) + \left(17z^3 - 4z^2\right) = 23z^3 - 4z^2 + z$

15. $\left(-4x^2 - 6x + 1\right) - \left(-3x^2 - 7x + 1\right) = -4x^2 - 6x + 1 + 3x^2 + 7x - 1 = -x^2 + x$

16. $\left(3x^3 - 5x + 7\right) - \left(8x^3 + x^2 - 2x + 1\right) = 3x^3 - 5x + 7 - 8x^3 - x^2 + 2x - 1 = -5x^3 - x^2 - 3x + 6$

17. $f(-1) = 2(-1)^2 - 3(-1) + 2 = 2 + 3 + 2 = 7$

18. $f(4) = 1 - (4) - 4(4)^3 = 1 - 4 - 256 = -259$

19.   When $x = 2$, $y = -4$ and so $f(2) = -4$.

20. $(f+g)(-2) = \left((-2)^2 - 1\right) + \left((-2) - 3(-2)^2\right) = 4 - 1 + (-2) - 3(4) = -11;$

$(f-g)(x) = \left(x^2 - 1\right) - \left(x - 3x^2\right) = x^2 - 1 - x + 3x^2 = 4x^2 - x - 1$

Section 5.2

21. $5(3x-4) = 15x - 20$

22. $-2x\left(1 + x - 4x^2\right) = -2x - 2x^2 + 8x^3$

23. $x^3 \cdot x^5 = x^{3+5} = x^8$

24. $-2x^3 \cdot 3x = -6x^{3+1} = -6x^4$

25. $-7xy^7 \cdot 6xy = -42x^{1+1} \cdot y^{7+1} = -42x^2 y^8$

26. $12xy^4 \cdot 5x^2 y = 60x^{1+2} \cdot y^{4+1} = 60x^3 y^5$

27. $(x+4)(x+5) = x^2 + 5x + 4x + 20 = x^2 + 9x + 20$

28. $(x-7)(x-8) = x^2 - 8x - 7x + 56 = x^2 - 15x + 56$

29. $(6x+3)(2x-9) = 12x^2 - 54x + 6x - 27 = 12x^2 - 48x - 27$

30. $\left(y - \dfrac{1}{3}\right)\left(y + \dfrac{1}{3}\right) = y^2 - \left(\dfrac{1}{3}\right)^2 = y^2 - \dfrac{1}{9}$

31. $4x^2\left(2x^2 - 3x - 1\right) = 8x^4 - 12x^3 - 4x^2$

32. $-x\left(4 + 5x - 7x^2\right) = -4x - 5x^2 + 7x^3$

33. $(4x+y)(4x-y) = 16x^2 - y^2$

34. $(x+3)^2 = x^2 + 6x + 9$

35. $(2y-5)^2 = 4y^2 - 20y + 25$

36. $(a-b)\left(a^2 + ab + b^2\right) = a^3 - b^3$

37. $\left(5m - 2n^4\right)^2 = 25m^2 - 20mn^4 + 4n^8$

38. $\left((r-1)+t\right)\left((r-1)-t\right) = (r-1)^2 - t^2 = r^2 - 2r + 1 - t^2$

Section 5.3

39. $25x^2 - 30x = 5x(5x - 6)$

40. $12x^4 + 8x^3 - 16x^2 = 4x^2\left(3x^2 + 2x - 4\right)$

41. $x^2 + 3x = 0 \Rightarrow x(x+3) = 0 \Rightarrow$ Either $x = 0$ or $x + 3 = 0 \Rightarrow x = -3$

42. $7x^4 = 28x^2 \Rightarrow 7x^4 - 28x^2 = 0 \Rightarrow 7x^2\left(x^2 - 4\right) = 0 \Rightarrow 7x^2(x-2)(x+2) \Rightarrow$ Either $x = 0$ or $x = 2$ or $x = -2$

43. $2t^2 - 3t + 1 = 0 \Rightarrow (2t-1)(t-1) = 0 \Rightarrow$ Either $2t-1 = 0 \Rightarrow t = \frac{1}{2}$ or $t-1 = 0 \Rightarrow t = 1$

44. $4z(z-3) + 4(z-3) = 0 \Rightarrow (z-3)(4z+4) = 0 \Rightarrow$ Either $z-3 = 0 \Rightarrow z = 3$ or $4z+4 = 0 \Rightarrow z = -1$

45. $2x^3 + 2x^2 - 3x - 3 = 2x^2(x+1) - 3(x+1) = (x+1)(2x^2 - 3)$

46. $z^3 + z^2 + z + 1 = z^2(z+1) + 1(z+1) = (z+1)(z^2 + 1)$

47. $ax - bx + ay - by = x(a-b) + y(a-b) = (a-b)(x+y)$

48. $(fg)(4) = (4+1)(4-3) = (5)(1) = 5; \quad (fg)(x) = (x+1)(x-3) = x^2 - 3x + x - 3 = x^2 - 2x - 3$

49.   (a) The graph crossed the x-axis at 0 and 3.

      (b) The solutions to $P(x) = 0$ are 0 and 3.

      (c) The zeros of $P(x)$ are 0 and 3.

50.   (a) The graph crossed the x-axis at -1 and 2.

      (b) The solutions to $P(x) = 0$ are -1 and 2.

      (c) The zeros of $P(x)$ are -1 and 2.

Section 5.4

51. $x^2 + 8x + 12 = (x+2)(x+6)$

52. $x^2 - 5x - 50 = (x-10)(x+5)$

53. $9x^2 + 25x - 6 = (x+3)(9x-2)$

54. $4x^2 - 22x + 10 = 2(2x^2 - 11x + 5) = 2(x-5)(2x-1)$

55. $x^3 - 4x^2 + 3x = x(x^2 - 4x + 3) = x(x-3)(x-1)$

56. $2x^4 + 14x^3 + 20x^2 = 2x^2(x^2 + 7x + 10) = 2x^2(x+2)(x+5)$

57. $5x^4 + 15x^3 - 90x^2 = 5x^2(x^2 + 3x - 18) = 5x^2(x-3)(x+6)$

58. $10x^3 - 90x^2 + 200x = 10x(x^2 - 9x + 20) = 10x(x-5)(x-4)$

59.   Since $Y_1 = 0$ when $x = -3$ and $x = 5$, the factors are $(x+3)(x-5)$. This checks using FOIL.

60.   Since $Y_1 = 0$ when $x = 11$ and $x = 13$, the factors are $(x-11)(x-13)$. This checks using FOIL.

61.   Since $x = -7$ and $x = 4$ are zeros of the polynomial, the factors are $(x+7)(x-4)$. This checks using FOIL.

62.   Since $x = 8$ and $x = 13$ are zeros of the polynomial, the solutions are 8 and 13. These solutions check.

Section 5.5

63. $t^2 - 49 = (t-7)(t+7)$

64. $4y^2 - 9x^2 = (2y - 3x)(2y + 3x)$

65. $x^2 + 4x + 4 = (x+2)^2$

66. $16x^2 - 8x + 1 = (4x - 1)^2$

67. $x^3 - 27 = (x-3)(x^2 + 3x + 9)$

68. $64x^3 + 27y^3 = (4x + 3y)(16x^2 - 12xy + 9y^2)$

69. $10y^3 - 10y = 10y(y^2 - 1) = 10y(y-1)(y+1)$

70. $4r^4 - t^6 = (2r^2 - t^3)(2r^2 + t^3)$

71. $m^4 - 16n^4 = (m^2 - 4n^2)(m^2 + 4n^2) = (m - 2n)(m + 2n)(m^2 + 4n^2)$

72. $n^3 - 2n^2 - n + 2 = n^2(n-2) - 1(n-2) = (n^2 - 1)(n-2) = (n-1)(n+1)(n-2)$

73. $25a^2 - 30ab + 9b^2 = (5a - 3b)^2$

74. $2r^3 - 12r^2 t + 18rt^2 = 2r(r^2 - 6rt + 9t^2) = 2r(r - 3t)^2$

75. $a^6 + 27b^3 = (a^2 + 3b)(a^4 - 3a^2 b + 9b^2)$

76. $8p^6 - q^3 = (2p^2 - q)(4p^4 + 2p^2 q + q^2)$

Section 5.6

77. $5x^3 - 10x^2 = 5x^2(x - 2)$

78. $-2x^3 + 32x = -2x(x^2 - 16) = -2x(x - 4)(x + 4)$

79. $x^4 - 16y^4 = (x^2 - 4y^2)(x^2 + 4y^2) = (x - 2y)(x + 2y)(x^2 + 4y^2)$

80. $4x^3 + 8x^2 - 12x = 4x(x^2 + 2x - 3) = 4x(x + 3)(x - 1)$

81. $-2x^3 + 11x^2 - 12x = -x(2x^2 - 11x + 12) = -x(2x - 3)(x - 4)$

82. $x^4 - 8x^2 - 9 = (x^2 + 1)(x^2 - 9) = (x^2 + 1)(x - 3)(x + 3)$

83. $64a^3 + b^3 = (4a + b)(16a^2 - 4ab + b^2)$

84. $8 - y^3 = (2 - y)(4 + 2y + y^2)$

85. $(z + 3)^2 - 16 = ((z + 3) - 4)((z + 3) + 4) = (z - 1)(z + 7)$

86. $x^4 - 5x^3 - 4x^2 + 20x = x(x^3 - 5x^2 - 4x + 20) = x(x^2(x-5) - 4(x-5))$

$= x(x^2 - 4)(x-5) = x(x-2)(x+2)(x-5)$

Section 5.7

87. $x^2 - 16 = 0 \Rightarrow (x-4)(x+4) = 0 \Rightarrow x = 4$ or $x = -4$. The solutions are -4 and 4.

The graph of $y = x^2 - 16$ (not shown) crosses the *x*-axis at $x = -4$ and $x = 4$. The solutions are -4 and 4.

88. $x^2 - 2x - 3 = 0 \Rightarrow (x-3)(x+1) = 0 \Rightarrow x = 3$ or $x = -1$. The solutions are -1 and 3.

The graph of $y = x^2 - 2x - 3$ (not shown) crosses the *x*-axis at $x = -1$ and $x = 3$. The solutions are

-1 and 3.

89. $4x^2 - 28x + 49 = 0 \Rightarrow (2x-7)^2 = 0 \Rightarrow 2x - 7 = 0 \Rightarrow x = \dfrac{7}{2}$

90. $x^2 + 8 = 0 \Rightarrow x^2 = -8$. There are no solutions.

91. $3x^2 = 2x + 5 \Rightarrow 3x^2 - 2x - 5 = 0 \Rightarrow (x+1)(3x-5) = 0 \Rightarrow x = -1$ or $x = \dfrac{5}{3}$

92. $4x^2 + 5x = 6 \Rightarrow 4x^2 + 5x - 6 = 0 \Rightarrow (x+2)(4x-3) = 0 \Rightarrow x = -2$ or $x = \dfrac{3}{4}$

93. $x^3 = x \Rightarrow x^3 - x = 0 \Rightarrow x(x^2-1) = 0 \Rightarrow x(x-1)(x+1) = 0 \Rightarrow x = 0$ or $x = 1$ or $x = -1$

94. $x^3 - 6x^2 + 11x = 6 \Rightarrow x^3 - 6x^2 + 11x - 6 = 0$; the graph of $y = x^3 - 6x^2 + 11x - 6$ (not shown) crosses the *x*-

axis at $x = 1$, $x = 2$ and $x = 3$.

The solutions are 1, 2 and 3.

95. $x^3 + x^2 - 72x = 0 \Rightarrow x(x^2 + x - 72) = 0 \Rightarrow x(x-8)(x+9) = 0 \Rightarrow x = 0$ or $x = 8$ or $x = -9$

96. $x^4 - 15x^3 + 56x^2 = 0 \Rightarrow x^2(x^2 - 15x + 56) = 0 \Rightarrow x^2(x-8)(x-7) = 0 \Rightarrow x = 0$ or $x = 8$ or $x = 7$

97. $x^4 = 16 \Rightarrow x^4 - 16 = 0 \Rightarrow (x^2-4)(x^2+4) = 0 \Rightarrow (x-2)(x+2)(x^2+4) = 0 \Rightarrow$

$x = 2$ or $x = -2$, $(x^2 = -4$ is not possible$)$. The only solutions are -2 and 2.

98. $x^4 + 5x^2 = 36 \Rightarrow x^4 + 5x^2 - 36 = 0 \Rightarrow (x^2-4)(x^2+9) = 0 \Rightarrow (x-2)(x+2)(x^2+9) = 0 \Rightarrow$

$x = 2$ or $x = -2$, $(x^2 = -9$ is not possible$)$. The only solutions are -2 and 2.

Applications

99.   (a) $R(x) = 15x$

(b) $P(x) = R(x) - C(x) = 15x - (3x + 9000) = 12x - 9000$

(c) $P(4000) = 12(4000) - 9000 = 48,000 - 9000 = 39,000$; the profit is \$39,000 for selling 4000 DVDs.

100. (a) Height $= x$, width $= x+5$, length $= x+5+5 = x+10$; volume $= x(x+5)(x+10)$

(b) $x(x+5)(x+10) = 168$

(c) $x(x^2 +15x+50)-168 = 0 \Rightarrow x^3 +15x^2 +50x-168 = 0$

$\Rightarrow (x-2)(x^2 +17x+84) = 0 \Rightarrow x = 2$, since $x^2 +17x+84 > 0$ for all x.

The dimensions are 2 inches by 7 inches by 12 inches.

101. Let x represent the height of the picture, then the overall area of the picture and frame is

$(x+6)(x+2+6) = (x+6)(x+8)$

$= 224 \Rightarrow x^2 +14x+48-224 = 0 \Rightarrow x^2 +14x-176 = 0$

$\Rightarrow (x-8)(x+22) = 0 \Rightarrow x = 8$, since $x = -22$ has no meaning in this context. The dimensions of the picture

are 8 inches by 10 inches.

102. $\frac{1}{2}(x+2)(x-3) = \frac{1}{2}(x^2 -x-6) = \frac{1}{2}x^2 -\frac{1}{2}x-3$

103. (a) In May, $x = 5$; $f(5) = -1.466(5)^2 +20.25(5)+9 = 73.6°F$.

(b) Table $Y_1 = -1.466X^2 +20.25X+9$ with TblStart = 1 and ΔTbl = 1. See Figure 103b. It is greatest in

July.

(c) Graph $Y_1 = -1.466X^2 +20.25X+9$ in [1, 12, 1] by [30, 90, 10]. See Figure 103c. The temperature

increases from January to July and then decreases from July to December.

[1, 12, 1] by [30, 90, 10]

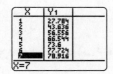   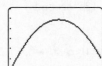

Figure 103b   Figure 103c

104. (a) $100\left(1-\dfrac{x}{100}\right)^2 = 100\left(1-2(1)\left(\dfrac{x}{100}\right)+\left(\dfrac{x}{100}\right)^2\right) = 100\left(1-\dfrac{x}{50}+\dfrac{x^2}{10,000}\right) = 100-2x+\dfrac{x^2}{100}$

$= \dfrac{x^2}{100}-2x+100$

(b) $100\left(1-\dfrac{70}{100}\right)^2 = 100(0.3)^2 = 9\%$ and $\dfrac{1}{100}(70)^2 -2(70)+100 = 49-140+100 = 9\%$

105. If x is the largest of three consecutive integers then the smallest is $x-2$ and the middle integer is $x-1$.

The product is given by $(x-2)(x-1)x = (x^2 -3x+2)x = x^3 -3x^2 +2x$.

106.   See Figure 106.

[–25, 20, 5] by [–160, 100, 10]

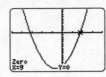

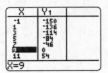

Figure 106   Figure 107a   Figure 107b

107.   Let x represent the width of the building. Then $x + 7$ represents the length of the building. The equation

needed is given by $x(x+7) = 144 \Rightarrow x^2 + 7x - 144 = 0$.

(a)   Graph $Y_1 = X^2 + 7X - 144$ in [–25, 20, 5] by [–160, 100, 10]. See Figure 107a.

Here $y = 0$ when $x = 9$. The dimensions are 9×16 feet.

(b)   Table $Y_1 = X^2 + 7X - 144$ with TblStart = –1 and ΔTbl = 2. See Figure 107b.

Here $Y_1 = 0$ when $x = 9$. The dimensions are 9×16 feet.

(c) $x^2 + 7x - 144 = 0 \Rightarrow (x-9)(x+16) = 0 \Rightarrow$ Either $x - 9 = 0 \Rightarrow x = 9$ or $x + 16 = 0 \Rightarrow x = -16$. Since

$x = -16$ has no meaning in this problem, we know that $x = 9$. The dimensions are 9×16 feet.

108.   Let l represent the length of the rectangle. Then since the perimeter is 50 feet,

$2l + 2x = 50 \Rightarrow l = 25 - x$. The area is $A = (25 - x)x = 25x - x^2$.

109.   (a) $-16t^2 + 66t = 0 \Rightarrow 8t^2 - 33t = 0 \Rightarrow t(8t - 33) = 0 \Rightarrow t = 0$ or $8t - 33 = 0 \Rightarrow t = 0$ or $t = \dfrac{33}{8}$

The ball strikes the ground at $\dfrac{33}{8} = 4.125$ seconds.

(b) $-16t^2 + 66t = 50 \Rightarrow -16t^2 + 66t - 50 = 0 \Rightarrow 8t^2 - 33t + 25 = 0 \Rightarrow (t-1)(8t - 25) \Rightarrow t = 1$ or $t = \dfrac{25}{8}$. The

ball is 50 feet high at 1 second and $\dfrac{25}{8} = 3.125$ seconds.

110.   (a) $R(x) = (50 - x)(600 + 20x)$

(b) $(50 - x)(600 + 20x) = 32,000 \Rightarrow 30,000 + 400x - 20x^2 = 32,000 \Rightarrow x^2 - 20x + 100 = 0 \Rightarrow$

$(x - 10)^2 = 0 \Rightarrow x - 10 = 0 \Rightarrow x = 10$.

When $x = 10$ the ticket price is $50 - 10 = \$40$.

111. $-0.0001x^2 + 500 = 400 \Rightarrow 0.0001x^2 - 100 = 0 \Rightarrow (0.01x - 10)(0.01x + 10) = 0 \Rightarrow$

$0.01x - 10 = 0$ or $0.01 + 10 = 0 \Rightarrow x = 1000$ or $x = -1000$. The values are -1000 and 1000.

Chapter 5 Test

1. $x^2y^2 - 4x + 9x - 5x^2y^2 = 5x - 4x^2y^2$

2. $\left(-2x^3 - 6x + 1\right) - \left(5x^3 - x^2 + x - 10\right) = -2x^3 - 6x + 1 - 5x^3 + x^2 - x + 10 = -7x^3 + x^2 - 7x + 11$

3. $f(-2) = 2(-2)^3 - (-2)^2 - 5(-2) + 2 = -16 - 4 + 10 + 2 = -8$

4. Since $y = -2$ when $x = 2$, $f(2) = -2$.

5. The degree is $3 + 1 = 4$. The coefficient is 3.

6. $8xyz$

7. $-(-2)^2(3) + 3(-2)(3)^2 = -(4)(3) + 3(-2)(9)$

    $= -12 - 54 = -66$

8. $(f - g)(4) = 2(4) - 5 - \left(1 - 4^3\right) = 8 - 5 - (1 - 64)$

    $= 8 - 5 + 63 = 66$

    $(f + g)(x) = 2x - 5 + 1 - x^3 = -x^3 + 2x - 4$

9. $-\dfrac{2}{5}x^2(10x - 5) = -4x^3 + 2x^2$

10. $2xy^7 \cdot 7xy = 14 \cdot x^{1+1} \cdot y^{7+1} = 14x^2y^8$

11. $(2x + 1)(5x - 7) = 10x^2 - 14x + 5x - 7 = 10x^2 - 9x - 7$

12. $(5 - 3x)^2 = 25 - 30x + 9x^2 = 9x^2 - 30x + 25$

13. $(5x - 4y)(5x + 4y) = (5x)^2 - (4y)^2 = 25x^2 - 16y^2$

14. $-2x^2\left(x^2 - 3x + 2\right) = -2x^2 \cdot x^2 + 2x^2 \cdot 3x - 2x^2 \cdot 2$

    $= -2x^4 + 6x^3 - 4x^2$

15. $(x - 2y)\left(x^2 + 2xy + 4y^2\right) = x \cdot x^2 + x \cdot 2xy + x \cdot 4y^2$

    $-2y \cdot x^2 - 2y \cdot 2xy - 2y \cdot 4y^2 = x^3 + 2x^2y + 4xy^2 - 2x^2y$

    $-4xy^2 - 8y^3 = x^3 - 8y^3$

16. $2x^2(x - 1)(x + 1) = 2x^2\left(x^2 - 1\right) = 2x^2 \cdot x^2 - 2x^2 \cdot 1 = 2x^4 - 2x^2$

17. $x^2 - 3x - 10 = (x - 5)(x + 2)$

18. $2x^3 + 6x = 2x\left(x^2 + 3\right)$

19. $3x^2 + 7x - 20 = (x + 4)(3x - 5)$

20. $5x^4 - 5x^2 = 5x^2\left(x^2 - 1\right) = 5x^2(x - 1)(x + 1)$

21. $2x^3 + x^2 - 10x - 5 = x^2(2x+1) - 5(2x+1) = (2x+1)(x^2 - 5)$

22. $49x^2 - 14x + 1 = (7x-1)^2$

23. $x^3 + 8 = (x+2)(x^2 - 2x + 4)$

24. $4x^2y^4 + 8x^4y^2 = 4x^2y^2(y^2 + 2x^2)$

25. $a^2 - 3ab + 2b^2 = (a-b)(a-2b)$

26. The degree is 3. The leading coefficient is 1.

27. $(2m^3 - 4n^2)^2 = 4m^6 - 16m^3n^2 + 16n^4$

28. Since $x = -8$ and $x = 6$ are zeros of the polynomial, the factors are $(x+8)(x-6)$.

29. If x is an even integer, the next consecutive even integer is $x + 2$. The product is $x(x+2)$.

30. $5x^2 = 15x \Rightarrow 5x^2 - 15x = 0 \Rightarrow 5x(x-3) = 0 \Rightarrow$ Either $5x = 0 \Rightarrow x = 0$ or $x - 3 = 0 \Rightarrow x = 3$

31. $4t^2 + 19t - 5 = 0 \Rightarrow (4t-1)(t+5) = 0 \Rightarrow$ Either $4t - 1 = 0 \Rightarrow t = \dfrac{1}{4}$ or $t + 5 = 0 \Rightarrow t = -5$

32. $2z^4 - 8z^2 = 0 \Rightarrow 2z^2(z^2 - 4) = 0 \Rightarrow 2z^2(z-2)(z+2) = 0 \Rightarrow z = -2,\ 0,$ or 2

33. $x^4 - 2x^2 + 1 = 0 \Rightarrow (x^2 - 1)(x^2 - 1) = 0 \Rightarrow x^2 - 1 = 0$

     $\Rightarrow (x+1)(x-1) = 0 \Rightarrow$ Either $x = -1$ or $x = 1$

34. $x(x-3) + (x-3) = 0 \Rightarrow (x+1)(x-3) = 0 \Rightarrow$ Either

     $x = -1$ or $x = 3$

35. Let x represent the height of the frame. Then $x + 4$ represents the width of the frame.

     (a) The equation needed is given by $x(x+4) = 221$.

     (b) $x(x+4) = 221 \Rightarrow x^2 + 4x - 221 = 0 \Rightarrow (x-13)(x+17) = 0 \Rightarrow x = 13$ or $x = -17$.

          Since $x = -17$ has no meaning in this problem, the height is 13 inches and the width is 17 inches.

     (c) The perimeter is $P = 2(13) + 2(17) = 26 + 34 = 60$ inches.

36. (a) For May, $x = 5;\ f(5) = -0.091(5)^3 + 0.66(5)^2 + 5.78(5) + 23.5 \approx 57.5°\text{F}$

     (b) The dew point starts at about 30°F in January and increases to a maximum of about 65°F in July
          Then it decreases to 30°F by the end of December.

37.

     $(x+2)(x+3) = x^2 + 2x + 3x + 6 = x^2 + 5x + 6$

38. $-16t^2 + 96t + 3 = 131 \Rightarrow -16t^2 + 96t - 128 = 0$

      $\Rightarrow -16\left(t^2 - 6t + 8\right) = 0 \Rightarrow (t - 2)(t - 4) = 0$

      $\Rightarrow$ Either $t = 2$ or $t = 4$

      The baseball was 131 feet in the air at 2 seconds and 4 seconds.

Chapter 5 Extended and Discovery Exercises

1.   (a) The population of lynx oscillated during the recorded time period at nearly 10-year intervals.

      (b) The population of snowshoe hares oscillated during the recorded time period at nearly 10-year intervals.

      (c) As the number of snowshoe hares increased, the lynx would have more plentiful prey and hence the lynx population would begin to increase. As the lynx population increased, the snowshoe hares would be killed in increasing numbers causing their population to begin to decrease. With a limited number of snowshoe hares available for food, the lynx population would decline. Now with fewer predators, the snowshoe hare population could begin to rise once again. This cycle repeated itself three times during the recorded time period.

2.   (a) Plot the data in [1, 10, 1] by [0, 30, 3]. See Figure 2a. The data appear to be nonlinear.

      (b) Graph $Y_1 = X^{\wedge}2.5$ in [1, 10, 1] by [0, 30, 3]. See Figure 2b. The model is only accurate for the first 3 planets.

      (c) By trial and error, the model $y = x^{1.5}$ fits quite well.

      (d) The orbit for Neptune is $y = (30.1)^{1.5} \approx 165.1$ years. The orbit for Pluto is $y = (39.4)^{1.5} \approx 247.3$ years.

      [1, 10, 1] by [0, 30, 3]   [1, 10, 1] by [0, 30, 3]

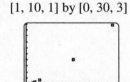   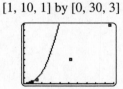

         Figure 2a   Figure 2b

3.   (a) Plot the data in [1890, 2000, 20] by [0, 60, 10]. See Figure 3. The data appear to be nonlinear.

      (b) By trial and error the value is $k \approx 0.006$.

      (c) In 2005 the number of women in the work force will be $y = 0.006(2005 - 1900)^2 + 5.3 \approx 71.5$ million.

4.   (a) Plot the data in [1993, 2005, 1] by [45, 115, 10]. See Figure 4.

(b) By trial and error the value is $k \approx 0.7$.

(c) In 2006 the number of Americans over 100 will be $y = 0.7(2006-1994)^2 + 50 \approx 150.8$ thousand.

[1890, 2000, 20] by [0, 60, 10]   [1993, 2005, 1] by [45, 115, 10]

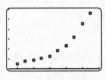   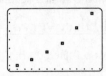

Figure 3   Figure 4

5.   (a) $\begin{bmatrix} 10,000^2 & 10,000 & 1 & | & 412 \\ 40,000^2 & 40,000 & 1 & | & 843 \\ 100,000^2 & 100,000 & 1 & | & 2550 \end{bmatrix}$

(b) The results from a graphing calculator are $a \approx 1.56 \times 10^{-7}$, $b \approx 0.007$, and $c \approx 330.9$.

(c) $1.56 \times 10^{-7}(20,000)^2 + 0.007(20,000) + 330.9 \approx \$533.30$.

This is close to the actual value. Note that unrounded values for a, b, and c give $\$524.37$.

Chapters 1-5 Cumulative Review Exercises

1. $A = \frac{1}{2}(5)\left(\frac{3}{2}\right) = \frac{15}{4}$

2. $\frac{b-a}{a}$

3.   (a) $\left(\frac{x^{-3}}{y}\right)^2 = \frac{x^{-6}}{y^2} = \frac{1}{x^6 y^2}$

(b) $\frac{\left(3r^{-1}t\right)^{-4}}{r^2\left(t^2\right)^{-2}} = \frac{3^{-4}r^4t^{-4}}{r^2t^{-4}} = \frac{1}{3^4} \cdot r^{4-2} \cdot t^{-4-(-4)}$

$= \frac{1}{81} \cdot r^2 \cdot t^0 = \frac{r^2}{81}$

(c) $\left(ab^{-2}\right)^4\left(ab^3\right)^{-1} = a^4b^{-8}a^{-1}b^{-3} = a^3b^{-11} = \frac{a^3}{b^{11}}$

4. $5.859 \times 10^4 = 58,590$

5. $f(-3) = -2(-3)^2 + 3(-3) = -2(9) + 3(-3) = -18 - 9 = -27$

6.   The radicand must be greater than or equal to zero.

$D : x \geq 4$

7.

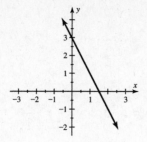

8.

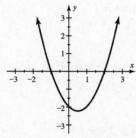

9. (a) Yes. It passes the vertical line test.

(b) D: all real numbers; $R : y \geq -4$

(c) $f(1) = -4; f(2) = -3$

(d) The graph crosses the x-axis at -1 and 3.

(e) The function has the value 0 where the graph crosses the x-axis at -1 and 3; $x = -1, 3$.

(f) $f(x) = (x+1)(x-3)$

10. $2x - y = 4 \Rightarrow y = 2x - 4$; so a line parallel to this line has slope $m = 2$. The line through $(-2, 5)$ with slope

$m = 2$ has equation $y - 5 = 2(x - (-2)) \Rightarrow y = 2x + 4 + 5$

$\Rightarrow y = 2x + 9$.

11. Two points on the graph of f are $(-2, 12)$ and $(-1, 7)$.

$m = \dfrac{y_2 - y_1}{x_2 - x_1} = \dfrac{7 - 12}{-1 - (-2)} = \dfrac{-5}{1} = -5;\ y - 12 = -5(x - (-2))$

$\Rightarrow y = -5x - 10 + 12 \Rightarrow y = -5x + 2$

12. $-2(5x - 1) = 1 - (5 - x) \Rightarrow -10x + 2 = 1 - 5 + x$

$\Rightarrow -11x = -6 \Rightarrow x = \dfrac{6}{11}$

13. $3x + 4 \leq x - 1 \Rightarrow 2x \leq -5 \Rightarrow x \leq -\dfrac{5}{2} \Rightarrow \left(-\infty, -\dfrac{5}{2}\right]$

14. $|5x - 10| > 5 \Rightarrow 5x - 10 < -5$ or $5x - 10 > 5 \Rightarrow$

$5x < 5$ or $5x > 15 \Rightarrow x < 1$ or $x > 3 \Rightarrow (-\infty, 1) \cup (3, \infty)$

15. $-2 < 2 - 7x \le 2 \Rightarrow -4 < -7x \le 0 \Rightarrow \dfrac{4}{7} > x \ge 0$

$\Rightarrow 0 \le x < \dfrac{4}{7} \Rightarrow \left[0, \ \dfrac{4}{7}\right)$

16.   Note that $4x + 2y = 10 \Rightarrow 2y = -4x + 10 \Rightarrow y = -2x + 5$ and $-x + 5y = 3 \Rightarrow 5y = x + 3 \Rightarrow y = \dfrac{1}{5}x + \dfrac{3}{5}.$

Graph $Y_1 = -2x + 5$ and $Y_2 = 1/5 \, x + 3/5$ in $[-5, 5, 1]$ by $[-5, 5, 1]$. The solution is $(2, \ 1)$.

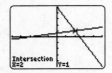

Figure 16

17.   (a) $-x - 2y = 5 \Rightarrow -2y - 5 = x.$ Substitute $x = -2y - 5$ into the second equation and solve for y:

$2(-2y - 5) + 4y = -10 \Rightarrow -4y - 10 + 4y = -10 \Rightarrow -10 = -10.$

This is an identity, so the equations are dependent.

The solutions are $\{(x, \ y) | x + 2y = -5\}.$

(b) Add the two equations to eliminate the variable y.

$\begin{aligned} 3x + 2y &= 7 \\ 2x - 2y &= 3 \\ \hline 5x \phantom{{}+ 2y} &= 10 \end{aligned}$

$\Rightarrow x = 2.$ Substitute $x = 2$ into the first equation and solve for

$y: 3(2) + 2y = 7 \Rightarrow 2y = 1$

$\Rightarrow y = \dfrac{1}{2}.$ The solution is $\left(2, \ \dfrac{1}{2}\right).$

18.   Note that $2x + y \le 4 \Rightarrow y \le -2x + 4$ and $x - 2y < -2 \Rightarrow -2y < -x - 2 \Rightarrow y > \dfrac{1}{2}x + 1$

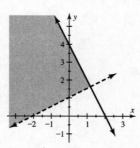

19.   Add the first two equations to eliminate the variables y and z.

$\begin{aligned} x + y - z &= -2 \\ x - y + z &= -6 \\ \hline 2x \phantom{{}+ y - z} &= -8 \end{aligned} \quad \Rightarrow x = -4$

Multiply the second equation by -1 and add to the third equation to eliminate the variables x and y.

$$\begin{array}{r} -x+y\ -z =6 \\ \underline{x-y-2z=3} \\ -3z=9 \Rightarrow z=-3 \end{array} \quad \Rightarrow z=-3$$

Substitute $x=-4$ and $z=-3$ into the first equation and solve for $y: -4+y-(-3)=-2 \Rightarrow y=-1$.

The solution is $(-4,\ -1,\ -3)$.

20. $\begin{bmatrix} 1 & -1 & -1 & | & -2 \\ 2 & 1 & 1 & | & 8 \\ -1 & -1 & 2 & | & 0 \end{bmatrix} \begin{matrix} R_2+R_1 \to \\ \\ 2R_3+R_2 \to \end{matrix} \begin{bmatrix} 3 & 0 & 0 & | & 6 \\ 2 & 1 & 1 & | & 8 \\ 0 & -1 & 5 & | & 8 \end{bmatrix} \frac{1}{3}R_1 \to \begin{bmatrix} 1 & 0 & 0 & | & 2 \\ 2 & 1 & 1 & | & 8 \\ 0 & -1 & 5 & | & 8 \end{bmatrix} -2R_1+R_2 \to$

$\begin{bmatrix} 1 & 0 & 0 & | & 2 \\ 0 & 1 & 1 & | & 4 \\ 0 & -1 & 5 & | & 8 \end{bmatrix} R_2+R_3 \to \begin{bmatrix} 1 & 0 & 0 & | & 2 \\ 0 & 1 & 1 & | & 4 \\ 0 & 0 & 6 & | & 12 \end{bmatrix} \frac{1}{6}R_3 \to \begin{bmatrix} 1 & 0 & 0 & | & 2 \\ 0 & 1 & 1 & | & 4 \\ 0 & 0 & 1 & | & 2 \end{bmatrix} -R_3+R_2 \to$

$\begin{bmatrix} 1 & 0 & 0 & | & 2 \\ 0 & 1 & 0 & | & 2 \\ 0 & 0 & 1 & | & 2 \end{bmatrix} \Rightarrow (2,2,2)$

21. $0(0 \cdot 0 - 1(-1)) - 1(1 \cdot 0 - 3(-1)) + (-3)(1 \cdot 1 - 3 \cdot 0)$

$= 0 - (0+3) - 3(1-0) = -3 - 3(1) = -3 - 3 = -6$

22. $x^2 + 4 = 0 \Rightarrow x^2 = -4$, which is not possible for any real number x. There are no solutions.

23. $-2x(x^2 - 2x + 5) = -2x^3 + 4x^2 - 10x$

24. $(2a+b)(2a-b) = 4a^2 - 2ab + 2ab - b^2 = 4a^2 - b^2$

25. $(x-1)(x+1)(x+3) = (x^2-1)(x+3) = x^3 + 3x^2 - x - 3$

26. $(4x+9)(2x-1) = 8x^2 - 4x + 18x - 9 = 8x^2 + 14x - 9$

27. $(x^2+3y^3)^2 = (x^2+3y^3)(x^2+3y^3) = x^4 + 3x^2y^3 + 3x^2y^3 + 9y^6$

$= x^4 + 6x^2y^3 + 9y^6$

28. $-2x(1-x^2) = -2x + 2x^3 = 2x^3 - 2x$

29. $x^2 - 8x - 33 = (x+3)(x-11)$

30. $10x^3 + 65x^2 - 35x = 5x(2x^2 + 13x - 7) = 5x(2x-1)(x+7)$

31. $4x^2 - 100 = 4(x^2 - 25) = 4(x-5)(x+5)$

32. $49x^2 - 70x + 25 = (7x-5)^2$

33. $r^4 - r = r(r^3 - 1) = r(r-1)(r^2 + r + 1)$

34. $x^3 + 2x^2 + x + 2 = x^2(x+2) + 1(x+2) = (x^2+1)(x+2)$

35. $4x^2 - 1 = 0 \Rightarrow (2x-1)(2x+1) = 0 \Rightarrow 2x-1 = 0$ or $2x+1 = 0$

$\Rightarrow x = \dfrac{1}{2}$ or $x = -\dfrac{1}{2}$. The solutions are $-\dfrac{1}{2}$ and $\dfrac{1}{2}$.

36. $3x^2 + 14x - 5 = 0 \Rightarrow (3x-1)(x+5) = 0 \Rightarrow 3x-1 = 0$ or $x+5 = 0 \Rightarrow x = \dfrac{1}{3}$ or $x = -5$. The solutions are

-5 and $\dfrac{1}{3}$.

37. $x^3 + 4x = 4x^2 \Rightarrow x^3 - 4x^2 + 4x = 0 \Rightarrow x(x^2 - 4x + 4) = 0$

$\Rightarrow x(x-2)^2 = 0 \Rightarrow x = 0$ or $x - 2 = 0 \Rightarrow x = 0$ or $x = 2$.

The solutions are 0 and 2.

38. $x^4 = x^2 \Rightarrow x^4 - x^2 = 0 \Rightarrow x^2(x^2 - 1) = 0 \Rightarrow x^2(x-1)(x+1) = 0$

$\Rightarrow x^2 = 0$ or $x - 1 = 0$ or $x + 1 = 0 \Rightarrow x = 0$ or $x = 1$ or $x = -1$.

The solutions are $-1, 0,$ and 1.

39.   Graph $Y_1 = \sqrt{(2)}X - 1.1(X - \pi)$ and $Y_2 = 1 - 2X$ in $[-5, 5, 1]$ by $[-5, 5, 1]$. The solution is approximately

$x = -1.1$.

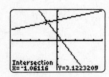

Figure 39

40.   Graph $Y_1 = (\pi - 1) \times {^\wedge} 2 - \sqrt{(3)}$ and $Y_2 = 5 - 1.3X$ in $[-10, 10, 1]$ by $[-10, 10, 1]$. The solutions are

approximately $x = -2.1$ and $x = 1.5$.

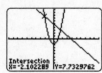

Figure 40a   Figure 40b

Applications

41.   Let x represent the amount invested at 6% interest. Then

$0.06x + 0.07(4000 - x) = 257 \Rightarrow 0.06x + 280 - 0.07x = 257$

$\Rightarrow -0.01x = -23 \Rightarrow x = 2300$. So $2300 is invested at 6% and $1700 is invested at 7%.

42.   (a) $C(1995) = 750(1995 - 1990) + 6800 = 750(5) + 6800 = 10{,}550;$

In 1995 the car cost $10,550.

(b)   Slope = 750; The cost increased, on average, by $750 per year from 1990 to 2005.

43.   Let x represent the length of a side of the square base.

$$x^2(x-5)=1008 \Rightarrow x^3-5x^2-1008=0$$

$$\Rightarrow (x-12)(x^2+7x+84)=0 \Rightarrow x=12, \text{ since } x^2+7x+84$$

is never zero. The dimensions are 12 inches by 12 inches by 7 inches.

44.   Let x, y, and z represent the measures of the angles of the triangle from smallest to largest. Then the system needed is

$$\begin{array}{ll} x+y+z=180 & x+y+z=180 \\ z=x+y-20 \Rightarrow & -x-y+z=-20 \\ y+z=x+100 & -x+y+z=100 \end{array}$$

Add the first two equations to eliminate the variables x and y.

$$\begin{array}{l} x+y+z=180 \\ \underline{-x-y+z=-20} \Rightarrow z=80 \\ 2z=160 \end{array}$$

Add the first and third equations to eliminate the variable x.

$$\begin{array}{l} x+y+z=180 \\ \underline{-x+y+z=100} \Rightarrow 2y+2(80)=280 \Rightarrow 2y=120 \Rightarrow y=60 \\ 2y+2z=280 \end{array}$$

Then $x+y+z=180 \Rightarrow x+60+80=180 \Rightarrow x=40$. The angles are $40°, 60°$, and $80°$.

Critical Thinking Solutions for Chapter 5

Section 5.1

•   The volume of a cube is $V=l\cdot w\cdot h$. So one cube's volume is given by y^3. Volume for 4 cubes is $4y^3$.

•   No, for example $(2x^2+x)+(1-2x^2)=x+1$.

Section 5.2

•   If we let $x=1$ the two expressions are not equal. *Answers may vary.*

Section 5.3

• $xy-4x-3y+12=0 \Rightarrow x(y-4)-3(y-4)=0 \Rightarrow (y-4)(x-3)=0 \Rightarrow y=4 \text{ or } x=3$

Section 5.4

•   Graph $Y_1=100X(100-X)$ in [0, 100, 10] by [0, 300,000, 30,000]. This graph is not shown.

    The maximum occurs when $x=50$. The price that maximizes revenue is $50 per ticket.

•   No. The trinomials shown cannot be factored using the methods discussed in this section.

Chapter 6: Rational Expressions and Functions

6.1: Introduction to Rational Functions and Equations

Concepts

1. A rational expression is a polynomial divided by a nonzero polynomial. Example: $\dfrac{x^2+1}{3x}$ *Answers may vary.*

2. $x = 4$

3. Multiply each side by $x + 7$.

4. No, the equation is undefined when $x = 5$.

5. No, it simplifies to $5 + x$.

6. 1

7. a

8. 0

9. Yes

10. Yes

11. Yes

12. No, $|x|$ is not a polynomial.

13. No, $\sqrt{x}$ is not a polynomial.

14. Yes

Rational Functions

15. $f(x) = \dfrac{x}{x+1}$

17. $f(x) = \dfrac{x^2}{x-2}$

19. The denominator cannot equal zero, so $D = \{x \mid x \neq -2\}$.

21. The denominator cannot equal zero, so $D = \left\{x \mid x \neq \frac{1}{3}\right\}$.

23. The denominator cannot equal zero, and
$$t^2 - 4 = 0 \Rightarrow (t-2)(t+2) = 0 \Rightarrow t = 2 \text{ or } t = -2, \text{ so } D = \{t \mid t \neq -2, t \neq 2\}.$$

25. The denominator cannot equal zero, and
$$x^2 - 3x + 2 = 0 \Rightarrow (x-2)(x-1) = 0 \Rightarrow x = 2 \text{ or } x = 1, \text{ so } D = \{x \mid x \neq 1, x \neq 2\}.$$

27. The denominator cannot equal zero, and $x^3 - 4x = 0 \Rightarrow x(x^2 - 4) = 0 \Rightarrow x(x-2)(x+2) = 0 \Rightarrow x = 0 \text{ or } x = 2$
or $x = -2$, so $D = \{x \mid x \neq -2, x \neq 0, x \neq 2\}$.

29.   See Figure 29. The denominator cannot equal zero, so $x-1 \neq 0$. The domain is $\{x \mid x \neq 1\}$.

31.   See Figure 31. The denominator cannot equal zero, so $2x \neq 0$. The domain is $\{x \mid x \neq 0\}$.

33.   See Figure 33. The denominator cannot equal zero, so $x+2 \neq 0$. The domain is $\{x \mid x \neq -2\}$.

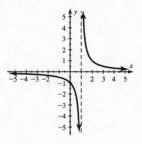
Figure 29

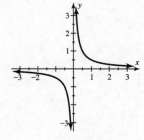
Figure 31

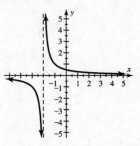
Figure 33

35.   See Figure 35. This denominator will never equal zero. The domain is $\{x \mid -\infty < x < \infty\}$.

37.   See Figure 37. The denominator cannot equal zero, so $2x-3 \neq 0$. The domain is $\left\{x \mid x \neq \dfrac{3}{2}\right\}$.

39.   The denominator cannot equal zero, so $x^2 - 1 = 0 \Rightarrow (x-1)(x+1) = 0 \Rightarrow x = 1$ or $x = -1$. The domain is

$\{x \mid x \neq -1, \ x \neq 1\}$.

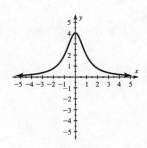
Figure 35

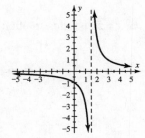
Figure 37

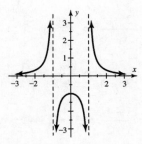
Figure 39

41. $f(-2) = \dfrac{1}{(-2)-1} = -\dfrac{1}{3}$.

43. $f(-3) = \dfrac{(-3)+1}{(-3)-1} = \dfrac{-2}{-4} = \dfrac{1}{2}$.

45. $f(-2) = \dfrac{(-2)^2 - 3(-2) + 5}{(-2)^2 + 1} = \dfrac{15}{5} = 3$.

47. $f(-3) = 2$ and $f(1) = 0$; The equation on the vertical asymptote is $x = -1$.

49. $f(-1) = 0$ and $f(2)$ is undefined; The equations on the vertical asymptotes are $x = -2$ and $x = 2$.

51.

x	-2	-1	0	1	2
$f(x) = \dfrac{1}{x-1}$	$-\dfrac{1}{3}$	$-\dfrac{1}{2}$	-1	—	1

When $x=1, \dfrac{1}{x-1}=\dfrac{1}{1-1}=\dfrac{1}{0}$, so f(1) is undefined.

Rational Equations

53. $\dfrac{3}{x}=5 \Rightarrow 3=5x \Rightarrow \dfrac{3}{5}=x \Rightarrow x=\dfrac{3}{5}$

55. $\dfrac{1}{x-2}=-1 \Rightarrow 1=-1(x-2) \Rightarrow 1=-x+2 \Rightarrow -1=-x \Rightarrow x=1$

57. $\dfrac{x}{x+1}=2 \Rightarrow x=2(x+1) \Rightarrow x=2x+2 \Rightarrow -x=2 \Rightarrow x=-2$

59. $\dfrac{2x+1}{3x-2}=1 \Rightarrow 2x+1=3x-2 \Rightarrow 1=x-2 \Rightarrow 3=x \Rightarrow x=3$

61. $\dfrac{3}{x+2}=x \Rightarrow 3=x(x+2) \Rightarrow x^2+2x-3=0 \Rightarrow (x+3)(x-1)=0 \Rightarrow x=-3 \text{ or } x=1$

63. $\dfrac{6}{x+1}=3x \Rightarrow \dfrac{2}{x+1}=x \Rightarrow 2=x(x+1) \Rightarrow x^2+x-2=0 \Rightarrow (x-1)(x+2) \Rightarrow x=1 \text{ or } x=-2$

65. $\dfrac{1}{x^2-1}=-1 \Rightarrow 1=-1(x^2-1) \Rightarrow 1=-x^2+1 \Rightarrow 0=-x^2 \Rightarrow 0=x^2 \Rightarrow 0=x \Rightarrow x=0$

67. $\dfrac{x}{x-5}=\dfrac{2x-5}{x-5} \Rightarrow x=2x-5 \Rightarrow -x=-5 \Rightarrow x=5$, but this solution does not check. No solution.

69. $\dfrac{4x}{x+2}=\dfrac{-8}{2x+4} \Rightarrow \dfrac{4x}{x+2}=\dfrac{2(-4)}{2(x+2)} \Rightarrow \dfrac{4x}{x+2}=\dfrac{-4}{x+2} \Rightarrow 4x=-4 \Rightarrow x=-1$

71. $\dfrac{2x}{x+2}=\dfrac{x-4}{x+2} \Rightarrow 2x=x-4 \Rightarrow x=-4$

73. Graph $Y_1=(4+X)/(2X)$ and $Y_2=-0.5$ in $[-4.7, 4.7, 1]$ by $[-3.1, 3.1, 1]$. See Figure 73. $x=-2$

75. Table $Y_1=(2X)/(X^2-4)$ and $Y_2=-2/3$ with TblStart = -5 and ΔTbl = 1. See Figure 75. $x=-4 \text{ or } x=1$

77. Table $Y_1=1/(X-1)$ and $Y_2=X-1$ with TblStart = -3 and ΔTbl = 1. See Figure 77. $x=0 \text{ or } x=2$

$[-4.7, 4.7, 1]$ by $[-3.1, 3.1, 1]$

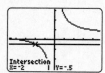

Figure 73   Figure 75   Figure 77

Using More Than One Method

79. (a) Symbolic: $\dfrac{1}{x+2} = 1 \Rightarrow 1 = x+2 \Rightarrow -1 = x \Rightarrow x = -1$

(b) Graphical: Graph $Y_1 = 1/(X+2)$ and $Y_2 = 1$ in $[-4.7, 4.7, 1]$ by $[-3.1, 3.1, 1]$. See Figure 79b. $x = -1$

(c) Numerical: Table $Y_1 = 1/(X+2)$ and $Y_2 = 1$ with TblStart $= -3$ and ΔTbl $= 1$. See Figure 79c. $x = -1$

$[-4.7, 4.7, 1]$ by $[-3.1, 3.1, 1]$

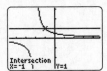   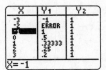

Figure 79b   Figure 79c

81. (a) Symbolic: $\dfrac{x}{2x+1} = \dfrac{2}{5} \Rightarrow 5x = 2(2x+1) \Rightarrow 5x = 4x+2 \Rightarrow x = 2$

(b) Graphical: Graph $Y_1 = X/(2X+1)$ and $Y_2 = 2/5$ in $[-2.35, 2.35, 1]$ by $[-1.55, 1.55, 1]$. See Figure 81b.

$x = 2$

(c) Numerical: Table $Y_1 = X/(2X+1)$ and $Y_2 = 2/5$ with TblStart $= 0$ and ΔTbl $= 1$. See Figure 81c. $x = 2$

$[-2.35, 2.35, 1]$ by $[-1.55, 1.55, 1]$

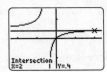   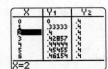

Figure 81b   Figure 81c

83. Graph $Y_1 = X^3/(X-1)$ and $Y_2 = -1/(X+1)$ in $[-5, 5, 1]$ by $[-5, 5, 1]$. The solutions are approximately

$x = -1.62$ and $x = 0.62$

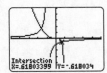   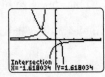

Figure 83a   Figure 83b

85. $a = \dfrac{x}{b} \Rightarrow ba = b \cdot \dfrac{x}{b} \Rightarrow ab = x$

Operations on Functions

87. (a) $(f+g)(3) = 5(3) + (3+1) = 15 + 4 = 19$

(b) $(f-g)(-2) = 5(-2) - (-2+1) = -10 - (-1) = -9$

(c) $(fg)(5) = (5 \cdot 5)(5+1) = 25(6) = 150$

(d) $(f/g)(0) = (5 \cdot 0)/(0+1) = \dfrac{0}{1} = 0$

89. (a) $(f+g)(3) = (2 \cdot 3 - 1) + (4 \cdot 3^2) = 6 - 1 + 4 \cdot 9 = 5 + 36 = 41$

(b) $(f-g)(-2) = (2(-2)-1)-(4(-2)^2) = -4-1-4(4) = -5-16 = -21$

(c) $(fg)(5) = (2 \cdot 5 - 1)(4 \cdot 5^2) = (9)(100) = 900$

(d) $(f/g)(0) = (2 \cdot 0 - 1)/(4 \cdot 0^2) = \dfrac{-1}{0}$, which is undefined

91. (a) $(f+g)(x) = (x+1)+(x+2) = 2x+3$

(b) $(f-g)(x) = (x+1)-(x+2) = -1$

(c) $(fg)(x) = (x+1)(x+2) = x^2+3x+2$

(d) $(f/g)(x) = \dfrac{x+1}{x+2}$

93. (a) $(f+g)(x) = (1-x)+(x^2) = x^2-x+1$

(b) $(f-g)(x) = (1-x)-(x^2) = 1-x-x^2$

(c) $(fg)(x) = (1-x)(x^2) = x^2-x^3$

(d) $(f/g)(x) = \dfrac{1-x}{x^2}$

Applications

95. The graph should increase quickly at first and then continue to increase at a slower rate. The answer is graph c.

97. The graph should increase slowly at first and then continue to increase at a faster rate. The answer is graph d.

99. (a) $f(400) = \dfrac{2540}{400} = 6.35$;   A curve with a radius of 400 feet will have an outer rail elevation of 6.35 inches.

(b) Table $Y_1 = 2540/X$ with TblStart = 100 and ΔTbl = 50. See Figure 99.

(c) If the radius of the curve doubles, the outer rail elevation is halved.

(d) $\dfrac{2540}{r} = 5 \Rightarrow 2540 = 5r \Rightarrow r = \dfrac{2540}{5} = 508$ feet

Figure 99

101. (a) $D(0.05) = \dfrac{900}{10.5+30(0.05)} = \dfrac{900}{12} = 75$; The braking distance is 75 feet when the uphill grade is 0.05.

(b) $60 = \dfrac{900}{10.5+30x} \Rightarrow 60(10.5+30x) = 900 \Rightarrow 630+1800x = 900 \Rightarrow 1800x = 270 \Rightarrow x = 0.15$

103. (a) $T(4) = \dfrac{1}{5-(4)} = \dfrac{1}{1} = 1;$ When cars leave the ramp at a rate of 4 vehicles per minute, the wait is 1 minute.

(b) As more cars try to exit, the waiting time increases. This agrees with intuition.

(c) $3 = \dfrac{1}{5-x} \Rightarrow 3(5-x) = 1 \Rightarrow 15 - 3x = 1 \Rightarrow -3x = -14 \Rightarrow x = 4.\overline{6}$ vehicles per minute.

105. (a) $P(1) = \dfrac{(1)-1}{(1)} = \dfrac{0}{1} = 0$ and $P(50) = \dfrac{(50)-1}{(50)} = \dfrac{49}{50} = 0.98$

When there is only 1 ball there is no chance of losing. With 50 balls there is a 98% chance of losing.

(b) Graph $Y_1 = (X-1)/X$ in [0, 100, 10] by [0, 1, 0.1]. See Figure 105.

[0, 100, 10] by [0, 1, 0.1]

Figure 105

(c) It increases. This agrees with intuition since there are more balls without the winning number.

(d) $0.975 = \dfrac{x-1}{x} \Rightarrow 0.975x = x - 1 \Rightarrow -0.025x = -1 \Rightarrow x = 40$ balls

6.2: Multiplication and Division of Rational Expressions

Concepts

1. 1

2. $\dfrac{x-3}{3-x} = \dfrac{x-3}{-1(x-3)} = \dfrac{1}{-1} = -1$

3. No, $\dfrac{x^2-1}{x-1} = \dfrac{(x-1)(x+1)}{x-1} = x+1.$

4. No, it is equal to neither $\dfrac{1}{x}$ nor $1+\dfrac{3}{x}.$

5. We multiply $\dfrac{2}{3}$ by $\dfrac{7}{5}.$

6. $\dfrac{ac}{bd}$

7. $\dfrac{ad}{bc}$

8. $\dfrac{a}{b}$

Review of Fractions

9. $\dfrac{1}{2} \cdot \dfrac{4}{5} = \dfrac{1}{1} \cdot \dfrac{2}{5} = \dfrac{2}{5}$

11. $\dfrac{7}{8} \cdot \dfrac{4}{3} \cdot (-3) = \dfrac{7}{2} \cdot \dfrac{1}{3} \cdot \dfrac{-3}{1} = \dfrac{7}{2} \cdot \dfrac{1}{1} \cdot \dfrac{-1}{1} = -\dfrac{7}{2}$

13. $\dfrac{3}{8} \cdot 2 = \dfrac{3}{8} \cdot \dfrac{2}{1} = \dfrac{3}{4} \cdot \dfrac{1}{1} = \dfrac{3}{4}$

15. $-\dfrac{7}{11} \div 14 = -\dfrac{7}{11} \cdot \dfrac{1}{14} = -\dfrac{1}{11} \cdot \dfrac{1}{2} = -\dfrac{1}{22}$

17. $\dfrac{5}{7} \div \dfrac{15}{14} = \dfrac{5}{7} \cdot \dfrac{14}{15} = \dfrac{1}{1} \cdot \dfrac{2}{3} = \dfrac{2}{3}$

19. $6 \div \left(-\dfrac{1}{3}\right) = 6 \cdot (-3) = -18$

Simplifying Rational Expressions

21. $\dfrac{5x}{x^2} = \dfrac{5}{x}$

23. $\dfrac{3z+6}{z+2} = \dfrac{3(z+2)}{z+2} = 3$

25. $\dfrac{2z+2}{3z+3} = \dfrac{2(z+1)}{3(z+1)} = \dfrac{2}{3}$

27. $\dfrac{(x-1)(x+1)}{x-1} = x+1$

29. $\dfrac{x^2-4}{x+2} = \dfrac{(x-2)(x+2)}{x+2} = x-2$

31. $\dfrac{x(x-1)}{(x+1)(x-1)} = \dfrac{x}{x+1}$

33. $\dfrac{(3x+1)(x+2)}{(x+2)(5x-2)} = \dfrac{3x+1}{5x-2}$

35. $\dfrac{x+5}{x^2+2x-15} = \dfrac{x+5}{(x+5)(x-3)} = \dfrac{1}{x-3}$

37. $\dfrac{x^2+2x}{x^2+3x+2} = \dfrac{x(x+2)}{(x+1)(x+2)} = \dfrac{x}{x+1}$

39. $\dfrac{6x^2+7x-5}{2x^2-11x+5} = \dfrac{(3x+5)(2x-1)}{(x-5)(2x-1)} = \dfrac{3x+5}{x-5}$

41. $\dfrac{a^2-b^2}{a-b} = \dfrac{(a-b)(a+b)}{a-b} = a+b$

43. $\dfrac{m^3+n^3}{m+n} = \dfrac{(m+n)\left(m^2-mn+n^2\right)}{m+n} = m^2-mn+n^2$

45. $-\dfrac{4-t}{t-4} = \dfrac{-1(4-t)}{t-4} = \dfrac{t-4}{t-4} = 1$

47. $\dfrac{4m-n}{-4m+n} = \dfrac{4m-n}{-1(4m-n)} = \dfrac{1}{-1} = -1$

49. $\dfrac{5-y}{y-5} = \dfrac{-1(y-5)}{y-5} = -1$

Reciprocals

51. $\dfrac{1}{4x}$

53. $\dfrac{5b}{2a}$

55. $\dfrac{5-x}{3-x}$

57. $\dfrac{x^2+1}{1} = x^2+1$

Multiplication and Division of Rational Expressions

59. $\dfrac{2}{x} \cdot \dfrac{x-1}{3x} = \dfrac{2(x-1)}{3x^2}$

61. $\dfrac{x-2}{x} \cdot \dfrac{x-3}{x+4} = \dfrac{(x-2)(x-3)}{x(x+4)}$

63. $\dfrac{1}{2x} \cdot \dfrac{4x}{2} = \dfrac{1}{2x} \cdot \dfrac{2x}{1} = \dfrac{2x}{2x} = 1$

65. $\dfrac{5a}{4} \cdot \dfrac{12}{5a} = \dfrac{1}{1} \cdot \dfrac{3}{1} = 3$

67. $\dfrac{9x^2y^4}{8xy^6} \cdot \dfrac{\left(2xy^2\right)^3}{3(xy)^4} = \dfrac{9x^2y^4}{8xy^6} \cdot \dfrac{8x^3y^6}{3x^4y^4} = \dfrac{9\cdot 8 \cdot x^5y^{10}}{8\cdot 3 \cdot x^5y^{10}} = \dfrac{9}{3} = 3$

69. $\dfrac{x+1}{2x-5} \cdot \dfrac{2x-5}{x} = \dfrac{x+1}{1} \cdot \dfrac{1}{x} = \dfrac{x+1}{x}$

71. $\dfrac{b^2+1}{b^2-1} \cdot \dfrac{b-1}{b+1} = \dfrac{b^2+1}{(b-1)(b+1)} \cdot \dfrac{b-1}{b+1} = \dfrac{b^2+1}{b+1} \cdot \dfrac{1}{b+1} = \dfrac{b^2+1}{(b+1)^2}$

73. $\dfrac{x^2-2x-35}{2x^3-3x^2} \cdot \dfrac{x^3-x^2}{2x-14} = \dfrac{(x-7)(x+5)}{x^2(2x-3)} \cdot \dfrac{x^2(x-1)}{2(x-7)} = \dfrac{x+5}{2x-3} \cdot \dfrac{x-1}{2} = \dfrac{(x-1)(x+5)}{2(2x-3)}$

75. $\dfrac{3n-9}{n^2-9} \cdot \left(n^3+27\right) = \dfrac{3(n-3) \cdot (n+3)\left(n^2-3n+9\right)}{(n+3)(n-3)} = \dfrac{3\left(n^2-3n+9\right)}{1} = 3\left(n^2-3n+9\right)$

77. $\dfrac{3n-9}{n^2-9} \cdot \dfrac{n^3+27}{12} = \dfrac{3(n-3) \cdot (n+3)\left(n^2-3n+9\right)}{(n+3)(n-3) \cdot 12} = \dfrac{n^2-3n+9}{4}$

79. $\dfrac{x}{y} \cdot \dfrac{2y}{x} \cdot \dfrac{2}{xy} = \dfrac{x \cdot 2y \cdot 2}{y \cdot x \cdot xy} = \dfrac{4}{xy}$

81. $\dfrac{x-1}{y} \cdot \dfrac{y(x+y)}{2} \cdot \dfrac{y}{x+y} = \dfrac{(x-1) \cdot y^2(x+y)}{2y \cdot (x+y)} = \dfrac{y(x-1)}{2}$

83. $\dfrac{3x}{2} \div \dfrac{2x}{5} = \dfrac{3x}{2} \cdot \dfrac{5}{2x} = \dfrac{3}{2} \cdot \dfrac{5}{2} = \dfrac{15}{4}$

85. $\dfrac{8a^4}{3b} \div \dfrac{a^5}{9b^2} = \dfrac{8a^4}{3b} \cdot \dfrac{9b^2}{a^5} = \dfrac{8}{1} \cdot \dfrac{3b}{a} = \dfrac{24b}{a}$

87. $(2n+4) \div \dfrac{n+2}{n-1} = \dfrac{2(n+2)}{1} \cdot \dfrac{n-1}{n+2} = 2(n-1)$

89. $\dfrac{6b}{b+2} \div \dfrac{3b^4}{2b+4} = \dfrac{6b}{b+2} \cdot \dfrac{2(b+2)}{3b^4} = \dfrac{2}{1} \cdot \dfrac{2}{b^3} = \dfrac{4}{b^3}$

91. $\dfrac{3a+1}{a^7} \div \dfrac{a+1}{3a^8} = \dfrac{3a+1}{a^7} \cdot \dfrac{3a^8}{a+1} = \dfrac{3a+1}{1} \cdot \dfrac{3a}{a+1} = \dfrac{3a(3a+1)}{a+1}$

93. $\dfrac{x+5}{x-x^3} \div \dfrac{25-x^2}{x^3} = \dfrac{x+5}{x(1-x^2)} \cdot \dfrac{x^3}{(5-x)(5+x)} = \dfrac{1}{1-x^2} \cdot \dfrac{x^2}{5-x} = \dfrac{x^2}{(x-5)(x^2-1)}$

95. $\dfrac{x^2-3x+2}{x^2+5x+6} \div \dfrac{x^2+x-2}{x^2+2x-3} = \dfrac{(x-2)(x-1)}{(x+2)(x+3)} \cdot \dfrac{(x+3)(x-1)}{(x+2)(x-1)} = \dfrac{(x-2)(x-1)}{(x+2)^2}$

97. $\dfrac{x^2-4}{x^2+x-2} \div \dfrac{x-2}{x-1} = \dfrac{(x-2)(x+2)}{(x+2)(x-1)} \cdot \dfrac{x-1}{x-2} = 1$

99. $\dfrac{3y}{x^2} \div \dfrac{y^2}{x} \div \dfrac{y}{5x} = \dfrac{3y}{x^2} \cdot \dfrac{x}{y^2} \cdot \dfrac{5x}{y} = \dfrac{15x^2y}{x^2y^3} = \dfrac{15}{y^2}$

101. $\dfrac{x-3}{x-1} \div \dfrac{x^2}{x-1} \div \dfrac{x-3}{x} = \dfrac{x-3}{x-1} \cdot \dfrac{x-1}{x^2} \cdot \dfrac{x}{x-3} = \dfrac{1}{x}$

Operations on Functions

103. $(fg)(x) = \dfrac{x+2}{x^2-3x+2} \cdot \dfrac{x-1}{x+2} = \dfrac{(x+2)(x-1)}{(x-2)(x-1)(x+2)} = \dfrac{1}{x-2}$

105. $(f/g)(x) = \dfrac{x^2-1}{x+1} = \dfrac{(x-1)(x+1)}{x+1} = x-1$

Geometry

107. (a) $\dfrac{5x^2+12x+4}{x+2}=\dfrac{(5x+2)(x+2)}{x+2}=5x+2$

 (b) If the width is 8 feet, $x+2=8 \Rightarrow x=6$. The length is $5(6)+2=32$ feet.

109. (a) $\dfrac{4x^3+4x^2+x}{x}=4x^2+4x+1$

 (b) Since $4x^2+4x+1=(2x+1)^2$, a side of the bottom is $2(10)+1=21$. The dimensions are 21 by 21 by 10.

Checking Basic Concepts for Sections 6.1 & 6.2

1. (a) $f(2)=\dfrac{(2)}{(2)-1}=\dfrac{2}{1}=2$

 (b) The denominator cannot equal zero, so $x-1\neq 0$. The domain is $\{x\,|\,x\neq 1\}$.

 (c) See Figure 1. Vertical asymptote: $x=1$.

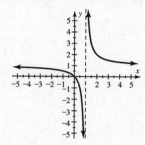

Figure 1

2. (a) $\dfrac{6}{2x+3}=3 \Rightarrow 6=3(2x+3) \Rightarrow 6=6x+9 \Rightarrow 6x=-3 \Rightarrow x=-\dfrac{1}{2}$

 (b) $\dfrac{2}{x-1}=x \Rightarrow 2=x(x-1) \Rightarrow x^2-x-2=0 \Rightarrow (x+1)(x-2)=0 \Rightarrow x=-1 \text{ or } x=2$

3. $\dfrac{x^2-6x-7}{x^2-1}=\dfrac{(x-7)(x+1)}{(x-1)(x+1)}=\dfrac{x-7}{x-1}$

4. (a) $\dfrac{2x^2}{x^2-1}\cdot\dfrac{x+1}{4x}=\dfrac{2x^2}{(x-1)(x+1)}\cdot\dfrac{x+1}{4x}=\dfrac{x}{x-1}\cdot\dfrac{1}{2}=\dfrac{x}{2(x-1)}$

 (b) $\dfrac{1}{x-2}\div\dfrac{3}{(x-2)(x+3)}=\dfrac{1}{x-2}\cdot\dfrac{(x-2)(x+3)}{3}=\dfrac{1}{1}\cdot\dfrac{x+3}{3}=\dfrac{x+3}{3}$

6.3: Addition and Subtraction of Rational Expressions

Concepts

1. 18

2. 18

3. $(x-5)(x+5)$

4. $(x-5)(x+5)$

5. A common denominator

6. A common denominator

7. $\dfrac{a+b}{c}$

8. $\dfrac{a-b}{c}$

Least Common Multiples

9. $10 = 2 \cdot 5$ and $15 = 3 \cdot 5$, the LCM is $2 \cdot 3 \cdot 5 = 30$.

11. $34 = 2 \cdot 17$ and $51 = 3 \cdot 17$, the LCM is $2 \cdot 3 \cdot 17 = 102$.

13. $6a = 2 \cdot 3 \cdot a$ and $9a^2 = 3 \cdot 3 \cdot a \cdot a$, the LCM is $2 \cdot 3 \cdot 3 \cdot a \cdot a = 18a^2$.

15. $10x^2 = 2 \cdot 5 \cdot x \cdot x$ and $25\left(x^2 - x\right) = 5 \cdot 5 \cdot x \cdot (x-1)$, the LCM is $2 \cdot 5 \cdot 5 \cdot x \cdot x \cdot (x-1) = 50x^2\left(x-1\right)$.

17. $x^2 + 2x + 1 = (x+1) \cdot (x+1)$ and $x^2 - 4x - 5 = (x-5) \cdot (x+1)$, the LCM is $(x-5)(x+1)^2$.

19. The LCM is $(x+y)(x-y)$.

Review of Fractions

21. $\dfrac{1}{7} + \dfrac{4}{7} = \dfrac{5}{7}$

23. $\dfrac{2}{3} + \dfrac{5}{6} + \dfrac{1}{4} = \dfrac{8}{12} + \dfrac{10}{12} + \dfrac{3}{12} = \dfrac{21}{12} = \dfrac{7}{4}$

25. $\dfrac{1}{10} - \dfrac{3}{10} = \dfrac{-2}{10} = -\dfrac{1}{5}$

27. $\dfrac{3}{2} - \dfrac{1}{8} = \dfrac{12}{8} - \dfrac{1}{8} = \dfrac{11}{8}$

Addition and Subtraction of Rational Expressions

29. $\dfrac{1}{x} + \dfrac{3}{x} = \dfrac{4}{x}$

31. $\dfrac{2}{x^2 - 4} - \dfrac{x+1}{x^2 - 4} = \dfrac{2 - (x+1)}{x^2 - 4} = \dfrac{1 - x}{x^2 - 4}$

33. $\dfrac{4}{x^2} + \dfrac{5}{x^2} = \dfrac{9}{x^2}$

35. $\dfrac{4}{xy} - \dfrac{7}{xy} = -\dfrac{3}{xy}$

37. $\dfrac{x}{x+1} + \dfrac{1}{x+1} = \dfrac{x+1}{x+1} = 1$

39. $\dfrac{2z}{4-z} - \dfrac{3z-4}{4-z} = \dfrac{2z-(3z-4)}{4-z} = \dfrac{-z+4}{4-z} = \dfrac{4-z}{4-z} = 1$

41. $\dfrac{4r}{5t^2} + \dfrac{r}{5t^2} = \dfrac{4r+r}{5t^2} = \dfrac{5r}{5t^2} = \dfrac{r}{t^2}$

43. $\dfrac{3t}{t^2-t-6} + \dfrac{2-2t}{t^2-t-6} = \dfrac{3t+2-2t}{t^2-t-6} = \dfrac{t+2}{(t-3)(t+2)} = \dfrac{1}{t-3}$

45. $\dfrac{5b}{3a} - \dfrac{7b}{5a} = \dfrac{5b\cdot 5}{15a} - \dfrac{7b\cdot 3}{15a} = \dfrac{25b-21b}{15a} = \dfrac{4b}{15a}$

47. $\dfrac{4}{n-4} + \dfrac{3}{2-n} = \dfrac{4}{n-4} - \dfrac{3}{n-2} = \dfrac{4(n-2)}{(n-4)(n-2)} - \dfrac{3(n-4)}{(n-4)(n-2)} = \dfrac{4n-8-3n+12}{(n-4)(n-2)} = \dfrac{n+4}{(n-4)(n-2)}$

49. $\dfrac{x}{x+4} - \dfrac{x+1}{x} = \dfrac{x^2}{x(x+4)} - \dfrac{(x+1)(x+4)}{x(x+4)} = \dfrac{x^2-(x^2+5x+4)}{x(x+4)} = \dfrac{-5x-4}{x(x+4)}$

51. $\dfrac{2}{x^2} - \dfrac{4x-1}{x} = \dfrac{2}{x^2} - \dfrac{x(4x-1)}{x^2} = \dfrac{2-(4x^2-x)}{x^2} = \dfrac{-4x^2+x+2}{x^2}$

53. $\dfrac{x+3}{x-5} + \dfrac{5}{x-3} = \dfrac{(x+3)(x-3)}{(x-5)(x-3)} + \dfrac{5(x-5)}{(x-5)(x-3)} = \dfrac{x^2-9+5x-25}{(x-5)(x-3)} = \dfrac{x^2+5x-34}{(x-5)(x-3)}$

55. $\dfrac{4n}{n^2-9} - \dfrac{8}{n-3} = \dfrac{4n}{(n-3)(n+3)} - \dfrac{8(n+3)}{(n-3)(n+3)} = \dfrac{4n-8n-24}{(n-3)(n+3)} = -\dfrac{4(n+6)}{(n-3)(n+3)}$

57. $\dfrac{x}{x^2-9} + \dfrac{5x}{x-3} = \dfrac{x}{(x-3)(x+3)} + \dfrac{5x(x+3)}{(x-3)(x+3)} = \dfrac{x+5x^2+15x}{(x-3)(x+3)} = \dfrac{x(5x+16)}{(x-3)(x+3)}$

59. $\dfrac{b}{2b-4} - \dfrac{b-1}{b-2} = \dfrac{b}{2(b-2)} - \dfrac{2(b-1)}{2(b-2)} = \dfrac{b-(2b-2)}{2(b-2)} = \dfrac{-(b-2)}{2(b-2)} = -\dfrac{1}{2}$

61. $\dfrac{2x}{x-5} + \dfrac{2x-1}{3x^2-16x+5} = \dfrac{2x(3x-1)}{(x-5)(3x-1)} + \dfrac{2x-1}{(x-5)(3x-1)} = \dfrac{6x^2-2x+2x-1}{(x-5)(3x-1)} = \dfrac{6x^2-1}{(x-5)(3x-1)}$

63. $\dfrac{4x}{x-y} - \dfrac{9}{x+y} = \dfrac{4x(x+y)}{(x-y)(x+y)} - \dfrac{9(x-y)}{(x-y)(x+y)} = \dfrac{4x^2+4xy-(9x-9y)}{(x-y)(x+y)} = \dfrac{4x^2+4xy-9x+9y}{(x-y)(x+y)}$

65. $\dfrac{3}{(x-1)(x-2)} + \dfrac{4x}{(x+1)(x-2)} = \dfrac{3(x+1)}{(x-2)(x-1)(x+1)} + \dfrac{4x(x-1)}{(x-2)(x-1)(x+1)}$

$= \dfrac{3x+3+4x^2-4x}{(x-2)(x-1)(x+1)} = \dfrac{4x^2-x+3}{(x-2)(x-1)(x+1)}$

67. $\dfrac{3}{x^2-x-6} - \dfrac{2}{x^2+5x+6} = \dfrac{3}{(x-3)(x+2)} - \dfrac{2}{(x+3)(x+2)}$

$= \dfrac{3(x+3)}{(x-3)(x+2)(x+3)} - \dfrac{2(x-3)}{(x-3)(x+2)(x+3)} = \dfrac{3x+9-2x+6}{(x-3)(x+2)(x+3)}$

$$= \frac{x+15}{(x-3)(x+2)(x+3)}$$

69. $\dfrac{3}{x^2-2x+1}+\dfrac{1}{x^2-3x+2}=\dfrac{3(x-2)}{(x-2)(x-1)^2}+\dfrac{x-1}{(x-2)(x-1)^2}=\dfrac{3x-6+x-1}{(x-2)(x-1)^2}=\dfrac{4x-7}{(x-2)(x-1)^2}$

71. $\dfrac{3x}{x^2+2x-3}+\dfrac{1}{x^2-2x+1}=\dfrac{3x(x-1)}{(x+3)(x-1)(x-1)}+\dfrac{1(x+3)}{(x+3)(x-1)(x-1)}=\dfrac{3x^2-2x+3}{(x+3)(x-1)^2}$

73. $\dfrac{4c}{ab}+\dfrac{3b}{ac}-\dfrac{2a}{bc}=\dfrac{4c^2}{abc}+\dfrac{3b^2}{abc}-\dfrac{2a^2}{abc}=\dfrac{4c^2+3b^2-2a^2}{abc}=\dfrac{-2a^2+3b^2+4c^2}{abc}$

75. $5-\dfrac{6}{n^2-36}+\dfrac{3}{n-6}=\dfrac{5(n-6)(n+6)}{(n-6)(n+6)}-\dfrac{6}{(n-6)(n+6)}+\dfrac{3(n+6)}{(n-6)(n+6)}$

$$= \frac{5n^2-180-6+3n+18}{(n-6)(n+6)}=\frac{5n^2+3n-168}{(n-6)(n+6)}$$

77. $\dfrac{3}{x-5}-\dfrac{1}{x-3}-\dfrac{2x}{x-5}=\dfrac{3(x-3)}{(x-5)(x-3)}-\dfrac{x-5}{(x-5)(x-3)}-\dfrac{2x(x-3)}{(x-5)(x-3)}=\dfrac{-2(x^2-4x+2)}{(x-5)(x-3)}$

79. $\dfrac{5}{2x-3}+\dfrac{x}{x+1}-\dfrac{x}{2x-3}=\dfrac{5(x+1)}{(x+1)(2x-3)}+\dfrac{x(2x-3)}{(x+1)(2x-3)}-\dfrac{x(x+1)}{(x+1)(2x-3)}$

$$= \frac{5x+5+2x^2-3x-x^2-x}{(x+1)(2x-3)}=\frac{x^2+x+5}{(x+1)(2x-3)}$$

81. $\dfrac{1}{x-1}-\dfrac{2}{x+1}+\dfrac{x}{x^2-1}=\dfrac{1(x+1)}{(x-1)(x+1)}-\dfrac{2(x-1)}{(x-1)(x+1)}+\dfrac{x}{(x-1)(x+1)}$

$$= \frac{x+1-2x+2+x}{(x-1)(x+1)}=\frac{3}{(x-1)(x+1)}$$

83. $\dfrac{a-b}{a+b}+\dfrac{a+b}{a-b}=\dfrac{(a-b)(a-b)}{(a+b)(a-b)}+\dfrac{(a+b)(a+b)}{(a+b)(a-b)}=\dfrac{a^2-2ab+b^2+a^2+2ab+b^2}{(a+b)(a-b)}$

$$= \frac{2a^2+2b^2}{(a+b)(a-b)}$$

Addition and Subtraction of Functions

85. $(f+g)(x)=\dfrac{1}{x+1}+\dfrac{x}{x+1}=\dfrac{1+x}{x+1}=1$

$(f-g)(x)=\dfrac{1}{x+1}-\dfrac{x}{x+1}=\dfrac{1-x}{x+1}$

87. $(f+g)(x)=\dfrac{1}{x+2}+\dfrac{1}{x-2}=\dfrac{x-2}{(x-2)(x+2)}+\dfrac{x+2}{(x-2)(x+2)}=\dfrac{2x}{(x-2)(x+2)}$

$(f-g)(x)=\dfrac{1}{x+2}-\dfrac{1}{x-2}=\dfrac{x-2}{(x-2)(x+2)}-\dfrac{x+2}{(x-2)(x+2)}=\dfrac{-4}{(x-2)(x+2)}$

Applications

89. Find the LCM of 12 and $30:12 = 2 \cdot 2 \cdot 3$ and $30 = 2 \cdot 3 \cdot 5$, so the LCM is $2 \cdot 2 \cdot 3 \cdot 5 = 60$. The planets will

be in alignment in 60 years.

91. $\dfrac{1}{R} = \dfrac{1}{80} + \dfrac{1}{300} = \dfrac{30}{2400} + \dfrac{8}{2400} = \dfrac{38}{2400} = \dfrac{19}{1200}$ and so $R = \dfrac{1200}{19} \approx 63.2$ ohms

93. $\dfrac{1}{S} = \dfrac{1}{0.1} - \dfrac{1}{10} \Rightarrow \dfrac{1}{S} = 10 - 0.1 \Rightarrow \dfrac{1}{S} = 9.9 \Rightarrow S = \dfrac{1}{9.9} \approx 0.101$ feet, or about 1.2 inches.

95. $\dfrac{8}{d^2} + \dfrac{8}{d^2} = \dfrac{16}{d^2}$ W/m^2

6.4: Rational Equations

Concepts

1. Multiply each side by $x + 2$.

2. Multiply each side by the LCD.

3. $\dfrac{1}{2}$

4. bc

5. LCD

6. intersect

Solving Rational Equations

7. The factored denominators are x and 5. The LCD is $5x$.

9. The factored denominators are $x - 1$ and $(x-1)(x+1)$. The LCD is $(x-1)(x+1) = x^2 - 1$.

11. The factored denominators are 2 and $2x + 1$ and $2(x-2)$. The LCD is $2(2x+1)(x-2)$.

13. The LCD is 6. The first step is to multiply both sides of the equation by the LCD.

$$6 \cdot \left(\dfrac{x}{3} + \dfrac{1}{2} \right) = \dfrac{5}{6} \cdot 6 \Rightarrow 2x + 3 = 5 \Rightarrow 2x = 2 \Rightarrow x = \dfrac{2}{2} = 1$$

15. The LCD is $15x$. The first step is to multiply both sides of the equation by the LCD.

$$15x \cdot \left(\dfrac{2}{x} - \dfrac{7}{3} \right) = -\dfrac{29}{15} \cdot 15x \Rightarrow 30 - 35x = -29x \Rightarrow 30 = 6x \Rightarrow x = \dfrac{30}{6} = 5$$

17. The LCD is $3(x-1)$. The first step is to multiply both sides of the equation by the LCD.

$$3(x-1) \cdot \dfrac{x}{x-1} = \dfrac{4}{3} \cdot 3(x-1) \Rightarrow 3x = 4x - 4 \Rightarrow x = 4$$

19. The LCD is $6x$. The first step is to multiply both sides of the equation by the LCD.

$$6x \cdot \dfrac{1}{2x} - 6x \cdot \dfrac{5}{3x} = 1 \cdot 6x \Rightarrow 3 - 10 = 6x \Rightarrow -7 = 6x \Rightarrow x = -\dfrac{7}{6}$$

21. The LCD is $x + 1$. The first step is to multiply both sides of the equation by the LCD.

$$(x+1)\cdot\frac{1}{x+1}-(x+1)(1)=\frac{3}{x+1}\cdot(x+1)\Rightarrow 1-x-1=3\Rightarrow -x=3\Rightarrow x=-3$$

23. The LCD is $x-3$. The first step is to multiply both sides of the equation by the LCD.

$$(x-3)\cdot\frac{1}{x-3}+(x-3)\cdot\frac{x}{x-3}=\frac{2x}{x-3}\cdot(x-3)\Rightarrow 1+x=2x\Rightarrow x=1$$

25. The LCD is $(x-1)(x+4)$. The first step is to multiply both sides of the equation by the LCD.

$$(x-1)(x+4)\cdot\frac{3}{x-1}=\frac{6}{x+4}\cdot(x-1)(x+4)\Rightarrow 3x+12=6x-6\Rightarrow 18=3x\Rightarrow x=\frac{18}{3}=6$$

27. The LCD is $(3z+4)(2z-5)$. The first step is to multiply both sides of the equation by the LCD.

$$(3z+4)(2z-5)\cdot\frac{6}{3z+4}=\frac{4}{2z-5}\cdot(3z+4)(2z-5)\Rightarrow 12z-30=12z+16\Rightarrow -30=16\Rightarrow \text{ No solutions.}$$

29. The LCD is $(t-1)(t+2)$. The first step is to multiply both sides of the equation by the LCD.

$$(t-1)(t+2)\cdot\frac{5}{t-1}+(t-1)(t+2)\cdot\frac{2}{t+2}=\frac{15}{t^2+t-2}\cdot(t-1)(t+2)\Rightarrow$$

$$5t+10+2t-2=15\Rightarrow 7t+8=15\Rightarrow 7t=7\Rightarrow t=1$$

Some expressions in the original equation are not defined when $t=1$. No solutions.

31. The LCD is $6n$. The first step is to multiply both sides of the equation by the LCD.

$$6n\cdot\frac{1}{3n}-6n\cdot 2=\frac{1}{2n}\cdot 6n\Rightarrow 2-12n=3\Rightarrow -12n=1\Rightarrow n=-\frac{1}{12}$$

33. The LCD is x^2. The first step is to multiply both sides of the equation by the LCD.

$$x^2\cdot\left(\frac{1}{x}+\frac{1}{x^2}\right)=2\cdot x^2\Rightarrow x+1=2x^2\Rightarrow 2x^2-x-1=0\Rightarrow(2x+1)(x-1)=0\Rightarrow x=-\frac{1}{2}\text{ or }1$$

35. The LCD is $(x+2)(x-3)$. The first step is to multiply both sides of the equation by the LCD.

$$(x+2)(x-3)\cdot\frac{x}{x+2}=\frac{4}{x-3}\cdot(x+2)(x-3)\Rightarrow x^2-3x=4x+8\Rightarrow x^2-7x-8=0\Rightarrow$$

$$(x+1)(x-8)=0\Rightarrow x=-1\text{ or }8$$

37. $$\frac{2w+1}{3w}-\frac{4w-3}{w}=0\Rightarrow 3w\cdot\frac{2w+1}{3w}-3w\cdot\frac{(4w-3)}{w}=0\cdot 3w\Rightarrow 2w+1-12w+9=0\Rightarrow$$

$$-10w+10=0\Rightarrow -10w=-10\Rightarrow w=1$$

39. $$\frac{3}{2y}+\frac{2y}{y-4}=-\frac{11}{2}\Rightarrow 2y(y-4)\cdot\frac{3}{2y}+2y(y-4)\cdot\frac{2y}{y-4}=-\frac{11}{2}\cdot 2y(y-4)\Rightarrow$$

$$3y-12+4y^2=-11y^2+44y\Rightarrow 15y^2-41y-12=0\Rightarrow(15y+4)(y-3)=0\Rightarrow y=-\frac{4}{15}\text{ or }y=3$$

41. $$\frac{1}{(x-1)^2}+\frac{3}{x^2-1}=\frac{5}{x^2-1}\Rightarrow 1(x+1)+3(x-1)=5(x-1)\Rightarrow x+1+3x-3=5x-5\Rightarrow 4x-2=5x-5\Rightarrow x=3$$

43. $\dfrac{x^2+x-2}{x-2}-\dfrac{4}{x-2}=0\Rightarrow\dfrac{x^2+x-2-4}{x-2}=0\Rightarrow\dfrac{x^2+x-6}{x-2}=0$

$\Rightarrow\dfrac{(x+3)(x-2)}{x-2}=0\Rightarrow x+3=0\Rightarrow x=-3$

45. $\dfrac{4}{x+3}=\dfrac{x}{3-x}=\dfrac{18}{x^2-9}\Rightarrow(x-3)(x+3)\cdot\dfrac{4}{x+3}-(x-3)(x+3)\dfrac{-x}{x-3}$

$=\dfrac{18}{(x-3)(x+3)}\cdot(x-3)(x+3)\Rightarrow4(x-3)-(-x)(x+3)=18$

$\Rightarrow4x-12+x^2+3x=18\Rightarrow x^2+7x-30=0\Rightarrow(x+10)(x-3)=0$

$\Rightarrow x=-10$ ($x=3$ makes expressions in the original equation undefined)

47. $\dfrac{1}{x}-4x=0\Rightarrow x\cdot\dfrac{1}{x}-x\cdot4x=0\cdot x\Rightarrow1-4x^2=0$

$\Rightarrow(1-2x)(1+2x)=0\Rightarrow x=\dfrac{1}{2}\text{ or }x=-\dfrac{1}{2}$

49. $\dfrac{a}{x}-\dfrac{b}{x}=c\Rightarrow\dfrac{a-b}{x}=c\Rightarrow x\cdot\dfrac{a-b}{x}=c\cdot x$

$\Rightarrow a-b=cx\Rightarrow\dfrac{a-b}{c}=x$

51. $y_1=y_2$ when $x=1$. By substitution of $x=1$, this answer checks.

53. $y_1=y_2$ when $x=2$. By substitution of $x=2$, this answer checks

55.   Graph $Y_1=1/(X-1)$ and $Y_2=1/2$ in [–4.7, 4.7, 1] by [–3.1, 3.1, 1]. See Figure 55.

The solution is the x-coordinate of the intersection point, $x=3$.

[–4.7, 4.7, 1] by [–3.1, 3.1, 1]

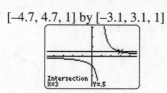

Figure 55

57.   Graph $Y_1=3/(X+2)$ and $Y_2=X$ in [–4.7, 4.7, 1] by [–6.2, 6.2, 1]. See Figures 57a & 57b.

The solutions are the x-coordinates of the intersection points, $x=-3$ or $x=1$.

[–4.7, 4.7, 1] by [–6.2, 6.2, 1]   [–4.7, 4.7, 1] by [–6.2, 6.2, 1]

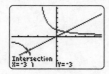   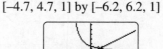

Figure 57a   Figure 57b

59.   Graph $Y_1 = 1/(X-2)$ and $Y_2 = 2$ in $[-4.7, 4.7, 1]$ by $[-3.1, 3.1, 1]$. See Figure 59.

The solution is the x-coordinate of the intersection point, $x = 2.5$.

$[-4.7, 4.7, 1]$ by $[-3.1, 3.1, 1]$

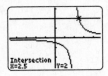

Figure 59

61.   Graph $Y_1 = 1/X + 1/X^2$ and $Y_2 = 15/4$ in $[-4.7, 4.7, 1]$ by $[-6.2, 6.2, 1]$. See Figures 61a & 61b.

The solutions are the x-coordinates of the intersection points, $x = -0.4$ or $x \approx 0.67$.

$[-4.7, 4.7, 1]$ by $[-6.2, 6.2, 1]$ $[-4.7, 4.7, 1]$ by $[-6.2, 6.2, 1]$

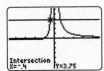   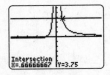

Figure 61a   Figure 61b

63.   Graph $Y_1 = 1/(X+2) - 1/(X-2)$ and $Y_2 = 4/3$ in $[-4.7, 4.7, 1]$ by $[-3.1, 3.1, 1]$. See Figures 63a & 63b.

The solutions are the x-coordinates of the intersection points, $x = -1$ or $x = 1$.

$[-4.7, 4.7, 1]$ by $[-3.1, 3.1, 1]$ $[-4.7, 4.7, 1]$ by $[-3.1, 3.1, 1]$

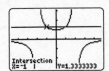   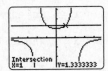

Figure 63a   Figure 63b

Using More Than One Method

65.   Symbolic: The LCD is $3x$. The first step is to multiply both sides of the equation by the LCD.

$$3x \cdot \left(\frac{1}{x} + \frac{2}{3} \right) = 1 \cdot 3x \Rightarrow 3 + 2x = 3x \Rightarrow 3 = x \Rightarrow x = 3$$

Graphical: Graph $Y_1 = 1/X + 2/3$ and $Y_2 = 1$ in $[-4.7, 4.7, 1]$ by $[-1.55, 1.55, 1]$. See Figure 65a.

The solution is the x-coordinate of the intersection point, $x = 3$.

Numerical: Table $Y_1 = 1/X + 2/3$ and $Y_2 = 1$ with TblStart $= 0$ and ΔTbl $= 1$. See Figure 65b.

The solution is the x-value when $Y_1 = Y_2$, $x = 3$.

$[-4.7, 4.7, 1]$ by $[-1.55, 1.55, 1]$

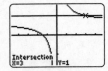   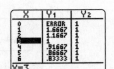

Figure 65a   Figure 65b

67.   Symbolic: The LCD is $3x(x+2)$. The first step is to multiply both sides of the equation by the LCD.

$$3x(x+2)\cdot\left(\frac{1}{x}+\frac{1}{x+2}\right)=\frac{4}{3}\cdot 3x(x+2)\Rightarrow 3x+6+3x=4x^2+8x\Rightarrow 4x^2+2x-6=0\Rightarrow$$

$$2(2x+3)(x-1)=0\Rightarrow x=-\frac{3}{2}\text{ or }x=1$$

Graphical: Graph $Y_1=1/X+1/(X+2)$ and $Y_2=4/3$ in $[-4.7, 4.7, 1]$ by $[-3.1, 3.1, 1]$. See Figures 67a & 67b.

The solutions are the x-coordinates of the intersection points, $x=-\frac{3}{2}$ or $x=1$.

Numerical: Table $Y_1=1/X+1/(X+2)$ and $Y_2=4/3$ with TblStart $=-2$ and ΔTbl $=0.5$. See Figure 67c.

The solutions are the x-values when $Y_1=Y_2$, $x=-\frac{3}{2}$ or $x=1$.

$[-4.7, 4.7, 1]$ by $[-3.1, 3.1, 1]$ $[-4.7, 4.7, 1]$ by $[-3.1, 3.1, 1]$

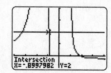   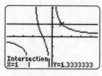   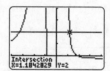

Figure 67a   Figure 67b   Figure 67c

69.   Graph $Y_1=X/(X^2-1)-3/(X+2)$ and $Y_2=2$ in $[-4, 4, 1]$ by $[-1, 4, 1]$. The solutions are the x-coordinates

of the intersection points, approximately $x=-3.28$, $x=-0.90$, or $x=1.18$.

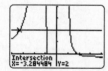      

Figure 69a   Figure 69b   Figure 69c

Solving an Equation for a Variable

71. $t=\dfrac{d}{r}\Rightarrow rt=d\Rightarrow r=\dfrac{d}{t}\Rightarrow$

73. $h=\dfrac{2A}{b}\Rightarrow bh=2A\Rightarrow b=\dfrac{2A}{h}$

75. $\dfrac{1}{2a}=\dfrac{1}{b}\Rightarrow 2a=b\Rightarrow a=\dfrac{b}{2}$

77. $\dfrac{1}{R}=\dfrac{1}{R_1}+\dfrac{1}{R_2}\Rightarrow R_1R_2=RR_2+RR_1\Rightarrow R_1R_2-RR_1=RR_2\Rightarrow R_1(R_2-R)=RR_2\Rightarrow R_1=\dfrac{RR_2}{R_2-R}$

79. $T=\dfrac{1}{15-x}\Rightarrow T(15-x)=1\Rightarrow 15-x=\dfrac{1}{T}\Rightarrow -x=\dfrac{1}{T}-15\Rightarrow x=15-\dfrac{1}{T}$

81. $\dfrac{1}{r}=\dfrac{1}{t+1}\Rightarrow t+1=r\Rightarrow t=r-1$

83. $\dfrac{1}{r} = \dfrac{a}{a+b} \Rightarrow a+b = ar \Rightarrow b = ar - a \Rightarrow b = a(r-1)$

Applications

85.   (a) $N = 8 \Rightarrow 8 = \dfrac{x^2}{3600 - 60x} \Rightarrow (3600 - 60x)(8) = \dfrac{x^2}{3600 - 60x} \cdot (3600 - 60x)$

$\Rightarrow 28,800 - 480x = x^2 \Rightarrow x^2 + 480x - 28,800 = 0$

Graph $Y_1 = X^2 + 480X - 28,800$ and $Y_2 = 0$ in [0, 75, 5] by [–10,000, 10,000, 1000]. The solution is the

x-coordinate of the intersection point, $x \approx 54$ cars. See Figure 85.

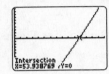

Figure 85

(b)   If cars arrive, on average, at 54 per hour, the line will be, on average, 8 cars long.

87.   (a) $\dfrac{x}{5} + \dfrac{x}{2} = 1$

(b)   The LCD is 10. The first step is to multiply both sides of the equation by the LCD.

$10 \cdot \left(\dfrac{x}{5} + \dfrac{x}{2} \right) = 1 \cdot 10 \Rightarrow 2x + 5x = 10 \Rightarrow 7x = 10 \Rightarrow x = \dfrac{10}{7}$ hours.

(c)   Graph $Y_1 = X/5 + X/2$ and $Y_2 = 1$ in [–4.7, 4.7, 1] by [–3.1, 3.1, 1]. See Figure 87.

The solution is the x-coordinate of the intersection point, $x \approx 1.43$ hours.

[–4.7, 4.7, 1] by [–3.1, 3.1, 1]

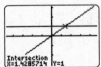

Figure 87

89.   Let x represent the speed of the winner. Then $x - 2$ represents the speed of the second-place runner.

Since $t = \dfrac{d}{r}$, the time needed for the winner to finish is $\dfrac{5}{x}$ and the time for the second-place runner is $\dfrac{5}{x-2}$.

Since the winner finished 7.5 minutes $\left(\dfrac{1}{8} \text{ hour} \right)$ ahead of second place, the equation is $\dfrac{5}{x-2} - \dfrac{5}{x} = \dfrac{1}{8}$.

The LCD is $8x(x-2)$. The first step is to multiply both sides of the equation by the LCD.

$8x(x-2) \cdot \left(\dfrac{5}{x-2} - \dfrac{5}{x} \right) = \dfrac{1}{8} \cdot 8x(x-2) \Rightarrow 40x - 40x + 80 = x^2 - 2x \Rightarrow x^2 - 2x - 80 = 0 \Rightarrow$

$(x-10)(x+8) = 0 \Rightarrow x = 10$ or $x = -8$

Since the value $x = -8$ has no meaning, the solutions are winner: 10 mph, second-place: $10 - 2 = 8$ mph.

91. $\dfrac{1}{10,000} = \dfrac{0.625}{H} \Rightarrow H = 0.625(10,000) = 6250$ feet.

93.   Let x represent the speed of the current. Then $15 - x$ represents the boat's upstream speed and $15 + x$

represents the boat's downstream speed. Since $t = \dfrac{d}{r}$, the equation is $\dfrac{36}{15-x} + \dfrac{36}{15+x} = 5$.

The LCD is $(15-x)(15+x)$. The first step is to multiply both sides of the equation by the LCD.

$$(15-x)(15+x) \cdot \left(\dfrac{36}{15-x} + \dfrac{36}{15+x} \right) = 5 \cdot (15-x)(15+x) \Rightarrow 540 + 36x + 540 - 36x = 1125 - 5x^2$$

$$\Rightarrow 5x^2 - 45 = 0 \Rightarrow 5x^2 = 45 \Rightarrow x^2 = 9 \Rightarrow x = \pm\sqrt{9} \Rightarrow x = 3 \text{ mph (the value } x = -3 \text{ has no meaning)}$$

95.   Let x represent the plane's speed with no wind. Then $x + 50$ represents the plane's speed with the wind and

$x - 50$ represents the plane's speed into the wind. Since $t = \dfrac{d}{r}$, the equation is $\dfrac{675}{x+50} = \dfrac{450}{x-50}$.

The LCD is $(x+50)(x-50)$. The first step is to multiply both sides of the equation by the LCD.

$$(x+50)(x-50) \cdot \dfrac{675}{x+50} = \dfrac{450}{x-50} \cdot (x+50)(x-50) \Rightarrow 675x - 33,750 = 450x + 22,500 \Rightarrow$$

$225x = 56,250 \Rightarrow x = 250$ mph

97.   Let x and $x + 3$ represent the time needed by the faster employee and the slower employee respectively.

$$\dfrac{1}{x} + \dfrac{1}{x+3} = \dfrac{1}{2} \Rightarrow 2(x+3) + 2x = x(x+3) \Rightarrow 2x + 6 + 2x = x^2 + 3x \Rightarrow x^2 - x - 6 = 0 \Rightarrow$$

$(x-3)(x+2) = 0 \Rightarrow x = 3$ or $x = -2$. The value $x = -2$ has no physical meaning.

The faster employee can mow the field in 3 hours. The slower employee takes 6 hours.

99.   Let x and $x - 1$ represent the speed of the faster person and the slower person respectively.

$$\dfrac{12}{x} = \dfrac{9}{x-1} \Rightarrow 12x - 12 = 9x \Rightarrow 3x = 12 \Rightarrow x = 4$$

The faster parson walks 4 mph. The slower person walks 3 mph.

101.   Let x represent the time needed to empty the pool.

$$\dfrac{x}{40} - \dfrac{x}{60} = 1 \Rightarrow 3x - 2x = 120 \Rightarrow x = 120 \text{ hours. The pool is empty in 120 hours.}$$

103.   Yes; the value is slightly greater than 2.

Checking Basic Concepts for Sections 6.3 & 6.4

1.   (a) $\dfrac{x}{x^2-1} + \dfrac{1}{x^2-1} = \dfrac{x+1}{(x-1)(x+1)} = \dfrac{1}{x-1}$

(b) $\dfrac{1}{x-2} - \dfrac{3}{x} = \dfrac{x}{x(x-2)} - \dfrac{3(x-2)}{x(x-2)} = \dfrac{x-(3x-6)}{x(x-2)} = \dfrac{-2x+6}{x(x-2)} = \dfrac{-2(x-3)}{x(x-2)} = -\dfrac{2(x-3)}{x(x-2)}$

(c) $\dfrac{1}{x(x-1)}+\dfrac{1}{x^2-1}-\dfrac{2}{x(x+1)}=\dfrac{1(x+1)+1x-2(x-1)}{x(x-1)(x+1)}=\dfrac{3}{x(x-1)(x+1)}$

2. (a) $\dfrac{6}{x}-\dfrac{1}{2}=1\Rightarrow 12-x=2x\Rightarrow 12=3x\Rightarrow x=4$

(b) $\dfrac{3}{2x-1}=\dfrac{2}{x+1}\Rightarrow 3(x+1)=2(2x-1)\Rightarrow 3x+3=4x-2\Rightarrow x=5$

(c) $\dfrac{2}{x-1}+\dfrac{3}{x+2}=\dfrac{x}{x^2+x-2}\Rightarrow 2(x+2)+3(x-1)=x\Rightarrow 5x+1=x\Rightarrow 4x=-1\Rightarrow x=-\dfrac{1}{4}$

3. $\dfrac{x}{12}+\dfrac{x}{28}=1\Rightarrow 7x+3x=84\Rightarrow 10x=84\Rightarrow x=8.4\ \text{min}$

4. $\dfrac{1}{S}=\dfrac{1}{F}-\dfrac{1}{D}\Rightarrow DF=SD-SF\Rightarrow DF=S(D-F)\Rightarrow \dfrac{DF}{D-F}=S\Rightarrow S=\dfrac{DF}{D-F}$

6.5: Complex Fractions

Concepts

1. $\dfrac{5}{7}\cdot\dfrac{11}{3}=\dfrac{55}{21}$

2. $\dfrac{a}{b}\cdot\dfrac{d}{c}$

3. Multiply both the numerator and the denominator by $x-1$. *Answers may vary.*

4. A complex fraction is a rational expression with fractions in its numerator, denominator or both.

5. $\dfrac{z+\dfrac{3}{4}}{z-\dfrac{3}{4}}$

6. $\dfrac{\dfrac{a}{b}}{\dfrac{a-b}{a+b}}$

Simplifying Complex Fractions

7. $\dfrac{\dfrac{1}{5}}{\dfrac{4}{7}}=\dfrac{1}{5}\cdot\dfrac{7}{4}=\dfrac{7}{20}$

9. $\dfrac{1+\dfrac{1}{3}}{1-\dfrac{1}{3}}=\dfrac{1+\dfrac{1}{3}}{1-\dfrac{1}{1}}\cdot\dfrac{3}{3}=\dfrac{3+1}{3-1}=\dfrac{4}{2}=2$

11. $\dfrac{2+\dfrac{2}{3}}{2-\dfrac{1}{4}} = \dfrac{2+\dfrac{2}{3}}{2-\dfrac{1}{4}} \cdot \dfrac{12}{12} = \dfrac{24+8}{24-3} = \dfrac{32}{21}$

13. $\dfrac{\dfrac{a}{b}}{\dfrac{3a}{2b^2}} = \dfrac{a}{b} \cdot \dfrac{2b^2}{3a} = \dfrac{2b}{3}$

15. $\dfrac{\dfrac{x}{2y}}{\dfrac{2x}{3y}} = \dfrac{x}{2y} \cdot \dfrac{3y}{2x} = \dfrac{3}{4}$

17. $\dfrac{\dfrac{8}{n+1}}{\dfrac{4}{n-1}} = \dfrac{8}{n+1} \cdot \dfrac{n-1}{4} = \dfrac{2(n-1)}{n+1}$

19. $\dfrac{\dfrac{2k+3}{k}}{\dfrac{k-4}{k}} = \dfrac{2k+3}{k} \cdot \dfrac{k}{k-4} = \dfrac{2k+3}{k-4}$

21. $\dfrac{\dfrac{3}{z^2-4}}{\dfrac{z}{z^2-4}} = \dfrac{3}{z^2-4} \cdot \dfrac{z^2-4}{z} = \dfrac{3}{z}$

23. $\dfrac{\dfrac{x}{x^2-16}}{\dfrac{1}{x-4}} = \dfrac{x}{x^2-16} \cdot \dfrac{x-4}{1} = \dfrac{x(x-4)}{x^2-16} = \dfrac{x(x-4)}{(x-4)(x+4)} = \dfrac{x}{x+4}$

25. $\dfrac{1+\dfrac{1}{x}}{x+1} = \dfrac{1+\dfrac{1}{x}}{x+1} \cdot \dfrac{x}{x} = \dfrac{x+1}{x(x+1)} = \dfrac{1}{x}$

27. $\dfrac{\dfrac{1}{x-3}}{\dfrac{1}{x}-\dfrac{3}{x-3}} = \dfrac{\dfrac{1}{x-3}}{\dfrac{1}{x}-\dfrac{3}{x-3}} \cdot \dfrac{x(x-3)}{x(x-3)} = \dfrac{x}{x-3-3x} = \dfrac{x}{-2x-3} = -\dfrac{x}{2x+3}$

29. $\dfrac{\dfrac{1}{x}+\dfrac{2}{x^2}}{\dfrac{3}{x}-\dfrac{1}{x^2}} = \dfrac{\dfrac{1}{x}+\dfrac{2}{x^2}}{\dfrac{3}{x}-\dfrac{1}{x^2}} \cdot \dfrac{x^2}{x^2} = \dfrac{x+2}{3x-1}$

31. $\dfrac{\dfrac{1}{x+3}+\dfrac{2}{x-3}}{2-\dfrac{1}{x-3}} = \dfrac{\dfrac{1}{x+3}+\dfrac{2}{x-3}}{2-\dfrac{1}{x-3}} \cdot \dfrac{(x-3)(x+3)}{(x-3)(x+3)} = \dfrac{x-3+2(x+3)}{2(x-3)(x+3)-(x+3)} = \dfrac{3(x+1)}{(x+3)(2x-7)}$

33. $\dfrac{\dfrac{4}{x-5}}{\dfrac{1}{x+5}+\dfrac{1}{x}} = \dfrac{\dfrac{4}{x-5}}{\dfrac{1}{x+5}+\dfrac{1}{x}} \cdot \dfrac{x(x-5)(x+5)}{x(x-5)(x+5)} = \dfrac{4x(x+5)}{x(x-5)+(x-5)(x+5)} = \dfrac{4x(x+5)}{(x-5)(2x+5)}$

35. $\dfrac{\dfrac{1}{p^2q}+\dfrac{1}{pq^2}}{\dfrac{1}{p^2q}-\dfrac{1}{pq^2}} = \dfrac{\dfrac{1}{p^2q}+\dfrac{1}{pq^2}}{\dfrac{1}{p^2q}-\dfrac{1}{pq^2}} \cdot \dfrac{p^2q^2}{p^2q^2} = \dfrac{q+p}{q-p} = \dfrac{p+q}{q-p}$

37. $\dfrac{\dfrac{1}{a}+\dfrac{1}{b}}{\dfrac{1}{b}-\dfrac{1}{a}} = \dfrac{\dfrac{1}{a}+\dfrac{1}{b}}{\dfrac{1}{b}-\dfrac{1}{a}} \cdot \dfrac{ab}{ab} = \dfrac{b+a}{a-b} = \dfrac{a+b}{a-b}$

39. $\dfrac{\dfrac{1}{x}+\dfrac{1}{x+1}}{\dfrac{2}{x+1}-\dfrac{1}{x+1}} = \dfrac{\dfrac{1}{x}+\dfrac{1}{x+1}}{\dfrac{1}{x+1}} = \dfrac{\dfrac{1}{x}+\dfrac{1}{x+1}}{\dfrac{1}{x+1}} \cdot \dfrac{x(x+1)}{x(x+1)} = \dfrac{x+1+x}{x} = \dfrac{2x+1}{x}$

$= \dfrac{-2(2x-3)(x-2)}{2x-3} = -2(x-2)$

41. $\dfrac{3^{-1}-4^{-1}}{5^{-1}+4^{-1}} = \dfrac{\dfrac{1}{3}-\dfrac{1}{4}}{\dfrac{1}{5}+\dfrac{1}{4}} = \dfrac{\dfrac{4}{12}-\dfrac{3}{12}}{\dfrac{4}{20}+\dfrac{5}{20}} = \dfrac{\dfrac{1}{12}}{\dfrac{9}{20}} = \dfrac{1}{12} \cdot \dfrac{20}{9} = \dfrac{5}{27}$

43. $\dfrac{m^{-1}-2n^{-2}}{1+(mn)^{-2}} = \dfrac{\dfrac{1}{m}-\dfrac{2}{n^2}}{1+\dfrac{1}{m^2n^2}} = \dfrac{\dfrac{1}{m}-\dfrac{2}{n^2}}{1+\dfrac{1}{m^2n^2}} \cdot \dfrac{m^2n^2}{m^2n^2} = \dfrac{mn^2-2m^2}{m^2n^2+1}$

45. $\dfrac{1-(2n+1)^{-1}}{1+(2n+1)^{-1}} = \dfrac{1-\dfrac{1}{2n+1}}{1+\dfrac{1}{2n+1}} = \dfrac{1-\dfrac{1}{2n+1}}{1+\dfrac{1}{2n+1}} \cdot \dfrac{2n+1}{2n+1} = \dfrac{2n+1-1}{2n+1+1} = \dfrac{2n}{2(n+1)} = \dfrac{n}{n+1}$

47. $\dfrac{\dfrac{x}{x^2-4}-\dfrac{1}{x^2-4}}{\dfrac{1}{x+4}} = \dfrac{\dfrac{x-1}{x^2-4}}{\dfrac{1}{x+4}} = \dfrac{x-1}{x^2-4} \cdot \dfrac{x+4}{1} = \dfrac{(x-1)(x+4)}{x^2-4}$

Applications

49. $\dfrac{P\left(1+\dfrac{r}{12}\right)^{24}-P}{\dfrac{r}{12}}$

51. $R = \dfrac{1}{\dfrac{1}{R_1}+\dfrac{1}{R_2}} = \dfrac{1}{\dfrac{1}{R_1}+\dfrac{1}{R_2}} \cdot \dfrac{R_1R_2}{R_1R_2} = \dfrac{R_1R_2}{R_2+R_1} \Rightarrow R = \dfrac{R_1R_2}{R_1+R_2}$

6.6: Modeling with Proportions and Variation

Concepts

1. A proportion is a statement that two ratios are equal.

2. $\dfrac{5}{6} = \dfrac{x}{7}$

3. It doubles.

4. It is reduced by half.

5. constant

6. constant

7. kxy

8. $kx^2 y^3$

9. It would vary directly. If the number of people doubled, the food bill would double.

10. It would vary inversely. Increasing the number of painters would decrease the time needed to paint the building.

Proportions

11. $\dfrac{x}{14} = \dfrac{5}{7} \Rightarrow 7x = 70 \Rightarrow x = \dfrac{70}{10} = 10$

13. $\dfrac{8}{x} = \dfrac{2}{3} \Rightarrow 2x = 24 \Rightarrow x = \dfrac{24}{2} = 12$

15. $\dfrac{6}{13} = \dfrac{h}{156} \Rightarrow 13h = 936 \Rightarrow h = \dfrac{936}{13} = 72$

17. $\dfrac{7}{4z} = \dfrac{5}{3} \Rightarrow 20z = 21 \Rightarrow z = \dfrac{21}{20}$

19. $\dfrac{2}{3x+1} = \dfrac{5}{x} \Rightarrow 2x = 5(3x+1) \Rightarrow 2x = 15x + 5 \Rightarrow -13x = 5 \Rightarrow x = -\dfrac{5}{13}$

21. $\dfrac{4}{x} = \dfrac{x}{9} \Rightarrow 36 = x^2 \Rightarrow x^2 - 36 = 0 \Rightarrow (x-6)(x+6) = 0 \Rightarrow x = 6 \text{ or } x = -6$

23. (a) $\dfrac{7}{9} = \dfrac{10}{x}$

    (b) $\dfrac{7}{9} = \dfrac{10}{x} \Rightarrow 7x = 90 \Rightarrow x = \dfrac{90}{7}$

25. (a) $\dfrac{5}{3} = \dfrac{x}{6}$

    (b) $\dfrac{5}{3} = \dfrac{x}{6} \Rightarrow 3x = 30 \Rightarrow x = \dfrac{30}{3} = 10$

27. (a) $\dfrac{78}{6} = \dfrac{x}{8}$

(b) $\dfrac{78}{6} = \dfrac{x}{8} \Rightarrow 6x = 624 \Rightarrow x = \dfrac{624}{6} = \$104$

29. (a) $\dfrac{2}{90} = \dfrac{5}{x}$

    (b) $\dfrac{2}{90} = \dfrac{5}{x} \Rightarrow 2x = 450 \Rightarrow x = \dfrac{450}{2} = 225 \text{ min}$

Variation

31. (a) $y = kx \Rightarrow 6 = 3k \Rightarrow k = 2$

    (b) $y = 2x \Rightarrow y = 2(7) = 14$

33. (a) $y = kx \Rightarrow 5 = 2k \Rightarrow k = 2.5$

    (b) $y = 2.5x \Rightarrow y = 2.5(7) \Rightarrow 17.5$

35. (a) $y = kx \Rightarrow -120 = 16k \Rightarrow k = -7.5$

    (b) $y = -7.5x \Rightarrow y = -7.5(7) = -52.5$

37. (a) $y = \dfrac{k}{x} \Rightarrow 5 = \dfrac{k}{4} \Rightarrow k = 20$

    (b) $y = \dfrac{20}{x} \Rightarrow y = \dfrac{20}{10} = 2$

39. (a) $y = \dfrac{k}{x} \Rightarrow 100 = \dfrac{k}{\frac{1}{2}} \Rightarrow k = 50$

    (b) $y = \dfrac{50}{x} \Rightarrow y = \dfrac{50}{10} = 5$

41. (a) $y = \dfrac{k}{x} \Rightarrow 20 = \dfrac{k}{20} \Rightarrow k = 400$

    (b) $y = \dfrac{400}{x} \Rightarrow y = \dfrac{400}{10} = 40$

43. (a) $z = kxy \Rightarrow 6 = 3(8)k \Rightarrow 6 = 24k \Rightarrow k = 0.25$

    (b) $z = 0.25xy \Rightarrow z = 0.25(5)(7) = 8.75$

45. (a) $z = kxy \Rightarrow 5775 = 25(21)k \Rightarrow 5775 = 525k \Rightarrow k = 11$

    (b) $z = 11xy \Rightarrow z = 11(5)(7) = 385$

47. (a) $z = kxy \Rightarrow 25 = \dfrac{1}{2}(5)k \Rightarrow 25 = \dfrac{5}{2}k \Rightarrow k = 10$

    (b) $z = 10xy \Rightarrow z = 10(5)(7) = 350$

49.   (a) Direct. The ratios $\dfrac{y}{x}$ always equal 1.5.

(b) $y = kx \Rightarrow 3 = 2k \Rightarrow k = 1.5$; The equation is $y = 1.5x$.

(c) See Figure 49.

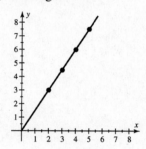

Figure 49

51.   (a) Neither. The products xy do not remain constant.

(b) N/A

(c) N/A

53.   (a) Neither. The ratios $\dfrac{y}{x}$ do not remain constant.

(b) N/A

(c) N/A

55.   (a) Direct. The ratios $\dfrac{y}{x}$ always equal $-2b$.

(b) $y = kx \Rightarrow -2b = 1k \Rightarrow k = -2b$. The equation is $y = -2bx$.

57.   Direct. The ratios $\dfrac{y}{x}$ always equal 1. $y = kx \Rightarrow 1 = 1k \Rightarrow k = 1$

59.   Neither. The ratios $\dfrac{y}{x}$ do not remain constant.

61.   Direct. The ratios $\dfrac{y}{x}$ always equal 2. $y = kx \Rightarrow 2 = 1k \Rightarrow k = 2$

Applications

63. $\dfrac{6}{7} = \dfrac{x}{27} \Rightarrow 7x = 162 \Rightarrow x = \dfrac{162}{7} \approx 23.1$ feet

65. $\dfrac{8}{1} = \dfrac{11}{x} \Rightarrow 8x = 11 \Rightarrow x = \dfrac{11}{8} = 1.375$ inches

67.   Let x be the total number of largemouth bass in the population.

$$\dfrac{300}{x} = \dfrac{17}{112} \Rightarrow 17x = 33,600 \Rightarrow x \approx 2000$$

69. $d = kx^2$; $300 = k(60)^2 \Rightarrow 300 = 3600k \Rightarrow k = \dfrac{1}{12}$; $d = \dfrac{1}{12}x^2$, $x = 30 \Rightarrow d = \dfrac{1}{12}(30)^2 = 75$ feet

71.   (a) Direct; The ratios $\dfrac{R}{W}$ always equal 0.012.

(b) $R = 0.012W$. See Figure 71.

(c) $R = 0.012(3200) = 38.4$ lb

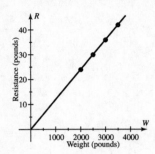

Figure 71

73.   (a) Direct; The ratios $\dfrac{G}{A}$ always equal 27.

(b) $G = 27A$. See Figure 73.

(c) For each square inch increase in the cross-sectional area of the hose, the flow increases by 27 gal/min.

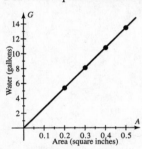

Figure 73

75.   (a) $F = \dfrac{k}{L} \Rightarrow 150 = \dfrac{k}{8} \Rightarrow k = 1200$ and so $F = \dfrac{1200}{L}$

(b) $F = \dfrac{1200}{20} = 60$ lb

77.   (a) Direct

(b) $y = kx \Rightarrow -95 = 5k \Rightarrow k = -19$ and so $y = -19x$

(c) It is negative. For each one mile increase in altitude the temperature decreases by 19°F.

(d) $y = -19(3.5) = -66.5$; The temperature decreases by 66.5°F.

79. $\dfrac{35}{2} = \dfrac{25}{x} \Rightarrow 35x = 50 \Rightarrow x \approx 1.43$ ohms

81. $z = kx^2 y^3 \Rightarrow 31.9 = (2)^2 (2.5)^3 k \Rightarrow 31.9 = 62.5k \Rightarrow k = 0.5104$ and so $z = 0.5104x^2 y^3$

83. $S = kwt^2 \Rightarrow 300 = k(5)(3)^2 \Rightarrow 300 = 45k \Rightarrow k = \dfrac{20}{3}$ and so $S = \dfrac{20}{3}wt^2$

When $w = 5$ and $t = 2$, $S = \dfrac{20}{3}(5)(2)^2 \approx 133$ lb.

85. $W_m = kW_E \Rightarrow 28 = 175k \Rightarrow k = 0.16$; so $W_m = 0.16W_E$

   A 220-pound person would weigh $W_m = 0.16(220) = 35.2$ pounds on the moon.

87. $V = kIR \Rightarrow 220 = (10)(22)k \Rightarrow k = 1$; so $V = IR$

   When $I = 15$ and $R = 50$ the voltage is $V = (15)(50) = 750$.

Checking Basic Concepts for Sections 6.5 & 6.6

1. (a) $\dfrac{3 - \dfrac{1}{x^2}}{3 + \dfrac{1}{x^2}} = \dfrac{3 - \dfrac{1}{x^2}}{3 + \dfrac{1}{x^2}} \cdot \dfrac{x^2}{x^2} = \dfrac{3x^2 - 1}{3x^2 + 1}$

   (b) $\dfrac{\dfrac{2}{x-1} - \dfrac{2}{x+1}}{\dfrac{4}{x^2 - 1}} = \dfrac{\dfrac{2}{x-1} - \dfrac{2}{x+1}}{\dfrac{4}{x^2 - 1}} \cdot \dfrac{(x-1)(x+1)}{(x-1)(x+1)} = \dfrac{2(x+1) - 2(x-1)}{4} = \dfrac{4}{4} = 1$

2. (a) $y = kx \Rightarrow 6 = 8k \Rightarrow k = 0.75$

   (b) $y = 0.75x \Rightarrow y = 0.75(11) = 8.25$

3. (a) Direct. The ratios $\dfrac{y}{x}$ always equal 0.4. The equation is $y = 0.4x$.

   (b) Inverse. The products xy always equal 120. The equation is $y = \dfrac{120}{x}$.

6.7: Division of Polynomials

Concepts

1. term

2. $\dfrac{a}{c} + \dfrac{b}{c}$

3. $5 \cdot 4 + 1$

4. $(3x - 5) \cdot (x - 2) + (-2)$

5. No, the divisor is not of the form $x - k$.

6. The remainder is the last number in the third row.

Division by a Monomial

7. $\dfrac{4x - 6}{2} = \dfrac{4x}{2} - \dfrac{6}{2} = 2x - 3$

   Table $Y_1 = (4X - 6)/2$ and $Y_2 = 2X - 3$ with TblStart $= 0$ and ΔTbl $= 1$. See Figure 7.

9. $\dfrac{6x^3-9x}{3x}=\dfrac{6x^3}{3x}-\dfrac{9x}{3x}=2x^2-3$

Table $Y_1=\left(6X^{\wedge}3-9X\right)/(3X)$ and $Y_2=2X^2-3$ with TblStart = 0 and ΔTbl = 1. See Figure 9.

11. $\left(4x^2-x+1\right)\div 2x^2=\dfrac{4x^2}{2x^2}-\dfrac{x}{2x^2}+\dfrac{1}{2x^2}=2-\dfrac{1}{2x}+\dfrac{1}{2x^2}$

Table $Y_1=\left(4X^2-X+1\right)/\left(2X^2\right)$ and $Y_2=2-1/(2X)+1/\left(2X^2\right)$ with TblStart = 0 and ΔTbl = 1. See Figure 11.

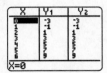   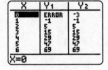   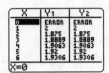

Figure 7   Figure 9   Figure 11

13. $\dfrac{9x^2-12x-3}{3}=\dfrac{9x^2}{3}-\dfrac{12x}{3}-\dfrac{3}{3}=3x^2-4x-1$

15. $\dfrac{12a^3-18a}{6a}=\dfrac{12a^3}{6a}-\dfrac{18a}{6a}=2a^2-3$

17. $\left(16x^3-24x\right)\div(12x)=\dfrac{16x^3}{12x}-\dfrac{24x}{12x}=\dfrac{4}{3}x^2-2$

19. $\left(a^2b^2-4ab+ab^2\right)\div ab=\dfrac{a^2b^2}{ab}-\dfrac{4ab}{ab}+\dfrac{ab^2}{ab}=ab-4+b$

21. $\dfrac{6m^4n^4+3m^2n^2-12}{3m^2n^2}=\dfrac{6m^4n^4}{3m^2n^2}+\dfrac{3m^2n^2}{3m^2n^2}-\dfrac{12}{3m^2n^2}=2m^2n^2+1-\dfrac{4}{m^2n^2}$

Division by a Polynomial

23.
$$
\begin{array}{r}
3x-7 \\
x-3\overline{\smash{\big)}\,3x^2-16x+21} \\
\underline{3x^2-9x} \\
-7x+21 \\
\underline{-7x+21} \\
0
\end{array}
$$

The solution is: $3x-7$

25.
$$
\begin{array}{r}
x^2-1 \\
2x+3\overline{\smash{\big)}\,2x^3+3x^2-2x-2} \\
\underline{2x^3+3x^2} \\
-2x-2 \\
\underline{-2x-3} \\
1
\end{array}
$$

The solution is: $x^2-1+\dfrac{1}{2x+3}$

27.

$$\begin{array}{r} 10x+10 \\ x-1\overline{\smash{\big)}\,10x^2+0x-5} \\ \underline{10x^2-10x} \\ 10x-5 \\ \underline{10x-10} \\ 5 \end{array}$$

The solution is: $10x+10+\dfrac{5}{x-1}$

29.

$$\begin{array}{r} 4x^2-1 \\ x+2\overline{\smash{\big)}\,4x^3+8x^2-x-2} \\ \underline{4x^3+8x^2} \\ 0-x-2 \\ \underline{-x-2} \\ 0 \end{array}$$

The solution is: $4x^2-1$

31.

$$\begin{array}{r} x^2-4x+19 \\ x+4\overline{\smash{\big)}\,x^3+0x^2+3x-4} \\ \underline{x^3+4x^2} \\ -4x^2+3x \\ \underline{-4x^2-16x} \\ 19x-4 \\ \underline{19x+76} \\ -80 \end{array}$$

The solution is: $x^2-4x+19-\dfrac{80}{x+4}$

33.

$$\begin{array}{r} x^2+3x-6 \\ 3x-1\overline{\smash{\big)}\,3x^3+8x^2-21x+7} \\ \underline{3x^3-x^2} \\ 9x^2-21x \\ \underline{9x^2-3x} \\ -18x+7 \\ \underline{-18x+6} \\ 1 \end{array}$$

The solution is: $x^2+3x-6+\dfrac{1}{3x-1}$

35.

$$\begin{array}{r} a^3-a \\ 2a+5\overline{\smash{\big)}\,2a^4+5a^3-2a^2-5a} \\ \underline{2a^4+5a^3} \\ 0-2a^2-5a \\ \underline{-2a^2-5a} \\ 0 \end{array}$$

The solution is: a^3-a

37.

$$
\begin{array}{r}
3x+4 \\
x^2+0x-4\overline{\smash{\big)}3x^3+4x^2-12x-16} \\
\underline{3x^3+0x^2-12x} \\
4x^2+0x-16 \\
\underline{4x^2+0x-16} \\
0
\end{array}
$$

The solution is: $3x+4$

39.

$$
\begin{array}{r}
x+1 \\
x^2+0x-1\overline{\smash{\big)}x^3+x^2-x+0} \\
\underline{x^3+0x^2-x} \\
x^2+0x+0 \\
\underline{x^2+0x-1} \\
1
\end{array}
$$

The solution is: $x+1+\dfrac{1}{x^2-1}$

41.

$$
\begin{array}{r}
2a^2-a+3 \\
a^2-a+5\overline{\smash{\big)}2a^4-3a^3+14a^2-8a+10} \\
\underline{2a^4-2a^3+10a^2} \\
-a^3+4a^2-8a+10 \\
\underline{-a^3+a^2-5a} \\
3a^2-3a+10 \\
\underline{3a^2-3a+15} \\
-5
\end{array}
$$

The solution is: $2a^2-a+3-\dfrac{5}{a^2-a+5}$

43. $\left(a^2+ab+b^2\right)(a-b)+1=a^3+a^2b+ab^2-a^2b-ab^2-b^3+1$

$= a^3-b^3+1$

Synthetic Division

45.

$$
\begin{array}{r}
1\underline{|} \quad\; 1 \quad 3 \quad -1 \\
\quad\quad 1 \quad\; 4 \\
\overline{\quad 1 \quad 4 \quad\;\; 3}
\end{array}
$$

The solution is: $x+4+\dfrac{3}{x-1}$

47.

$$
\begin{array}{r}
7\underline{|} \quad\; 3 \quad -22 \quad 7 \\
\quad\quad 21 \quad -7 \\
\overline{\quad 3 \quad\; -1 \quad\; 0}
\end{array}
$$

The solution is: $3x-1$

49.

$$
\begin{array}{r}
-4\underline{|} \quad\; 1 \quad 7 \quad 14 \quad 8 \\
\quad\quad -4 \quad -12 \quad -8 \\
\overline{\quad 1 \quad 3 \quad\;\; 2 \quad\;\; 0}
\end{array}
$$

The solution is: x^2+3x+2

51.

$$
\begin{array}{r}
2\underline{|} \quad\; 2 \quad 1 \quad 0 \quad -1 \\
\quad\quad 4 \quad 10 \quad 20 \\
\overline{\quad 2 \quad 5 \quad 10 \quad 19}
\end{array}
$$

The solution is: $2x^2+5x+10+\dfrac{19}{x-2}$

53.

$$
\begin{array}{r}
-2\underline{|} \quad\; 2 \quad 0 \quad 3 \quad 0 \quad -4 \\
\quad\quad -4 \quad 8 \quad -22 \quad 44 \\
\overline{\quad 2 \quad -4 \quad 11 \quad -22 \quad 40}
\end{array}
$$

The solution is: $2x^3-4x^2+11x-22+\dfrac{40}{x+2}$

55. $1\rfloor$ 1   0   0   0   −1   The solution is: $b^3 + b^2 + b + 1$
          1   1   1   1
        ‾‾‾‾‾‾‾‾‾‾‾‾‾‾‾‾
         1   1   1   1   0

Geometry

57. Use polynomial division to find $\left(6x^2 + 7x - 5\right) \div \left(2x - 1\right) = 3x + 5$. That is, $L = 3x + 5$.

When $x = 8$, $L = 3(8) + 5 = 29$ feet.

59. The volume of a box is $V = lwh$, so $h = \dfrac{V}{lw}$. Divide the volume by the length times the width.

Use polynomial division to find $\left(x^3 + 7x^2 + 14x + 8\right) \div \left(x^2 + 5x + 4\right) = x + 2$. That is, $h = x + 2$.

61. (a) By synthetic division the remainder for $\left(x^2 - 3x + 2\right) \div \left(x - 1\right)$ is 0.

$$p(1) = (1)^2 - 3(1) + 2 = 0$$

(b) By synthetic division the remainder for $\left(3x^2 + 5x - 2\right) \div \left(x + 2\right)$ is 0.

$$p(-2) = 3(-2)^2 + 5(-2) - 2 = 0$$

(c) By synthetic division the remainder for $\left(3x^3 - 4x^2 - 5x + 3\right) \div \left(x - 2\right)$ is 1.

$$p(2) = 3(2)^3 - 4(2)^2 - 5(2) + 3 = 1$$

(d) By synthetic division the remainder for $\left(x^3 - 2x^2 - x + 2\right) \div \left(x + 1\right)$ is 0.

$$p(-1) = (-1)^3 - 2(-1)^2 - (-1) + 2 = 0$$

(e) By synthetic division the remainder for $\left(x^4 - 5x^2 - 1\right) \div \left(x - 3\right)$ is 35.

$$p(3) = (3)^4 - 5(3)^2 - 1 = 35$$

63. (a) $p(2) = (2)^2 + (2) - 6 = 0$

Yes, it is a factor because by synthetic division $\left(x^2 + x - 6\right) \div \left(x - 2\right) = x + 3$.

(b) $p(1) = (1)^2 + 4(1) - 5 = 0$

Yes, it is a factor because by synthetic division $\left(x^2 + 4x - 5\right) \div \left(x - 1\right) = x + 5$.

(c) $p(-2) = (-2)^2 + 8(-2) + 11 = -1$

No, it is not a factor because by synthetic division $\left(x^2 + 8x + 11\right) \div \left(x + 2\right) = x + 6$ Remainder -1.

(d) $p(-1) = (-1)^3 + (-1)^2 + (-1) + 1 = 0$

Yes, it is a factor because by synthetic division $\left(x^3 + x^2 + x + 1\right) \div \left(x + 1\right) = x^2 + 1$.

(e) $p(2) = (2)^3 - 3(2)^2 - (2) - 3 = -9$

No, it is not a factor because $\left(x^3 - 3x^2 - x - 3\right) \div \left(x - 2\right) = x^2 - x - 3$ Remainder -9.

Checking Basic Concepts for Section 6.7

1.
(a) $\dfrac{2x - x^2}{x^2} = \dfrac{2x}{x^2} - \dfrac{x^2}{x^2} = \dfrac{2}{x} - 1$

(b) $\dfrac{6a^2 - 9a + 15}{3a} = \dfrac{6a^2}{3a} - \dfrac{9a}{3a} + \dfrac{15}{3a} = 2a - 3 + \dfrac{5}{a}$

2.
$$\begin{array}{r} 5x - 3 \\ 2x+1{\overline{\smash{\big)}\,10x^2 - x + 4}} \\ \underline{10x^2 + 5x} \\ -6x + 4 \\ \underline{-6x - 3} \\ 7 \end{array}$$
The solution is: $5x - 3 + \dfrac{7}{2x+1}$

3.
(a)
$$\begin{array}{r} 1| \quad 2 \quad -5 \quad\;\; 0 \quad -1 \\ \underline{\quad\quad\;\; 2 \quad -3 \quad -3} \\ 2 \quad -3 \quad -3 \quad -4 \end{array}$$
The solution is: $2x^2 - 3x - 3 - \dfrac{4}{x-1}$

(b)
$$\begin{array}{r} -2| \quad 2 \quad -5 \quad\;\; 0 \quad\;\; -1 \\ \underline{\quad\quad\;\; -4 \quad 18 \quad -36} \\ 2 \quad -9 \quad 18 \quad -37 \end{array}$$
The solution is: $2x^2 - 9x + 18 - \dfrac{37}{x+2}$

Chapter 6 Review Exercises

Section 6.1

1.   Symbolic: $f(x) = \dfrac{1}{x-1}$

2.   Symbolic: $f(x) = \dfrac{x-3}{x}$

3.   See Figure 3. Because the denominator cannot be equal to 0, the domain is $\{x \mid x \neq -2\}$.

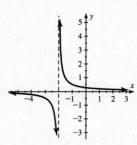

Figure 3

4. $f(3) = \dfrac{1}{(3)^2 - 1} = \dfrac{1}{9-1} = \dfrac{1}{8}$; The values $x = \pm 1$ are not in the domain of f.

5. $f(-3) = \dfrac{3}{(-3)+2} = \dfrac{3}{-1} = -3$ and $f(2) = \dfrac{3}{(2)+2} = \dfrac{3}{4} = 0.75$

6. $f(-2) = \dfrac{2(-2)}{(-2)^2-4} = \dfrac{-4}{0}$ is undefined and $f(3) = \dfrac{2(3)}{(3)^2-4} = \dfrac{6}{5} = 1.2$

7. $f(0) = 3$ and $f(2) = 1$; The equation of the vertical asymptote is $x = 1$.

8. $f(-3) = 2$ and $f(-2)$ is undefined; The equation of the vertical asymptote is $x = -2$.

9. The LCD is x. The first step is to multiply both sides of the equation by the LCD.

$x \cdot \left(\dfrac{3-x}{x}\right) = 2 \cdot x \Rightarrow 3 - x = 2x \Rightarrow 3 = 3x \Rightarrow x = 1$

10. The LCD is $5(x+3)$. The first step is to multiply both sides of the equation by the LCD.

$5(x+3) \cdot \left(\dfrac{1}{x+3}\right) = \dfrac{1}{5} \cdot 5(x+3) \Rightarrow 5 = x+3 \Rightarrow x = 2$

11. The LCD is $x-2$. The first step is to multiply both sides of the equation by the LCD.

$(x-2) \cdot \left(\dfrac{x}{x-2}\right) = (x-2) \cdot \dfrac{2x-2}{x-2} \Rightarrow x = 2x-2 \Rightarrow x = 2$

Some expressions in the original equation are not defined when $x = 2$. No solutions.

12. The LCD is $2-3x$. The first step is to multiply both sides of the equation by the LCD.

$(2-3x) \cdot \left(\dfrac{4}{2-3x}\right) = -1 \cdot (2-3x) \Rightarrow 4 = 3x-2 \Rightarrow 3x = 6 \Rightarrow x = 2$

13. (a) $(f+g)(3) = \left(2(3)^2-3(3)\right)+(2(3)-3) = (2(9)-9)+(6-3) = (18-9)+3 = 9+3 = 12$

(b) $(fg)(3) = \left(2(3)^2-3(3)\right)(2(3)-3) = (2(9)-9)(6-3) = (18-9)(3) = 9(3) = 27$

14. (a) $(f-g)(x) = \left(x^2-1\right)-(x-1) = x^2-x$

(b) $(f/g)(x) = \dfrac{x^2-1}{x-1} = \dfrac{(x-1)(x+1)}{x-1} = x+1$

Section 6.2

15. $\dfrac{4a}{6a^4} = \dfrac{2}{3a^3}$

16. $\dfrac{(x-3)(x+2)}{(x+1)(x-3)} = \dfrac{x+2}{x+1}$

17. $\dfrac{x^2-4}{x-2} = \dfrac{(x-2)(x+2)}{x-2} = x+2$

18. $\dfrac{x^2-6x-7}{2x^2-x-3} = \dfrac{(x-7)(x+1)}{(2x-3)(x+1)} = \dfrac{x-7}{2x-3}$

19. $\dfrac{8-x}{x-8} = \dfrac{-1(x-8)}{x-8} = -1$

20. $-\dfrac{3-2x}{2x-3} = \dfrac{-1(3-2x)}{2x-3} = \dfrac{2x-3}{2x-3} = 1$

21. $\dfrac{1}{2y} \cdot \dfrac{4y^2}{8} = \dfrac{1}{2} \cdot \dfrac{y}{2} = \dfrac{y}{4}$

22. $\dfrac{x+2}{x-5} \cdot \dfrac{x-5}{x+1} = \dfrac{x+2}{1} \cdot \dfrac{1}{x+1} = \dfrac{x+2}{x+1}$

23. $\dfrac{x^2+1}{x^2-1} \cdot \dfrac{x-1}{x+1} = \dfrac{x^2+1}{(x-1)(x+1)} \cdot \dfrac{x-1}{x+1} = \dfrac{x^2+1}{x+1} \cdot \dfrac{1}{x+1} = \dfrac{x^2+1}{(x+1)^2}$

24. $\dfrac{x^2+2x+1}{x^2-9} \cdot \dfrac{x+3}{x+1} = \dfrac{(x+1)(x+1)}{(x-3)(x+3)} \cdot \dfrac{x+3}{x+1} = \dfrac{x+1}{x-3} \cdot \dfrac{1}{1} = \dfrac{x+1}{x-3}$

25. $\dfrac{1}{3y} \div \dfrac{1}{9y^4} = \dfrac{1}{3y} \cdot \dfrac{9y^4}{1} = \dfrac{1}{1} \cdot \dfrac{3y^3}{1} = 3y^3$

26. $\dfrac{2x+2}{3x-3} \div \dfrac{x+1}{x-1} = \dfrac{2(x+1)}{3(x-1)} \cdot \dfrac{x-1}{x+1} = \dfrac{2}{3} \cdot \dfrac{1}{1} = \dfrac{2}{3}$

27. $\dfrac{x^2+2x}{x^2-25} \div \dfrac{x+2}{x+5} = \dfrac{x(x+2)}{(x-5)(x+5)} \cdot \dfrac{x+5}{x+2} = \dfrac{x}{x-5} \cdot \dfrac{1}{1} = \dfrac{x}{x-5}$

28. $\dfrac{x^2+2x-15}{x^2+4x+3} \div \dfrac{x-3}{x+1} = \dfrac{(x+5)(x-3)}{(x+3)(x+1)} \cdot \dfrac{x+1}{x-3} = \dfrac{x+5}{x+3} \cdot \dfrac{1}{1} = \dfrac{x+5}{x+3}$

29. $\dfrac{x^2+x-2}{2x^2-7x+3} \cdot \dfrac{x^2-3x}{x^2+2x-3} = \dfrac{(x+2)(x-1)}{(2x-1)(x-3)} \cdot \dfrac{x(x-3)}{(x+3)(x-1)} = \dfrac{x(x+2)}{(2x-1)(x+3)}$

30. $\dfrac{x^2+3x+2}{x^2+7x+12} \div \dfrac{x^2+4x+4}{x^2+4x+3} = \dfrac{(x+2)(x+1)}{(x+3)(x+4)} \cdot \dfrac{(x+3)(x+1)}{(x+2)(x+2)} = \dfrac{(x+1)^2}{(x+2)(x+4)}$

Section 6.3

31. $36 = 2 \cdot 2 \cdot 3 \cdot 3$ and $24 = 2 \cdot 2 \cdot 2 \cdot 3$ and $16 = 2 \cdot 2 \cdot 2 \cdot 2$. The LCM is $2 \cdot 2 \cdot 2 \cdot 2 \cdot 3 \cdot 3 = 144$.

32. $4ab = 2 \cdot 2 \cdot a \cdot b$ and $a^2b = a \cdot a \cdot b$. The LCM is $2 \cdot 2 \cdot a \cdot a \cdot b = 4a^2b$.

33. $9x^2y = 3 \cdot 3 \cdot x \cdot x \cdot y$ and $6xy^3 = 2 \cdot 3 \cdot x \cdot y \cdot y \cdot y$. The LCM is $2 \cdot 3 \cdot 3 \cdot x \cdot x \cdot y \cdot y \cdot y = 18x^2y^3$.

34. $x-1 = (x-1)$ and $x+2 = (x+2)$. The LCM is $(x-1)(x+2)$.

35. $x^2-9 = (x-3)(x+3)$ and $x(x+3) = x(x+3)$. The LCM is $x(x-3)(x+3)$.

36. $x^2 = x \cdot x$ and $x-3 = (x-3)$ and $x^2-6x+9 = (x-3)^2$. The LCM is $x^2(x-3)^2$.

37. $\dfrac{1}{x+4} + \dfrac{3}{x+4} = \dfrac{1+3}{x+4} = \dfrac{4}{x+4}$

38. $\dfrac{2}{x} + \dfrac{x-3}{x} = \dfrac{2+x-3}{x} = \dfrac{x-1}{x}$

39. $\dfrac{1}{x+1} - \dfrac{x}{x+1} = \dfrac{1-x}{x+1}$

40. $\dfrac{2x}{x-2} - \dfrac{2}{x-2} = \dfrac{2x-2}{x-2}$

41. $\dfrac{4}{1-t} + \dfrac{t}{t-1} = \dfrac{4}{-1(t-1)} + \dfrac{t}{t-1} = \dfrac{-4}{t-1} + \dfrac{t}{t-1} = \dfrac{t-4}{t-1}$

42. $\dfrac{2}{y-2} - \dfrac{2}{y+2} = \dfrac{2(y+2)}{(y-2)(y+2)} - \dfrac{2(y-2)}{(y-2)(y+2)} = \dfrac{2y+4-2y+4}{(y-2)(y+2)} = \dfrac{8}{(y-2)(y+2)}$

43. $\dfrac{4b}{a^2 c} - \dfrac{3a}{b^2 c} = \dfrac{4b^3}{a^2 b^2 c} - \dfrac{3a^3}{a^2 b^2 c} = \dfrac{4b^3 - 3a^3}{a^2 b^2 c}$

44. $\dfrac{r}{5t^2} + \dfrac{t}{5r^2} = \dfrac{r^3}{5r^2 t^2} + \dfrac{t^3}{5r^2 t^2} = \dfrac{r^3 + t^3}{5r^2 t^2}$

45. $\dfrac{4}{a^2 - b^2} - \dfrac{2}{a+b} = \dfrac{4}{(a-b)(a+b)} - \dfrac{2(a-b)}{(a-b)(a+b)} = \dfrac{4-2a+2b}{(a-b)(a+b)} = -\dfrac{2(a-b-2)}{(a-b)(a+b)}$

46. $\dfrac{a}{a-b} + \dfrac{b}{a+b} = \dfrac{a(a+b)}{(a-b)(a+b)} + \dfrac{b(a-b)}{(a-b)(a+b)} = \dfrac{a^2 + 2ab - b^2}{(a-b)(a+b)}$

47. $\dfrac{1}{x^2 - 3x + 2} + \dfrac{1}{x^2 + x - 2} = \dfrac{1}{(x-2)(x-1)} + \dfrac{1}{(x+2)(x-1)}$

$= \dfrac{x+2}{(x-2)(x-1)(x+2)} + \dfrac{x-2}{(x-2)(x-1)(x+2)} = \dfrac{x+2+x-2}{(x-2)(x-1)(x+2)}$

$= \dfrac{2x}{(x-2)(x-1)(x+2)}$

48. $\dfrac{1}{x^2 - 5x + 6} - \dfrac{1}{x^2 + x - 6} = \dfrac{1}{(x-2)(x-3)} - \dfrac{1}{(x+3)(x-2)}$

$= \dfrac{x+3}{(x-3)(x-2)(x+3)} - \dfrac{x-3}{(x-3)(x-2)(x+3)} = \dfrac{x+3-x+3}{(x-3)(x-2)(x+3)}$

$= \dfrac{6}{(x-3)(x-2)(x+3)}$

49. $\dfrac{1}{x-2} + \dfrac{2}{x+2} - \dfrac{x}{x-2} = \dfrac{x+2}{(x-2)(x+2)} + \dfrac{2(x-2)}{(x-2)(x+2)} - \dfrac{x(x+2)}{(x-2)(x+2)}$

$= \dfrac{x+2+2x-4-x^2-2x}{(x-2)(x+2)} = \dfrac{-x^2 + x - 2}{(x-2)(x+2)}$

50.
$$\frac{2x}{2x-1}-\frac{3}{2x+1}-\frac{1}{2x-1}=\frac{2x(2x+1)}{(2x-1)(2x+1)}-\frac{3(2x-1)}{(2x-1)(2x+1)}-\frac{2x+1}{(2x-1)(2x+1)}$$

$$=\frac{4x^2+2x-6x+3-2x-1}{(2x-1)(2x+1)}=\frac{4x^2-6x+2}{(2x-1)(2x+1)}=\frac{2(x-1)(2x-1)}{(2x-1)(2x+1)}$$

$$=\frac{2x-2}{2x+1}$$

Section 6.4

51.   The LCD is $2x$. The first step is to multiply both sides of the equation by the LCD.

$$\frac{4}{x}-\frac{5}{2x}=\frac{1}{2}\Rightarrow 2x\cdot\frac{4}{x}-2x\cdot\frac{5}{2x}=\frac{1}{2}\cdot 2x\Rightarrow 8-5=x\Rightarrow x=3$$

52.   Cross multiply to obtain the following.

$$\frac{1}{x-4}=\frac{3}{2x-1}\Rightarrow 3(x-4)=1(2x-1)\Rightarrow 3x-12=2x-1\Rightarrow x=11$$

53.   The LCD is x^2. The first step is to multiply both sides of the equation by the LCD.

$$x^2\cdot\left(\frac{2}{x^2}-\frac{1}{x}\right)=1\cdot x^2\Rightarrow 2-x=x^2\Rightarrow x^2+x-2=0\Rightarrow (x+2)(x-1)=0\Rightarrow x=-2\text{ or }x=1$$

54.   The LCD is $x(x+4)$. The first step is to multiply both sides of the equation by the LCD.

$$x(x+4)\cdot\left(\frac{1}{x+4}-\frac{1}{x}\right)=1\cdot x(x+4)\Rightarrow x-(x+4)=x^2+4x\Rightarrow x^2+4x+4=0\Rightarrow (x+2)(x+2)=0\Rightarrow x=-2$$

55.   The LCD is $3(x-1)(x+1)$. The first step is to multiply both sides of the equation by the LCD.

$$3(x-1)(x+1)\cdot\left(\frac{1}{x^2-1}-\frac{1}{x-1}\right)=\frac{2}{3}\cdot 3(x-1)(x+1)\Rightarrow 3-3(x+1)=2(x-1)(x+1)\Rightarrow$$

$$3-3x-3=2x^2-2\Rightarrow 2x^2+3x-2=0\Rightarrow (x+2)(2x-1)=0\Rightarrow x=-2\text{ or }x=\frac{1}{2}$$

56.   The LCD is $(x-3)(x+3)$. The first step is to multiply both sides of the equation by the LCD.

$$(x^2-9)\cdot\left(\frac{1}{x-3}+\frac{1}{x+3}\right)=\frac{-5}{x^2-9}\cdot(x^2-9)\Rightarrow x+3+x-3=-5\Rightarrow 2x=-5\Rightarrow x=-\frac{5}{2}$$

57.   The LCD is $(x+2)(x-2)^2$. The first step is to multiply both sides of the equation by the LCD.

$$(x+2)(x-2)^2\cdot\frac{1}{(x-2)^2}-(x+2)(x-2)^2\cdot\frac{1}{x^2-4}=\frac{2}{(x-2)^2}\cdot(x+2)(x-2)^2\Rightarrow$$

$$x+2-(x-2)=2(x+2)\Rightarrow 4=2x+4\Rightarrow 2x=0\Rightarrow x=0$$

58.   The LCD is $2(x+1)(x+2)$. The first step is to multiply both sides of the equation by the LCD.

$$2(x+1)(x+2)\cdot\frac{1}{x^2+3x+2}+2(x+1)(x+2)\cdot\frac{x}{x+2}=\frac{1}{2}\cdot2(x+1)(x+2)\Rightarrow$$

$$2+2x(x+1)=(x+1)(x+2)\Rightarrow2x^2+2x+2=x^2+3x+2\Rightarrow x^2-x=0\Rightarrow x=0\text{ or }x=1$$

59. $\dfrac{-3x-3}{x-3}+\dfrac{x^2+x}{x-3}=7x+7\Rightarrow\dfrac{-3x-3+x^2+x}{x-3}=7x+7$

$\Rightarrow\dfrac{x^2-2x-3}{x-3}=7x+7\Rightarrow\dfrac{(x-3)(x+1)}{x-3}=7x+7$

$\Rightarrow x+1=7x+7\Rightarrow-6x=6\Rightarrow x=-1$

60. $\dfrac{1}{x-4}+\dfrac{x}{x+4}=\dfrac{8}{x^2-4}\Rightarrow(x-4)(x+4)\cdot\dfrac{1}{x-4}+(x-4)(x+4)\cdot\dfrac{x}{x+4}$

$=\dfrac{8}{(x-4)(x+4)}\cdot(x-4)(x+4)\Rightarrow x+4+x(x-4)=8$

$\Rightarrow x+4+x^2-4x=8\Rightarrow x^2-3x-4=0\Rightarrow(x-4)(x+1)=0$

$\Rightarrow x-4=0\text{ or }x+1=0\Rightarrow x=-1$ ($x=4$ is not a solution because this value makes expressions in the

original equation undefined)

61. $y_1=y_2$ when $x=2$. By substitution of $x=2$, this answer checks.

62. $y_1=y_2$ when $x=-2$ and $x=3$. By substitution of $x=-2$ and $x=3$, these answers check.

63. $m=\dfrac{y_2-y_1}{x_2-x_1}\Rightarrow m(x_2-x_1)=y_2-y_1\Rightarrow m(x_2-x_1)+y_1=y_2\Rightarrow y_2=m(x_2-x_1)+y_1$

64. $T=\dfrac{a}{b+2}\Rightarrow T(b+2)=a\Rightarrow a=T(b+2)$

65. $\dfrac{1}{f}=\dfrac{1}{p}+\dfrac{1}{q}\Rightarrow pq=fq+fp\Rightarrow pq-fp=fq\Rightarrow p(q-f)=fq\Rightarrow p=\dfrac{fq}{q-f}$

66. $I=\dfrac{2a+3b}{ab}\Rightarrow abI=2a+3b\Rightarrow abI-3b=2a\Rightarrow b(aI-3)=2a\Rightarrow b=\dfrac{2a}{aI-3}$

Section 6.5

67. $\dfrac{\dfrac{3}{5}}{\dfrac{10}{13}}=\dfrac{3}{5}\cdot\dfrac{13}{10}=\dfrac{39}{50}$

68. $\dfrac{\dfrac{4}{ab}}{\dfrac{2}{bc}}=\dfrac{4}{ab}\cdot\dfrac{bc}{2}=\dfrac{2c}{a}$

69. $\dfrac{\dfrac{2n-1}{n}}{\dfrac{3n}{2n+1}}=\dfrac{2n-1}{n}\cdot\dfrac{2n+1}{3n}=\dfrac{4n^2-1}{3n^2}$

70. $\dfrac{\dfrac{1}{x-y}}{\dfrac{1}{x^2-y^2}} = \dfrac{1}{x-y} \cdot \dfrac{x^2-y^2}{1} = \dfrac{(x-y)(x+y)}{x-y} = x+y$

71. $\dfrac{2+\dfrac{3}{x}}{2-\dfrac{3}{x}} = \dfrac{2+\dfrac{3}{x}}{2-\dfrac{3}{x}} \cdot \dfrac{x}{x} = \dfrac{2x+3}{2x-3}$

72. $\dfrac{\dfrac{1}{x}+\dfrac{1}{2}}{\dfrac{1}{4}-\dfrac{2}{x}} = \dfrac{\dfrac{1}{x}+\dfrac{1}{2}}{\dfrac{1}{4}-\dfrac{2}{x}} \cdot \dfrac{4x}{4x} = \dfrac{4+2x}{x-8} = \dfrac{2(2+x)}{x-8}$

73. $\dfrac{\dfrac{4}{x}+\dfrac{1}{x-1}}{\dfrac{1}{x}-\dfrac{2}{x-1}} = \dfrac{\dfrac{4}{x}+\dfrac{1}{x-1}}{\dfrac{1}{x}-\dfrac{2}{x-1}} \cdot \dfrac{x(x-1)}{x(x-1)} = \dfrac{4(x-1)+x}{x-1-2x} = \dfrac{5x-4}{-x-1} = \dfrac{5x-4}{-1(x+1)} = \dfrac{4-5x}{x+1}$

74. $\dfrac{\dfrac{1}{x+3}-\dfrac{1}{x-3}}{\dfrac{4}{x+3}-\dfrac{2}{x-3}} = \dfrac{\dfrac{1}{x+3}-\dfrac{1}{x-3}}{\dfrac{4}{x+3}-\dfrac{2}{x-3}} \cdot \dfrac{(x+3)(x-3)}{(x+3)(x-3)} = \dfrac{x-3-(x+3)}{4(x-3)-2(x+3)} = \dfrac{-6}{2x-18} = \dfrac{-3}{x-9}$

Section 6.6

75. $\dfrac{x}{6} = \dfrac{6}{20} \Rightarrow 20x = 36 \Rightarrow x = \dfrac{36}{20} \Rightarrow x = \dfrac{9}{5}$

76. $\dfrac{11}{x} = \dfrac{5}{7} \Rightarrow 5x = 77 \Rightarrow x = \dfrac{77}{5}$

77. $\dfrac{x+1}{5} = \dfrac{x}{3} \Rightarrow 5x = 3x+3 \Rightarrow 2x = 3 \Rightarrow x = \dfrac{3}{2}$

78. $\dfrac{3}{7} = \dfrac{4}{x-1} \Rightarrow 3x-3 = 28 \Rightarrow 3x = 31 \Rightarrow x = \dfrac{31}{3}$

79. $\dfrac{x+1}{5} = \dfrac{10}{15} \Rightarrow 15x+15 = 50 \Rightarrow 15x = 35 \Rightarrow x = \dfrac{35}{15} = \dfrac{7}{3}$

80. $\dfrac{7}{8} = \dfrac{11}{x} \Rightarrow 7x = 88 \Rightarrow x = \dfrac{88}{7} \approx 12.6$

81. $y = kx \Rightarrow 8 = 2k \Rightarrow k = 4 \Rightarrow y = 4x$, so when $x = 7$, $y = 4(7) = 28$

82. $y = \dfrac{k}{x} \Rightarrow 5 = \dfrac{k}{10} \Rightarrow k = 50 \Rightarrow y = \dfrac{50}{x}$, so when $x = 25$, $y = \dfrac{50}{25} = 2$

83. $z = kxy \Rightarrow 483 = k(23)(7) \Rightarrow 483 = 161k \Rightarrow k = \dfrac{483}{161} \Rightarrow k = 3$

84. $z = kxy^2 \Rightarrow 891 = k(22)(3)^2 \Rightarrow 891 = 198k \Rightarrow k = \dfrac{891}{198} \Rightarrow k = \dfrac{9}{2} \Rightarrow z = \dfrac{9}{2}xy^2$

So when $x = 10$ and $y = 4$, $z = \dfrac{9}{2}(10)(4)^2 = 720$

85.   Inverse. The products xy always equal 200. $y = \dfrac{k}{x} \Rightarrow 100 = \dfrac{k}{2} \Rightarrow k = 200$; The equation is $y = \dfrac{200}{x}$.

86.   Direct. The ratios $\dfrac{y}{x}$ always equal 3. $y = kx \Rightarrow 9 = 3k \Rightarrow k = 3$; The equation is $y = 3x$.

87.   Direct. The ratios $\dfrac{y}{x}$ always equal $\dfrac{1}{2}$. The constant of variation is $k = \dfrac{1}{2}$.

88.   Inverse. The products xy always equal 6. The constant of variation is $k = 6$.

Section 6.7

89. $\dfrac{10x+15}{5} = \dfrac{10x}{5} + \dfrac{15}{5} = 2x+3$

90. $\dfrac{2x^2+x}{x} = \dfrac{2x^2}{x} + \dfrac{x}{x} = 2x+1$

91. $\left(4x^3 - x^2 + 2x\right) \div 2x^2 = \dfrac{4x^3 - x^2 + 2x}{2x^2} = \dfrac{4x^3}{2x^2} - \dfrac{x^2}{2x^2} + \dfrac{2x}{2x^2} = 2x - \dfrac{1}{2} + \dfrac{1}{x}$

92. $\left(4a^3b - 6ab^2\right) \div \left(2a^2b^2\right) = \dfrac{4a^3b - 6ab^2}{2a^2b^2} = \dfrac{4a^3b}{2a^2b^2} - \dfrac{6ab^2}{2a^2b^2} = \dfrac{2a}{b} - \dfrac{3}{a}$

93.

$$
\begin{array}{r}
2x-3 \\
x+1\overline{\smash{\big)}\,2x^2 - x - 2} \\
\underline{2x^2 + 2x} \\
-3x-2 \\
\underline{-3x-3} \\
1
\end{array}
$$

The solution is: $2x - 3 + \dfrac{1}{x+1}$

94.

$$
\begin{array}{r}
x^2-x+2 \\
x+1\overline{\smash{\big)}\,x^3 + 0x^2 + x - 1} \\
\underline{x^3 + x^2} \\
-x^2+x \\
\underline{-x^2-x} \\
2x-1 \\
\underline{2x+2} \\
-3
\end{array}
$$

The solution is: $x^2 - x + 2 - \dfrac{3}{x+1}$

95.

$$
\begin{array}{r}
2x^2 - x + 1 \\
3x-2\overline{\smash{\big)}\,6x^3 - 7x^2 + 5x - 1} \\
\underline{6x^3 - 4x^2} \\
-3x^2 + 5x \\
\underline{-3x^2 + 2x} \\
3x - 1 \\
\underline{3x - 2} \\
1
\end{array}
$$

The solution is: $2x^2 - x + 1 + \dfrac{1}{3x-2}$

96.

$$
\begin{array}{r}
2x - 9 \\
x^2 + 0x - 2\overline{\smash{\big)}\,2x^3 - 9x^2 + 21x - 21} \\
\underline{2x^3 + 0x^2 - 4x} \\
-9x^2 + 25x - 21 \\
\underline{-9x^2 + 0x + 18} \\
25x - 39
\end{array}
$$

The solution is: $2x - 9 + \dfrac{25x - 39}{x^2 - 2}$

97.

$$
\begin{array}{r|rrr}
3 & 2 & -11 & 13 \\
 & & 6 & -15 \\
\hline
 & 2 & -5 & -2
\end{array}
$$

The solution is: $2x - 5 - \dfrac{2}{x-3}$

98.

$$
\begin{array}{r|rrrr}
-4 & 3 & 10 & -4 & 19 \\
 & & -12 & 8 & -16 \\
\hline
 & 3 & -2 & 4 & 3
\end{array}
$$

The solution is: $3x^2 - 2x + 4 + \dfrac{3}{x+4}$

Applications

99. (a) $D(0) = \dfrac{1600}{9.6 - 30(0)} = \dfrac{1600}{9.6} = 166.\overline{6}; \; D(0.1) = \dfrac{1600}{9.6 - 30(0.1)} = \dfrac{1600}{6.6} = 242.\overline{42}$

The downhill grade adds $242.\overline{42} - 166.\overline{6} \approx 75.8$ feet to the braking distance.

(b) $200 = \dfrac{1600}{9.6 - 30x} \Rightarrow 200(9.6 - 30x) = 1600 \Rightarrow 9.6 - 30x = 8 \Rightarrow -30x = -1.6 \Rightarrow x = 0.05\overline{3}$

100. (a) Table $Y_1 = 1/(10 - X)$ with TblStart = 9 and ΔTbl = 0.1. See Figure 100.

(b) As x approaches 10 the waiting time increases. As vehicles arrive at a faster rate, the wait in line increases.

101. (a) $N(14) = \dfrac{14^2}{225 - 15(14)} = \dfrac{196}{15} \approx 13$ cars

(b) As x increases the number of cars in line increases. This agrees with intuition.

102. (a) $P(0) = \dfrac{5}{(0)^2 + 1} = \dfrac{5}{1} = 5$; Initially there were 5000 fish in the lake.

(b) Graph $Y_1 = 5/(X^2 + 1)$ in [0, 6, 1] by [0, 6, 1]. See Figure 102.

(c) Over this 6-year period, the fish population decreased.

(d) $1 = \dfrac{5}{x^2 + 1} \Rightarrow x^2 + 1 = 5 \Rightarrow x^2 = 4 \Rightarrow x = 2$ (the value $x = -2$ has no physical meaning)

The population was 1 thousand after 2 years.

103.   (a) $\dfrac{x}{2} + \dfrac{x}{3} = 1$

(b) $6 \cdot \left(\dfrac{x}{2} + \dfrac{x}{3} \right) = 1 \cdot 6 \Rightarrow 3x + 2x = 6 \Rightarrow 5x = 6 \Rightarrow x = \dfrac{6}{5} = 1.2$ hours

(c)   Graph $Y_1 = X/2 + X/3$ and $Y_2 = 1$ in [0, 2, 1] by [0, 2, 1]. See Figure 103. $x = 1.2$ hours

[0, 6, 1] by [0, 6, 1]

[0, 2, 1] by [0, 2, 1]

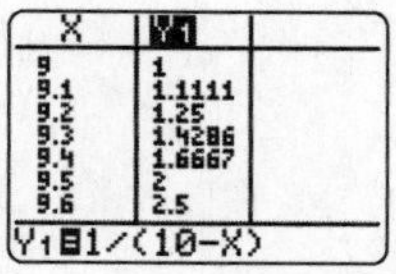

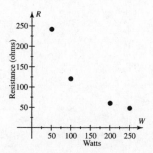

Figure 100

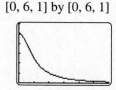

Figure 102

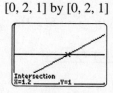

Figure 103

104.   Let x represent the speed of the current. Then $12 - x$ represents the boat's upstream speed and $12 + x$ represents the boat's downstream speed. Since $t = \dfrac{d}{r}$, the equation is $\dfrac{48}{12 - x} + \dfrac{48}{12 + x} = 9$.

The LCD is $(12 - x)(12 + x)$. The first step is to multiply both sides of the equation by the LCD.

$(12 - x)(12 + x) \cdot \left(\dfrac{48}{12 - x} + \dfrac{48}{12 + x} \right) = 9 \cdot (12 - x)(12 + x) \Rightarrow 576 + 48x + 576 - 48x = 1296 - 9x^2$

$\Rightarrow 9x^2 - 144 = 0 \Rightarrow 9x^2 = 144 \Rightarrow x^2 = 16 \Rightarrow x = \pm\sqrt{16} \Rightarrow x = 4$ mph ($x = -4$ has no meaning)

105. $\dfrac{650}{74} = \dfrac{387}{x} \Rightarrow 650x = 28{,}638 \Rightarrow x = \dfrac{28{,}638}{650} \approx 44$ minutes

106. $\dfrac{5}{3} = \dfrac{x}{26} \Rightarrow 3x = 130 \Rightarrow x = \dfrac{130}{3} \approx 43.3$ feet

107.   (a)   Plot the data in [0, 300, 50] by [0, 250, 50]. See Figure 107. The data represent inverse variation.

(b) $R = \dfrac{k}{x} \Rightarrow 242 = \dfrac{k}{50} \Rightarrow k = 12{,}100$; Equation: $R = \dfrac{12{,}100}{W}$

(c) $R = \dfrac{12{,}100}{55} = 220$ ohms

Figure 107

108. (a) $D = kW \Rightarrow 0.3 = 5k \Rightarrow k = 0.06$; Equation: $D = 0.06W$

   (b) $D = 0.06(7) = 0.42$ inches

109. The change in temperature is $18°F$ so $\dfrac{18}{4000} = \dfrac{x}{6000} \Rightarrow 4000x = 108,000 \Rightarrow x = 27\,°$ change at 6000 feet.

   The temperature at 6000 feet is $80 - 27 = 53°F$.

110. (a) Direct. The ratios $\dfrac{y}{x}$ always equal $\dfrac{7}{2}$.

   (b) $y = kx \Rightarrow 3.5 = 1k \Rightarrow k = 3.5$, Equation: $y = 3.5x$

   (c) $y = 3.5(2.3) = 8.05\%$

Chapter 6 Test

1. $f(x) = \dfrac{x}{x+2}$

2. (a) $f(-2) = \dfrac{1}{4(-2)^2 - 1} = \dfrac{1}{16 - 1} = \dfrac{1}{15}$

   (b) Here $4x^2 - 1 \neq 0 \Rightarrow x \neq -\dfrac{1}{2}$ and $x \neq \dfrac{1}{2}$. The domain is $\left\{ x \,\middle|\, x \neq \dfrac{1}{2},\, x \neq -\dfrac{1}{2} \right\}$.

3. See Figure 3.

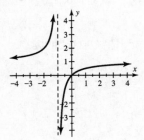

Figure 3

4. $\dfrac{2a^3}{4a^2} = \dfrac{a}{2}$

5. $\dfrac{1 - 2t}{2t - 1} = \dfrac{-1(2t - 1)}{2t - 1} = -1$

6. $\dfrac{x^2 - 2x - 15}{2x^2 - x - 21} = \dfrac{(x-5)(x+3)}{(2x-7)(x+3)} = \dfrac{x-5}{2x-7}$

7. $\dfrac{x^2 + 4}{x^2 - 4} \cdot \dfrac{x-2}{x+2} = \dfrac{x^2 + 4}{(x-2)(x+2)} \cdot \dfrac{x-2}{x+2} = \dfrac{x^2 + 4}{x+2} \cdot \dfrac{1}{x+2} = \dfrac{x^2 + 4}{(x+2)^2}$

8. $\dfrac{1}{4y^2} \div \dfrac{1}{8y^4} = \dfrac{1}{4y^2} \cdot \dfrac{8y^4}{1} = \dfrac{1}{1} \cdot \dfrac{2y^2}{1} = 2y^2$

9. $\dfrac{x}{x+5}+\dfrac{1-x}{x+5}=\dfrac{x+1-x}{x+5}=\dfrac{1}{x+5}$

10. $\dfrac{2x}{x-2}-\dfrac{1}{x+2}=\dfrac{2x(x+2)}{(x-2)(x+2)}-\dfrac{x-2}{(x-2)(x+2)}=\dfrac{2x(x+2)-(x-2)}{(x-2)(x+2)}$

$=\dfrac{2x^2+4x-x+2}{(x-2)(x+2)}=\dfrac{2x^2+3x+2}{(x-2)(x+2)}$

11. $\dfrac{a^2}{3b}-\dfrac{2b^3}{5a}=\dfrac{a^2}{3b}\cdot\dfrac{5a}{5a}-\dfrac{2b^3}{5a}\cdot\dfrac{3b}{3b}=\dfrac{5a^3}{15ab}-\dfrac{6b^4}{15ab}$

$=\dfrac{5a^3-6b^4}{15ab}$

12. $\dfrac{x}{x-2}-\dfrac{4}{x^2}-\dfrac{2}{x}=\dfrac{x}{(x-2)}\cdot\dfrac{x^2}{x^2}-\dfrac{4}{x^2}\cdot\dfrac{x-2}{x-2}-\dfrac{2}{x}\cdot\dfrac{x(x-2)}{x(x-2)}$

$=\dfrac{x^3}{x^2(x-2)}-\dfrac{4x-8}{x^2(x-2)}-\dfrac{2x^2-4x}{x^2(x-2)}$

$=\dfrac{x^3-4x+8-2x^2+4x}{x^2(x-2)}=\dfrac{x^3-2x^2+8}{x^2(x-2)}$

13. $\dfrac{3+\dfrac{3}{x}}{3-\dfrac{3}{x}}=\dfrac{3+\dfrac{3}{x}}{3-\dfrac{3}{x}}\cdot\dfrac{x}{x}=\dfrac{3x+3}{3x-3}=\dfrac{3(x+1)}{3(x-1)}=\dfrac{x+1}{x-1}$

14. $\dfrac{1}{z+4}-\dfrac{z}{(z+4)^2}=\dfrac{z+4}{(z+4)(z+4)}-\dfrac{z}{(z+4)(z+4)}=\dfrac{z+4-z}{(z+4)(z+4)}=\dfrac{4}{(z+4)^2}$

15. $\dfrac{\dfrac{1}{x-2}+\dfrac{x}{x-2}}{\dfrac{1}{3}-\dfrac{5}{x-2}}=\dfrac{\dfrac{1}{x-2}+\dfrac{x}{x-2}}{\dfrac{1}{3}-\dfrac{5}{x-2}}\cdot\dfrac{3(x-2)}{3(x-2)}=\dfrac{3+3x}{x-2-15}=\dfrac{3x+3}{x-17}=\dfrac{3(x+1)}{x-17}$

16. The LCD is $7(5t+1)$. The first step is to multiply both sides of the equation by the LCD.

$7(5t+1)\cdot\dfrac{t}{5t+1}=\dfrac{2}{7}\cdot7(5t+1)\Rightarrow 7t=10t+2\Rightarrow-3t=2\Rightarrow t=-\dfrac{2}{3}$

17. The LCD is $(2x-1)(x+4)$. The first step is to multiply both sides of the equation by the LCD.

$(2x-1)(x+4)\cdot\dfrac{x}{2x-1}=\dfrac{x+2}{x+4}\cdot(2x-1)(x+4)$

$\Rightarrow(x+4)x=(x+2)(2x-1)\Rightarrow x^2+4x=2x^2+3x-2$

$\Rightarrow 0=x^2-x-2\Rightarrow 0=(x+1)(x-2)\Rightarrow$ Either $x=-1$ or

$x=2$

18. The LCD is $2-x$. The first step is to multiply both sides of the equation by the LCD.

$$(2-x) \cdot \left(\frac{x+4}{2-x} - \frac{2x-1}{2-x} \right) = 0 \cdot (2-x)$$

$$\Rightarrow x+4-(2x-1)=0 \Rightarrow x+4-2x+1=0 \Rightarrow x=5$$

19. The LCD is $(x-2)(x+2)$. The first step is to multiply both sides of the equation by the LCD.

$$(x-2)(x+2) \cdot \left(\frac{1}{x^2-4} - \frac{1}{x-2} \right) = \frac{1}{x+2} \cdot (x-2)(x+2) \Rightarrow 1-(x+2)=x-2 \Rightarrow$$

$$-x-1=x-2 \Rightarrow -2x=-1 \Rightarrow x=\frac{1}{2}$$

20. $F = \dfrac{Gm}{r^2} \Rightarrow Fr^2 = Gm \Rightarrow m = \dfrac{Fr^2}{G}$

21. Graph $Y_1 = \sqrt{2} \big/ \left(X^2 + 2 \right) - (1.7X) \big/ (X-1.2)$ and $Y_2 = \sqrt{3}$ in [–5, 5, 1] by [–5, 5, 1].

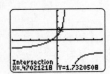

Figure 21

The solution is approximately 0.47.

22. $(f-g)(2) = (2-1) - \left(2^2 - 2(2) + 1 \right) = 1 - (4-4+1) = 1-1 = 0;$

$$(f/g)(x) = \frac{x-1}{x^2-2x+1} = \frac{x-1}{(x-1)(x-1)} = \frac{1}{x-1}$$

23. $\dfrac{7}{12} = \dfrac{x}{20} \Rightarrow 12x = 140 \Rightarrow x = \dfrac{140}{12} = \dfrac{35}{3}$

24. $y = kx \Rightarrow 8 = 23k \Rightarrow k = \dfrac{8}{23} \Rightarrow y = \dfrac{8}{23}x$, so when $x=10$, $y = \dfrac{8}{23}(10) = \dfrac{80}{23}$

25. Inverse. The products xy always equal 100. $y = \dfrac{k}{x} \Rightarrow 50 = \dfrac{k}{2} \Rightarrow k = 100;$ The equation is $y = \dfrac{100}{x}$.

26. Direct. The ratios $\dfrac{y}{x}$ always equal 1.5. $y = kx \Rightarrow 3 = 2k \Rightarrow k = 1.5;$ The equation is $y = 1.5x$.

27. $\dfrac{4a^3 + 10a}{2a} = \dfrac{4a^3}{2a} + \dfrac{10a}{2a} = 2a^2 + 5$

28. Using synthetic division:

$$\begin{array}{r|rrrr} -2 & 3 & 5 & 0 & -2 \\ & & -6 & 2 & -4 \\ \hline & 3 & -1 & 2 & -6 \end{array}$$

The solution is: $3x^2 - x + 2 - \dfrac{6}{x+2}$

29. $\dfrac{73}{50} = \dfrac{x}{180} \Rightarrow 50x = 13{,}140 \Rightarrow x = \dfrac{13{,}140}{50} = 262.8$ inches, or $\dfrac{262.8}{12} = 21.9$ feet

30. (a) Graph $Y_1 = 1/(25-X)$ in [0, 25, 5] by [0, 2, 0.5]. Vertical asymptote: $x = 25$. See Figure 30.

(b) $1 = \dfrac{1}{25-x} \Rightarrow 25 - x = 1 \Rightarrow -x = -24 \Rightarrow x = 24$ vehicles per minute

[0, 25, 5] by [0, 2, 0.5]

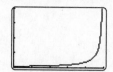

Figure 30

31. (a) $\dfrac{x}{24} + \dfrac{x}{30} = 1$

(b) The LCD is 120. The first step is to multiply both sides of the equation by the LCD.

$$120 \cdot \left(\frac{x}{24} + \frac{x}{30} \right) = 1 \cdot 120 \Rightarrow 5x + 4x = 120 \Rightarrow 9x = 120 \Rightarrow x = \frac{120}{9} = \frac{40}{3} \approx 13.3 \text{ hours}$$

32. The change in dew point is 11°F so $\dfrac{11}{10,000} = \dfrac{x}{7500} \Rightarrow 10,000x = 82,500 \Rightarrow x = 8.25°$ change at 7500 feet.

The temperature at 7500 feet is $80 - 8.25 = 41.75°F$.

Chapter 6 Extended and Discovery Exercises

1. (a) The matching graph is iii. Graph $Y_1 = \sqrt{(X)}$ in [0, 15, 1] by [0, 8, 1]. See Figure 1a.

Graph $Y_1 = \left(10X^2 + 80X + 32\right) / \left(X^2 + 40X + 80\right)$ in [0, 15, 1] by [0, 8, 1]. See Figure 1-iii.

[0, 15, 1] by [0, 8, 1]   [0, 15, 1] by [0, 8, 1]   [0, 15, 1] by [0, 8, 1]   [0, 15, 1] by [0, 8, 1]

     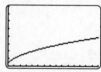   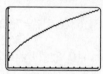   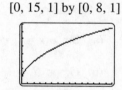

Figure 1a   Figure 1-iii   Figure 1b   Figure 1-ii

(b) The matching graph is ii. Graph $Y_1 = \sqrt{(4X+1)}$ in [0, 15, 1] by [0, 8, 1]. See Figure 1b.

Graph $Y_1 = \left(15X^2 + 75X + 33\right) / \left(X^2 + 23X + 31\right)$ in [0, 15, 1] by [0, 8, 1]. See Figure 1-ii.

(c) The matching graph is iv. Graph $Y_1 = \sqrt[3]{(X)}$ in [0, 15, 1] by [0, 8, 1]. See Figure 1c.

Graph $Y_1 = \left(7X\wedge 3 + 42X^2 + 30X + 2\right) / \left(2X\wedge 3 + 30X^2 + 42X + 7\right)$ in [0, 15, 1] by [0, 8, 1]. See Figure

1-iv.

[0, 15, 1] by [0, 8, 1]   [0, 15, 1] by [0, 8, 1]   [0, 15, 1] by [−1, 1, 1]   [0, 15, 1] by [−1, 1, 1]

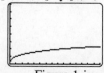   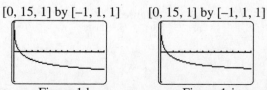

Figure 1c   Figure 1-iv   Figure 1d   Figure 1-i

(d) The matching graph is i. Graph $Y_1 = \left(1 - \sqrt{(X)}\right) / \left(1 + \sqrt{(X)}\right)$ in [0, 15, 1] by [–1, 1, 1]. See Figure 1d.

Graph $Y_1 = \left(2 - 2X^2\right) / \left(3X^2 + 10X + 3\right)$ in [0, 15, 1] by [–1, 1, 1]. See Figure 1-i.

2.   It quadruples.

3.   It doubles.

4.   (a) $1.125 = k(2.1)^{1.25} \Rightarrow k = \dfrac{1.125}{2.1^{1.25}} \approx 0.445$

(b) $0.8 = 0.445x^{1.25} \Rightarrow x^{1.25} = \dfrac{0.8}{0.445} \Rightarrow \left(x^{1.25}\right)^{0.8} = \left(\dfrac{0.8}{0.445}\right)^{0.8} \Rightarrow x \approx 1.60$ grams

5. $V = kr^2 \Rightarrow 150 = k(5)^2 \Rightarrow 150 = 25k \Rightarrow k = \dfrac{150}{25} = 6 \Rightarrow V = 6r^2$ so when $r = 7, V = 6(7)^2 = 294$ in^3

6.   It is reduced to $\frac{1}{4}$ of its value.

7.   It is reduced to $\frac{1}{8}$ of its value.

8.   (a) $200 = \dfrac{k}{4000^2} \Rightarrow k = 200\left(4000^2\right) = 3.2 \times 10^9$

(b) Graph $Y_1 = \left(3.2 \times 10^\wedge 9\right) / X^2$ in [0, 10,000, 1000] by [0, 200, 50]. See Figure 8.

Fifty pounds is $\frac{1}{4}$ of 200 pounds and so the person should be twice as far or 8000 miles from Earth's

center.

(c) One percent is $\frac{1}{100}$ of the weight of the object on Earth's surface, thus the object must be 10 times as far

or 40,000 miles from Earth's center.

9.   (a) Since $k = I \cdot d^2, k = 31.68(0.5)^2 = 7.92$.

(b) Graph $Y_1 = 7.92 / X^2$ in [0, 5, 1] by [0, 30, 5]. See Figure 9.

As the distance from the light bulb increases the intensity decreases.

(c) When the distance is doubled, the intensity is reduced to $\frac{1}{4}$ of its value.

(d) $1 = \dfrac{7.92}{d^2} \Rightarrow d^2 = 7.92 \Rightarrow d = \sqrt{7.92} \approx 2.81$ meters

[0, 10,000, 1000] by [0, 200, 50]   [0, 5, 1] by [0, 30, 5]

Figure 8   Figure 9

Chapters 1-6 Cumulative Review Exercises

1. $y = \sqrt{74 + (7)} = \sqrt{81} = 9$

2. $r = 16 - (-4)^2 = 16 - 16 = 0$

3. Natural: 1; Whole: 0, 1; Integer: $-\dfrac{12}{4}, 0, 1$; Rational: $-\dfrac{12}{4}, 0, 1, 2.\overline{11}, \dfrac{13}{2}$; Irrational: $\sqrt{3}$

4. By substitution of values, the formula that best fits the data is (i).

5. $\dfrac{18x^{-2}y^3}{3x^2 y^{-3}} = 6x^{-2-2} y^{3-(-3)} = 6x^{-4} y^6 = \dfrac{6y^6}{x^4}$

6. $\left(\dfrac{2c^2}{3d^3}\right)^{-2} = \left(\dfrac{3d^3}{2c^2}\right)^2 = \dfrac{3^2 (d^3)^2}{2^2 (c^2)^2} = \dfrac{9d^6}{4c^4}$

7. Move the decimal point 10 places to the left: $67,300,000,000 = 6.73 \times 10^{10}$.

8. The domain corresponds to the x-coordinates and the range corresponds to the y-coordinates.

    $D = \{-3, -1, 0, 4\}; R = \{-2, 0, 1, 5\}$

9. When $x = -2$, $y = 3(-2) - 1 = -7$. The other ordered pairs can be found similarly.

    The ordered pairs are $(-2, -7), (-1, -4), (0, -1), (1, 2), (2, 5)$. See Figure 9.

10. When $x = -2$, $y = \dfrac{4 - (-2)^2}{2} = \dfrac{4 - 4}{2} = \dfrac{0}{2} = 0$. The other ordered pairs can be found similarly.

    The ordered pairs are $(-2, 0), \left(-1, \dfrac{3}{2}\right), (0, 2), \left(1, \dfrac{3}{2}\right), (2, 0)$. See Figure 10.

11. See Figure 11.

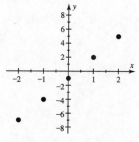

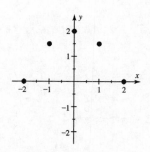

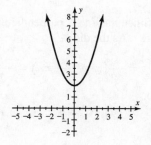

Figure 9   Figure 10   Figure 11

12. The function is defined for all values of the variable except 1. The domain is $\{x \mid x \ne 1\}$.

13. The slope is $m = \dfrac{3 - (-5)}{2 - (-2)} = \dfrac{8}{4} = 2$. Since $f(x) = -1$ when $x = 0$ the y-intercept is -1. Here $f(x) = 2x - 1$.

14.   See Figure 14.

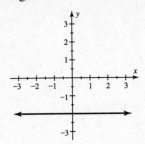

Figure 14

15.   The equation is in the form $f(x) = mx + b$. The slope is -3 and the y-intercept is 2.

16. $m = \dfrac{9-(-3)}{2-6} = \dfrac{12}{-4} = -3$

17.   See Figure 17.

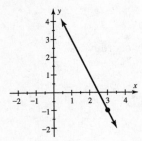

Figure 17

18.   The graph falls 1 unit for each 2 units of run. The slope is $-\dfrac{1}{2}$. The y-intercept is 2. So $y = -\dfrac{1}{2}x + 2$.

19.   Since the line is parallel to $y = 2x + 3$, the slope is $m = 2$. Using the point-slope form gives

$$y = 2(x-1) + 4 \Rightarrow y = 2x - 2 + 4 \Rightarrow y = 2x + 2$$

20.   Since the line is perpendicular to $y = \dfrac{2}{3}x - 2$, the slope is $m = -\dfrac{3}{2}$. Using the point-slope form gives

$$y = -\dfrac{3}{2}(x-2) + 1 \Rightarrow y = -\dfrac{3}{2}x + 3 + 1 \Rightarrow y = -\dfrac{3}{2}x + 4$$

21. $\dfrac{2}{5}(x+1) - 6 = -4 \Rightarrow \dfrac{2}{5}(x+1) = 2 \Rightarrow x+1 = 5 \Rightarrow x = 4$

22. $\dfrac{1}{4}\left(\dfrac{t-5}{3}\right) - 6 = \dfrac{2}{3}t - (3t+7) \Rightarrow t - 5 - 72 = 8t - 12(3t+7) \Rightarrow t - 77 = -28t - 84 \Rightarrow 29t = -7 \Rightarrow t = -\dfrac{7}{29}$

23.   The graphs intersect at the point $(3, -2)$, so $y_1 = y_2$ when $x = 3$.

24. $\dfrac{2x+3}{4} \le \dfrac{1}{3} \Rightarrow 6x + 9 \le 4 \Rightarrow 6x \le -5 \Rightarrow x \le -\dfrac{5}{6}$. The interval is $\left(-\infty, -\dfrac{5}{6}\right]$.

25. $\dfrac{1}{2}z - 5 > \dfrac{3}{4}z - (2z+5) \Rightarrow 2z - 20 > 3z - 4(2z+5) \Rightarrow 7z > 0 \Rightarrow z > 0$. The interval is $(0, \infty)$.

26. $-6 \le -\dfrac{2}{3}x - 4 < -2 \Rightarrow -2 \le -\dfrac{2}{3}x < 2 \Rightarrow 3 \ge x > -3 \Rightarrow -3 < x \le 3.$

      The interval is $(-3, 3]$. See Figure 26.

27. $3x - 1 \le 5$ or $2x + 5 > 13 \Rightarrow 3x \le 6$ or $2x > 8 \Rightarrow x \le 2$ or $x > 4.$

      The interval is $(-\infty, 2] \cup (4, \infty)$. See Figure 27.

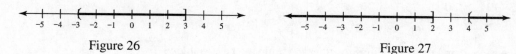

                     Figure 26   Figure 27

28. $\left| \dfrac{1}{3}x + 6 \right| = 4 \Rightarrow \dfrac{1}{3}x + 6 = -4 \Rightarrow \dfrac{1}{3}x = -10 \Rightarrow x = -30$ or $\dfrac{1}{3}x + 6 = 4 \Rightarrow \dfrac{1}{3}x = -2 \Rightarrow x = -6$

29.   The solutions to $|2x - 3| < 11$ satisfy $c < x < d$ where c and d are the solutions to $|2x - 3| = 11$.

      $|2x - 3| = 11$ is equivalent to $2x - 3 = -11 \Rightarrow x = -4$ and $2x - 3 = 11 \Rightarrow x = 7$. The interval is $(-4, 7)$.

30.   First divide each side of $-3|t - 5| \le -18$ by -3 to obtain $|t - 5| \ge 6$.

      The solutions to $|t - 5| \ge 6$ satisfy $t \le c$ or $t \ge d$ where c and d are the solutions to $|t - 5| = 6$.

      $|t - 5| = 6$ is equivalent to $t - 5 = -6 \Rightarrow t = -1$ and $t - 5 = 6 \Rightarrow t = 11$.

      The interval is $(-\infty, -1] \cup [11, \infty)$.

31.   By substitution, $(-8, -1)$ is a solution to the given system of equations.

32.   See Figure 32.

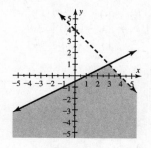

         Figure 32

33.   Multiplying the second equation by 4 and adding the two equations will eliminate the variable y.

      $\begin{array}{r} 2x - 8y = 5 \\ 16x + 8y = 4 \\ \hline 18x = 9 \end{array}$

                Thus, $x = \dfrac{1}{2}$. And so $2\left(\dfrac{1}{2}\right) - 8y = 5 \Rightarrow y = -\dfrac{1}{2}$. The solution is $\left(\dfrac{1}{2}, -\dfrac{1}{2}\right)$.

34.   Multiplying the second equation by 2 and adding the two equations will eliminate the variable x.

      $\begin{array}{r} 2x - 3y = 12 \\ -2x + 4y = -12 \\ \hline y = 0 \end{array}$ Thus, $y = 0$. And so $2x - 3(0) = 12 \Rightarrow x = 6$. The solution is $(6, 0)$.

35.   From the graph of the region of feasible solutions (not shown), the vertices are $(0, 0)$, $(0, 3)$, $(2, 2)$, and $(3, 0)$.

The maximum value of R occurs at one of the vertices. For $(0, 0)$, $R = 2(0) + 3(0) = 0$.

For $(0, 3)$, $R = 2(0) + 3(3) = 9$. For $(2, 2)$, $R = 2(2) + 3(2) = 10$. For $(3, 0)$, $R = 2(3) + 3(0) = 6$.

The maximum value is $R = 10$.

36. Multiply the second equation by 3 and add the first and second equations to eliminate the variable y.

$$\begin{array}{r} 2x + 3y - z = 3 \\ 9x - 3y + 12z = 30 \\ \hline 11x + 11z = 33 \end{array} \quad \text{or} \quad x + z = 3$$

Add the second and third equations together to eliminate the variable y.

$$\begin{array}{r} 3x - y + 4z = 10 \\ 2x + y - 2z = -1 \\ \hline 5x + 2z = 9 \end{array}$$

Multiply the first *new* equation by -2 and add the first *new* equation and second *new* equation to eliminate z.

$$\begin{array}{r} -2x - 2z = -6 \\ 5x + 2z = 9 \\ \hline 3x = 3 \end{array}$$

And so $x = 1$. Substitute $x = 1$ into the first *new* equation: $(1) + z = 3 \Rightarrow z = 2$

Substitute $x = 1$ and $z = 2$ into the *original* first equation: $2(1) + 3y - (2) = 3 \Rightarrow y = 1$

The solution is $(1, 1, 2)$.

37. $$\begin{bmatrix} 1 & 1 & -1 & | & 4 \\ -1 & -1 & -1 & | & 0 \\ 1 & -2 & 1 & | & -9 \end{bmatrix} \begin{matrix} R_2 + R_1 \to \\ R_3 - R_1 \to \end{matrix} \begin{bmatrix} 1 & 1 & -1 & | & 4 \\ 0 & 0 & -2 & | & 4 \\ 0 & -3 & 2 & | & -13 \end{bmatrix} \begin{matrix} Exchange \\ R_2 \leftrightarrow R_3 \end{matrix} \begin{bmatrix} 1 & 1 & -1 & | & 4 \\ 0 & -3 & 2 & | & -13 \\ 0 & 0 & -2 & | & 4 \end{bmatrix}$$

$$(-1/2)R_3 \to \begin{bmatrix} 1 & 1 & -1 & | & 4 \\ 0 & -3 & 2 & | & -13 \\ 0 & 0 & 1 & | & -2 \end{bmatrix} \begin{matrix} R_1 + R_3 \to \\ R_2 - 2R_3 \to \end{matrix} \begin{bmatrix} 1 & 1 & 0 & | & 2 \\ 0 & -3 & 0 & | & -9 \\ 0 & 0 & 1 & | & -2 \end{bmatrix} \begin{matrix} R_1 + (1/3)R_2 \to \\ (-1/3)R_2 \to \end{matrix} \begin{bmatrix} 1 & 0 & 0 & | & -1 \\ 0 & 1 & 0 & | & 3 \\ 0 & 0 & 1 & | & -2 \end{bmatrix}$$

The solution is $(-1, 3, -2)$.

38. $\det A = 4(3) - 1(-2) = 12 + 2 = 14$

39. The triangle has vertices $(-3, 2)$, $(2, 1)$ and $(1, -2)$. The matrix needed is $A = \begin{bmatrix} -3 & 2 & 1 \\ 2 & 1 & -2 \\ 1 & 1 & 1 \end{bmatrix}$.

The area is $D = \left| \dfrac{1}{2} \det([A]) \right| = 8 \text{ in}^2$.

40. $E = \det \begin{bmatrix} 8 & 7 \\ 18 & 5 \end{bmatrix} = 40 - 126 = -86; \quad F = \det \begin{bmatrix} 6 & 8 \\ -8 & 18 \end{bmatrix} = 108 + 64 = 172$

$D = \det \begin{bmatrix} 6 & 7 \\ -8 & 5 \end{bmatrix} = 30 + 56 = 86;$ The solution is $x = \dfrac{E}{D} = \dfrac{-86}{86} = -1$ and $y = \dfrac{F}{D} = \dfrac{172}{86} = 2$.

41. $3x^2 (x^3 + 5x - 2) = 3x^5 + 15x^3 - 6x^2$

42. $(2x - 5)(x + 3) = 2x^2 + 6x - 5x - 15 = 2x^2 + x - 15$

43. $6x + 3x^2 = 0 \Rightarrow 3x(2+x) = 0 \Rightarrow$ Either $3x = 0 \Rightarrow x = 0$ or $2 + x = 0 \Rightarrow x = -2$

44. $3a^3 - a^2 + 15a - 5 = a^2(3a-1) + 5(3a-1) = (3a-1)(a^2+5)$

45. $4x^2 + 5x - 6 = (4x-3)(x+2)$

46. $6x^3 - 9x^2 - 6x = 3x(2x^2 - 3x - 2) = 3x(x-2)(2x+1)$

47. $9a^2 - 4b^2 = (3a-2b)(3a+2b)$

48. $64t^3 + 27 = (4t+3)(16t^2 - 12t + 9)$

49. $3x^2 - 11x = 4 \Rightarrow 3x^2 - 11x - 4 = 0 \Rightarrow (3x+1)(x-4) = 0 \Rightarrow x = -\dfrac{1}{3}$ or $x = 4$

50. $x^4 - x^3 - 30x^2 = 0 \Rightarrow x^2(x^2 - x - 30) = 0 \Rightarrow x^2(x+5)(x-6) = 0 \Rightarrow x = 0$ or $x = -5$ or $x = 6$

51. $\dfrac{x^2 + 3x - 10}{x^2 - 4} \cdot \dfrac{x-2}{x+5} = \dfrac{(x+5)(x-2)(x-2)}{(x-2)(x+2)(x+5)} = \dfrac{x-2}{x+2}$

52. $\dfrac{x^2 + 2x - 24}{x^2 + 3x - 18} \div \dfrac{x+3}{x^2-9} = \dfrac{x^2 + 2x - 24}{x^2 + 3x - 18} \cdot \dfrac{x^2 - 9}{x+3} = \dfrac{(x+6)(x-4)(x-3)(x+3)}{(x+6)(x-3)(x+3)} = x - 4$

53. $\dfrac{2}{t+2} - \dfrac{t}{t^2-4} = \dfrac{2(t-2)}{(t-2)(t+2)} - \dfrac{t}{(t-2)(t+2)} = \dfrac{2t-4-t}{(t-2)(t+2)} = \dfrac{t-4}{t^2-4}$

54. $\dfrac{4a}{3ab^2} + \dfrac{b}{a^2c} = \dfrac{4a^2c}{3a^2b^2c} + \dfrac{3b^3}{3a^2b^2c} = \dfrac{4a^2c + 3b^3}{3a^2b^2c}$

55. $\dfrac{1}{x+4} + \dfrac{1}{x-4} = \dfrac{-7}{x^2-16} \Rightarrow 1(x-4) + 1(x+4) = -7 \Rightarrow 2x = -7 \Rightarrow x = -\dfrac{7}{2}$

56. $\dfrac{1}{y^2 + 3y - 4} - \dfrac{y}{y-1} = -1 \Rightarrow 1 - y(y+4) = -1(y+4)(y-1) \Rightarrow 1 - y^2 - 4y = -y^2 - 3y + 4 \Rightarrow$

    $1 - 4y = 4 - 3y \Rightarrow y = -3$

57. $J = \dfrac{y+z}{z} \Rightarrow Jz = y + z \Rightarrow Jz - z = y \Rightarrow z(J-1) = y \Rightarrow z = \dfrac{y}{J-1}$

58. $\dfrac{\dfrac{4}{x^2} + \dfrac{1}{x}}{\dfrac{4}{x^2} - \dfrac{1}{x}} = \dfrac{\dfrac{4}{x^2} + \dfrac{1}{x}}{\dfrac{4}{x^2} - \dfrac{1}{x}} \cdot \dfrac{x^2}{x^2} = \dfrac{4+x}{4-x}$

59. $y = \dfrac{k}{x} \Rightarrow 2 = \dfrac{k}{4} \Rightarrow k = 8 \Rightarrow y = \dfrac{8}{x}$, so when $x = 16$, $y = \dfrac{8}{16} = \dfrac{1}{2}$

60.

$$x+2\overline{)x^3+0x^2+2x+11}$$

$$\begin{array}{r} x^2-2x+6 \\ \underline{x^3+2x^2} \\ -2x^2+2x \\ \underline{-2x^2-4x} \\ 6x+11 \\ \underline{6x+12} \\ -1 \end{array}$$

The solution is: $x^2-2x+6-\dfrac{1}{x+2}$

61. $(f+g)(3)=\left(3^2-3(3)+2\right)+(3-2)=(9-9+2)+1\ =2+1=3$

62. $(f/g)(x)=\dfrac{x^2-3x+2}{x-2}=\dfrac{(x-2)(x-1)}{x-2}=x-1$

63.   (a) See Figure 63. These data are linear.

(b) $m=\dfrac{100-0}{212-32}=\dfrac{100}{180}=\dfrac{5}{9}$. Letting $h=32$ and $k=0$, the function is $f(x)=\dfrac{5}{9}(x-32)$.

(c) $f(104)=\dfrac{5}{9}(104-32)=\dfrac{5}{9}(72)=40°C$

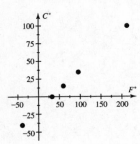

Figure 63

64.   Let x represent the length of each of the shorter sides. Then $\dfrac{1}{2}x+7$ represents the length of the longer side.

The perimeter is 22 so $x+x+\dfrac{1}{2}x+7=22\Rightarrow\dfrac{5}{2}x=15\Rightarrow x=6$. The sides are 6, 6, and 10 inches.

65.   Let x and y represent the time spent on the stair climber and the bicycle respectively. Then the system needed is $x+y=1.5$ and $690x+540y=885$. Multiplying the first equation by -540 and adding the two equations will eliminate the variable y.

$$\begin{array}{r} -540x-540y=-810 \\ 690x+540y=\ \ \ 885 \\ \hline 150x=75 \end{array}$$

Thus, $x=0.5$. And so $(0.5)+y=1.5\Rightarrow y=1$.

The athlete spent 0.5 hour on the stair climber and 1 hour on the stationary bicycle.

66.   Let x and y represent the number of $2 and $5 tickets respectively. Then the system needed is $x+y=30$ and $2x+5y=78$. Multiplying the first equation by -2 and adding the two equations will eliminate the x.

$$-2x-2y=-60$$
$$\underline{2x+5y=\ \ 78}$$
$$3y=18$$ Thus, $y=6$. And so $x+(6)=30 \Rightarrow x=24$. There were 24 children and 6 adults.

67.   Let x represent the width of the rectangle. Then $x+4$ represents the length.

$$x(x+4)=165 \Rightarrow x^2+4x-165=0 \Rightarrow (x-11)(x+15)=0 \Rightarrow x=11 \text{ or } x=-15$$

The solution $x=-15$ has no physical meaning. The rectangle is 11 feet by 15 feet.

68.   (a) $-16t^2+44t=0 \Rightarrow 4t^2-11t=0 \Rightarrow t(4t-11)=0 \Rightarrow t=0 \text{ or } 4t-11=0 \Rightarrow t=0 \text{ or } t=\dfrac{11}{4}$

       The ball strikes the ground at $\dfrac{11}{4}=2.75$ seconds.

     (b) $-16t^2+44t=18 \Rightarrow -16t^2+44t-18=0 \Rightarrow 8t^2-22t+9=0 \Rightarrow (2t-1)(4t-9) \Rightarrow t=\dfrac{1}{2} \text{ or } t=\dfrac{9}{4}$. The

       ball is 18 feet high at $\dfrac{1}{2}$ second and $\dfrac{9}{4}=2.25$ seconds.

69.   (a) $\dfrac{x}{15}+\dfrac{x}{10}=1$

     (b) $\dfrac{x}{15}+\dfrac{x}{10}=1 \Rightarrow 2x+3x=30 \Rightarrow 5x=30 \Rightarrow x=6$ hours

70. $\dfrac{6}{4}=\dfrac{x}{32} \Rightarrow 4x=192 \Rightarrow x=48$ feet

Critical Thinking Solutions for Chapter 6

Section 6.3

- No, for example $\dfrac{1}{2}=\dfrac{1}{4}+\dfrac{1}{4}$, but $2 \neq 4+4$.

Section 6.4

- At 6 miles per hour the bicyclist rides 1 mile up hill in $\dfrac{1}{6}$ hour. At 12 miles per hour the bicyclist rides 1 mile

     down hill in $\dfrac{1}{12}$ hour. Thus the bicyclist rides a total of 2 miles in $\dfrac{1}{6}+\dfrac{1}{12}=\dfrac{2}{12}+\dfrac{1}{12}=\dfrac{3}{12}=\dfrac{1}{4}$ hour.

     The average speed of the bicyclist is $2 \div \dfrac{1}{4}=8$ miles per hour.

Section 6.5

- No, $\dfrac{\dfrac{x}{y}}{2+\dfrac{x}{y}}=\dfrac{\dfrac{x}{y}}{2+\dfrac{x}{y}} \cdot \dfrac{y}{y}=\dfrac{x}{2y+x}$. No, $\dfrac{2+\dfrac{x}{y}}{\dfrac{x}{y}}=\dfrac{2+\dfrac{x}{y}}{\dfrac{x}{y}} \cdot \dfrac{y}{y}=\dfrac{2y+x}{x}$.

Section 6.6

• If the width doubles the strength doubles, whereas if the thickness doubles, the strength quadruples. If both triple, the strength increases by 27 times.

Section 6.7

• 1. Since $x - k$ is a factor, $p(x) = (x-k)(\text{some other factor})$. So $P(k) = (k-k)(\text{some other factor}) = 0$.

  2. Since $p(k) = 0$, an x-intercept on the graph must be k.

Chapter 7: Radical Expressions and Functions

7.1: Radical Expressions and Rational Exponents

Concepts

1. ± 3

2. 3

3. 2

4. Yes

5. b

6. $\sqrt{x} = x^{1/2}$

7. $\sqrt[3]{a^4} = a^{4/3}$

8. $\sqrt[n]{a}$

9. $\left(\sqrt[n]{a}\right)^m$ or $\sqrt[n]{a^m}$

10. Yes

Radical Expressions

11. $\sqrt{9} = 3$

13. $\sqrt{0.36} = 0.6$

15. $\sqrt{\dfrac{16}{25}} = \dfrac{4}{5}$

17. $\sqrt{x^2} = x$ since $x > 0$

19. $\sqrt[3]{27} = 3$

21. $\sqrt[3]{-64} = -4$

23. $\sqrt[3]{\dfrac{8}{27}} = \dfrac{2}{3}$

25. $-\sqrt[3]{x^9} = -\sqrt[3]{\left(x^3\right)^3} = -x^3$

27. $\sqrt[3]{(2x)^6} = \sqrt[3]{\left((2x)^2\right)^3} = (2x)^2 = 4x^2$

29. $\sqrt[4]{81} = 3$

31. $\sqrt[5]{-243} = \sqrt[5]{(-3)^5} = -5$

33. $\sqrt[4]{-16}$ is not possible to evaluate over the real numbers since the index is even and the radicand is negative.

35. $-\sqrt{5} \approx -2.24$

37. $\sqrt[3]{5} \approx 1.71$

39. $\sqrt[5]{-7} \approx -1.48$

41. $16^{1/5} \approx 1.74$

43. $5^{1/3} \approx 1.71$

45. $9^{3/5} \approx 3.74$

47. $4^{-3/7} \approx 0.55$

49. $9^{1/2} = \sqrt{9} = 3$

51. $8^{1/3} = \sqrt[3]{8} = 2$

53. $\left(\dfrac{4}{9}\right)^{1/2} = \sqrt{\dfrac{4}{9}} = \dfrac{2}{3}$

55. $(-8)^{2/3} = \sqrt[3]{(-8)^2} = \sqrt[3]{64} = 4$

57. $\left(\dfrac{1}{8}\right)^{-1/3} = 8^{1/3} = \sqrt[3]{8} = 2$

59. $16^{-3/4} = \dfrac{1}{\sqrt[4]{16^3}} = \dfrac{1}{\left(\sqrt[4]{16}\right)^3} = \dfrac{1}{(2)^3} = \dfrac{1}{8}$

61. $\left(4^{1/2}\right)^{-3} = \dfrac{1}{\left(\sqrt{4}\right)^3} = \dfrac{1}{2^3} = \dfrac{1}{8}$

63. $z^{1/4} = \sqrt[4]{z}$

65. $y^{-2/5} = \dfrac{1}{\sqrt[5]{y^2}}$

67. $(3x)^{1/3} = \sqrt[3]{3x}$

69. $\sqrt{y} = y^{1/2}$

71. $\sqrt{x} \cdot \sqrt{x} = \left(x^{1/2}\right)^2 = x$

73. $\sqrt[3]{8x^2} = \sqrt[3]{8} \cdot \sqrt[3]{x^2} = 2x^{2/3}$

75. $\dfrac{\sqrt{49x}}{\sqrt[3]{x^2}} = \dfrac{\sqrt{49} \cdot \sqrt{x}}{x^{2/3}} = 7x^{1/2-2/3} = 7x^{-1/6} = \dfrac{7}{x^{1/6}}$

77. $\left(b^{1/a}\, b^{2/a}\right)^a = \left(b^{1/a+2/a}\right)^a = \left(b^{3/a}\right)^a = b^{\frac{3}{a} \cdot a} = b^3$

79. $\left(x^2\right)^{3/2} = x^{2 \cdot 3/2} = x^3$

81. $\sqrt[3]{x^3 y^6} = \left(x^3 y^6\right)^{1/3} = x^{3 \cdot 1/3} \cdot y^{6 \cdot 1/3} = xy^2$

83. $\sqrt{\dfrac{y^4}{x^2}} = \left(\dfrac{y^4}{x^2}\right)^{1/2} = \dfrac{y^{4\cdot 1/2}}{x^{2\cdot 1/2}} = \dfrac{y^2}{x}$

85. $\sqrt{y^3} \cdot \sqrt[3]{y^2} = (y^3)^{1/2} \cdot (y^2)^{1/3} = y^{3\cdot 1/2} \cdot y^{2\cdot 1/3} = y^{3/2} \cdot y^{2/3} = y^{3/2 + 2/3} = y^{13/6}$

87. $\left(\dfrac{x^6}{27}\right)^{2/3} = \dfrac{x^{6\cdot 2/3}}{27^{2/3}} = \dfrac{x^4}{\left(\sqrt[3]{27}\right)^2} = \dfrac{x^4}{3^2} = \dfrac{x^4}{9}$

89. $\left(\dfrac{x^2}{y^6}\right)^{-1/2} = \left(\dfrac{y^6}{x^2}\right)^{1/2} = \dfrac{y^{6\cdot 1/2}}{x^{2\cdot 1/2}} = \dfrac{y^3}{x}$

91. $\sqrt{\sqrt{y}} = \left(y^{1/2}\right)^{1/2} = y^{1/2\cdot 1/2} = y^{1/4}$

93. $\left(a^{-1/2}\right)^{4/3} = a^{-1/2\cdot 4/3} = a^{-2/3} = \dfrac{1}{a^{2/3}}$

95. $\dfrac{\left(k^{1/2}\right)^{-3}}{\left(k^2\right)^{1/4}} = \dfrac{k^{-3/2}}{k^{1/2}} = k^{-3/2 - 1/2} = k^{-4/2} = k^{-2} = \dfrac{1}{k^2}$

97. $\sqrt{b} \cdot \sqrt[4]{b} = b^{1/2} \cdot b^{1/4} = b^{1/2 + 1/4} = b^{3/4}$

99. $p^{1/2}\left(p^{3/2} + p^{1/2}\right) = p^{1/2 + 3/2} + p^{1/2 + 1/2} = p^2 + p$

101. $\sqrt[3]{x}\left(\sqrt{x} - \sqrt[3]{x^2}\right) = x^{1/3}\left(x^{1/2} - x^{2/3}\right) = x^{1/3 + 1/2} - x^{1/3 + 2/3} = x^{5/6} - x$

103. $\sqrt{(-4)^2} = \sqrt{16} = 4$

105. $\sqrt{y^2} = |y|$

107. $\sqrt{(x-5)^2} = |x-5|$

109. $\sqrt{x^2 - 2x + 1} = \sqrt{(x-1)^2} = |x-1|$

111. $\sqrt[4]{y^4} = \left(y^4\right)^{1/4} = y^{4\cdot 1/4} = |y|$

113. $\sqrt[4]{x^{12}} = \left(x^{12}\right)^{1/4} = x^{12\cdot 1/4} = \left|x^3\right|$

115. $\sqrt[5]{x^5} = \left(x^5\right)^{1/5} = x^{5\cdot 1/5} = x$

Applications

117. $15\pi R^2 = 65 \;\Rightarrow\; R^2 = \dfrac{65}{15\pi} \;\Rightarrow\; R = \sqrt{\dfrac{65}{15\pi}} \approx 1.17$ miles

119. $N(h) = 1.6h^{-1/2} = 1.6(2.5)^{-1/2} = \dfrac{1.6}{\sqrt{2.5}} \approx 1.01;$ The stepping frequency is about 1 step per second.

121. $s = \dfrac{1}{2}(3+4+5) = 6 \;\Rightarrow\; A = \sqrt{6(6-3)(6-4)(6-5)} = \sqrt{6(3)(2)(1)} = \sqrt{36} = 6$

7.2: Simplifying Radical Expressions

Concepts

1.   Yes

2.   No

3. $\sqrt[3]{ab}$

4. $\dfrac{a}{b}$

5. $\dfrac{a}{b}$

6. $\dfrac{\sqrt{3}}{\sqrt{27}} = \sqrt{\dfrac{3}{27}} = \sqrt{\dfrac{1}{9}} = \dfrac{1}{3}$

7.   No; $\sqrt{50} = \sqrt{25 \cdot 2} = \sqrt{25} \cdot \sqrt{2} = 5\sqrt{2}$ and $\sqrt{25} + \sqrt{25}$

   $= 5 + 5 = 2(5)$ or 10

8.   Yes; $\sqrt{50} = \sqrt{25 \cdot 2} = \sqrt{25} \cdot \sqrt{2} = 5\sqrt{2}$

9.   No, because $1^3 \neq 3$.

10.   Yes, because $4^3 = 64$.

Multiplying and Dividing

11. $\sqrt{3} \cdot \sqrt{3} = \sqrt{3 \cdot 3} = \sqrt{9} = 3$

13. $\sqrt{2} \cdot \sqrt{50} = \sqrt{2 \cdot 50} = \sqrt{100} = 10$

15. $\sqrt[3]{4} \cdot \sqrt[3]{16} = \sqrt[3]{4 \cdot 16} = \sqrt[3]{64} = 4$

17. $\sqrt{\dfrac{9}{25}} = \dfrac{\sqrt{9}}{\sqrt{25}} = \dfrac{3}{5}$

19. $\sqrt{\dfrac{1}{2}} \cdot \sqrt{\dfrac{1}{8}} = \sqrt{\dfrac{1 \cdot 1}{2 \cdot 8}} = \sqrt{\dfrac{1}{16}} = \dfrac{\sqrt{1}}{\sqrt{16}} = \dfrac{1}{4}$

21. $\sqrt[3]{\dfrac{2}{3}} \cdot \sqrt[3]{\dfrac{4}{3}} \cdot \sqrt[3]{\dfrac{1}{3}} = \sqrt[3]{\dfrac{2}{3} \cdot \dfrac{4}{3} \cdot \dfrac{1}{3}} = \sqrt[3]{\dfrac{8}{27}} = \dfrac{2}{3}$

23. $\sqrt{x^3} \cdot \sqrt{x^3} = \sqrt{x^3 \cdot x^3} = \sqrt{x^6} = x^{6/2} = x^3$

25. $\sqrt[3]{\dfrac{7}{27}} = \dfrac{\sqrt[3]{7}}{\sqrt[3]{27}} = \dfrac{\sqrt[3]{7}}{3}$

27. $\sqrt[4]{\dfrac{x}{81}} = \dfrac{\sqrt[4]{x}}{\sqrt[4]{81}} = \dfrac{\sqrt[4]{x}}{3}$

29. $\sqrt{\dfrac{9}{z^2}} = \dfrac{\sqrt{9}}{\sqrt{z^2}} = \dfrac{3}{z}$

31. $\sqrt{\dfrac{x}{2}} \cdot \sqrt{\dfrac{x}{8}} = \sqrt{\dfrac{x \cdot x}{2 \cdot 8}} = \sqrt{\dfrac{x^2}{16}} = \dfrac{\sqrt{x^2}}{\sqrt{16}} = \dfrac{x}{4}$

33. $\dfrac{\sqrt{45}}{\sqrt{5}} = \sqrt{\dfrac{45}{5}} = \sqrt{9} = 3$

35. $\sqrt[3]{-4} \cdot \sqrt[3]{-16} = \sqrt[3]{-4 \cdot (-16)} = \sqrt[3]{64} = 4$

37. $\sqrt[4]{9} \cdot \sqrt[4]{9} = \sqrt[4]{9 \cdot 9} = \sqrt[4]{81} = 3$

39. $\dfrac{\sqrt[5]{64}}{\sqrt[5]{-2}} = \sqrt[5]{\dfrac{64}{-2}} = \sqrt[5]{-32} = -2$

41. $\dfrac{\sqrt{a^2 b}}{\sqrt{b}} = \sqrt{\dfrac{a^2 b}{b}} = \sqrt{a^2} = a$

43. $\dfrac{\sqrt[3]{54}}{\sqrt[3]{2}} = \sqrt[3]{\dfrac{54}{2}} = \sqrt[3]{27} = 3$

45. $\sqrt{4x^4} = \sqrt{4} \cdot \sqrt{\left(x^2\right)^2} = 2x^2$

47. $\sqrt[3]{-5a^6} = \sqrt[3]{-5} \cdot \sqrt[3]{\left(a^2\right)^3} = \sqrt[3]{-5} \cdot a^2 = -a^2 \sqrt[3]{5}$

49. $\sqrt[4]{16x^4 y} = \sqrt[4]{16} \cdot \sqrt[4]{x^4} \cdot \sqrt[4]{y} = 2x\sqrt[4]{y}$

51. $\sqrt{3x} \cdot \sqrt{12x} = \sqrt{3 \cdot 12 \cdot x \cdot x} = \sqrt{36x^2} = \sqrt{36} \cdot \sqrt{x^2} = 6x$

53. $\sqrt[3]{8x^6 y^3 z^9} = \sqrt[3]{8} \cdot \sqrt[3]{\left(x^2\right)^3} \cdot \sqrt[3]{y^3} \cdot \sqrt[3]{\left(z^3\right)^3} = 2x^2 yz^3$

55. $\sqrt[4]{\dfrac{3}{4}} \cdot \sqrt[4]{\dfrac{27}{4}} = \sqrt[4]{\dfrac{3}{4} \cdot \dfrac{27}{4}} = \sqrt[4]{\dfrac{81}{16}} = \dfrac{\sqrt[4]{81}}{\sqrt[4]{16}} = \dfrac{3}{2}$

57. $\sqrt[3]{12} \cdot \sqrt[3]{ab} = \sqrt[3]{12ab}$

59. $\sqrt[4]{25z} \cdot \sqrt[4]{25z} = \sqrt[4]{625z^2} = \sqrt[4]{625} \cdot \sqrt[4]{z^2} = 5\sqrt{z}$

61. $\sqrt[5]{\dfrac{7a}{b^2}} \cdot \sqrt[5]{\dfrac{b^2}{7a^6}} = \sqrt[5]{\dfrac{7ab^2}{7a^6 b^2}} = \sqrt[5]{\dfrac{1}{a^5}} = \dfrac{1}{a}$

63. $\sqrt{x+4} \cdot \sqrt{x-4} = \sqrt{(x+4)(x-4)} = \sqrt{x^2 - 16}$

65. $\sqrt[3]{a+1} \cdot \sqrt[3]{a^2 - a + 1} = \sqrt[3]{(a+1)\left(a^2 - a + 1\right)} = \sqrt[3]{a^3 + 1}$

67. $\dfrac{\sqrt{x^2 + 2x + 1}}{\sqrt{x+1}} = \sqrt{\dfrac{x^2 + 2x + 1}{x+1}} = \sqrt{\dfrac{(x+1)(x+1)}{x+1}} = \sqrt{x+1}$

69. $\sqrt{500} = \sqrt{100 \cdot 5} = \sqrt{100} \cdot \sqrt{5} = 10\sqrt{5}$; the answer is 10.

71. $\sqrt{8} = \sqrt{4 \cdot 2} = \sqrt{4} \cdot \sqrt{2} = 2\sqrt{2}$; the answer is 2.

73. $\sqrt{45} = \sqrt{9 \cdot 5} = \sqrt{9} \cdot \sqrt{5} = 3\sqrt{5}$; the answer is 3.

75. $\sqrt{200} = \sqrt{100 \cdot 2} = \sqrt{100} \cdot \sqrt{2} = 10\sqrt{2}$

77. $\sqrt[3]{81} = \sqrt[3]{27 \cdot 3} = \sqrt[3]{27} \cdot \sqrt[3]{3} = 3\sqrt[3]{3}$

79. $\sqrt[4]{64} = \sqrt[4]{16 \cdot 4} = \sqrt[4]{16} \cdot \sqrt[4]{4} = 2\sqrt[4]{4} = 2\sqrt[4]{2^2} = 2\sqrt{2}$

81. $\sqrt[5]{-64} = \sqrt[5]{-2^6} = \sqrt[5]{-2^5 \cdot 2} = \sqrt[5]{-2^5} \cdot \sqrt[5]{2} = -2\sqrt[5]{2}$

83. $\sqrt{b^5} = \sqrt{\left(b^2\right)^2 \cdot b} = \sqrt{\left(b^2\right)^2} \cdot \sqrt{b} = b^2\sqrt{b}$

85. $\sqrt{8n^3} = \sqrt{\left(2n\right)^2 \cdot 2n} = \sqrt{\left(2n\right)^2} \cdot \sqrt{2n} = 2n\sqrt{2n}$

87. $\sqrt{12a^2b^5} = \sqrt{\left(2ab^2\right)^2 \cdot 3b} = \sqrt{\left(2ab^2\right)^2} \cdot \sqrt{3b} = 2ab^2\sqrt{3b}$

89. $\sqrt[3]{-125x^4y^5} = \sqrt[3]{\left(-5xy\right)^3 \cdot xy^2} = \sqrt[3]{\left(-5xy\right)^3} \cdot \sqrt[3]{xy^2} = -5xy\sqrt[3]{xy^2}$

91. $\sqrt[3]{5t} \cdot \sqrt[3]{125t} = \sqrt[3]{625t^2} = \sqrt[3]{5^4t^2} = \sqrt[3]{5^3 \cdot 5t^2} = \sqrt[3]{5^3} \cdot \sqrt[3]{5t^2} = 5\sqrt[3]{5t^2}$

93. $\sqrt[4]{\dfrac{9t^5}{r^8}} \cdot \sqrt[4]{\dfrac{9r}{5t}} = \sqrt[4]{\dfrac{81rt^5}{5r^8t}} = \sqrt[4]{\dfrac{81t^4}{5r^7}} = \dfrac{\sqrt[4]{\left(3t\right)^4}}{\sqrt[4]{r^4 \cdot 5r^3}} = \dfrac{3t}{r\sqrt[4]{5r^3}}$

95. $\sqrt[3]{\dfrac{27x^2}{y^3}} = \dfrac{\sqrt[3]{27} \cdot \sqrt[3]{x^2}}{\sqrt[3]{y^3}} = \dfrac{3\sqrt[3]{x^2}}{y}$

97. $\sqrt{\dfrac{7a^2}{27}} \cdot \sqrt{\dfrac{7a}{3}} = \sqrt{\dfrac{49a^3}{81}} = \dfrac{\sqrt{49} \cdot \sqrt{a^3}}{\sqrt{81}} = \dfrac{7\sqrt{a^2} \cdot \sqrt{a}}{9}$

$= \dfrac{7a\sqrt{a}}{9}$

99. $\left(\sqrt[mn]{a^mb^m}\right)^n = \left(\sqrt[m]{a^mb^m}\right)^{n/n} = \sqrt[m]{(ab)^m} = (ab)^{m/m} = ab$

101. $\sqrt{3} \cdot \sqrt[3]{3} = 3^{1/2} \cdot 3^{1/3} = 3^{1/2+1/3} = 3^{5/6} = \sqrt[6]{3^5}$

103. $\sqrt[4]{8} \cdot \sqrt[3]{4} = \sqrt[4]{2^3} \cdot \sqrt[3]{2^2} = 2^{3/4} \cdot 2^{2/3} = 2^{3/4+2/3} = 2^{17/12} = 2^{12/12+5/12} = 2 \cdot 2^{5/12} = 2\sqrt[12]{2^5}$

105. $\sqrt[4]{27} \cdot \sqrt[3]{9} \cdot \sqrt{3} = \sqrt[4]{3^3} \cdot \sqrt[3]{3^2} \cdot \sqrt{3} = 3^{3/4} \cdot 3^{2/3} \cdot 3^{1/2} = 3^{3/4+2/3+1/2} = 3^{23/12} = 3^{12/12} \cdot 3^{11/12} = 3\sqrt[12]{3^{11}}$

107. $\sqrt[4]{x^3} \cdot \sqrt[3]{x} = x^{3/4} \cdot x^{1/3} = x^{3/4+1/3} = x^{13/12} = x^{12/12} \cdot x^{1/12} = x\sqrt[12]{x}$

109. $\sqrt[4]{rt} \cdot \sqrt[3]{r^2t} = (rt)^{1/4} \cdot \left(r^2t\right)^{1/3} = r^{1/4}t^{1/4} \cdot r^{2/3}t^{1/3} = r^{1/4+2/3}t^{1/4+1/3} = r^{11/12}t^{7/12} = \sqrt[12]{r^{11}t^7}$

Applications

111.   (a) $A = 100\sqrt[3]{8^2} = 100\sqrt[3]{64} = 100 \cdot 4 = 400$ square inches

(b) $A = 100\sqrt[3]{W^2} \Rightarrow A = 100W^{2/3}$

Checking Basic Concepts for Sections 7.1 & 7.2

1.   (a) ± 7

(b) 7

2.   (a) $\sqrt[3]{-8} = -2$

(b) $-\sqrt[4]{81} = -3$

3.   (a) $x^{3/2} = \sqrt{x^3}$ or $\left(\sqrt{x}\right)^3$

(b) $x^{2/3} = \sqrt[3]{x^2}$ or $\left(\sqrt[3]{x}\right)^2$

(c) $x^{-2/5} = \dfrac{1}{\sqrt[5]{x^2}}$ or $\dfrac{1}{\left(\sqrt[5]{x}\right)^2}$

4. $\sqrt{(x-1)^2} = |x-1|$

5.   (a) $\left(64^{-3/2}\right)^{1/3} = 64^{-3/2 \cdot 1/3} = 64^{-1/2} = \dfrac{1}{64^{1/2}} = \dfrac{1}{\sqrt{64}} = \dfrac{1}{8}$

(b) $\sqrt{5} \cdot \sqrt{20} = \sqrt{5 \cdot 20} = \sqrt{100} = 10$

(c) $\sqrt[3]{-8x^4 y} = \sqrt[3]{(-2x)^3 \cdot xy} = -2x\sqrt[3]{xy}$

(d) $\sqrt{\dfrac{4b}{5}} \cdot \sqrt{\dfrac{4b^3}{5}} = \sqrt{\dfrac{4b \cdot 4b^3}{5 \cdot 5}} = \dfrac{\sqrt{16b^4}}{\sqrt{25}} = \dfrac{4b^2}{5}$

6. $\sqrt[3]{7} \cdot \sqrt{7} = 7^{1/3} \cdot 7^{1/2} = 7^{1/3+1/2} = 7^{5/6} = \sqrt[6]{7^5}$

7.3: Operations on Radical Expressions

Concepts

1. $2\sqrt{a}$

2. $3\sqrt[3]{b}$

3.   like

4.   Yes, these are like radicals. $4\sqrt{15} - 3\sqrt{15} = (4-3)\sqrt{15} = \sqrt{15}$

5.   No; 6 and $3\sqrt{5}$ are not like radicals.

6. $\dfrac{\sqrt{7}}{\sqrt{7}}$

7. $\sqrt{t} + 5$

8.   The conjugate of the denominator over itself: $\dfrac{5+\sqrt{2}}{5+\sqrt{2}}$

Like Radicals

9.   Not possible, since $\sqrt{12} = 2\sqrt{3}$ and $\sqrt{24} = 2\sqrt{6}$.

11.   Since $\sqrt{28} = \sqrt{4 \cdot 7} = \sqrt{4} \cdot \sqrt{7} = 2\sqrt{7}$ and $\sqrt{63} = \sqrt{9 \cdot 7} = \sqrt{9} \cdot \sqrt{7} = 3\sqrt{7}$, the like radicals are

$\sqrt{7}, 2\sqrt{7}$, and $3\sqrt{7}$.

13.   Since $\sqrt[3]{16} = \sqrt[3]{8 \cdot 2} = \sqrt[3]{8} \cdot \sqrt[3]{2} = 2\sqrt[3]{2}$ and $\sqrt[3]{-54} = \sqrt[3]{-27 \cdot 2} = \sqrt[3]{-27} \cdot \sqrt[3]{2} = -3\sqrt[3]{2}$, the like radicals are

$2\sqrt[3]{2}$ and $-3\sqrt[3]{2}$.

15.   Not possible, since $\sqrt{x^2 y} = x\sqrt{y}$ and $\sqrt{4y^2} = 2y$.

17.   Since $\sqrt[3]{8xy} = \sqrt[3]{8} \cdot \sqrt[3]{xy} = 2\sqrt[3]{xy}$ and $\sqrt[3]{x^4 y^4} = \sqrt[3]{(xy)^3 \cdot xy} = \sqrt[3]{(xy)^3} \cdot \sqrt[3]{xy} = xy\sqrt[3]{xy}$, the like radicals are

$2\sqrt[3]{xy}$ and $xy\sqrt[3]{xy}$.

Addition and Subtraction of Radicals

19. $2\sqrt{3} + 7\sqrt{3} = 9\sqrt{3}$

21. $4\sqrt[3]{5} + 2\sqrt[3]{5} = 6\sqrt[3]{5}$

23.   Not possible, since 7 and $4\sqrt{7}$ are not like radicals

25.   Not possible, since $2\sqrt{3}$ and $3\sqrt{2}$ are not like radicals

27.   Not possible, since $\sqrt{3}$ and $\sqrt[3]{3}$ are not like radicals

29. $\sqrt[3]{16} + 3\sqrt[3]{2} = \sqrt[3]{8 \cdot 2} + 3\sqrt[3]{2} = \sqrt[3]{8} \cdot \sqrt[3]{2} + 3\sqrt[3]{2} = 2\sqrt[3]{2} + 3\sqrt[3]{2} = 5\sqrt[3]{2}$

31. $\sqrt{2} + \sqrt{18} + \sqrt{32} = \sqrt{2} + \sqrt{9 \cdot 2} + \sqrt{16 \cdot 2} = \sqrt{2} + \sqrt{9} \cdot \sqrt{2} + \sqrt{16} \cdot \sqrt{2} = \sqrt{2} + 3\sqrt{2} + 4\sqrt{2} = 8\sqrt{2}$

33. $11\sqrt{11} - 5\sqrt{11} = 6\sqrt{11}$

35. $\sqrt{x} + \sqrt{x} - \sqrt{y} = 2\sqrt{x} - \sqrt{y}$

37. $\sqrt[3]{z} + \sqrt[3]{z} = 2\sqrt[3]{z}$

39. $2\sqrt[3]{6} - 7\sqrt[3]{6} = -5\sqrt[3]{6}$

41. $\sqrt[3]{y^6} - \sqrt[3]{y^3} = \sqrt[3]{(y^2)^3} - \sqrt[3]{y^3} = y^2 - y$

43. $3\sqrt{28} + 3\sqrt{7} = 3\sqrt{4 \cdot 7} + 3\sqrt{7} = 3 \cdot 2\sqrt{7} + 3\sqrt{7} = 9\sqrt{7}$

45. $\sqrt[4]{48} + 4\sqrt[4]{3} = \sqrt[4]{16 \cdot 3} + 4\sqrt[4]{3} = \sqrt[4]{16} \cdot \sqrt[4]{3} + 4\sqrt[4]{3} = 2\sqrt[4]{3} + 4\sqrt[4]{3} = 6\sqrt[4]{3}$

47. $\sqrt{9x} + \sqrt{16x} = \sqrt{9} \cdot \sqrt{x} + \sqrt{16} \cdot \sqrt{x} = 3\sqrt{x} + 4\sqrt{x} = 7\sqrt{x}$

49. $3\sqrt{2k} + \sqrt{8k} + \sqrt{18k} = 3\sqrt{2k} + \sqrt{4 \cdot 2k} + \sqrt{9 \cdot 2k} = 3\sqrt{2k} + \sqrt{4} \cdot \sqrt{2k} + \sqrt{9} \cdot \sqrt{2k}$

$$= 3\sqrt{2k} + 2\sqrt{2k} + 3\sqrt{2k} = 8\sqrt{2k}$$

51. $\sqrt{44} - 4\sqrt{11} = \sqrt{4 \cdot 11} - 4\sqrt{11} = 2\sqrt{11} - 4\sqrt{11} = -2\sqrt{11}$

53. $2\sqrt[3]{16} + \sqrt[3]{2} - \sqrt{2} = 2\sqrt[3]{8 \cdot 2} + \sqrt[3]{2} - \sqrt{2} = 2 \cdot 2\sqrt[3]{2} + \sqrt[3]{2} - \sqrt{2} = 5\sqrt[3]{2} - \sqrt{2}$

55. $\sqrt[3]{xy} - 2\sqrt[3]{xy} = -\sqrt[3]{xy}$

57. $\sqrt{4x+8} + \sqrt{x+2} = \sqrt{4(x+2)} + \sqrt{x+2} = 2\sqrt{x+2} + \sqrt{x+2} = 3\sqrt{x+2}$

59. $\sqrt{25x^3} - \sqrt{x^3} = \sqrt{25} \cdot \sqrt{x^2} \cdot \sqrt{x} - \sqrt{x^2} \cdot \sqrt{x} = 5x\sqrt{x} - x\sqrt{x} = 4x\sqrt{x}$

61. $\sqrt[3]{\dfrac{7x}{8}} - \dfrac{\sqrt[3]{7x}}{3} = \dfrac{\sqrt[3]{7x}}{\sqrt[3]{8}} - \dfrac{\sqrt[3]{7x}}{3} = \dfrac{\sqrt[3]{7x}}{2} - \dfrac{\sqrt[3]{7x}}{3} = \dfrac{3\sqrt[3]{7x}}{6} - \dfrac{2\sqrt[3]{7x}}{6} = \dfrac{\sqrt[3]{7x}}{6}$

63. $\dfrac{4\sqrt{3}}{3} + \dfrac{\sqrt{3}}{6} = \dfrac{4\sqrt{3}}{3} \cdot \dfrac{2}{2} + \dfrac{\sqrt{3}}{6} = \dfrac{8\sqrt{3}}{6} + \dfrac{\sqrt{3}}{6} = \dfrac{8\sqrt{3} + \sqrt{3}}{6} = \dfrac{9\sqrt{3}}{6} = \dfrac{3\sqrt{3}}{2}$

65. $\dfrac{15\sqrt{8}}{4} - \dfrac{2\sqrt{2}}{5} = \dfrac{15 \cdot 2\sqrt{2}}{4} \cdot \dfrac{5}{5} - \dfrac{2\sqrt{2}}{5} \cdot \dfrac{4}{4} = \dfrac{150\sqrt{2}}{20} - \dfrac{8\sqrt{2}}{20} = \dfrac{150\sqrt{2} - 8\sqrt{2}}{20} = \dfrac{142\sqrt{2}}{20} = \dfrac{71\sqrt{2}}{10}$

67. $2\sqrt[4]{64} - \sqrt[4]{324} + \sqrt[4]{4} = 2\sqrt[4]{16 \cdot 4} - \sqrt[4]{81 \cdot 4} + \sqrt[4]{4} = 4\sqrt[4]{4} - 3\sqrt[4]{4} + \sqrt[4]{4} = 2\sqrt[4]{4} = 2\sqrt{2}$

69. $5\sqrt[4]{x^5} - \sqrt[4]{x} = 5\sqrt[4]{x^4 \cdot x} - \sqrt[4]{x} = 5x\sqrt[4]{x} - \sqrt[4]{x} = (5x-1)\sqrt[4]{x}$

71. $\sqrt{64x^3} - \sqrt{x} + 3\sqrt{x} = \sqrt{(8x)^2 \cdot x} - \sqrt{x} + 3\sqrt{x} = 8x\sqrt{x} - \sqrt{x} + 3\sqrt{x} = 2\sqrt{x}(4x+1)$

73. $\sqrt[4]{81a^5b^5} - \sqrt[4]{ab} = \sqrt[4]{(3ab)^4 \cdot ab} - \sqrt[4]{ab} = 3ab\sqrt[4]{ab} - \sqrt[4]{ab} = (3ab-1)\sqrt[4]{ab}$

75. $5\sqrt[3]{\dfrac{n^4}{125}} - 2\sqrt[3]{n} = 5\sqrt[3]{\dfrac{n^3}{125} \cdot n} - 2\sqrt[3]{n} = 5 \cdot \dfrac{n}{5}\sqrt[3]{n} - 2\sqrt[3]{n} = n\sqrt[3]{n} - 2\sqrt[3]{n} = (n-2)\sqrt[3]{n}$

Multiplying Binomials Containing Radicals

77. $(\sqrt{x} - 3)(\sqrt{x} + 2) = (\sqrt{x})^2 + 2\sqrt{x} - 3\sqrt{x} - 6 = x - \sqrt{x} - 6$

79. $(3 + \sqrt{7})(3 - \sqrt{7}) = 3^2 - (\sqrt{7})^2 = 9 - 7 = 2$

81. $(11 - \sqrt{2})(11 + \sqrt{2}) = 11^2 - (\sqrt{2})^2 = 121 - 2 = 119$

83. $(\sqrt{x} + 8)(\sqrt{x} - 8) = (\sqrt{x})^2 - 8^2 = x - 64$

85. $(\sqrt{ab} - \sqrt{c})(\sqrt{ab} + \sqrt{c}) = (\sqrt{ab})^2 - (\sqrt{c})^2 = ab - c$

87. $(\sqrt{x} - 7)(\sqrt{x} + 8) = (\sqrt{x})^2 + 8\sqrt{x} - 7\sqrt{x} - 56 = x + \sqrt{x} - 56$

Rationalizing the Denominator

89. $\dfrac{1}{\sqrt{7}} = \dfrac{1}{\sqrt{7}} \cdot \dfrac{\sqrt{7}}{\sqrt{7}} = \dfrac{\sqrt{7}}{(\sqrt{7})^2} = \dfrac{\sqrt{7}}{7}$

91. $\dfrac{4}{\sqrt{3}} = \dfrac{4}{\sqrt{3}} \cdot \dfrac{\sqrt{3}}{\sqrt{3}} = \dfrac{4\sqrt{3}}{\left(\sqrt{3}\right)^2} = \dfrac{4\sqrt{3}}{3}$

93. $\dfrac{5}{3\sqrt{5}} = \dfrac{5}{3\sqrt{5}} \cdot \dfrac{\sqrt{5}}{\sqrt{5}} = \dfrac{5\sqrt{5}}{3\left(\sqrt{5}\right)^2} = \dfrac{5\sqrt{5}}{3\cdot 5} = \dfrac{\sqrt{5}}{3}$

95. $\sqrt{\dfrac{b}{12}} = \dfrac{\sqrt{b}}{\sqrt{12}} \cdot \dfrac{\sqrt{12}}{\sqrt{12}} = \dfrac{\sqrt{12b}}{\left(\sqrt{12}\right)^2} = \dfrac{\sqrt{4\cdot 3b}}{12} = \dfrac{\sqrt{4}\cdot\sqrt{3b}}{12} = \dfrac{2\sqrt{3b}}{12} = \dfrac{\sqrt{3b}}{6}$

97. $\dfrac{rt}{2\sqrt{r^3}} = \dfrac{rt}{2\sqrt{r^2}\cdot\sqrt{r}} \cdot \dfrac{\sqrt{r}}{\sqrt{r}} = \dfrac{rt\sqrt{r}}{2r\left(\sqrt{r}\right)^2} = \dfrac{t\sqrt{r}}{2r}$

99. $\dfrac{1}{3-\sqrt{2}} = \dfrac{1}{3-\sqrt{2}} \cdot \dfrac{3+\sqrt{2}}{3+\sqrt{2}} = \dfrac{3+\sqrt{2}}{9-2} = \dfrac{3+\sqrt{2}}{7}$

101. $\dfrac{\sqrt{2}}{\sqrt{5}+2} = \dfrac{\sqrt{2}}{\sqrt{5}+2} \cdot \dfrac{\sqrt{5}-2}{\sqrt{5}-2} = \dfrac{\sqrt{10}-2\sqrt{2}}{5-4} = \dfrac{\sqrt{10}-2\sqrt{2}}{1} = \sqrt{10}-2\sqrt{2}$

103. $\dfrac{\sqrt{7}-2}{\sqrt{7}+2} = \dfrac{\sqrt{7}-2}{\sqrt{7}+2} \cdot \dfrac{\sqrt{7}-2}{\sqrt{7}-2} = \dfrac{7-4\sqrt{7}+4}{7-4} = \dfrac{11-4\sqrt{7}}{3}$

105. $\dfrac{1}{\sqrt{7}-\sqrt{6}} = \dfrac{1}{\sqrt{7}-\sqrt{6}} \cdot \dfrac{\sqrt{7}+\sqrt{6}}{\sqrt{7}+\sqrt{6}} = \dfrac{\sqrt{7}+\sqrt{6}}{7-6} = \dfrac{\sqrt{7}+\sqrt{6}}{1} = \sqrt{7}+\sqrt{6}$

107. $\dfrac{\sqrt{z}}{\sqrt{z}-3} = \dfrac{\sqrt{z}}{\sqrt{z}-3} \cdot \dfrac{\sqrt{z}+3}{\sqrt{z}+3} = \dfrac{z+3\sqrt{z}}{z-9}$

109. $\dfrac{\sqrt{a}+\sqrt{b}}{\sqrt{a}-\sqrt{b}} = \dfrac{\sqrt{a}+\sqrt{b}}{\sqrt{a}-\sqrt{b}} \cdot \dfrac{\sqrt{a}+\sqrt{b}}{\sqrt{a}+\sqrt{b}} = \dfrac{a+2\sqrt{ab}+b}{a-b}$

111. $\dfrac{1}{\sqrt{x+1}-\sqrt{x}} = \dfrac{1}{\sqrt{x+1}-\sqrt{x}} \cdot \dfrac{\sqrt{x+1}+\sqrt{x}}{\sqrt{x+1}+\sqrt{x}} = \dfrac{\sqrt{x+1}+\sqrt{x}}{x+1-x} = \dfrac{\sqrt{x+1}+\sqrt{x}}{1} = \sqrt{x+1}+\sqrt{x}$

113. $\dfrac{3}{\sqrt[3]{x}} = \dfrac{3}{x^{1/3}} \cdot \dfrac{x^{2/3}}{x^{2/3}} = \dfrac{3x^{2/3}}{x} = \dfrac{3\sqrt[3]{x^2}}{x}$

115. $\dfrac{1}{\sqrt[3]{x^2}} = \dfrac{1}{x^{2/3}} \cdot \dfrac{x^{1/3}}{x^{1/3}} = \dfrac{x^{1/3}}{x} = \dfrac{\sqrt[3]{x}}{x}$

Geometry

117. $\sqrt{27} + \sqrt{48} + \sqrt{75} = \sqrt{9\cdot 3} + \sqrt{16\cdot 3} + \sqrt{25\cdot 3} = 3\sqrt{3} + 4\sqrt{3} + 5\sqrt{3} = 12\sqrt{3} \approx 20.8$ cm

7.4: Radical Functions

Concepts

1. See Figure 1.

2. See Figure 2.

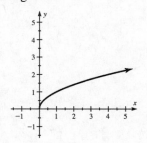

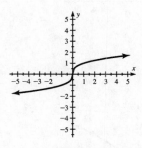

Figure 1   Figure 2

3. $\{x \mid x \geq 0\}$

4. All real numbers

5. $f(x) = x^p$, where p is rational

6. $f(x) = \sqrt[n]{x}$, where n is an integer and $n \geq 2$

7. The variable cannot be negative. The domain is $\{x \mid x \geq 0\}$.

8. All real numbers.

Root Functions

9. $f(10) = \sqrt{10-1} = \sqrt{9} = 3$; $f(0)$ is not possible because $\sqrt{0-1} = \sqrt{-1}$

11. $f(-1) = \sqrt{3-3(-1)} = \sqrt{3+3} = \sqrt{6}$; $f(5)$ is not possible because $\sqrt{3-3(5)} = \sqrt{3-15} = \sqrt{-12}$

13. $f(-4) = \sqrt{(-4)^2 - (-4)} = \sqrt{16+4} = \sqrt{20} = \sqrt{4 \cdot 5} = 2\sqrt{5}$;

$f(3) = \sqrt{3^2 - 3} = \sqrt{9-3} = \sqrt{6}$

15. $f(-3) = \sqrt[3]{(-3)^2 - 8} = \sqrt[3]{9-8} = \sqrt[3]{1} = 1$; $f(4) = \sqrt[3]{4^2 - 8} = \sqrt[3]{16-8} = \sqrt[3]{8} = 2$

17. $f(1) = \sqrt[3]{1-9} = \sqrt[3]{-8} = -2$; $f(10) = \sqrt[3]{10-9} = \sqrt[3]{1} = 1$

19. $f(-2) = \sqrt[3]{3 - (-2)^2} = \sqrt[3]{3-4} = \sqrt[3]{-1} = -1$;

$f(3)\sqrt[3]{3 - 3^2} = \sqrt[3]{3-9} = \sqrt[3]{-6}$ or $-\sqrt[3]{6}$

21. $T(64) = \frac{1}{2}\sqrt{64} = \frac{1}{2}(8) = 4$

23. $f(4) = \sqrt{4+5} + \sqrt{4} = \sqrt{9} + \sqrt{4} = 3+2 = 5$

25. $f(x) = \sqrt{x} + 1$

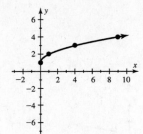

x	$\sqrt{x}+1$
–1	—
0	1
1	2
4	3
9	4

27. $f(x) = \sqrt{3x}$

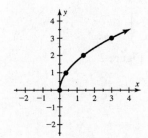

x	$\sqrt{3x}$
–1	—
0	0
$\frac{1}{3}$	1
$\frac{4}{3}$	2
3	3

29. $f(x) = 2\sqrt[3]{x}$

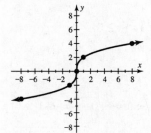

x	$2\sqrt[3]{x}$
–8	–4
–1	–2
0	0
1	2
8	4

31. $f(x) = \sqrt[3]{x} - 1$

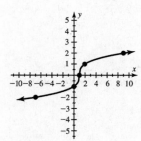

x	$\sqrt[3]{x}-1$
–7	–2
0	–1
1	0
2	1
9	2

33. $x + 2 \geq 0 \Rightarrow x \geq -2 \Rightarrow$ Domain: $[-2, \infty)$

35. $x - 2 \geq 0 \Rightarrow x \geq 2 \Rightarrow$ Domain: $[2, \infty)$

37. $2x - 4 \geq 0 \Rightarrow 2x \geq 4 \Rightarrow x \geq 2 \Rightarrow$ Domain: $[2, \infty)$

39. $1 - x \geq 0 \Rightarrow -x \geq -1 \Rightarrow x \leq 1 \Rightarrow$ Domain: $(-\infty, 1]$

41. $8 - 5x \geq 0 \Rightarrow -5x \geq -8 \Rightarrow x \leq \dfrac{8}{5} \Rightarrow$ Domain: $\left(-\infty, \dfrac{5}{8}\right]$

43. $3x^2 + 4 \geq 0 \Rightarrow 3x^2 \geq -4 \Rightarrow x^2 \geq -\dfrac{4}{3} \Rightarrow$ Domain: $(-\infty, \infty)$

45. $2x + 1 > 0 \Rightarrow 2x > -1 \Rightarrow x > -\dfrac{1}{2} \Rightarrow$ Domain: $\left(-\dfrac{1}{2}, \infty\right)$

47.   See Figure 47. This graph is shifted 2 units left.

49.   See Figure 49. This graph is shifted 2 units upward.

51.   See Figure 51. This graph is shifted 2 units left.

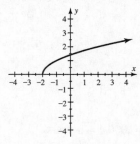
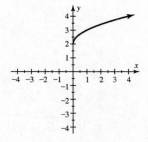
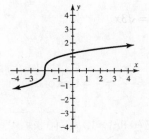

     Figure 47   Figure 49   Figure 51

Power Functions

53. $f(x) = x^{1/2} = \sqrt{x}$

55. $f(x) = x^{2/3} = \sqrt[3]{x^2}$

57. $f(x) = x^{-1/5} = \dfrac{1}{x^{1/5}} = \dfrac{1}{\sqrt[5]{x}}$

59. $f(4) = 4^{5/2} = \left(\sqrt{4}\right)^5 = 2^5 = 32; \ f(5) = 5^{5/2} \approx 55.90$

61. $f(-32) = (-32)^{7/5} = \dfrac{1}{(-32)^{7/5}} = \dfrac{1}{\left(\sqrt[5]{-32}\right)^7} = \dfrac{1}{(-2)^7} = -\dfrac{1}{128} \approx -0.01; \ f(10) = 10^{-7/5} = \dfrac{1}{10^{7/5}} \approx 0.04$

63. $f(256) = 256^{1/4} = \sqrt[4]{256} = 4; \ f(-10) = (-10)^{1/4} = \sqrt[4]{-10} \Rightarrow$ Not possible

65. $f(32) = 32^{2/5} = \left(\sqrt[5]{32}\right)^2 = 2^2 = 4; \ f(-32) = (-32)^{2/5} = \left(\sqrt[5]{-32}\right)^2 = (-2)^2 = 4$

67.   Graph $Y_1 = X \wedge (1/5)$ and $Y_2 = X \wedge (1/3)$ in $[0, 6, 1]$ by $[0, 6, 1]$. See Figure 67. Function $g(x)$ increases faster.

69.   Graph $Y_1 = X \wedge 1.2$ and $Y_2 = X \wedge 0.45$ in $[0, 6, 1]$ by $[0, 6, 1]$. See Figure 69. Function $f(x)$ increases faster.

     $[0, 6, 1]$ by $[0, 6, 1]$ $[0, 6, 1]$ by $[0, 6, 1]$

     Figure 67   Figure 69

71. $x^p > x^q$

73. (a) $(f+g)(2) = \sqrt{8 \cdot 2} + \sqrt{2 \cdot 2} = \sqrt{16} + \sqrt{4} = 4 + 2 = 6$

    (b) $(f-g)(x) = \sqrt{8x} - \sqrt{2x} = \sqrt{4 \cdot 2x} - \sqrt{2x} = 2\sqrt{2x} - \sqrt{2x} = \sqrt{2x}$

    (c) $(fg)(x) = \left(\sqrt{8x}\right)\left(\sqrt{2x}\right) = \sqrt{8x \cdot 2x} = \sqrt{16x^2} = 4|x|$

    (d) $(f/g)(x) = \dfrac{\sqrt{8x}}{\sqrt{2x}} = \sqrt{\dfrac{8x}{2x}} = \sqrt{4} = 2$

75. b

77. c

Applications

79. $T(4) = \dfrac{\sqrt{4}}{2} = \dfrac{2}{2} = 1$ second

81. (a) $T(0.8c) = 10\sqrt{1 - (0.8c/c)^2} = 10\sqrt{1 - 0.8^2} = 10\sqrt{1 - 0.64} = 10\sqrt{0.36} = 10 \cdot 0.6 = 6$ years

    (b) The twin in the spaceship will be 4 years younger than the twin on Earth.

83. (a)

[0, 5, 1] by [0, 0.5, 0.1]

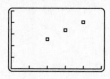

    (b) $A(2) = k(2)^{2/3} = 0.254 \Rightarrow k = \dfrac{0.254}{2^{2/3}} \Rightarrow k \approx 0.16$

    (c)

[0, 5, 1] by [0, 0.5, 0.1]

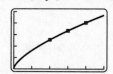

Yes

    (d) $A(2.5) = 0.16(2.5)^{2/3} \approx 0.295$ square meters

85.   (a) $L = kW^{1/3} \Rightarrow 0.422 = k \cdot 0.1^{1/3} \Rightarrow k = \dfrac{0.422}{0.1^{1/3}} \approx 0.91$

      (b)   Plot the data and graph $Y_1 = 0.91X \wedge (1/3)$ in [0, 1.5, 0.1] by [0, 1, 0.1]. See Figure 85. It increases.

      (c) $L = 0.91(0.7)^{1/3} \approx 0.808$ meters

      (d) $L = 0.91(0.65)^{1/3} \approx 0.788$;   A bird weighing 0.65 kg has a wing span of about 0.788 meters.

        [0, 1600, 400] by [0, 220, 20]

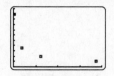

         Figure 85

87. $S(150) = 342(150)^{0.425} \approx 2877$ square inches

Checking Basic Concepts for Sections 7.3 & 7.4

1.   (a) $\sqrt{3} \cdot \sqrt{12} = \sqrt{3 \cdot 12} = \sqrt{36} = 6$

      (b) $\dfrac{\sqrt[3]{81}}{\sqrt[3]{3}} = \sqrt[3]{\dfrac{81}{3}} = \sqrt[3]{27} = 3$

      (c) $\sqrt{36x^6} = \sqrt{36} \cdot \sqrt{\left(x^3\right)^2} = 6x^3$

2.   (a) $5\sqrt{6} + 2\sqrt{6} + \sqrt{7} = 7\sqrt{6} + \sqrt{7}$

      (b) $8\sqrt[3]{x} - 3\sqrt[3]{x} = 5\sqrt[3]{x}$

      (c) $\sqrt{9x} - \sqrt{4x} = \sqrt{9} \cdot \sqrt{x} - \sqrt{4} \cdot \sqrt{x} = 3\sqrt{x} - 2\sqrt{x} = \sqrt{x}$

3.   (a) $\sqrt[3]{xy^4} - \sqrt[3]{x^4 y} = \sqrt[3]{y^3 \cdot xy} - \sqrt[3]{x^3 \cdot xy} = y\sqrt[3]{xy} - x\sqrt[3]{xy} = (y - x)\sqrt[3]{xy}$

      (b) $\left(4 - \sqrt{2}\right)\left(4 + \sqrt{2}\right) = 4^2 - \left(\sqrt{2}\right)^2 = 16 - 2 = 14$

4. $\dfrac{6}{2\sqrt{6}} = \dfrac{6}{2\sqrt{6}} \cdot \dfrac{\sqrt{6}}{\sqrt{6}} = \dfrac{6\sqrt{6}}{2 \cdot 6} = \dfrac{\sqrt{6}}{2}$

5. $\dfrac{2}{\sqrt{5} - 1} = \dfrac{2}{\sqrt{5} - 1} \cdot \dfrac{\sqrt{5} + 1}{\sqrt{5} + 1} = \dfrac{2\left(\sqrt{5} + 1\right)}{5 - 1} = \dfrac{2\left(\sqrt{5} + 1\right)}{4} = \dfrac{\sqrt{5} + 1}{2}$

6.   (a) See Figure 6a. $f(-1)$ is undefined

      (b) See Figure 6b. $f(-1) = -1$

      (c) See Figure 6c. $f(-1) = 1$

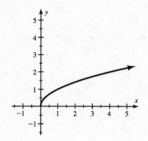

Figure 6a

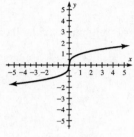

Figure 6b

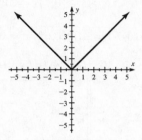

Figure 6c

7. $f(64) = 0.2(64)^{2/3} = 0.2\left(\sqrt[3]{64}\right)^2 = 0.2(4)^2 = 0.2 \cdot 16 = 3.2$

8. $x - 4 \geq 0 \Rightarrow x \geq 4 \Rightarrow$ Domain: $[4, \infty)$

7.5: Equations Involving Radical Expressions

Concepts

1.   Square each side.

2.   Cube each side.

3.   Yes

4.   We must check the answers.

5.   The Pythagorean theorem is used to find an unknown side of a right triangle. *Answers may vary*.

6. $c^2 = 3^2 + 4^2 \Rightarrow c^2 = 9 + 16 \Rightarrow c^2 = 25 \Rightarrow c = \sqrt{25} = 5$

7. $d = \sqrt{\left(x_2 - x_1\right)^2 + \left(y_2 - y_1\right)^2}$

8. $x^{1/2} + x^{3/4} = 2$

9. $\sqrt{2} \cdot \sqrt{2} = \sqrt{2 \cdot 2} = \sqrt{4} = 2$

10. $\sqrt{5} \cdot \sqrt{5} = \sqrt{5 \cdot 5} = \sqrt{25} = 5$

11. $\sqrt{x} \cdot \sqrt{x} = \sqrt{x \cdot x} = \sqrt{x^2} = x$

12. $\sqrt{2x} \cdot \sqrt{2x} = \sqrt{2x \cdot 2x} = \sqrt{4x^2} = 2x$

13. $\left(\sqrt{2x+1}\right)^2 = 2x + 1$

14. $\left(\sqrt{7x}\right)^2 = 7x$

15. $\left(\sqrt[3]{5x^2}\right)^3 = 5x^2$

16. $\left(\sqrt[3]{(2x-5)}\right)^3 = 2x-5$

Symbolic Solutions

17. $\sqrt{x} = 8 \Rightarrow \left(\sqrt{x}\right)^2 = 8^2 \Rightarrow x = 64$

19. $\sqrt[4]{x} = 3 \Rightarrow \left(\sqrt[4]{x}\right)^4 = 3^4 \Rightarrow x = 81$

21. $\sqrt{2t+4} = 4 \Rightarrow \left(\sqrt{2t+4}\right)^2 = 4^2 \Rightarrow 2t+4 = 16 \Rightarrow 2t = 12 \Rightarrow t = 6$

23. $\sqrt{x+1} - 3 = 4 \Rightarrow \sqrt{x+1} = 7 \Rightarrow x+1 = 7^2 \Rightarrow x = 49-1 \Rightarrow x = 48$

25. $2\sqrt{x-2} + 1 = 5 \Rightarrow 2\sqrt{x-2} = 4 \Rightarrow \sqrt{x-2} = 2 \Rightarrow x-2 = 2^2 \Rightarrow x = 4+2 \Rightarrow x = 6$

27. $\sqrt{x+6} = x \Rightarrow \left(\sqrt{x+6}\right)^2 = x^2 \Rightarrow x+6 = x^2 \Rightarrow x^2 - x - 6 = 0 \Rightarrow (x+2)(x-3) = 0 \Rightarrow x = -2$ or $x = 3$. The solution $x = -2$ does not check. The solution is $x = 3$.

29. $\sqrt[3]{x} = 3 \Rightarrow \left(\sqrt[3]{x}\right)^3 = 3^3 \Rightarrow x = 27$

31. $\sqrt[3]{2z-4} = -2 \Rightarrow 2z-4 = (-2)^3 \Rightarrow 2z-4 = -8 \Rightarrow 2z = -4 \Rightarrow z = -2$

33. $\sqrt[4]{t+1} = 2 \Rightarrow t+1 = 2^4 \Rightarrow t+1 = 16 \Rightarrow t = 15$

35. $\sqrt{5z-1} = \sqrt{z+1} \Rightarrow \left(\sqrt{5z-1}\right)^2 = \left(\sqrt{z+1}\right)^2 \Rightarrow 5z-1 = z+1 \Rightarrow 4z = 2 \Rightarrow z = \dfrac{1}{2}$

37. $\sqrt{1-x} = 1-x \Rightarrow \left(\sqrt{1-x}\right)^2 = (1-x)^2 \Rightarrow 1-x = 1-2x+x^2 \Rightarrow x^2 - x = 0 \Rightarrow x(x-1) = 0 \Rightarrow x = 0$ or $x = 1$

39. $\sqrt{b^2-4} = b-2 \Rightarrow \left(\sqrt{b^2-4}\right)^2 = (b-2)^2 \Rightarrow b^2-4 = b^2 - 4b + 4 \Rightarrow 4b = 8 \Rightarrow b = 2$

41. $\sqrt{1-2x} = x+7 \Rightarrow \left(\sqrt{1-2x}\right)^2 = (x+7)^2 \Rightarrow 1-2x = x^2 + 14x + 49 \Rightarrow x^2 + 16x + 48 = 0 \Rightarrow$
$(x+12)(x+4) = 0 \Rightarrow x = -12$ or $x = -4$. The solution $x = -12$ does not check. The solution is $x = -4$.

43. $\sqrt{x} = \sqrt{x-5} + 1 \Rightarrow \left(\sqrt{x}\right)^2 = \left(\sqrt{x-5}+1\right)^2 \Rightarrow x = (x-5) + 2\sqrt{x-5} + 1 \Rightarrow$
$2\sqrt{x-5} = 4 \Rightarrow \left(2\sqrt{x-5}\right)^2 = 4^2 \Rightarrow 4(x-5) = 16 \Rightarrow 4x - 20 = 16 \Rightarrow 4x = 36 \Rightarrow x = 9$

45. $\sqrt{2t-2} + \sqrt{t} = 7 \Rightarrow \sqrt{2t-2} = 7 - \sqrt{t} \Rightarrow \left(\sqrt{2t-2}\right)^2 = \left(7-\sqrt{t}\right)^2 \Rightarrow 2t-2 = 49 - 14\sqrt{t} + t \Rightarrow$
$14\sqrt{t} = 51 - t \Rightarrow \left(14\sqrt{t}\right)^2 = (51-t)^2 \Rightarrow 196t = 2601 - 102t + t^2 \Rightarrow t^2 - 298t + 2601 = 0 \Rightarrow$
$(t-9)(t-289) = 0 \Rightarrow t = 9$ or $t = 289$. The solution $t = 289$ does not check. The solution is $t = 9$.

47. $x^2 = 49 \Rightarrow \sqrt{x^2} = \sqrt{49} \Rightarrow |x| = 7 \Rightarrow x = \pm 7$

49. $2z^2 = 200 \Rightarrow z^2 = 100 \Rightarrow \sqrt{z^2} = \sqrt{100} \Rightarrow |z| = 10 \Rightarrow z = \pm 10$

51. $(t+1)^2 = 16 \Rightarrow \sqrt{(t+1)^2} = \sqrt{16} \Rightarrow |t+1| = 4 \Rightarrow t+1 = \pm 4$

    $\Rightarrow t = 3 \text{ or } t = -5$

53. $(4-2x)^2 = 100 \Rightarrow \sqrt{(4-2x)^2} = \sqrt{100} \Rightarrow |4-2x| = 10 \Rightarrow 4-2x = \pm 10$

    $\Rightarrow 4-2x = 10 \text{ or } 4-2x = -10 \Rightarrow -2x = 6 \text{ or } -2x = -14 \Rightarrow x = -3 \text{ or } x = 7$

55. $b^3 = 64 \Rightarrow \sqrt[3]{b^3} = \sqrt[3]{64} \Rightarrow b = 4$

57. $2t^3 = -128 \Rightarrow t^3 = -64 \Rightarrow \sqrt[3]{t^3} = \sqrt[3]{-64} \Rightarrow t = -4$

59. $(x+1)^3 = 8 \Rightarrow \sqrt[3]{(x+1)^3} = \sqrt[3]{8} \Rightarrow x+1 = 2 \Rightarrow x = 1$

61. $(2-5z)^3 = -125 \Rightarrow \sqrt[3]{(2-5z)^3} = \sqrt[3]{-125} \Rightarrow 2-5z = -5 \Rightarrow -5z = -7 \Rightarrow z = \dfrac{7}{5}$

63. $x^4 = 16 \Rightarrow \sqrt[4]{x^4} = \sqrt[4]{16} \Rightarrow |x| = 2 \Rightarrow x = \pm 2$

65. $x^5 = 12 \Rightarrow \sqrt[5]{x^5} = \sqrt[5]{12} \Rightarrow x = \sqrt[5]{12}$

67. $2(x+2)^4 = 162 \Rightarrow (x+2)^4 = 81 \Rightarrow \sqrt[4]{(x+2)^4} = \sqrt[4]{81}$

    $\Rightarrow |x+2| = 3 \Rightarrow x+2 = \pm 3 \Rightarrow x = 1 \text{ or } x = -5$

Graphical Solutions

69. Graphical: Graph $Y_1 = \sqrt[3]{(X+5)}$ and $Y_2 = 2$ in [–7, 7, 1] by [0, 4, 1]. See Figure 69. The solution is $x = 3$.

71. Graphical: Graph $Y_1 = \sqrt{(2X-3)}$ and $Y_2 = \sqrt{(X)} - (1/2)$ in [0, 3, 1] by [–2, 2, 1]. See Figure 71.

    The solution is $x \approx 1.88$.

[–7, 7, 1] by [0, 4, 1]   [0, 3, 1] by [–2, 2, 1]

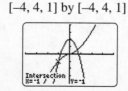
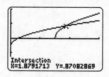

    Figure 69   Figure 71

73. Graphical: Graph $Y_1 = X \wedge (5/3)$ and $Y_2 = 2 - 3X^2$ in [–4, 4, 1] by [–4, 4, 1]. See Figures 73a & 73b.

    The solutions are $x = -1$ or $x \approx 0.70$.

[–4, 4, 1] by [–4, 4, 1]   [–4, 4, 1] by [–4, 4, 1]

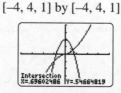

    Figure 73a   Figure 73b

75.   Graphical: Graph $Y_1 = X\wedge(1/3) - 1$ and $Y_2 = 2 - X$ in $[-3, 3, 1]$ by $[-3, 3, 1]$. See Figure 75.

The solution is $z \approx 1.79$.

77.   Graphical: Graph $Y_1 = \sqrt{(X+2)} + \sqrt{(3X+2)}$ and $Y_2 = 2$ in $[-2, 2, 1]$ by $[0, 5, 1]$. See Figure 77.

The solution is $y \approx -0.47$.

$[-3, 3, 1]$ by $[-3, 3, 1]$ $[-2, 2, 1]$ by $[0, 5, 1]$

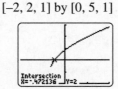

Figure 75   Figure 77

Using More Than One Method

79.   (a) $2\sqrt{x} = 8 \Rightarrow \left(2\sqrt{x}\right)^2 = 8^2 \Rightarrow 4x = 64 \Rightarrow x = 16$

(b) Graph $Y_1 = 2\sqrt{(X)}$ and $Y_2 = 8$ in $[0, 30, 5]$ by $[0, 10, 1]$. See Figure 79b. The solution is $x = 16$.

(c) Table $Y_1 = 2\sqrt{(X)}$ and $Y_2 = 8$ with TblStart $= 0$ and ΔTbl $= 4$. See Figure 79c. The solution is $x = 16$.

$[0, 30, 5]$ by $[0, 10, 1]$

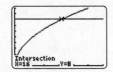   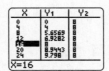

Figure 79b   Figure 79c

81.   (a) $\sqrt{6z - 2} = 8 \Rightarrow \left(\sqrt{6z-2}\right)^2 = 8^2 \Rightarrow 6z - 2 = 64 \Rightarrow 6z = 66 \Rightarrow z = 11$

(b) Graph $Y_1 = \sqrt{(6X - 2)}$ and $Y_2 = 8$ in $[0, 20, 2]$ by $[0, 10, 1]$. See Figure 81b. The solution is $z = 11$.

(c) Table $Y_1 = \sqrt{(6X - 2)}$ and $Y_2 = 8$ with TblStart $= 7$ and ΔTbl $= 1$. See Figure 81c. The solution is

$z = 11$.

$[0, 20, 2]$ by $[0, 10, 1]$

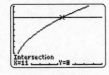   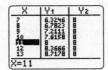

Figure 81b   Figure 81c

Solving an Equation for a Variable

83. $T = 2\pi\sqrt{\dfrac{L}{32}} \Rightarrow \dfrac{T}{2\pi} = \sqrt{\dfrac{L}{32}} \Rightarrow \left(\dfrac{T}{2\pi}\right)^2 = \dfrac{L}{32} \Rightarrow \dfrac{T^2}{4\pi^2} = \dfrac{L}{32} \Rightarrow 32 \cdot \dfrac{T^2}{4\pi^2} = L \Rightarrow L = \dfrac{8T^2}{\pi^2}$

85. $r = \sqrt{\dfrac{A}{\pi}} \Rightarrow r^2 = \dfrac{A}{\pi} \Rightarrow \pi r^2 = A \Rightarrow A = \pi r^2$

Pythagorean Theorem

87. Yes, since $6^2 + 8^2 = 10^2$. That is $36 + 64 = 100$.

89. Yes, since $\left(\sqrt{5}\right)^2 + \left(\sqrt{9}\right)^2 = \left(\sqrt{14}\right)^2$. That is $5 + 9 = 14$.

91. Yes, since $7^2 + 24^2 = 25^2$. That is $49 + 576 = 625$.

93. No, since $8^2 + 8^2 \neq 16^2$. That is $64 + 64 \neq 256$.

95. $4^2 + 4^2 = c^2 \Rightarrow c^2 = 16 + 16 \Rightarrow c^2 = 32 \Rightarrow c = \sqrt{32} = 4\sqrt{2}$

97. $24^2 + b^2 = 25^2 \Rightarrow 576 + b^2 = 625 \Rightarrow b^2 = 49 \Rightarrow b = \sqrt{49} = 7$

99. $3^2 + 4^2 = c^2 \Rightarrow c^2 = 9 + 16 \Rightarrow c^2 = 25 \Rightarrow c = \sqrt{25} = 5$

101. $\left(\sqrt{3}\right)^2 + b^2 = 8^2 \Rightarrow 3 + b^2 = 64 \Rightarrow b^2 = 61 \Rightarrow b = \sqrt{61}$

103. $a^2 + 48^2 = 50^2 \Rightarrow a^2 + 2304 = 2500 \Rightarrow a^2 = 196 \Rightarrow a = \sqrt{196} = 14$

Distance Formula

105. From $(-2, 1)$ to $(2, 3)$, $d = \sqrt{\left(2 - (-2)\right)^2 + \left(3 - 1\right)^2} = \sqrt{4^2 + 2^2} = \sqrt{16 + 4} = \sqrt{20} = 2\sqrt{5}$.

107. From $(10, 40)$ to $(30, -20)$,

$$d = \sqrt{\left(30 - 10\right)^2 + \left(-20 - 40\right)^2} = \sqrt{20^2 + \left(-60\right)^2} = \sqrt{400 + 3600} = \sqrt{4000} = 20\sqrt{10}.$$

109. $d = \sqrt{\left(4 - (-1)\right)^2 + \left(10 - 2\right)^2} = \sqrt{5^2 + 8^2} = \sqrt{25 + 64} = \sqrt{89}$.

111. $d = \sqrt{\left(4 - 0\right)^2 + \left(0 - (-3)\right)^2} = \sqrt{4^2 + 3^2} = \sqrt{16 + 9} = \sqrt{25} = 5$.

113. $\sqrt{\left(0 - x\right)^2 + \left(6 - 3\right)^2} = 5 \Rightarrow \sqrt{\left(-x\right)^2 + 3^2} = 5 \Rightarrow \sqrt{x^2 + 9} = 5 \Rightarrow \left(\sqrt{x^2 + 9}\right)^2 = 5^2 \Rightarrow$

$x^2 + 9 = 25 \Rightarrow x^2 = 16 \Rightarrow x = \sqrt{16} = \pm 4$. Since x is positive, $x = 4$.

115. $\sqrt{\left(62 - x\right)^2 + \left(6 - (-5)\right)^2} = 61 \Rightarrow \sqrt{\left(62 - x\right)^2 + 11^2} = 61 \Rightarrow \sqrt{\left(3844 - 124x + x^2\right) + 121} = 61 \Rightarrow$

$\sqrt{x^2 - 124x + 3965} = 61 \Rightarrow \left(\sqrt{x^2 - 124x + 3965}\right)^2 = 61^2 \Rightarrow x^2 - 124x + 3965 = 3721 \Rightarrow$

$x^2 - 124x + 244 = 0 \Rightarrow (x - 2)(x - 122) = 0 \Rightarrow x = 2$ or $x = 122$

Applications

117. $400 = 100\sqrt[3]{W^2} \Rightarrow \dfrac{400}{100} = \sqrt[3]{W^2} \Rightarrow 4 = \sqrt[3]{W^2} \Rightarrow 4^3 = \left(\sqrt[3]{W^2}\right)^3 \Rightarrow 64 = W^2 \Rightarrow W = 8 \text{ lb}$

119. $D(6) = 1.22\sqrt{6} \approx 2.988 \approx 3 \text{ miles}$

121. $1.22\sqrt{h} = 20 \Rightarrow \sqrt{h} = \dfrac{20}{1.22} \Rightarrow \left(\sqrt{h}\right)^2 = \left(\dfrac{20}{1.22}\right)^2 \Rightarrow h \approx 268.745 \approx 269 \text{ feet}$

123. $d^2 = 11.4^2 + 15.2^2 \Rightarrow d^2 = 129.96 + 231.04 \Rightarrow d^2 = 361 \Rightarrow d = \sqrt{361} = 19 \text{ inches}$

125. The height can be found using proportions: $\dfrac{16}{9} = \dfrac{29}{x} \Rightarrow 16x = 261 \Rightarrow x = \dfrac{261}{16} \approx 16.3$ inches.

Then $d^2 = 29^2 + 16.3^2 \Rightarrow d^2 = 841 + 265.69 \Rightarrow d^2 = 1106.69 \Rightarrow d = \sqrt{1106.69} \approx 33.3$ inches.

127. (a) $\dfrac{60}{11}\sqrt{d} = 60 \Rightarrow \sqrt{d} = 60\left(\dfrac{11}{60}\right) \Rightarrow \sqrt{d} = 11 \Rightarrow \left(\sqrt{d}\ \right)^2 = 11^2 \Rightarrow d = 121$ feet

(b) $\dfrac{60}{11}\sqrt{d} = 100 \Rightarrow \sqrt{d} = 100\left(\dfrac{11}{60}\right) \Rightarrow \sqrt{d} = \dfrac{55}{3} \Rightarrow \left(\sqrt{d}\ \right)^2 = \left(\dfrac{55}{3}\right)^2 \Rightarrow d \approx 336$ feet

129. (a) $V = 30\sqrt{\dfrac{285}{178}} \approx 38$ mph. The accident vehicle was traveling about 38 mph.

(b) $45\sqrt{\dfrac{D}{255}} = 60 \Rightarrow \sqrt{\dfrac{D}{255}} = \dfrac{60}{45} \Rightarrow \left(\sqrt{\dfrac{D}{255}}\right)^2 = \left(\dfrac{4}{3}\right)^2 \Rightarrow \dfrac{D}{255} = \dfrac{16}{9} \Rightarrow D = 255\left(\dfrac{16}{9}\right) \approx 453$ feet

131. (a) $W(2v) = 3.8(2v)^3 = 3.8 \cdot 8 \cdot v^3 = 8 \cdot \left(3.8v^3\right)$; The wattage generated increases by a factor of 8.

(b) $W = 3.8v^3 \Rightarrow \dfrac{W}{3.8} = v^3 \Rightarrow \sqrt[3]{\dfrac{W}{3.8}} = \sqrt[3]{v^3} \Rightarrow v = \sqrt[3]{\dfrac{W}{3.8}}$

(c) $v = \sqrt[3]{\dfrac{30,400}{3.8}} = \sqrt[3]{8000} = 20$ mph

133. $c^2 = a^2 + a^2 \Rightarrow c^2 = 2a^2 \Rightarrow c = \sqrt{2a^2} = a\sqrt{2}$

7.6: Complex Numbers

Concepts

1. $2 + 3i$; *Answers may vary.*

2. No, any real number a can be written as $a + 0i$.

3. i

4. -1

5. $i\sqrt{a}$

6. $10 - 7i$

7. $a + bi$

8. $\dfrac{2 + 4i}{2} = \dfrac{2}{2} + \dfrac{4}{2}i = 1 + 2i$

9. 4

10. -5

11. 0

12. pure

Complex Numbers

13. $\sqrt{-5} = i\sqrt{5}$

15. $\sqrt{-100} = i\sqrt{100} = i \cdot 10 = 10i$

17. $\sqrt{-144} = i\sqrt{144} = i \cdot 12 = 12i$

19. $\sqrt{-12} = i\sqrt{12} = i \cdot \sqrt{4 \cdot 3} = i \cdot \sqrt{4} \cdot \sqrt{3} = i \cdot 2 \cdot \sqrt{3} = 2i\sqrt{3}$

21. $\sqrt{-18} = i\sqrt{18} = i \cdot \sqrt{9 \cdot 2} = i \cdot \sqrt{9} \cdot \sqrt{2} = i \cdot 3 \cdot \sqrt{2} = 3i\sqrt{2}$

23. $(5+3i)+(-2-3i) = (5+(-2))+(3+(-3))i = 3+0i = 3$

25. $(2i)+(-8+5i) = (0+(-8))+(2+5)i = -8+7i$

27. $(2-7i)-(1+2i) = (2-1)+(-7-2)i = 1-9i$

29. $(5i)-(10-2i) = (0-10)+(5-(-2))i = -10+7i$

31. $(3+2i)(-1+5i) = -3+15i-2i+10i^2 = -3+13i+10(-1) = -3+13i-10 = -13+13i$

33. $4(5-3i) = 20-12i$

35. $(5+4i)(5-4i) = 25-16i^2 = 25-16(-1) = 25+16 = 41$

37. $(-4i)(5i) = -4 \cdot 5 \cdot i^2 = -20(-1) = 20$

39. $3i+(2-3i)-(1-5i) = 3i+2-3i-1+5i = 1+5i$

41. $(2+i)^2 = 4+4i+i^2 = 4+4i-1 = 3+4i$

43. $2i(-3+i) = -6i+2i^2 = -6i+2(-1) = -2-6i$

45. $i(1+i)^2 = i(1+2i+i^2) = i(1+2i-1) = i(2i) = 2i^2 = 2(-1) = -2$

47. $(a+3bi)(a-3bi) = a^2-9b^2i^2 = a^2-9b^2(-1) = a^2+9b^2$

49. When 11 is divided by 4, the result is 2 with remainder 3. Thus $i^{11} = i^3 = -i$.

51. When 21 is divided by 4, the result is 5 with remainder 1. Thus $i^{21} = i^1 = i$.

53. When 58 is divided by 4, the result is 14 with remainder 2. Thus $i^{58} = i^2 = -1$.

55. When 64 is divided by 4, the result is 16 with remainder 0. Thus $i^{64} = i^0 = 1$.

57. $3-4i$

59. Since $-6i = 0-6i$, the complex conjugate is $0+6i = 6i$.

61. $5+4i$

63. Since $-1 = -1+0i$, the complex conjugate is $-1-0i = -1$.

65. $\dfrac{2}{1+i} = \dfrac{2}{1+i} \cdot \dfrac{1-i}{1-i} = \dfrac{2(1-i)}{1^2-i^2} = \dfrac{2-2i}{1+1} = \dfrac{2-2i}{2} = \dfrac{2}{2}-\dfrac{2}{2}i = 1-i$

67. $\dfrac{3i}{5-2i}=\dfrac{3i}{5-2i}\cdot\dfrac{5+2i}{5+2i}=\dfrac{3i(5+2i)}{5^2-4i^2}=\dfrac{15i+6i^2}{25+4}=\dfrac{15i-6}{29}=-\dfrac{6}{29}+\dfrac{15}{29}i$

69. $\dfrac{8+9i}{5+2i}=\dfrac{8+9i}{5+2i}\cdot\dfrac{5-2i}{5-2i}=\dfrac{40-16i+45i-18i^2}{5^2-4i^2}=\dfrac{40+29i-18(-1)}{25+4}=\dfrac{58+29i}{29}=2+i$

71. $\dfrac{5+7i}{1-i}=\dfrac{5+7i}{1-i}\cdot\dfrac{1+i}{1+i}=\dfrac{5+5i+7i+7i^2}{1^2-i^2}=\dfrac{5+12i+7(-1)}{1+1}=\dfrac{-2+12i}{2}=-1+6i$

73. $\dfrac{2-i}{i}=\dfrac{2-i}{i}\cdot\dfrac{-i}{-i}=\dfrac{-2i+i^2}{-i^2}=\dfrac{-2i+(-1)}{-(-1)}=\dfrac{-2i-1}{1}=-1-2i$

75. $\dfrac{1}{i}+\dfrac{1}{2i}=\dfrac{2}{2i}+\dfrac{1}{2i}=\dfrac{3}{2i}=\dfrac{3}{2i}\cdot\dfrac{-2i}{-2i}=\dfrac{-6i}{-4i^2}=\dfrac{-6i}{-4(-1)}=\dfrac{-6i}{4}=-\dfrac{3}{2}i$

77. $\dfrac{1}{-1+i}-\dfrac{2}{i}=\dfrac{i}{i(-1+i)}-\dfrac{2(-1+i)}{i(-1+i)}=\dfrac{i}{-1-i}-\dfrac{-2+2i}{-1-i}=\dfrac{i+2-2i}{-1-i}=\dfrac{2-i}{-1-i}$

$=\dfrac{2-i}{-1-i}\cdot\dfrac{-1+i}{-1+i}=\dfrac{-2+2i+i-i^2}{1-i^2}=\dfrac{-2+3i-(-1)}{1-(-1)}=\dfrac{-1+3i}{2}=-\dfrac{1}{2}+\dfrac{3}{2}i$

Applications

79. $Z=\dfrac{40+70i}{2+3i}=\dfrac{40+70i}{2+3i}\cdot\dfrac{2-3i}{2-3i}=\dfrac{80-120i+140i-210i^2}{2^2-9i^2}=\dfrac{290+20i}{13}=\dfrac{290}{13}+\dfrac{20}{13}i$

Checking Basic Concepts for Sections 7.5 & 7.6

1. (a) $\sqrt{2x-4}=2\Rightarrow\left(\sqrt{2x-4}\right)^2=2^2\Rightarrow 2x-4=4\Rightarrow 2x=8\Rightarrow x=4$

(b) $\sqrt[3]{x-1}=3\Rightarrow\left(\sqrt[3]{x-1}\right)^3=3^3\Rightarrow x-1=27\Rightarrow x=28$

(c) $\sqrt{3x}=1+\sqrt{x+1}\Rightarrow\left(\sqrt{3x}\right)^2=\left(1+\sqrt{x+1}\right)^2\Rightarrow 3x=1+2\sqrt{x+1}+x+1\Rightarrow$

$3x=x+2+2\sqrt{x+1}\Rightarrow 2x-2=2\sqrt{x+1}\Rightarrow x-1=\sqrt{x+1}\Rightarrow(x-1)^2=\left(\sqrt{x+1}\right)^2\Rightarrow$

$x^2-2x+1=x+1\Rightarrow x^2-3x=0\Rightarrow x(x-3)=0\Rightarrow x=0\text{ or }x=3$

The solution $x=0$ does not check. The solution is $x=3$.

2. $d=\sqrt{(2-(-3))^2+(-7-5)^2}=\sqrt{5^2+(-12)^2}=\sqrt{25+144}=\sqrt{169}=13$

3. $h^2+12.8^2=16^2\Rightarrow h^2+163.84=256\Rightarrow h^2=92.16\Rightarrow h=\sqrt{92.16}=9.6\text{ inches}$

4. $(x+1)^4=16\Rightarrow\sqrt[4]{(x+1)^4}=\sqrt[4]{16}\Rightarrow|x+1|=2\Rightarrow x+1=\pm2\Rightarrow x=-3\text{ or }x=1$

5. (a) $\sqrt{-64}=i\sqrt{64}=i(8)=8i$

(b) $\sqrt{-17}=i\sqrt{17}$

6.   (a) $(2-3i)+(1-i)=(2+1)+\bigl(-3+(-1)\bigr)i=3-4i$

(b) $4i-(2+i)=(0-2)+(4-1)i=-2+3i$

(c) $(3-2i)(1+i)=3+3i-2i-2i^2=3+i-2(-1)=3+i+2=5+i$

(d) $\dfrac{3}{2-2i}=\dfrac{3}{2-2i}\cdot\dfrac{2+2i}{2+2i}=\dfrac{3(2+2i)}{2^2-4i^2}=\dfrac{6+6i}{4+4}=\dfrac{6+6i}{8}=\dfrac{6}{8}+\dfrac{6}{8}i=\dfrac{3}{4}+\dfrac{3}{4}i$

Chapter 7 Review Exercises

Section 7.1

1. $\sqrt{4}=2$

2. $\sqrt{36}=6$

3. $\sqrt{9x^2}=\sqrt{9}\cdot\sqrt{x^2}=3|x|$

4. $\sqrt{(x-1)^2}=|x-1|$

5. $\sqrt[3]{-64}=-4$

6. $\sqrt[3]{-125}=-5$

7. $\sqrt[3]{x^6}=\sqrt[3]{\left(x^2\right)^3}=x^2$

8. $\sqrt[3]{27x^3}=\sqrt[3]{27}\cdot\sqrt[3]{x^3}=3x$

9. $\sqrt[4]{16}=2$

10. $\sqrt[5]{-1}=-1$

11. $\sqrt[4]{x^8}=\sqrt[4]{\left(x^2\right)^4}=x^2$

12. $\sqrt[5]{(x+1)^5}=x+1$

13. $14^{1/2}=\sqrt{14}$

14. $(-5)^{1/3}=\sqrt[3]{-5}$

15. $\left(\dfrac{x}{y}\right)^{3/2}=\left(\sqrt{\dfrac{x}{y}}\right)^3$ or $\sqrt{\left(\dfrac{x}{y}\right)^3}$

16. $(xy)^{-2/3}=\dfrac{1}{(xy)^{2/3}}=\dfrac{1}{\sqrt[3]{(xy)^2}}$ or $\dfrac{1}{\left(\sqrt[3]{xy}\right)^2}$

17. $(-27)^{2/3}=\left(\sqrt[3]{-27}\right)^2=(-3)^2=9$

18. $16^{1/4}=\sqrt[4]{16}=2$

19. $16^{3/2} = \left(\sqrt{16}\ \right)^3 = 4^3 = 64$

20. $81^{3/4} = \left(\sqrt[4]{81}\ \right)^3 = 3^3 = 27$

21. $\left(z^3\right)^{2/3} = z^{3\cdot 2/3} = z^2$

22. $\left(x^2 y^4\right)^{1/2} = x^{2\cdot 1/2} \cdot y^{4\cdot 1/2} = xy^2$

23. $\left(\dfrac{x^2}{y^6}\right)^{3/2} = \dfrac{x^{2\cdot 3/2}}{y^{6\cdot 3/2}} = \dfrac{x^3}{y^9}$

24. $\left(\dfrac{x^3}{y^6}\right)^{-1/3} = \dfrac{x^{3\cdot(-1/3)}}{y^{6\cdot(-1/3)}} = \dfrac{x^{-1}}{y^{-2}} = \dfrac{y^2}{x}$

Section 7.2

25. $\sqrt{2} \cdot \sqrt{32} = \sqrt{64} = 8$

26. $\sqrt[3]{-4} \cdot \sqrt[3]{2} = \sqrt[3]{-8} = -2$

27. $\sqrt[3]{x^4} \cdot \sqrt[3]{x^2} = \sqrt[3]{x^6} = \sqrt[3]{\left(x^2\right)^3} = x^2$

28. $\dfrac{\sqrt{80}}{\sqrt{20}} = \sqrt{\dfrac{80}{20}} = \sqrt{4} = 2$

29. $\sqrt[3]{-\dfrac{x}{8}} = -\dfrac{\sqrt[3]{x}}{\sqrt[3]{8}} = -\dfrac{\sqrt[3]{x}}{2}$

30. $\sqrt{\dfrac{1}{3}} \cdot \sqrt{\dfrac{1}{3}} = \left(\sqrt{\dfrac{1}{3}}\right)^2 = \dfrac{1}{3}$

31. $\sqrt{48} = \sqrt{16\cdot 3} = \sqrt{16}\cdot\sqrt{3} = 4\sqrt{3}$

32. $\sqrt{54} = \sqrt{9\cdot 6} = \sqrt{9}\cdot\sqrt{6} = 3\sqrt{6}$

33. $\sqrt[3]{\dfrac{3}{x}} \cdot \sqrt[3]{\dfrac{9}{x^2}} = \sqrt[3]{\dfrac{3\cdot 9}{x\cdot x^2}} = \sqrt[3]{\dfrac{27}{x^3}} = \dfrac{\sqrt[3]{27}}{\sqrt[3]{x^3}} = \dfrac{3}{x}$

34. $\sqrt{32a^3 b^2} = \sqrt{\left(4ab\right)^2\cdot 2a} = \sqrt{\left(4ab\right)^2}\cdot\sqrt{2a} = 4ab\sqrt{2a}$

35. $\sqrt{3xy} \cdot \sqrt{27xy} = \sqrt{3\cdot 27\cdot xy\cdot xy} = \sqrt{81\left(xy\right)^2} = \sqrt{\left(9xy\right)^2} = 9xy$

36. $\sqrt[3]{-25z^2} \cdot \sqrt[3]{-5z^2} = \sqrt[3]{-25\cdot(-5)\cdot z^2\cdot z^2} = \sqrt[3]{125z^4} = \sqrt[3]{\left(5z\right)^3\cdot z} = \sqrt{\left(5z\right)^3}\cdot\sqrt[3]{z} = 5z\sqrt[3]{z}$

37. $\sqrt{x^2 + 2x + 1} = \sqrt{\left(x+1\right)^2} = x+1$

38. $\sqrt[4]{\dfrac{2a^2}{b}} \cdot \sqrt[4]{\dfrac{8a^3}{b^3}} = \sqrt[4]{\dfrac{16a^5}{b^4}} = \sqrt[4]{\left(\dfrac{2a}{b}\right)^4 \cdot a} = \sqrt[4]{\left(\dfrac{2a}{b}\right)^4} \cdot \sqrt[4]{a} = \dfrac{2a\sqrt[4]{a}}{b}$

39. $2\sqrt{x} \cdot \sqrt[3]{x} = 2x^{1/2} \cdot x^{1/3} = 2x^{1/2+1/3} = 2x^{5/6} = 2\sqrt[6]{x^5}$

40. $\sqrt[3]{rt} \cdot \sqrt[4]{r^2t^4} = (rt)^{1/3} \cdot \left(r^2t^4\right)^{1/4} = r^{1/3}t^{1/3} \cdot r^{1/2}t = r^{1/3+1/2}t^{1/3+1} = r^{5/6}t^{4/3} = r^{5/6}t^{8/6} = \sqrt[6]{r^5t^8}$ or $t\sqrt[6]{r^5t^2}$

Section 7.3

41. $3\sqrt{3} + \sqrt{3} = 4\sqrt{3}$

42. $\sqrt[3]{x} + 2\sqrt[3]{x} = 3\sqrt[3]{x}$

43. $3\sqrt[3]{5} - 6\sqrt[3]{5} = -3\sqrt[3]{5}$

44. $\sqrt[4]{y} - 2\sqrt[4]{y} = -\sqrt[4]{y}$

45. $2\sqrt{12} + 7\sqrt{3} = 2\sqrt{4 \cdot 3} + 7\sqrt{3} = 2\sqrt{4} \cdot \sqrt{3} + 7\sqrt{3} = 4\sqrt{3} + 7\sqrt{3} = 11\sqrt{3}$

46. $3\sqrt{18} - 2\sqrt{2} = 3\sqrt{9 \cdot 2} - 2\sqrt{2} = 3\sqrt{9} \cdot \sqrt{2} - 2\sqrt{2} = 9\sqrt{2} - 2\sqrt{2} = 7\sqrt{2}$

47. $7\sqrt[3]{16} - \sqrt[3]{2} = 7\sqrt[3]{8 \cdot 2} - \sqrt[3]{2} = 7\sqrt[3]{8} \cdot \sqrt[3]{2} - \sqrt[3]{2} = 14\sqrt[3]{2} - \sqrt[3]{2} = 13\sqrt[3]{2}$

48. $\sqrt{4x+4} + \sqrt{x+1} = \sqrt{4(x+1)} + \sqrt{x+1} = \sqrt{4} \cdot \sqrt{x+1} + \sqrt{x+1} = 3\sqrt{x+1}$

49. $\sqrt{4x^3} - \sqrt{x} = \sqrt{(2x)^2 \cdot x} - \sqrt{x} = \sqrt{(2x)^2} \cdot \sqrt{x} - \sqrt{x} = 2x\sqrt{x} - \sqrt{x} = (2x-1)\sqrt{x}$

50. $\sqrt[3]{ab^4} + 2\sqrt[3]{a^4b} = \sqrt[3]{b^3 \cdot ab} + 2\sqrt[3]{a^3 \cdot ab} = \sqrt[3]{b^3} \cdot \sqrt[3]{ab} + 2\sqrt[3]{a^3} \cdot \sqrt[3]{ab} = (b+2a)\sqrt[3]{ab}$

51. $\left(1+\sqrt{2}\right)\left(3+\sqrt{2}\right) = 3 + \sqrt{2} + 3\sqrt{2} + \left(\sqrt{2}\right)^2 = 3 + 4\sqrt{2} + 2 = 5 + 4\sqrt{2}$

52. $\left(7-\sqrt{5}\right)\left(1+\sqrt{3}\right) = 7 + 7\sqrt{3} - \sqrt{5} + \sqrt{15}$

53. $\left(3+\sqrt{6}\right)\left(3-\sqrt{6}\right) = 3^2 - \left(\sqrt{6}\right)^2 = 9 - 6 = 3$

54. $\left(10-\sqrt{5}\right)\left(10+\sqrt{5}\right) = 10^2 - \left(\sqrt{5}\right)^2 = 100 - 5 = 95$

55. $\left(\sqrt{a}+\sqrt{2b}\right)\left(\sqrt{a}-\sqrt{2b}\right) = \left(\sqrt{a}\right)^2 - \left(\sqrt{2b}\right)^2 = a - 2b$

56. $\left(\sqrt{xy}-1\right)\left(\sqrt{xy}+2\right) = \left(\sqrt{xy}\right)^2 + 2\sqrt{xy} - \sqrt{xy} - 2 = xy + \sqrt{xy} - 2$

57. $\dfrac{4}{\sqrt{5}} = \dfrac{4}{\sqrt{5}} \cdot \dfrac{\sqrt{5}}{\sqrt{5}} = \dfrac{4\sqrt{5}}{5}$

58. $\dfrac{r}{2\sqrt{t}} = \dfrac{r}{2\sqrt{t}} \cdot \dfrac{\sqrt{t}}{\sqrt{t}} = \dfrac{r\sqrt{t}}{2t}$

59. $\dfrac{1}{\sqrt{2}+3} = \dfrac{1}{\sqrt{2}+3} \cdot \dfrac{\sqrt{2}-3}{\sqrt{2}-3} = \dfrac{\sqrt{2}-3}{2-9} = \dfrac{\sqrt{2}-3}{-7} = \dfrac{3-\sqrt{2}}{7}$

60. $\dfrac{2}{5-\sqrt{7}} = \dfrac{2}{5-\sqrt{7}} \cdot \dfrac{5+\sqrt{7}}{5+\sqrt{7}} = \dfrac{10+2\sqrt{7}}{25-7} = \dfrac{10+2\sqrt{7}}{18} = \dfrac{5+\sqrt{7}}{9}$

61. $\dfrac{1}{\sqrt{8}-\sqrt{7}} = \dfrac{1}{\sqrt{8}-\sqrt{7}} \cdot \dfrac{\sqrt{8}+\sqrt{7}}{\sqrt{8}+\sqrt{7}} = \dfrac{\sqrt{8}+\sqrt{7}}{8-7} = \dfrac{\sqrt{8}+\sqrt{7}}{1} = \sqrt{8}+\sqrt{7}$

62. $\dfrac{\sqrt{a}-\sqrt{b}}{\sqrt{a}+\sqrt{b}} = \dfrac{\sqrt{a}-\sqrt{b}}{\sqrt{a}+\sqrt{b}} \cdot \dfrac{\sqrt{a}-\sqrt{b}}{\sqrt{a}-\sqrt{b}} = \dfrac{a-2\sqrt{ab}+b}{a-b}$

Section 7.4

63. See Figure 63.

64. See Figure 64.

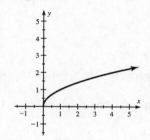
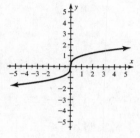

Figure 63   Figure 64

65. $f(x) = x^{1/2} = \sqrt{x};\ f(4) = 4^{1/2} = \sqrt{4} = 2$

66. $f(x) = x^{2/7} = \sqrt[7]{x^2};\ f(4) = 4^{2/7} = \sqrt[7]{4^2} = \sqrt[7]{16}$

67. See Figure 67. This graph is shifted 2 units downward.

68. See Figure 68. This graph is shifted 1 unit to the right.

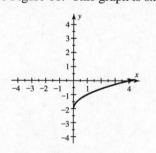
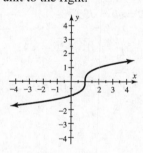

Figure 67   Figure 68

69. $x-1 \geq 0 \Rightarrow x \geq 1 \Rightarrow \text{Domain}: [1, \infty)$

70. $6-2x \geq 0 \Rightarrow -2x \geq -6 \Rightarrow x \leq 3 \Rightarrow \text{Domain}: (-\infty, 3]$

71. $x^2 + 1 \geq 0 \Rightarrow x^2 \geq -1 \Rightarrow \text{Domain}: (-\infty, \infty)$

72. $x+2 > 0 \Rightarrow x > -2 \Rightarrow \text{Domain}: (-2, \infty)$

Section 7.5

73. $\sqrt{x+2} = x \Rightarrow \left(\sqrt{x+2}\right)^2 = x^2 \Rightarrow x+2 = x^2 \Rightarrow x^2 - x - 2 = 0 \Rightarrow (x-2)(x+1) = 0 \Rightarrow x = 2 \text{ or } x = -1.$ The

solution $x = -1$ does not check. The solution is $x = 2$.

74. $\sqrt{2x-1} = \sqrt{x+3} \Rightarrow \left(\sqrt{2x-1}\right)^2 = \left(\sqrt{x+3}\right)^2 \Rightarrow 2x-1 = x+3 \Rightarrow x = 4$

75. $\sqrt[3]{x-1} = 2 \Rightarrow \left(\sqrt[3]{x-1}\right)^3 = 2^3 \Rightarrow x-1 = 8 \Rightarrow x = 9$

76. $\sqrt[3]{3x} = 3 \Rightarrow \left(\sqrt[3]{3x}\right)^3 = 3^3 \Rightarrow 3x = 27 \Rightarrow x = 9$

77. $\sqrt{2x} = x-4 \Rightarrow \left(\sqrt{2x}\right)^2 = (x-4)^2 \Rightarrow 2x = x^2 - 8x + 16 \Rightarrow x^2 - 10x + 16 = 0 \Rightarrow$

    $(x-2)(x-8) = 0 \Rightarrow x = 2$ or $x = 8$. The solution $x = 2$ does not check. The solution is $x = 8$.

78. $\sqrt{x} + 1 = \sqrt{x+2} \Rightarrow \left(\sqrt{x}+1\right)^2 = \left(\sqrt{x+2}\right)^2 \Rightarrow x + 2\sqrt{x} + 1 = x + 2 \Rightarrow 2\sqrt{x} = 1 \Rightarrow$

    $\left(2\sqrt{x}\right)^2 = 1^2 \Rightarrow 4x = 1 \Rightarrow x = \dfrac{1}{4}$

79. Graph $Y_1 = (2X-1)^{\wedge}(1/3)$ and $Y_2 = 2$ in $[-4, 6, 1]$ by $[-3, 3, 1]$. See Figure 79. The solution is $x = 4.5$.

80. Graph $Y_1 = X^{\wedge}(2/3)$ and $Y_2 = 3 - X$ in $[-5, 5, 1]$ by $[-5, 5, 1]$. See Figure 80. The solution is $x \approx 1.62$.

$[-4, 6, 1]$ by $[-3, 3, 1]$ $[-5, 5, 1]$ by $[-5, 5, 1]$

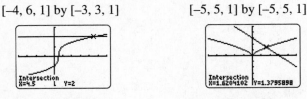

Figure 79   Figure 80

81. $c^2 = 4^2 + 7^2 \Rightarrow c^2 = 16 + 49 \Rightarrow c^2 = 65 \Rightarrow c = \sqrt{65}$

82. $5^2 + b^2 = 8^2 \Rightarrow 25 + b^2 = 64 \Rightarrow b^2 = 39 \Rightarrow b = \sqrt{39}$

83. $d = \sqrt{(2-(-2))^2 + (-2-3)^2} = \sqrt{4^2 + (-5)^2} = \sqrt{16+25} = \sqrt{41}$

84. $d = \sqrt{(-4-2)^2 + (1-(-3))^2} = \sqrt{(-6)^2 + 4^2} = \sqrt{36+16} = \sqrt{52} = 2\sqrt{13}$

85. $x^2 = 121 \Rightarrow x = \pm\sqrt{121} \Rightarrow x = \pm 11$

86. $2z^2 = 32 \Rightarrow z^2 = 16 \Rightarrow z = \pm\sqrt{16} \Rightarrow z = \pm 4$

87. $(x-1)^2 = 16 \Rightarrow x-1 = \pm\sqrt{16} \Rightarrow x-1 = \pm 4 \Rightarrow x = 1 \pm 4 \Rightarrow x = -3$ or $x = 5$

88. $x^3 = 64 \Rightarrow x = \sqrt[3]{64} \Rightarrow x = 4$

89. $(x-1)^3 = 8 \Rightarrow x-1 = \sqrt[3]{8} \Rightarrow x-1 = 2 \Rightarrow x = 3$

90. $(2x-1)^3 = 27 \Rightarrow 2x-1 = \sqrt[3]{27} \Rightarrow 2x-1 = 3 \Rightarrow 2x = 4 \Rightarrow x = 2$

91. $x^4 = 256 \Rightarrow \sqrt[4]{x^4} = \sqrt[4]{256} \Rightarrow |x| = 4 \Rightarrow x = \pm 4$

92. $x^5 = -1 \Rightarrow \sqrt[5]{x^5} = \sqrt[5]{-1} \Rightarrow x = -1$

93. $(x-3)^5 = -32 \Rightarrow \sqrt[5]{(x-3)^5} = \sqrt[5]{-32} \Rightarrow x-3 = -2 \Rightarrow x = 1$

94. $3(x+1)^4 = 3 \Rightarrow (x+1)^4 = 1 \Rightarrow \sqrt[4]{(x+1)^4} = \sqrt[4]{1}$

$\Rightarrow |x+1| = 1 \Rightarrow x+1 = \pm 1 \Rightarrow x = -2$ or $x = 0$

Section 7.6

95. $(1-2i)+(-3+2i) = (1+(-3))+(-2+2)i = -2+0i = -2$

96. $(1+3i)-(3-i) = (1-3)+(3-(-1))i = -2+4i$

97. $(1-i)(2+3i) = 2+3i-2i-3i^2 = 2+i-3(-1) = 2+i+3 = 5+i$

98. $\dfrac{3+i}{1-i} = \dfrac{3+i}{1-i} \cdot \dfrac{1+i}{1+i} = \dfrac{3+3i+i+i^2}{1^2-i^2} = \dfrac{3+4i-1}{1+1} = \dfrac{2+4i}{2} = \dfrac{2}{2}+\dfrac{4}{2}i = 1+2i$

99. $\dfrac{i(4+i)}{2-3i} = \dfrac{4i+i^2}{2-3i} = \dfrac{-1+4i}{2-3i} = \dfrac{-1+4i}{2-3i} \cdot \dfrac{2+3i}{2+3i} = \dfrac{-2-3i+8i+12i^2}{4-9i^2} = \dfrac{-2+5i+12(-1)}{4-9(-1)}$

$= \dfrac{-14+5i}{13} = -\dfrac{14}{13}+\dfrac{5}{13}i$

100. $(1-i)^2(1+i) = (1-2i+i^2)(1+i) = (1-2i-1)(1+i) = -2i(1+i) = -2i-2i^2 = 2-2i$

Applications

101. $\dfrac{\sqrt{h}}{2} = 4.6 \Rightarrow \sqrt{h} = 9.2 \Rightarrow (\sqrt{h})^2 = 9.2^2 \Rightarrow h = 84.64 \approx 85$ feet

102. $d^2 = 90^2 + 90^2 \Rightarrow d^2 = 8100 + 8100 \Rightarrow d^2 = 16,200 \Rightarrow d = \sqrt{16,200} \approx 127.3$ feet

103. $T = \dfrac{1}{4}\sqrt{10} \approx 0.79$ seconds

104. (a) $A = s^2 = (\sqrt{5})^2 = 5$ square units

(b) $V = s^3 = (\sqrt{5})^3 = 5\sqrt{5}$ cubic units

(c) $d^2 = (\sqrt{5})^2 + (\sqrt{5})^2 \Rightarrow d^2 = 5+5 \Rightarrow d^2 = 10 \Rightarrow d = \sqrt{10}$ units

(d) $d^2 = (\sqrt{5})^2 + (\sqrt{10})^2 \Rightarrow d^2 = 5+10 \Rightarrow d^2 = 15 \Rightarrow d = \sqrt{15}$ units

105. $2\pi\sqrt{\dfrac{L}{32.2}} = 1 \Rightarrow \sqrt{\dfrac{L}{32.2}} = \dfrac{1}{2\pi} \Rightarrow \left(\sqrt{\dfrac{L}{32.2}}\right)^2 = \left(\dfrac{1}{2\pi}\right)^2 \Rightarrow \dfrac{L}{32.2} = \dfrac{1}{4\pi^2} \Rightarrow L = \dfrac{32.2}{4\pi^2} \approx 0.82$ feet

106. $2\pi\sqrt{\dfrac{L}{5.1}} = 1 \Rightarrow \sqrt{\dfrac{L}{5.1}} = \dfrac{1}{2\pi} \Rightarrow \left(\sqrt{\dfrac{L}{5.1}}\right)^2 = \left(\dfrac{1}{2\pi}\right)^2 \Rightarrow \dfrac{L}{5.1} = \dfrac{1}{4\pi^2} \Rightarrow L = \dfrac{5.1}{4\pi^2} \approx 0.13$ feet. It is shorter.

107. $4(1+r)^{210} = 281 \Rightarrow (1+r)^{210} = \dfrac{281}{4} \Rightarrow \left((1+r)^{210}\right)^{1/210} = \left(\dfrac{281}{4}\right)^{1/210} \Rightarrow 1+r = \left(\dfrac{281}{4}\right)^{1/210} \Rightarrow$

$r = \left(\dfrac{281}{4}\right)^{1/210} - 1 \approx 0.02$.   From 1790 to 2000 the annual percentage growth rate was about 2%.

108. (a) $L = \sqrt{3.75(500)} = \sqrt{1875} \approx 43$ mph

(b) $L = 1.5\sqrt{500} \approx 34$ mph.   A steeper bank allows for a higher speed limit. This agrees with intuition.

109. $x^2 = 7 \Rightarrow x = \sqrt{7} \approx 2.65$ feet

110. $S = \pi(11)\sqrt{11^2 + 60^2} = 11\pi\sqrt{121 + 3600} = 11\pi\sqrt{3721} \approx 2108$ in^2

111. (a) $2^{-5700/5700} = 2^{-1} = \dfrac{1}{2}$

(b) $2^{-20,000/5700} = 2^{-200/57} \approx 0.09 = \dfrac{9}{100}$

Chapter 7 Test

1. $\sqrt[3]{-27} = \sqrt[3]{(-3)^3} = -3$

2. $\sqrt{(z+1)^2} = |z+1|$

3. $\sqrt{25x^4} = \sqrt{25} \cdot \sqrt{(x^2)^2} = 5x^2$

4. $\sqrt[3]{8z^6} = \sqrt[3]{8} \cdot \sqrt[3]{(z^2)^3} = 2z^2$

5. $\sqrt[4]{16x^4 y^5} = \sqrt[4]{(2xy)^4 \cdot y} = \sqrt[4]{(2xy)^4} \cdot \sqrt[4]{y} = 2xy\sqrt[4]{y}$

6. $\left(\sqrt{3} - \sqrt{2}\right)\left(\sqrt{3} + \sqrt{2}\right) = \left(\sqrt{3}\right)^2 - \left(\sqrt{2}\right)^2 = 3 - 2 = 1$

7. $7^{2/5} = \sqrt[5]{7^2}$ or $\left(\sqrt[5]{7}\right)^2$

8. $\left(\dfrac{x}{y}\right)^{-2/3} = \left(\dfrac{y}{x}\right)^{2/3} = \sqrt[3]{\left(\dfrac{y}{x}\right)^2}$ or $\left(\sqrt[3]{\dfrac{y}{x}}\right)^2$

9. $(-8)^{4/3} = \left(\sqrt[3]{-8}\right)^4 = (-2)^4 = 16$

10. $36^{-3/2} = \dfrac{1}{36^{3/2}} = \dfrac{1}{\left(\sqrt{36}\right)^3} = \dfrac{1}{6^3} = \dfrac{1}{216}$

11. $\sqrt[3]{x^4} = x^{4/3}$

12. $\sqrt{x} \cdot \sqrt[5]{x} = x^{1/2} \cdot x^{1/5} = x^{(1/2+1/5)} = x^{7/10}$

13. $4 - x \geq 0 \Rightarrow -x \geq -4 \Rightarrow x \leq 4 \Rightarrow$ Domain : $(-\infty, 4]$

14.   See Figure 14.

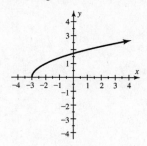

Figure 14

15. $\left(2z^{1/2}\right)^3 = 2^3 \cdot z^{1/2 \cdot 3} = 8z^{3/2}$

16. $\left(\dfrac{y^2}{z^3}\right)^{-1/3} = \left(\dfrac{z^3}{y^2}\right)^{1/3} = \dfrac{z^{3 \cdot 1/3}}{y^{2 \cdot 1/3}} = \dfrac{z}{y^{2/3}}$

17. $\sqrt{3} \cdot \sqrt{27} = \sqrt{81} = 9$

18. $\dfrac{\sqrt{y^3}}{\sqrt{4y}} = \sqrt{\dfrac{y^3}{4y}} = \sqrt{\dfrac{y^2}{4}} = \dfrac{\sqrt{y^2}}{\sqrt{4}} = \dfrac{y}{2}$

19. $7\sqrt{7} - 3\sqrt{7} + \sqrt{5} = 4\sqrt{7} + \sqrt{5}$

20. $7\sqrt[3]{x} - \sqrt[3]{x} = 6\sqrt[3]{x}$

21. $4\sqrt{18} + \sqrt{8} = 4(3)\sqrt{2} + 2\sqrt{2} = (12 + 2)\sqrt{2} = 14\sqrt{2}$

22. $\dfrac{\sqrt[3]{32}}{\sqrt[3]{4}} = \dfrac{\sqrt[3]{8 \cdot 4}}{\sqrt[3]{4}} = \dfrac{2\sqrt[3]{4}}{\sqrt[3]{4}} = 2$

23.   (a) $\sqrt{x-2} = 5 \Rightarrow \left(\sqrt{x-2}\right)^2 = 5^2 \Rightarrow x - 2 = 25 \Rightarrow x = 27$

       (b) $\sqrt[3]{x+1} = 2 \Rightarrow \left(\sqrt[3]{x+1}\right)^3 = 2^3 \Rightarrow x + 1 = 8 \Rightarrow x = 7$

       (c) $(x-1)^3 = 8 \Rightarrow \sqrt[3]{(x-1)^3} = \sqrt[3]{8} \Rightarrow x - 1 = 2 \Rightarrow x = 3$

       (d) $\sqrt{2x+2} = x - 11 \Rightarrow \left(\sqrt{2x+2}\right)^2 = (x-11)^2 \Rightarrow 2x + 2 = x^2 - 22x$

            $+121 \Rightarrow x^2 - 24x + 119 = 0 \Rightarrow (x-7)(x-17) = 0$

            $\Rightarrow x = 7$ or $x = 17$. The value $x = 7$ does not check. $x = 17$

24.   (a) $\dfrac{2}{3\sqrt{7}} = \dfrac{2}{3\sqrt{7}} \cdot \dfrac{\sqrt{7}}{\sqrt{7}} = \dfrac{2\sqrt{7}}{21}$

       (b) $\dfrac{1}{1+\sqrt{5}} = \dfrac{1}{1+\sqrt{5}} \cdot \dfrac{1-\sqrt{5}}{1-\sqrt{5}} = \dfrac{1-\sqrt{5}}{1-5} = \dfrac{1-\sqrt{5}}{-4} = \dfrac{-1+\sqrt{5}}{4}$

25. Graph $Y_1 = \sqrt{3X} - X + 1$ and $Y_2 = (X-1)^\wedge(1/3)$ in $[-5, 5, 1]$ by $[-5, 5, 1]$

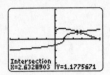

Figure 25

The solution is approximately $x = 2.63$.

26. $7^2 + b^2 = 13^2 \Rightarrow 49 + b^2 = 169 \Rightarrow b^2 = 120 \Rightarrow b = \sqrt{120} \approx 10.95$

27. $d = \sqrt{\left(-1-(-3)\right)^2 + (7-5)^2} = \sqrt{2^2 + 2^2} = \sqrt{4+4} = \sqrt{8} = 2\sqrt{2}$

28. $(-5+i) + (7-20i) = (-5+7) + (1+(-20))i = 2-19i$

29. $(3i) - (6-5i) = (0-6) + (3-(-5))i = -6+8i$

30. $\left(\dfrac{1}{2} - i\right)\left(\dfrac{1}{2} + i\right) = \dfrac{1}{4} + \dfrac{1}{2}i - \dfrac{1}{2}i - i^2 = \dfrac{1}{4} - (-1) = \dfrac{5}{4}$

31. $\dfrac{2i}{5+2i} = \dfrac{2i}{5+2i} \cdot \dfrac{5-2i}{5-2i} = \dfrac{2i(5-2i)}{5^2 - 4i^2} = \dfrac{10i - 4i^2}{25+4} = \dfrac{4+10i}{29} = \dfrac{4}{29} + \dfrac{10}{29}i$

32. (a) $V = \dfrac{4}{3}\pi r^3 \Rightarrow r^3 = \dfrac{3V}{4\pi} \Rightarrow \sqrt[3]{r^3} = \sqrt[3]{\dfrac{3V}{4\pi}} \Rightarrow r = \sqrt[3]{\dfrac{3V}{4\pi}}$

    (b) $r = \sqrt[3]{\dfrac{3(50)}{4\pi}} \approx 2.29$ inches

33. $27.4 W^{1/3} = 30 \Rightarrow W^{1/3} = \dfrac{30}{27.4} \Rightarrow \left(W^{1/3}\right)^3 = \left(\dfrac{30}{27.4}\right)^3 \Rightarrow W = \left(\dfrac{30}{27.4}\right)^3 \approx 1.31$ lb

Chapter 7 Extended and Discovery Exercises

1. By trial and error $\dfrac{m}{n} = \dfrac{3}{2}$. Plot the data and graph $Y_1 = 0.0002X^\wedge(3/2)$ in $[0, 2000, 500]$ by $[0, 20, 2]$. See Figure 1.

    $[0, 2000, 500]$ by $[0, 20, 2]$

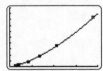

Figure 1

2. (a) $k(47)^{1.12}(11)^{1.98} = 11.4 \Rightarrow k = \dfrac{11.4}{(47)^{1.12}(11)^{1.98}} \approx 0.001325$

    (b) $V = 0.001325(105)^{1.12}(20)^{1.98} \approx 91.6$ ft^3

3. (a) $S = 15.7(154)^{0.425}(65)^{0.725} \approx 2753.963261 \approx 2754$ in^2

(b) It increases by a factor of $2^{0.425} \approx 1.34$.

(c) It increases by a factor of $2^{0.725} \approx 1.65$.

4. (a) Since the segment AB is on land, the expression is $30x$.

(b) The legs of right triangle BCD have lengths $CB = 1000 - x$ and $CD = 500$. Let $d = BD$.

$$d^2 = (1000 - x)^2 + 500^2 \Rightarrow d = \sqrt{(1000 - x)^2 + 500^2}.$$

(c) Since the segment BD is underwater, the expression is $50\sqrt{(1000 - x)^2 + 500^2}$.

(d) The expression is $30x + 50\sqrt{(1000 - x)^2 + 500^2}$.

(e) Graph $Y_1 = 30X + 50\sqrt{\left((100 - X)^2 + 500^2\right)}$ in [0, 1000, 100] by [40,000, 60,000, 5000]. See Figure 4.

The minimum cost is \$50,000 when $x = 625$ feet.

[0, 1000, 100] by [40,000, 60,000, 5000]

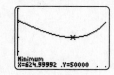

Figure 4

Chapters 1–7 Cumulative Review Exercises

1. $S = 4\pi(3)^2 = 4\pi(9) = 36\pi$

2. $D = \{-1,\ 0,\ 1\};\ R = \{2,\ 4\}$

3. (a) $\left(\dfrac{ab^2}{b^{-1}}\right)^{-3} = \dfrac{a^{-3}b^{-6}}{b^3} = a^{-3}b^{-6-3} = a^{-3}b^{-9} = \dfrac{1}{a^3 b^9}$

(b) $\dfrac{\left(x^2 y\right)^3}{x^2 \left(y^2\right)^{-3}} = \dfrac{x^6 y^3}{x^2 y^{-6}} = x^{6-2} y^{3-(-6)} = x^4 y^9$

(c) $(rt)^2 \left(r^2 t\right)^3 = r^2 t^2 r^6 t^3 = r^{2+6} t^{2+3} = r^8 t^5$

4. $0.00043 = 4.3 \times 10^{-4}$

5. $f(3) = \dfrac{3}{3-2} = \dfrac{3}{1} = 3$; the denominator cannot equal 0, so $x \neq 2$.

6. All real numbers

7.   See Figure 7.

8.   See Figure 8.

9.   See Figure 9.

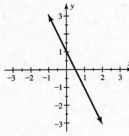   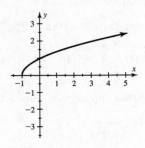   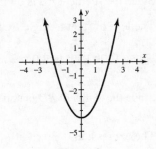

Figure 7   Figure 8   Figure 9

(a) D: All real numbers; $R: y \geq -4$

(b) $f(-2) = (-2)^2 - 4 = 4 - 4 = 0$

(c) The x-intercepts are -2 and 2.

(d) $x^2 - 4 = 0 \Rightarrow (x-2)(x+2) = 0 \Rightarrow x = 2$ or $x = -2$

10. The line perpendicular to $y = -2x$ has slope $m = \dfrac{1}{2}$.

$$y - 2 = \frac{1}{2}(x - (-1)) \Rightarrow y = \frac{1}{2}x + \frac{1}{2} + 2 \Rightarrow y = \frac{1}{2}x + \frac{5}{2}$$

11. $m = \dfrac{-5-7}{2-(-2)} = \dfrac{-12}{4} = -3;\ y - (-5) = -3(x-2)$

$\Rightarrow y = -3x + 6 - 5 \Rightarrow y = -3x + 1;\ f(x) = 1 - 3x$

12. See Figure 12.

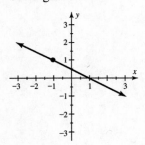

Figure 12

13. $5x - (3-x) = \dfrac{1}{2}x \Rightarrow 5x - 3 + x - \dfrac{1}{2}x = 0 \Rightarrow \dfrac{11}{2}x - 3 = 0 \Rightarrow \dfrac{11}{2}x = 3 \Rightarrow x = \dfrac{6}{11}$

14. $2x - 5 \leq 4 - x \Rightarrow 3x \leq 9 \Rightarrow x \leq 3; (-\infty, 3]$

15. $|x-2| \leq 3 \Rightarrow x - 2 \leq 3$ and $x - 2 \geq -3 \Rightarrow x \leq 5$ and $x \geq -1;\ [-1, 5]$

16. $-1 \leq 1 - 2x \leq 6 \Rightarrow -2 \leq -2x \leq 5 \Rightarrow 1 \geq x \geq -\dfrac{5}{2} \Rightarrow -\dfrac{5}{2} \leq x \leq 1;\ \left[-\dfrac{5}{2}, 1\right]$

17. (a) Add the equations.

$$2x - y = 4$$
$$\underline{x + y = 8}$$
$$3x \quad = 12 \Rightarrow x = 4$$

$$x + y = 8 \Rightarrow 4 + y = 8 \Rightarrow y = 4.$$

The solution is $(4, 4)$.

(b) Multiply the second equation by -3 and add to the first equation.

$$3x - 4y = 2$$
$$\underline{-3x + 4y = -3}$$
$$0 = -1 \qquad \text{contradiction; there are no solutions.}$$

18. Note that $x + 2y \le 2 \Rightarrow y \le -\dfrac{1}{2}x + 1$ and $-x + 3y \ge 3 \Rightarrow y \ge \dfrac{1}{3}x + 1$. See Figure 18.

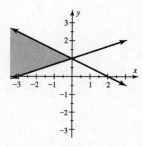

Figure 18

19. Add the first two equations to eliminate the variable z.

$$x + 2y - z = 6$$
$$\underline{x - 3y + z = -2}$$
$$2x - y \quad = 4$$

Add the first and third equations to eliminate the variable z.

$$x + 2y - z = 6$$
$$\underline{x + y + z = 6}$$
$$2x + 3y \quad = 12$$

Multiply this new equation by -1 and add it to the result of the first sum.

$$-2x - 3y = -12$$
$$\underline{2x - y = \quad 4}$$
$$-4y = -8 \qquad \Rightarrow y = 2; \, 2x - 2 = 4 \Rightarrow 2x = 6 \Rightarrow x = 3$$

Substitute $x = 3$ and $y = 2$ into the third equation to solve for z:

$$3 + 2 + z = 6 \Rightarrow 5 + z = 6 \Rightarrow z = 1. \text{ The solution is } (3, 2, 1).$$

20. $4\big(1 \cdot 1 - (-2) \cdot 0\big) - 2\big(2 \cdot 1 - 0 \cdot 0\big) + (-1)\big(2(-2) - 0 \cdot 1\big)$

$$= 4(1 - 0) - 2(2 - 0) - 1(-4 - 0) = 4(1) - 2(2) - 1(-4) = 4 - 4 + 4 = 4$$

21. $4x\left(4-x^3\right)=16x-4x^4=-4x^4+16x$

22. $(x-4)(x+4)=x^2-16$

23. $(5x+3)(x-2)=5x^2-10x+3x-6=5x^2-7x-6$

24. $(4x+9)^2=(4x)^2+2(4x)(9)+(9)^2=16x^2+72x+81$

25. $9x^2-16=(3x-4)(3x+4)$

26. $x^2-4x+4=(x-2)(x-2)=(x-2)^2$

27. $15x^3-9x^2=3x^2\left(5x-3\right)$

28. $12x^2-5x-3=(4x-3)(3x+1)$

29. $r^3-1=(r-1)\left(r^2+r+1\right)$

30. $x^3-3x^2+5x-15=x^2\left(x-3\right)+5\left(x-3\right)=\left(x-3\right)\left(x^2+5\right)$

31. $x^2-3x+2=0\Rightarrow(x-2)(x-1)=0\Rightarrow x=2\text{ or }x=1$

32. $x^3=4x\Rightarrow x^3-4x=0\Rightarrow x\left(x^2-4\right)=0\Rightarrow x(x-2)(x+2)=0$

    $\Rightarrow x=0\text{ or }x=2\text{ or }x=-2$

33. $\dfrac{x^2+3x+2}{x-3}\div\dfrac{x+1}{2x-6}=\dfrac{(x+2)(x+1)}{x-3}\cdot\dfrac{2(x-3)}{x+1}=2(x+2)$

34. $\dfrac{2}{x-1}+\dfrac{5}{x}=\dfrac{2}{x-1}\cdot\dfrac{x}{x}+\dfrac{5}{x}\cdot\dfrac{x-1}{x-1}=\dfrac{2x+5(x-1)}{x(x-1)}=\dfrac{2x+5x-5}{x(x-1)}=\dfrac{7x-5}{x(x-1)}$

35. $\sqrt{36x^2}=\sqrt{36}\cdot\sqrt{x^2}=6x$

36. $\sqrt[3]{64}=4$

37. $16^{-3/2}=\left(\sqrt{16}\right)^{-3}=4^{-3}=\dfrac{1}{4^3}=\dfrac{1}{64}$

38. $\sqrt[4]{625}=\sqrt[4]{(5)^4}=5$

39. $\sqrt{2x}\cdot\sqrt{8x}=\sqrt{(2x)(8x)}=\sqrt{16x^2}=4x$

40. $\sqrt{x}\cdot\sqrt[4]{x}=x^{1/2}\cdot x^{1/4}=x^{3/4}\text{ or }\sqrt[4]{x^3}$

41. $\dfrac{\sqrt[3]{16x^4}}{\sqrt[3]{2x}}=\sqrt[3]{\dfrac{16x^4}{2x}}=\sqrt[3]{8x^3}=2x$

42. $4\sqrt{12x}-2\sqrt{3x}=4\sqrt{4\cdot 3x}-2\sqrt{3x}=4\sqrt{4}\cdot\sqrt{3x}-2\sqrt{3x}$

    $=4(2)\sqrt{3x}-2\sqrt{3x}=8\sqrt{3x}-2\sqrt{3x}=6\sqrt{3x}$

43. $\left(2x+\sqrt{3}\right)\left(x-\sqrt{3}\right)=2x^2-2x\sqrt{3}+x\sqrt{3}-\left(\sqrt{3}\right)^2$

    $=2x^2-x\sqrt{3}-3$

44. The radicand must be greater than or equal to $0 \Rightarrow 1-x \geq 0 \Rightarrow x \leq 1;\ (-\infty,\ 1]$

45. See Figure 45.

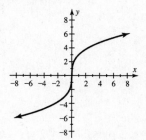

Figure 45

46. $d=\sqrt{(x_2-x_1)^2+(y_2-y_1)^2}=\sqrt{(1-(-2))^2+(2-3)^2}=\sqrt{3^2+(-1)^2}=\sqrt{9+1}=\sqrt{10}$

47. $(1-i)(2+3i)=2+3i-2i-3i^2=2+i-3(-1)=2+3+i=5+i$

48. Use the Pythagorean Theorem: $a^2+b^2=c^2 \Rightarrow 5^2+12^2=c^2 \Rightarrow 25+144=c^2 \Rightarrow 169=c^2 \Rightarrow c=13$

49. $2\sqrt{x+3}=x \Rightarrow \sqrt{x+3}=\dfrac{x}{2} \Rightarrow x+3=\left(\dfrac{x}{2}\right)^2 \Rightarrow x+3=\dfrac{x^2}{4}$

    $\Rightarrow 4(x+3)=4\left(\dfrac{x^2}{4}\right) \Rightarrow 4x+12=x^2 \Rightarrow x^2-4x-12=0$

    $\Rightarrow (x-6)(x+2)=0 \Rightarrow x=6\ (x=-2\text{ does not check}).$

50. $\sqrt[3]{x-1}=3 \Rightarrow x-1=3^3 \Rightarrow x-1=27 \Rightarrow x=28$

51. $\sqrt{x}+4=2\sqrt{x+5} \Rightarrow \left(\sqrt{x}+4\right)^2=4(x+5) \Rightarrow x+8\sqrt{x}+16=4x+20 \Rightarrow 8\sqrt{x}=3x+4 \Rightarrow 64x=(3x+4)^2$

    $\Rightarrow 64x=9x^2+24x+16 \Rightarrow 9x^2-40x+16=0 \Rightarrow (9x-4)(x-4)=0 \Rightarrow x=\dfrac{4}{9}\text{ or }x=4$

52. $\dfrac{1}{3}x^4=27 \Rightarrow x^4=81 \Rightarrow x=\sqrt[4]{81} \Rightarrow x=\pm 3$

53. Graph $Y_1=(X^2-2)^{(1/3)}+X$ and $Y_2=\sqrt{(X)}$ in $[-4.7,\ 4.7,\ 1]$ by $[-3.1,\ 3.1,\ 1]$. The solution is

    $x \approx 1.41.$

54. Graph $Y_1=X-X^{\left(1/3\right)}$ and $Y_2=\sqrt{(X+2)}$ in $[-6,\ 6,\ 1]$ by $[-4,\ 4,\ 1]$. The solution is $x \approx 4.06.$

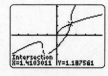 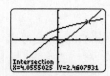

Figure 53   Figure 54

Applications

55.   Water is leaving the tank at 15 gallons per minute.

56.   See Figure 56.

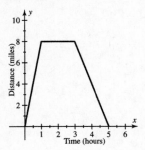

Figure 56

57.   Let x represent the amount invested at 5%. Then $0.05x + 0.04(2000 - x) = 93 \Rightarrow 0.05x + 80 - 0.04x = 93$

$\Rightarrow 0.01x = 13 \Rightarrow x = 1300$; $\$1300$ is invested at 5% and $\$700$ is invested at 4%.

58.   Let x represent the length of a side of the base. Then $x^2(x-4) = 256 \Rightarrow x^3 - 4x^2 - 256 = 0$

Table $Y_1 = X \wedge 3 - 4X \wedge 2 - 256$ with $\text{TblStart} = 1$ and $\Delta \text{Tbl} = 1$ (not shown). $x = 8$, so the dimensions are 8

inches by 8 inches by 4 inches.

59.   5 feet 3 inches $= 5.25$ feet; let h represent the height of the building; $\dfrac{5.25}{7.5} = \dfrac{h}{3.2}$

$\Rightarrow (5.25)(32) = 7.5h \Rightarrow 168 = 7.5h \Rightarrow h = 22.4$ feet

60.   Let x, y, z, represent the measures of the angles of the triangle from smallest to largest, respectively. The system needed is

$x + y + z = 180$
$z = x + y - 20$
$y + z = x + 90$

Substitute $z = x + y - 20$ into the third equation:

$y + x + y - 20 = x + 90 \Rightarrow 2y = 110 \Rightarrow y = 55$

Substitute $y = 55$ and $z = x + y - 20$ into the first equation:

$x + 55 + x + 55 - 20 = 180 \Rightarrow 2x + 90 = 180 \Rightarrow 2x = 90 \Rightarrow x = 45$

Then $z = x + y - 20 \Rightarrow z = 45 + 55 - 20 \Rightarrow z = 80$.

The angle measures are $45°$, $55°$, and $80°$.

Critical Thinking Solutions for Chapter 7

Section 7.1

• $\sqrt[6]{(-2)^6} = |-2| = 2$; $\sqrt[3]{(-2)^3} = -2$; n even $\Rightarrow \sqrt[n]{x^n} = |x|$ and n odd $\Rightarrow \sqrt[n]{x^n} = x$

Section 7.4

- Since hang time increases by a factor of $\sqrt{h}$, we would need 4 times the height to double the hang time.

    $4(50) = 200$ feet

Chapter 8: Quadratic Functions and Equations

8.1: Quadratic Functions and Their Graphs

Concepts

1.   parabola

2.   the vertex

3.   axis of symmetry

4. $(0, 0)$

5.   See Figure 5. *Answers may vary.*

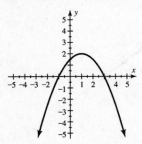

Figure 5

6. $-\dfrac{b}{2a}$

7.   narrower

8.   reflected

9. $ax^2 + bx + c$ with $a \neq 0$

10.   vertex

11. $f(-2) = 0$ and $f(0) = -4$

12. $f(-2) = 4$ and $f(2) = -4$

13. $f(-3) = -2$ and $f(1) = -2$

14. $f(-1) = -2$ and $f(2) = 1$

Graphs of Quadratic Functions

15.   The vertex is $(1, -2)$. The axis of symmetry is $x = 1$. The parabola opens upward.

   The graph is increasing when $x \geq 1$ and decreasing when $x \leq 1$.

17.   The vertex is $(-2, 3)$. The axis of symmetry is $x = -2$. The parabola opens downward.

   The graph is increasing when $x \leq -2$ and decreasing when $x \geq -2$.

19.   (a) See Figure 19.

(b) The vertex is $(0, 0)$. The axis of symmetry is $x = 0$.

(c) $f(-2) = \frac{1}{2}(-2)^2 = \frac{1}{2}(4) = 2$ and $f(3) = \frac{1}{2}(3)^2 = \frac{1}{2}(9) = 4.5$

21.   (a) See Figure 21.

(b) The vertex is $(0, -2)$. The axis of symmetry is $x = 0$.

(c) $f(-2) = (-2)^2 - 2 = 4 - 2 = 2$ and $f(3) = (3)^2 - 2 = 9 - 2 = 7$

23.   (a) See Figure 23.

(b) The vertex is $(0, 1)$. The axis of symmetry is $x = 0$.

(c) $f(-2) = -3(-2)^2 + 1 = -12 + 1 = -11$ and $f(3) = -3(3)^2 + 1 = -27 + 1 = -26$

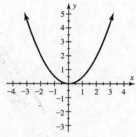

Figure 19

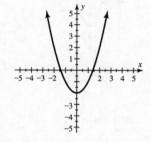

Figure 21

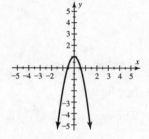

Figure 23

25.   (a) See Figure 25.

(b) The vertex is $(1, 0)$. The axis of symmetry is $x = 1$.

(c) $f(-2) = ((-2) - 1)^2 = (-3)^2 = 9$ and $f(3) = ((3) - 1)^2 = (2)^2 = 4$

27.   (a) See Figure 27.

(b) The vertex is $(-2, 0)$. The axis of symmetry is $x = -2$.

(c) $f(-2) = -(-2 + 2)^2 = -(0)^2 = 0$ and $f(3) = -(3 + 2)^2 = -(5)^2 = -25$

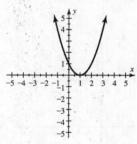

Figure 25

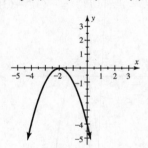

Figure 27

29. (a) See Figure 29.

     (b) The vertex is $(-0.5, -2.25)$. The axis of symmetry is $x = -0.5$.

     (c) $f(-2) = (-2)^2 + (-2) - 2 = 4 - 2 - 2 = 0$ and $f(3) = (3)^2 + (3) - 2 = 9 + 3 - 2 = 10$

31. (a) See Figure 31.

     (b) The vertex is $(0, -3)$. The axis of symmetry is $x = 0$.

     (c) $f(-2) = 2(-2)^2 - 3 = 8 - 3 = 5$ and $f(3) = 2(3)^2 - 3 = 18 - 3 = 15$

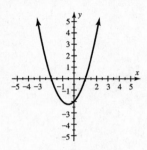

Figure 29   Figure 31

33. (a) See Figure 33.

     (b) The vertex is $(1, 1)$. The axis of symmetry is $x = 1$.

     (c) $f(-2) = 2(-2) - (-2)^2 = -4 - 4 = -8$ and $f(3) = 2(3) - (3)^2 = 6 - 9 = -3$

35. (a) See Figure 35.

     (b) The vertex is $(1, 1)$. The axis of symmetry is $x = 1$.

     (c) $f(-2) = -2(-2)^2 + 4(-2) - 1 = -8 - 8 - 1 = -17$ and $f(3) = -2(3)^2 + 4(3) - 1 = -18 + 12 - 1 = -7$

37. (a) See Figure 37.

     (b) The vertex is $(2, 4)$. The axis of symmetry is $x = 2$.

     (c) $f(-2) = \frac{1}{4}(-2)^2 - (-2) + 5 = 1 + 2 + 5 = 8$ and $f(3) = \frac{1}{4}(3)^2 - (3) + 5 = 2.25 - 3 + 5 = 4.25$

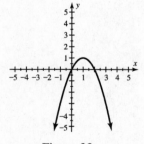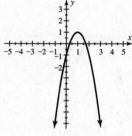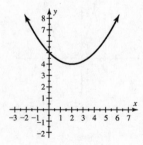

Figure 33   Figure 35   Figure 37

39. $x = -\dfrac{b}{2a} = -\dfrac{(-4)}{2(1)} = 2$, $f(2) = (2)^2 - 4(2) - 2 = 4 - 8 - 2 = -6$; The vertex is $(2, -6)$.

41. $x = -\dfrac{b}{2a} = -\dfrac{(-2)}{2(-\frac{1}{3})} = -3$, $f(-3) = -\dfrac{1}{3}(-3)^2 - 2(-3) + 1 = -3 + 6 + 1 = 4$; The vertex is $(-3, 4)$.

43. $x = -\dfrac{b}{2a} = -\dfrac{(0)}{2(-2)} = 0$, $f(0) = 3 - 2(0)^2 = 3 - 0 = 3$; The vertex is (0, 3).

45. $x = -\dfrac{b}{2a} = -\dfrac{(0.6)}{2(-0.3)} = 1$, $f(1) = -0.3(1)^2 + 0.6(1) + 1.1 = -0.3 + 0.6 + 1.1 = 1.4$; The vertex is (1, 1.4).

47. See Figure 47. Compared to $y = x^2$, the graph is reflected across the *x*-axis.

49. See Figure 49. Compared to $y = x^2$, the graph is narrower.

51. See Figure 51. Compared to $y = x^2$, the graph is wider.

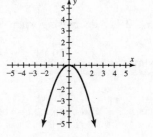

Figure 47

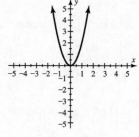

Figure 49

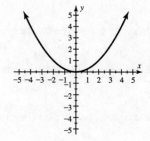

Figure 51

53. See Figure 53. Compared to $y = x^2$, the graph is reflected across the *x*-axis and is wider.

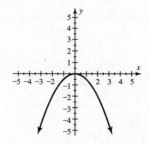

Figure 53

55. Because $-\dfrac{b}{2a} = -\dfrac{2}{2(1)} = -1$ and $f(-1) = (-1)^2 + 2(-1) - 1 = -2$, the vertex is $(-1, -2)$.

   The minimum *y*-value on the graph is -2. The graph is increasing when $x \geq -1$ and decreasing when $x \leq -1$.

57. Because $-\dfrac{b}{2a} = -\dfrac{-5}{2(1)} = \dfrac{5}{2}$ and $f\left(\dfrac{5}{2}\right) = \left(\dfrac{5}{2}\right)^2 - 5\left(\dfrac{5}{2}\right) = -\dfrac{25}{4}$, the vertex is $\left(\dfrac{5}{2}, -\dfrac{25}{4}\right)$.

   The minimum *y*-value on the graph is $-\dfrac{25}{4}$. The graph is increasing when $x \geq \dfrac{5}{2}$ and decreasing when $x \leq \dfrac{5}{2}$.

59. Because $-\dfrac{b}{2a} = -\dfrac{2}{2(2)} = -\dfrac{1}{2}$ and $f\left(-\dfrac{1}{2}\right) = 2\left(-\dfrac{1}{2}\right)^2 + 2\left(-\dfrac{1}{2}\right) - 3 = -\dfrac{7}{2}$, the vertex is $\left(-\dfrac{1}{2}, -\dfrac{7}{2}\right)$.

The minimum y-value on the graph is $-\dfrac{7}{2}$. The graph is increasing when $x \geq -\dfrac{1}{2}$ and decreasing when

$x \leq -\dfrac{1}{2}$.

61. Because $-\dfrac{b}{2a} = -\dfrac{2}{2(-1)} = 1$ and $f(1) = -(1)^2 + 2(1) + 5 = 6$, the vertex is $(1,\ 6)$.

The maximum y-value on the graph is 6. The graph is increasing when $x \leq 1$ and decreasing when $x \geq 1$.

63. Because $-\dfrac{b}{2a} = -\dfrac{4}{2(-1)} = 2$ and $f(2) = 4(2) - (2)^2 = 4$, the vertex is $(2,\ 4)$.

The maximum y-value on the graph is 4. The graph is increasing when $x \leq 2$ and decreasing when $x \geq 2$.

65. Because $-\dfrac{b}{2a} = -\dfrac{1}{2(-2)} = \dfrac{1}{4}$ and $f\left(\dfrac{1}{4}\right) = -2\left(\dfrac{1}{4}\right)^2 + \left(\dfrac{1}{4}\right) - 5 = -\dfrac{39}{8}$, the vertex is $\left(\dfrac{1}{4},\ -\dfrac{39}{8}\right)$.

The maximum y-value on the graph is $-\dfrac{39}{8}$. The graph is increasing when $x \leq \dfrac{1}{4}$ and decreasing when

$x \geq \dfrac{1}{4}$.

67. (a) $a = \dfrac{1}{2}$, so $a > 0$ and the graph opens upward. Because $0 < |a| < 1$, the graph is wider than the graph of

$y = x^2$.

(b) The axis of symmetry is $x = -\dfrac{b}{2a} = -\dfrac{1}{2\left(\frac{1}{2}\right)} = -\dfrac{1}{1} = -1$.

$f(-1) = \dfrac{1}{2}(-1)^2 + (-1) - \dfrac{3}{2} = \dfrac{1}{2}(1) - \dfrac{2}{2} - \dfrac{3}{2} = \dfrac{1}{2} - \dfrac{2}{2} - \dfrac{3}{2} = -\dfrac{4}{2} = -2$ so the vertex is $(-1, -2)$.

(c) The y-intercept is $c = -\dfrac{3}{2}$. To find the x-intercepts, let $y = 0$ and solve for x:

$\dfrac{1}{2}x^2 + x - \dfrac{3}{2} = 0 \Rightarrow 2\left(\dfrac{1}{2}x^2 + x - \dfrac{3}{2}\right) = 2(0) \Rightarrow x^2 + 2x - 3 = 0 \Rightarrow (x+3)(x-1) = 0 \Rightarrow x = -3 \text{ or } x = 1$.

(d) See Figure 67.

69. (a) $a = -1$, so $a < 0$ and the graph opens downward. Because $|a| = 1$, the graph has the same width as the

graph of $y = x^2$.

(b) The axis of symmetry is $x = -\dfrac{b}{2a} = -\dfrac{2}{2(-1)} = -\dfrac{2}{-2} = 1$. $f(1) = 2(1) - 1^2 = 2 - 1 = 1$, so the vertex is $(1, 1)$.

(c) The y-intercept is $c = 0$. To find the x-intercepts, let $y = 0$ and solve for x:

$2x - x^2 = 0 \Rightarrow x(2 - x) = 0 \Rightarrow x = 0 \text{ and } x = 2$.

(d) See Figure 69.

71. (a) $a = 2$, so $a > 0$ and the graph opens upward. Because $|a| > 1$, the graph is narrower than the graph of

$y = x^2$.

(b) The axis of symmetry is $x = -\dfrac{b}{2a} = -\dfrac{2}{2(2)} = -\dfrac{2}{4} = -\dfrac{1}{2}$.

$f\left(-\dfrac{1}{2}\right) = 2\left(-\dfrac{1}{2}\right)^2 + 2\left(-\dfrac{1}{2}\right) - 4 = 2\left(\dfrac{1}{4}\right) - 1 - 4 = \dfrac{1}{2} - \dfrac{2}{2} - \dfrac{8}{2} = -\dfrac{9}{2}$ so the vertex is $\left(-\dfrac{1}{2}, -\dfrac{9}{2}\right)$.

(c) The y-intercept is $c = -4$. To find the x-intercepts, let $y = 0$ and solve for x:

$2x^2 + 2x - 4 = 0 \Rightarrow 2(x^2 + x - 2) = 0 \Rightarrow 2(x + 2)(x - 1) = 0 \Rightarrow x = -2$ or $x = 1$.

(d) See Figure 71.

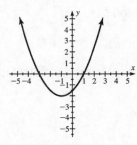

Figure 67

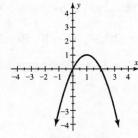

Figure 69

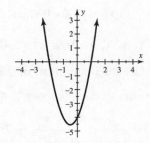

Figure 71

73. (a) Let x be one of the numbers, then $20 - x$ is the other number. $x(20 - x) = 20x - x^2$. The graph of

$y = -x^2 + 20x$ is a parabola that opens downward, so its vertex is the maximum point on the graph. The

x-value of the vertex is on the axis of symmetry, so $x = -\dfrac{b}{2a} = -\dfrac{20}{2(-1)} = -\dfrac{20}{-2} = 10$. So the numbers are

10 and 10.

Applications

75. d. The stone's distance from the ground would increase and then decrease.

77. a. The temperature would first decrease but after the repair, it would increase.

79. (a) When the ball is hit, $t = 0$. Then $h(0) = -16(0)^2 + 64(0) + 2 = 2$ feet.

(b) Find the x-coordinate of the vertex. $-\dfrac{b}{2a} = -\dfrac{64}{2(-16)} = 2$ seconds

(c) Find the y-coordinate of the vertex. $h(2) = -16(2)^2 + 64(2) + 2 = 66$ feet.

81. The x-coordinate of the vertex represents the time when the ball reaches its maximum height. Here we have

$x = \dfrac{-b}{2a} = \dfrac{-66}{2(-16)} = \dfrac{66}{32} \approx 2$ seconds. The maximum height is $h(2) = -16(2)^2 + 66(2) + 6 \approx 74$ feet.

83. (a) The revenue is increasing when $x \le 50$, and it is decreasing when $x \ge 50$.

(b) From the graph, the maximum revenue is $2500 when 50 tickets are sold.

(c) If x represents the number of tickets sold, then $100 - x$ represents the price of one ticket.

The total revenue is given by $f(x) = x(100 - x)$.

(d) Since $f(x) = -x^2 + 100x$, the number of tickets that should be sold to maximize revenue is

$$-\frac{b}{2a} = -\frac{100}{2(-1)} = 50.$$

The maximum revenue is $f(50) = 50(100 - 50) = 50(50) = \$2500$.

85.   Because there are 1200 feet of fence and the width of the enclosure is x, the length is given by $1200 - 2x$.

The area of the enclosure is $A(x) = x(1200 - 2x)$ or $A(x) = -2x^2 + 1200x$. The value of x that will maximize

the area is the x-coordinate of the vertex, $-\dfrac{b}{2a} = -\dfrac{1200}{2(-2)} = 300$. Thus the width is 300 feet and the length is

$1200 - 2(300) = 1200 - 600 = 600 \text{ feet}$. The enclosure measures 300 feet by 600 feet.

87.   (a) Graph $Y_1 = -0.095X^2 + 5.4X - 52.2$ in [20, 40, 5] by [0, 30, 5]. See Figure 87a.

(b) Since the x-coordinate of the vertex is $x \approx 28.4$, the temperature resulting in the greatest height for the

melon seedling is about 28.4°C. See Figure 87b.

(c) $x = -\dfrac{b}{2a} = -\dfrac{5.4}{2(-0.095)} \approx 28.4°C$

<table>
<tr><td>[20, 40, 5] by [0, 30, 5]</td><td>[20, 40, 5] by [0, 30, 5]</td></tr>
<tr><td>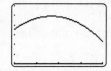</td><td>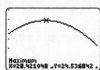</td></tr>
<tr><td>Figure 87a</td><td>Figure 87b</td></tr>
</table>

89. $S(8) = -0.227(8)^2 + 8.155(8) - 8.8 \approx 42 \text{ inches}$

8.2: Parabolas and Modeling

Concepts

1. $x^2 + 2$

2. $(x - 2)^2$

3. $(1,\ 2)$

4. $(-1,\ -2)$

5. $f(x) = ax^2 + bx + c$ or $f(x) = a(x - h)^2 + k$

6. $y = a(x - h)^2 + k;\ (h,\ k)$

7.   downward

8. $-\dfrac{b}{2a}$

Graphs of Parabolas

9.   (a) The vertex form of the function is $f(x) = (x-0)^2 + (-4)$. See Figure 9.

(b) The vertex is $(0, -4)$.

(c) Compared to the graph of $y = x^2$, the graph is shifted down 4 units.

11.   (a) The vertex form of the function is $f(x) = 2(x-0)^2 + 1$. See Figure 11.

(b) The vertex is $(0, 1)$.

(c) Compared to the graph of $y = x^2$, the graph is narrower and is shifted up 1 unit.

13.   (a) The vertex form of the function is $f(x) = (x-3)^2 + 0$. See Figure 13.

(b) The vertex is $(3, 0)$.

(c) Compared to the graph of $y = x^2$, the graph is shifted right 3 units.

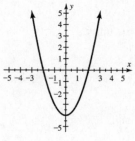 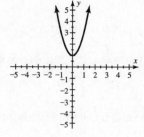 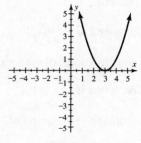

Figure 9   Figure 11   Figure 13

15.   (a) The vertex form of the function is $f(x) = -(x-0)^2 + 0$. See Figure 15.

(b) The vertex is $(0, 0)$.

(c) Compared to the graph of $y = x^2$, the graph is reflected across the x-axis.

17.   (a) The vertex form of the function is $f(x) = -(x-0)^2 + 2$. See Figure 17.

(b) The vertex is $(0, 2)$.

(c) Compared to the graph of $y = x^2$, the graph is reflected across the x-axis and shifted up 2 units.

19.   (a) The vertex form of the function is $f(x) = (x-(-2))^2 + 0$. See Figure 19.

(b) The vertex is $(-2, 0)$.

(c) Compared to the graph of $y = x^2$, the graph is shifted left 2 units.

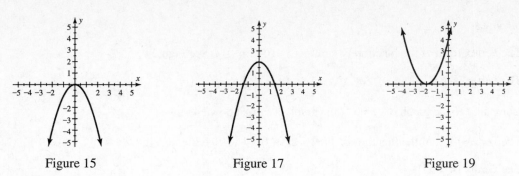

Figure 15   Figure 17   Figure 19

21. (a) The vertex form of the function is $f(x) = (x-(-1))^2 + (-2)$. See Figure 21.

     (b) The vertex is $(-1, -2)$.

     (c) Compared to the graph of $y = x^2$, the graph is shifted left 1 unit and down 2 units.

23. (a) The vertex form of the function is $f(x) = (x-1)^2 + 2$. See Figure 23.

     (b) The vertex is $(1, 2)$.

     (c) Compared to the graph of $y = x^2$, the graph is shifted right 1 unit and up 2 units.

25. (a) The vertex form of the function is $f(x) = 2(x-5)^2 + (-4)$. See Figure 25.

     (b) The vertex is $(5, -4)$.

     (c) Compared to the graph of $y = x^2$, the graph is narrower, shifted right 5 units and down 4 units.

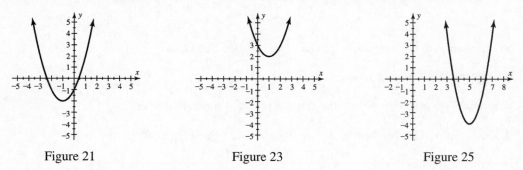

Figure 21   Figure 23   Figure 25

27. (a) The vertex form of the function is $f(x) = -\dfrac{1}{2}(x-(-3))^2 + 1$. See Figure 27.

     (b) The vertex is $(-3, 1)$.

     (c) Compared to the graph of $y = x^2$, the graph is wider, reflected across the *x*-axis, left 3 units and up 1 unit.

29. The graph of *f*(*x*) is the graph of $y = x^2$ translated 1 unit right and 2 units downward. The graph of *f*(*x*) is wider. See Figure 29.

31. The graph of *f*(*x*) is the graph of $y = x^2$ translated 1 unit left and 3 units upward. The graph of *f*(*x*) opens downward and is narrower. See Figure 31.

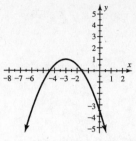

Figure 27

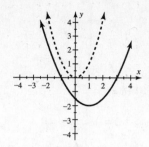

Figure 29

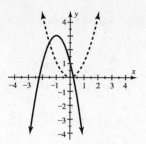

Figure 31

33. See Figure 33.

[–20, 20, 2] by [–20, 20, 2]

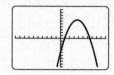

Figure 33

Vertex Form

35. Since $h = 3$, $k = 4$ and $a = 3$, the equation is $y = 3(x-3)^2 + 4$. Expand to obtain the other form:

$$y = 3(x^2 - 6x + 9) + 4 \implies y = 3x^2 - 18x + 27 + 4 \implies y = 3x^2 - 18x + 31$$

37. Since $h = 5$, $k = -2$ and $a = -\dfrac{1}{2}$, the equation is $y = -\dfrac{1}{2}(x-5)^2 - 2$. Expand to obtain the other form:

$$y = -\frac{1}{2}(x^2 - 10x + 25) - 2 \implies y = -\frac{1}{2}x^2 + 5x - \frac{25}{2} - 2 \implies y = -\frac{1}{2}x^2 + 5x - \frac{29}{2}$$

39. Since $h = 1$, $k = 2$ and $a = 1$, the equation is $y = (x-1)^2 + 2$.

41. Since $h = 0$, $k = -3$ and $a = -1$, the equation is $y = -(x-0)^2 - 3$ or $y = -x^2 - 3$.

43. Since $h = 0$, $k = -3$ and $a = 1$, the equation is $y = (x-0)^2 - 3$ or $y = x^2 - 3$.

45. Since $h = -1$, $k = 2$ and $a = -1$, the equation is $y = -(x+1)^2 + 2$.

47. $y = x^2 + 2x - 3 \implies y = (x^2 + 2x + 1) - 3 - 1 \implies y = (x+1)^2 - 4$. The vertex is $(-1,\ -4)$.

49. $y = x^2 - 4x + 5 \implies y = (x^2 - 4x + 4) + 5 - 4 \implies y = (x-2)^2 + 1$. The vertex is $(2,\ 1)$.

51. $y = x^2 + 3x - 2 \implies y = \left(x^2 + 3x + \dfrac{9}{4}\right) - 2 - \dfrac{9}{4} \implies y = \left(x + \dfrac{3}{2}\right)^2 - \dfrac{17}{4}$. The vertex is $\left(-\dfrac{3}{2},\ -\dfrac{17}{4}\right)$.

53. $y = x^2 - 7x + 1 \implies y = \left(x^2 - 7x + \dfrac{49}{4}\right) + 1 - \dfrac{49}{4} \implies y = \left(x - \dfrac{7}{2}\right)^2 - \dfrac{45}{4}$. The vertex is $\left(\dfrac{7}{2},\ -\dfrac{45}{4}\right)$.

55. $y = 3x^2 + 6x - 1 \implies y = 3(x^2 + 2x + 1) - 1 - 3 \implies y = 3(x+1)^2 - 4$. The vertex is $(-1,\ -4)$.

57. $y = 2x^2 - 3x \implies y = 2\left(x^2 - \dfrac{3}{2}x + \dfrac{9}{16}\right) - \dfrac{9}{8} \implies y = 2\left(x - \dfrac{3}{4}\right)^2 - \dfrac{9}{8}$. The vertex is $\left(\dfrac{3}{4},\ -\dfrac{9}{8}\right)$.

59. $y = -2x^2 - 8x + 5 \Rightarrow y = -2(x^2 + 4x + 4) + 5 + 8 \Rightarrow y = -2(x+2)^2 + 13$. The vertex is $(-2, 13)$.

Modeling Data

61.   Since $y = 2$ when $x = 1$, $2 = a(1)^2 \Rightarrow 2 = 1a \Rightarrow a = 2$.

63.   Since $y = 1.2$ when $x = 2$, $1.2 = a(2)^2 \Rightarrow 1.2 = 4a \Rightarrow a = 0.3$.

65.   A scatterplot of the data (not shown) indicates that the data point $(1, -3)$ is the vertex of the parabola.

       The function has the form $f(x) = a(x-1)^2 - 3$. By trial and error, $a = 2$. Thus $f(x) = 2(x-1)^2 - 3$.

67.   A scatterplot of the data (not shown) indicates that the data point $(1980, 6)$ is the vertex of the parabola.

       The function has the form $f(x) = a(x-1980)^2 + 6$. By trial and error, $a = 0.5$. So

       $f(x) = 0.5(x-1980)^2 + 6$.

69.   (a) See Figure 69.

       (b) Since $D = 12$ when $x = 12$, we have $12 = a(12)^2 \Rightarrow 12 = 144a \Rightarrow a = \dfrac{12}{144} \Rightarrow a = \dfrac{1}{12}$.

            The function is $D(x) = \dfrac{1}{12}x^2$.

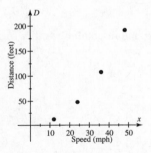

Figure 69

71.   (a) Graph the data (not shown) to see that the data point $(1982, 1)$ is near the vertex of a parabola. Graph

            $Y_1 = (X - 1982) \wedge 2 + 1$ to see if it is a reasonable model. A better model is

            $f(x) = 2(x-1982)^2 + 1$. *Answers may vary.*

       (b) Using the model, $f(1992) = 2(1992 - 1982)^2 + 1 = 2(10)^2 + 1 = 2(100) + 1 = 201$. The model predicts 201

            thousand deaths in 1992, which is close to the actual value of 202 thousand. *Answers may vary.*

Checking Basic Concepts for Sections 8.1 & 8.2

1.   (a) See Figure 1a. The vertex is $(0, -2)$, and the axis of symmetry is $x = 0$.

       (b) See Figure 1b. The vertex is $(1, -3)$, and the axis of symmetry is $x = 1$.

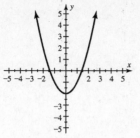

Figure 1a

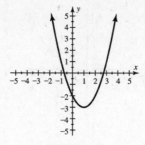

Figure 1b

2. The graph of y_1 opens upward whereas y_2 opens downward. Also, y_1 is narrower than y_2.

3. Because $-\dfrac{b}{2a} = -\dfrac{12}{2(-3)} = 2$ and $f(2) = -3(2)^2 + 12(2) - 5 = 7$, the vertex is $(2,\ 7)$.

   The maximum y-value on the graph is 7. The graph is increasing when $x \le 2$ and decreasing when $x \ge 2$.

4. (a) See Figure 4a. Compared to the graph of $y = x^2$, the graph is shifted right 1 unit and up 2 units.

   (b) See Figure 4b. Compared to the graph of $y = x^2$, the graph is reflected across the x-axis and left 3 units.

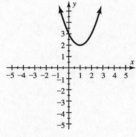

Figure 4a

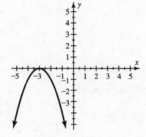

Figure 4b

5. (a) $y = x^2 + 14x - 7 \Rightarrow y = (x^2 + 14x + 49) - 7 - 49 \Rightarrow y = (x+7)^2 - 56$.

   (b) $y = 4x^2 + 8x - 2 \Rightarrow y = 4(x^2 + 2x + 1) - 2 - 4 \Rightarrow y = 4(x+1)^2 - 6$.

8.3: Quadratic Equations

Concepts

1. $x^2 + 3x - 2 = 0$; *Answers may vary.* A quadratic equation can have 0, 1 or 2 solutions.

2. Nonlinear

3. Factoring, square root property, completing the square

4. See Figure 4. *Answers may vary.*

5. See Figure 5. *Answers may vary.*

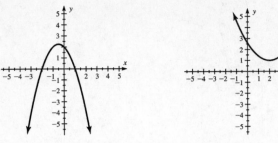

Figure 4   Figure 5

6.   The equation has two solutions. The solutions are the *x*-intercepts.

7. $x = \pm 8$; the square root property.

8. $\left(\dfrac{b}{2}\right)^2$

9.   Yes

10.   Yes

11.   No, there is no x^2 term.

12.   No, the degree of this polynomial is 3.

13.   Yes

14.   Yes

15.   No, the term $\sqrt{x}$ is not allowed in a quadratic equation.

16.   No, the variable *x* is located in the denominator.

17.   (a) $1 \pm \sqrt{7} \approx 1 \pm 2.65 = 3.65 \text{ or } -1.65$

      (b) $-2 \pm \sqrt{11} \approx -2 \pm 3.32 = 1.32 \text{ or } -5.32$

18.   (a) $\pm \dfrac{\sqrt{3}}{2} \approx \pm 0.87$

      (b) $\pm \dfrac{2\sqrt{5}}{7} \approx \pm 0.64$

19.   (a) $\dfrac{3 \pm \sqrt{13}}{5} \approx \dfrac{3 \pm 3.61}{5} = \dfrac{6.61}{5} \text{ or } -\dfrac{0.61}{5} \approx 1.32 \text{ or } -0.12$

      (b) $\dfrac{-5 \pm \sqrt{6}}{9} \approx \dfrac{-5 \pm 2.45}{9} = \dfrac{-2.55}{9} \text{ or } \dfrac{-7.45}{9} \approx -0.28 \text{ or } -0.83$

20.   (a) $\dfrac{2}{5} \pm \dfrac{\sqrt{5}}{5} \approx 0.4 \pm 0.45 = 0.85 \text{ or } -0.05$

      (b) $-\dfrac{3}{7} \pm \dfrac{\sqrt{3}}{7} \approx -\dfrac{3}{7} \pm 0.25 \approx -0.18 \text{ or } -0.68$

Solving Quadratic Equations

21. $-2, 1$

23.   No real solutions

25. $-2, 3$

27. -0.5

29. $x^2 - 4x - 5 = 0 \Rightarrow (x+1)(x-5) = 0 \Rightarrow x+1 = 0 \text{ or } x-5 = 0 \Rightarrow x = -1 \text{ or } 5$

A graph of $y = x^2 - 4x - 5$ (not shown) intersects the x-axis at -1 and 5.

31. $x^2 + 2x = 3 \Rightarrow x^2 + 2x - 3 = 0 \Rightarrow (x+3)(x-1) = 0 \Rightarrow x+3 = 0 \text{ or } x-1 = 0 \Rightarrow x = -3 \text{ or } 1$

A graph of $y = x^2 + 2x - 3$ (not shown) intersects the x-axis at -3 and 1.

33. $x^2 = 9 \Rightarrow x^2 - 9 = 0 \Rightarrow (x+3)(x-3) = 0 \Rightarrow x+3 = 0 \text{ or } x-3 = 0 \Rightarrow x = -3 \text{ or } 3$

A graph of $y = x^2 - 9$ (not shown) intersects the x-axis at -3 and 3.

35. $4x^2 - 4x - 3 = 0 \Rightarrow (2x+1)(2x-3) = 0 \Rightarrow 2x+1 = 0 \text{ or } 2x-3 = 0 \Rightarrow x = -\dfrac{1}{2} \text{ or } \dfrac{3}{2}$

A graph of $y = 4x^2 - 4x - 3$ (not shown) intersects the x-axis at $-\dfrac{1}{2}$ and $\dfrac{3}{2}$.

37. $x^2 + 2x = -1 \Rightarrow x^2 + 2x + 1 = 0 \Rightarrow (x+1)(x+1) = 0 \Rightarrow x+1 = 0 \Rightarrow x = -1$. A graph of $y = x^2 + 2x + 1$ (not shown) intersects the x-axis at -1.

39. $x^2 + 2 = 0$ has no real solutions. A graph of $y = x^2 + 2$ (not shown) does not intersect the x-axis.

41. $x^2 + 2x - 35 = 0 \Rightarrow (x+7)(x-5) = 0 \Rightarrow \text{Either } x+7 = 0 \Rightarrow x = -7 \text{ or } x-5 = 0 \Rightarrow x = 5$

43. $6x^2 - x - 1 = 0 \Rightarrow (3x+1)(2x-1) = 0 \Rightarrow \text{Either } 3x+1 = 0 \Rightarrow x = -\dfrac{1}{3} \text{ or } 2x-1 = 0 \Rightarrow x = \dfrac{1}{2}$

45. $2x^2 - 5x + 3 = 0 \Rightarrow (x-1)(2x-3) = 0 \Rightarrow \text{Either } x-1 = 0 \Rightarrow x = 1 \text{ or } 2x-3 = 0 \Rightarrow x = \dfrac{3}{2}$

47. $25x^2 - 350 = 125x \Rightarrow 25x^2 - 125x - 350 = 0 \Rightarrow 25(x^2 - 5x - 14) = 0 \Rightarrow 25(x-7)(x+2) = 0$

$\Rightarrow \text{Either } x-7 = 0 \Rightarrow x = 7 \text{ or } x+2 = 0 \Rightarrow x = -2$

49. $10x^2 - 27x + 18 = 0 \Rightarrow (5x-6)(2x-3) = 0 \Rightarrow \text{Either } 5x-6 = 0 \Rightarrow x = \dfrac{6}{5} \text{ or } 2x-3 = 0 \Rightarrow x = \dfrac{3}{2}$

51. $x^2 = 144 \Rightarrow x = \pm\sqrt{144} \Rightarrow x = \pm 12$

53. $5x^2 - 64 = 0 \Rightarrow 5x^2 = 64 \Rightarrow x^2 = \dfrac{64}{5} \Rightarrow x = \pm\sqrt{\dfrac{64}{5}} \Rightarrow x = \pm\dfrac{8}{\sqrt{5}} \text{ or } \pm\dfrac{8\sqrt{5}}{5}$

55. $(x+1)^2 = 25 \Rightarrow x+1 = \pm\sqrt{25} \Rightarrow x+1 = \pm 5 \Rightarrow x = -1 \pm 5 \Rightarrow x = -6 \text{ or } 4$

57. $(x-1)^2 = 64 \Rightarrow x-1 = \pm\sqrt{64} \Rightarrow x-1 = \pm 8 \Rightarrow x = 1 \pm 8 \Rightarrow x = -7 \text{ or } 9$

59. $(2x-1)^2 = 5 \Rightarrow 2x-1 = \pm\sqrt{5} \Rightarrow 2x = 1 \pm \sqrt{5} \Rightarrow x = \dfrac{1 \pm \sqrt{5}}{2}$

61. $10(x-5)^2 = 50 \Rightarrow (x-5)^2 = 5 \Rightarrow x-5 = \pm\sqrt{5} \Rightarrow x = 5 \pm \sqrt{5}$

Completing the Square

63. $\left(\dfrac{4}{2}\right)^2 = 2^2 = 4$

65. $\left(\dfrac{-5}{2}\right)^2 = \dfrac{25}{4}$

67. The term needed to complete the square is $\left(\dfrac{-8}{2}\right)^2 = (-4)^2 = 16$. The resulting perfect square is $(x-4)^2$.

69. The term needed to complete the square is $\left(\dfrac{9}{2}\right)^2 = \dfrac{81}{4}$. The resulting perfect square is $\left(x+\dfrac{9}{2}\right)^2$.

71. $x^2 - 2x = 24 \Rightarrow x^2 - 2x + 1 = 24 + 1 \Rightarrow (x-1)^2 = 25 \Rightarrow x - 1 = \pm\sqrt{25} \Rightarrow x - 1 = \pm 5 \Rightarrow x = 1 \pm 5 \Rightarrow x = -4 \text{ or } 6$

73. $x^2 + 6x - 2 = 0 \Rightarrow x^2 + 6x + 9 = 2 + 9 \Rightarrow (x+3)^2 = 11 \Rightarrow x + 3 = \pm\sqrt{11} \Rightarrow x = -3 \pm \sqrt{11}$

75. $x^2 - 3x = 5 \Rightarrow x^2 - 3x + \dfrac{9}{4} = 5 + \dfrac{9}{4} \Rightarrow \left(x - \dfrac{3}{2}\right)^2 = \dfrac{29}{4} \Rightarrow x - \dfrac{3}{2} = \pm\sqrt{\dfrac{29}{4}} \Rightarrow x = \dfrac{3}{2} \pm \dfrac{\sqrt{29}}{2} = \dfrac{3 \pm \sqrt{29}}{2}$

77. $x^2 - 5x + 1 = 0 \Rightarrow x^2 - 5x = -1 \Rightarrow x^2 - 5x + \dfrac{25}{4} = -1 + \dfrac{25}{4} \Rightarrow \left(x - \dfrac{5}{2}\right)^2 = \dfrac{21}{4} \Rightarrow$

$x - \dfrac{5}{2} = \pm\sqrt{\dfrac{21}{4}} \Rightarrow x = \dfrac{5}{2} \pm \dfrac{\sqrt{21}}{2} = \dfrac{5 \pm \sqrt{21}}{2}$

79. $x^2 - 4 = 2x \Rightarrow x^2 - 2x = 4 \Rightarrow x^2 - 2x + 1 = 4 + 1 \Rightarrow (x-1)^2 = 5 \Rightarrow x - 1 = \pm\sqrt{5} \Rightarrow x = 1 \pm \sqrt{5}$

81. $2x^2 - 3x = 4 \Rightarrow x^2 - \dfrac{3}{2}x = 2 \Rightarrow x^2 - \dfrac{3}{2}x + \dfrac{9}{16} = 2 + \dfrac{9}{16} \Rightarrow \left(x - \dfrac{3}{4}\right)^2 = \dfrac{41}{16} \Rightarrow x - \dfrac{3}{4} = \pm\sqrt{\dfrac{41}{16}} \Rightarrow$

$x = \dfrac{3}{4} \pm \dfrac{\sqrt{41}}{4} = \dfrac{3 \pm \sqrt{41}}{4}$

83. $4x^2 - 8x - 7 = 0 \Rightarrow x^2 - 2x - \dfrac{7}{4} = 0 \Rightarrow x^2 - 2x + 1 = \dfrac{7}{4} + 1 \Rightarrow (x-1)^2 = \dfrac{11}{4} \Rightarrow$

$x - 1 = \pm\sqrt{\dfrac{11}{4}} \Rightarrow x = 1 \pm \dfrac{\sqrt{11}}{2} \Rightarrow x = \dfrac{2 \pm \sqrt{11}}{2}$

85. $36x^2 + 18x + 1 = 0 \Rightarrow 36\left(x^2 + \dfrac{1}{2}x + \dfrac{1}{16}\right) = -1 + \dfrac{36}{16} \Rightarrow 36\left(x + \dfrac{1}{4}\right)^2 = \dfrac{20}{16} \Rightarrow$

$\left(x + \dfrac{1}{4}\right)^2 = \dfrac{20}{576} \Rightarrow x + \dfrac{1}{4} = \pm\sqrt{\dfrac{20}{576}} \Rightarrow x = -\dfrac{1}{4} \pm \dfrac{\sqrt{20}}{24} = \dfrac{-6 \pm 2\sqrt{5}}{24} = \dfrac{-3 \pm \sqrt{5}}{12}$

87. $3x^2 + 12x = 36 \Rightarrow 3x^2 + 12x - 36 = 0 \Rightarrow 3(x^2 + 4x - 12) = 0 \Rightarrow 3(x+6)(x-2) = 0 \Rightarrow$

Either $x + 6 = 0 \Rightarrow x = -6$ or $x - 2 = 0 \Rightarrow x = 2$

89. $x^2 + 4x = -2 \Rightarrow x^2 + 4x + 4 = -2 + 4 \Rightarrow (x+2)^2 = 2 \Rightarrow x + 2 = \pm\sqrt{2} \Rightarrow x = -2 \pm \sqrt{2}$

91. $3x^2 - 4 = 2 \Rightarrow 3x^2 = 6 \Rightarrow x^2 = 2 \Rightarrow x = \pm\sqrt{2}$

93. $-6x^2 + 70 = 16x \Rightarrow -6x^2 - 16x + 70 = 0 \Rightarrow -2(3x^2 + 8x - 35) = 0 \Rightarrow -2(3x - 7)(x + 5) = 0 \Rightarrow$

Either $3x - 7 = 0 \Rightarrow x = \dfrac{7}{3}$ or $x + 5 = 0 \Rightarrow x = -5$

95. $-3x(x - 8) = 6 \Rightarrow -3x^2 + 24x = 6 \Rightarrow x^2 - 8x = -2 \Rightarrow x^2 - 8x + 16 = -2 + 16 \Rightarrow (x - 4)^2 = 14$

$\Rightarrow x - 4 = \pm\sqrt{14} \Rightarrow x = 4 \pm \sqrt{14}$

97. $ax^2 - c = 0 \Rightarrow ax^2 = c \Rightarrow x^2 = \dfrac{c}{a} \Rightarrow x = \pm\sqrt{\dfrac{c}{a}}$

Solving Equations by More Than One Method

99.   (a) $x^2 - 3x - 18 = 0 \Rightarrow (x + 3)(x - 6) = 0 \Rightarrow$ Either $x + 3 = 0 \Rightarrow x = -3$ or $x - 6 = 0 \Rightarrow x = 6$

(b) Graph $Y_1 = X^2 - 3X - 18$ in [–5, 8, 1] by [–25, 5, 5]. See Figures 99a & 99b.

The solutions are the *x*-intercepts, $x = -3$ or $x = 6$.

(c) Table $Y_1 = X^2 - 3X - 18$ with TblStart = –6 and ΔTbl = 3. See Figure 99c.

Since $Y_1 = 0$ when $x = -3$ or when $x = 6$, the solutions are $x = -3$ or $x = 6$.

[–5, 8, 1] by [–25, 5, 5]   [–5, 8, 1] by [–25, 5, 5]

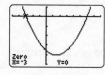   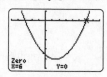   

Figure 99a   Figure 99b   Figure 99c

101.   (a) $x^2 - 8x + 15 = 0 \Rightarrow (x - 3)(x - 5) = 0 \Rightarrow$ Either $x - 3 = 0 \Rightarrow x = 3$ or $x - 5 = 0 \Rightarrow x = 5$

(b) Graph $Y_1 = X^2 - 8X + 15$ in [–10, 0, 1] by [–5, 10, 1]. See Figures 101a & 101b.

The solutions are the *x*-intercepts, $x = 3$ or $x = 5$.

(c) Table $Y_1 = X^2 - 8X + 15$ with TblStart = 1 and ΔTbl = 1. See Figure 101c.

Since $Y_1 = 0$ when $x = 3$ or when $x = 5$, the solutions are $x = 3$ or $x = 5$.

[–10, 0, 1] by [–5, 10, 1]   [–10, 0, 1] by [–5, 10, 1]

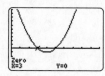   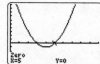   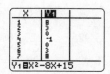

Figure 101a   Figure 101b   Figure 101c

103.   (a) $4(x^2 + 35) = 48x \Rightarrow x^2 + 35 = 12x \Rightarrow x^2 - 12x + 35 = 0 \Rightarrow (x - 5)(x - 7) = 0 \Rightarrow$

Either $x - 5 = 0 \Rightarrow x = 5$ or $x - 7 = 0 \Rightarrow x = 7$

(b) Graph $Y_1 = 4(X^2 + 35) - 48X$ in [–2, 10, 1] by [–6, 6, 1]. See Figures 103a & 103b.

The solutions are the *x*-intercepts, $x = 5$ or $x = 7$.

(c) Table $Y_1 = 4(X^2 + 35) - 48X$ with TblStart = 3 and ΔTbl = 1. See Figure 103c.

Since $Y_1 = 0$ when $x = 5$ or when $x = 7$, the solutions are $x = 5$ or $x = 7$.

[–3, 3, 1] by [–3, 3, 1]   [–3, 3, 1] by [–3, 3, 1]

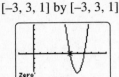   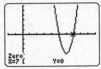   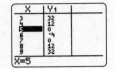

Figure 103a   Figure 103b   Figure 103c

Solving an Equation for a Variable

105. $x = y^2 - 1 \Rightarrow x + 1 = y^2 \Rightarrow \pm\sqrt{x+1} = y \Rightarrow y = \pm\sqrt{x+1}$

107. $K = \dfrac{1}{2}mv^2 \Rightarrow 2K = mv^2 \Rightarrow \dfrac{2K}{m} = v^2 \Rightarrow \pm\sqrt{\dfrac{2K}{m}} = v \Rightarrow v = \pm\sqrt{\dfrac{2K}{m}}$

109. $E = \dfrac{k}{r^2} \Rightarrow Er^2 = k \Rightarrow r^2 = \dfrac{k}{E} \Rightarrow r = \pm\sqrt{\dfrac{k}{E}}$

111. $LC = \dfrac{1}{(2\pi f)^2} \Rightarrow 4\pi^2 f^2 LC = 1 \Rightarrow f^2 = \dfrac{1}{4\pi^2 LC} \Rightarrow f = \pm\sqrt{\dfrac{1}{4\pi^2 LC}} \Rightarrow f = \pm\dfrac{1}{2\pi\sqrt{LC}}$

Applications

113. (a) $\dfrac{1}{2}x^2 = 450 \Rightarrow x^2 = 900 \Rightarrow x = \pm\sqrt{900} \Rightarrow x = 30$ mph ($x = -30$ has no physical meaning)

(b) $\dfrac{1}{2}x^2 = 800 \Rightarrow x^2 = 1600 \Rightarrow x = \pm\sqrt{1600} \Rightarrow x = 40$ mph ($x = -40$ has no physical meaning)

115. $60 - 16t^2 = 0 \Rightarrow -16t^2 = -60 \Rightarrow t^2 = \dfrac{60}{16} \Rightarrow t = \pm\sqrt{\dfrac{60}{16}} \Rightarrow t = \dfrac{\sqrt{60}}{4} \approx 1.9$ seconds. The value $t \approx -1.9$ seconds

has no physical meaning. The toy takes about 1.9 seconds to hit he ground. This is not twice the time it takes

to fall from a height of 30 feet.

117. $S = 42 \Rightarrow 42 = -0.227x^2 + 8.155x - 8.8$

$\Rightarrow -0.227x^2 + 8.155x = 50.8$

$\Rightarrow -0.227(x^2 - 35.925x) = 50.8$

$\Rightarrow x^2 - 35.925x = -223.789$

$\Rightarrow x^2 - 35.925x + 322.651 = -223.789 + 322.651$

$\Rightarrow (x - 17.9625)^2 = 98.862$

$\Rightarrow x - 17.9625 = \pm 9.943$

$\Rightarrow x = 17.9625 \pm 9.943$

$\Rightarrow x \approx 28$ or 8

28 feet from the screen is not reasonable, so the answer is about 8 feet.

119. Let x represent the number of hours the athletes run. The athletes are running along the legs of a right

triangle.

The distance between them is given by the length of the hypotenuse. One athlete runs $6x$ miles while the other runs $8x$ miles. By the Pythagorean Theorem we have the following

$$(6x)^2 + (8x)^2 = 20^2 \Rightarrow 36x^2 + 64x^2 = 400 \Rightarrow 100x^2 = 400 \Rightarrow x^2 = 4 \Rightarrow x = \pm 2.$$

The athletes are 20 miles apart after 2 hours.

121. (a) Let x represent the width of the lot. Then $x + 6$ represents the length. So $x(x+6) = 520$.

   The quadratic equation is $x^2 + 6x - 520 = 0$.

   (b) $x^2 + 6x - 520 = 0 \Rightarrow (x+26)(x-20) = 0 \Rightarrow x = -26$ or 20

   Since the value -26 has no physical meaning, the width is 20 feet.

123. The height of the seedling was about 22 centimeters when the temperature was about 23°C and 34°C.

125. (a) The rate of increase each 20-year period is not constant.

   (b) The vertex is (1800, 5). In 1800 the population was 5 million.

   (c) $0.0066(x-1800)^2 + 5 = 85 \Rightarrow 0.0066(x-1800)^2 = 80 \Rightarrow (x-1800)^2 \approx 12{,}121 \Rightarrow$

   $x - 1800 \approx \sqrt{12{,}121} \Rightarrow x - 1800 \approx 110 \Rightarrow x \approx 1910$. The population was 85 million in 1910.

8.4: The Quadratic Formula

Concepts

1. We use the quadratic formula to solve equations of the form $ax^2 + bx + c = 0$.

2. Completing the square

3. $b^2 - 4ac$

4. The quadratic equation has only one solution.

5. Factoring, square root property, completing the square and the quadratic formula

6. No. When $b^2 - 4ac < 0$, there are no real solutions.

7. $x^2 - 4 = 0 \Rightarrow x^2 = 4 \Rightarrow x = \pm\sqrt{4} \Rightarrow x = \pm 2$

8. $x^2 + 4 = 0 \Rightarrow x^2 = -4 \Rightarrow x = \pm\sqrt{-4} \Rightarrow x = \pm 2i$

The Quadratic Formula

9. $x = \dfrac{-11 \pm \sqrt{(11)^2 - 4(2)(-6)}}{2(2)} = \dfrac{-11 \pm \sqrt{169}}{4} = \dfrac{-11 \pm 13}{4} \Rightarrow x = -6$ or $x = \dfrac{1}{2}$

   Graph $Y_1 = 2X^2 + 11X - 6$ in $[-10, 5, 1]$ by $[-25, 10, 5]$. See Figures 9a & 9b. $x = -6$ or $x = \dfrac{1}{2}$

   Table $Y_1 = 2X^2 + 11X - 6$ with TblStart $= -7.3$ and ΔTbl $= 1.3$. See Figure 9c. $x = -6$ or $x = \dfrac{1}{2}$

[–10, 5, 1] by [–25, 10, 5]   [–10, 5, 1] by [–25, 10, 5]

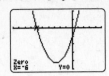   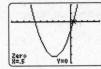   

Figure 9a   Figure 9b   Figure 9c

11. $x = \dfrac{-2 \pm \sqrt{(2)^2 - 4(-1)(-1)}}{2(-1)} = \dfrac{-2 \pm \sqrt{0}}{-2} = \dfrac{-2}{-2} \Rightarrow x = 1$

Graph $Y_1 = -X^2 + 2X - 1$ in [–5, 5, 1] by [–5, 5, 1]. See Figure 11a. $x = 1$

Table $Y_1 = -X^2 + 2X - 1$ with TblStart = –2 and ΔTbl = 1. See Figure 11b. $x = 1$

[–5, 5, 1] by [–5, 5, 1]

   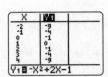

Figure 11a   Figure 11b

13. $x = \dfrac{-(-6) \pm \sqrt{(-6)^2 - 4(1)(-16)}}{2(1)} = \dfrac{6 \pm \sqrt{100}}{2} = \dfrac{6 \pm 10}{2} \Rightarrow x = -2 \text{ or } x = 8$

15. $x = \dfrac{-(-1) \pm \sqrt{(-1)^2 - 4(4)(-1)}}{2(4)} = \dfrac{1 \pm \sqrt{17}}{8}$

17. $x = \dfrac{-2 \pm \sqrt{(2)^2 - 4(-3)(-1)}}{2(-3)} = \dfrac{-2 \pm \sqrt{-8}}{-6} \Rightarrow$ No real solutions.

19. $x = \dfrac{-(-36) \pm \sqrt{(-36)^2 - 4(36)(9)}}{2(36)} = \dfrac{36 \pm \sqrt{0}}{72} = \dfrac{36}{72} = \dfrac{1}{2}$

21. $x = \dfrac{-(-6) \pm \sqrt{(-6)^2 - 4(2)(-2)}}{2(2)} = \dfrac{6 \pm \sqrt{52}}{4} = \dfrac{6 \pm 2\sqrt{13}}{4} = \dfrac{2(3 \pm \sqrt{13})}{2(2)} = \dfrac{3 \pm \sqrt{13}}{2}$

23. $x = \dfrac{-(-4) \pm \sqrt{(-4)^2 - 4(1)(1)}}{2(1)} = \dfrac{4 \pm \sqrt{12}}{2} = \dfrac{4 \pm 2\sqrt{3}}{2} = \dfrac{2(2 \pm \sqrt{3})}{2} = 2 \pm \sqrt{3}$

25. $x = \dfrac{-(-\frac{1}{2}) \pm \sqrt{(-\frac{1}{2})^2 - 4(\frac{3}{2})(-\frac{3}{2})}}{2(\frac{3}{2})} = \dfrac{\frac{1}{2} \pm \sqrt{\frac{37}{4}}}{3} = \dfrac{\frac{1}{2} \pm \frac{\sqrt{37}}{2}}{3} = \dfrac{1 \pm \sqrt{37}}{6}$

27. $x = \dfrac{-(-2) \pm \sqrt{(-2)^2 - 4(2)(-7)}}{2(2)} = \dfrac{2 \pm \sqrt{60}}{4} = \dfrac{2 \pm 2\sqrt{15}}{4} = \dfrac{2(1 \pm \sqrt{15})}{4} = \dfrac{1 \pm \sqrt{15}}{2}$

29. $x = \dfrac{-10 \pm \sqrt{(10)^2 - 4(-3)(-5)}}{2(-3)} = \dfrac{-10 \pm \sqrt{40}}{-6} = \dfrac{-10 \pm 2\sqrt{10}}{-6} = \dfrac{-2(5 \pm \sqrt{10})}{-6} = \dfrac{5 \pm \sqrt{10}}{3}$

The Discriminant

31. (a) Since the parabola opens upward, $a > 0$.

(b) The solutions are the x-intercepts, $x = -1$ or $x = 2$

(c) Since there are two real solutions, the discriminant is positive.

33. (a) Since the parabola opens upward, $a > 0$.

(b) Since there are no x-intercepts, there are no real solutions.

(c) Since there are no real solutions, the discriminant is negative.

35. (a) Since the parabola opens downward, $a < 0$.

(b) The solution is the x-intercept, $x = 2$

(c) Since there is one real solution, the discriminant is zero.

37. (a) $(1)^2 - 4(3)(-2) = 25$

(b) Since the discriminant is positive, there are two real solutions.

(c) Graph $Y_1 = 3X^2 + X - 2$ in [–3, 3, 1] by [–3, 3, 1]. See Figure 37. There are two x-intercepts.

39. (a) $(-4)^2 - 4(1)(4) = 0$

(b) Since the discriminant is zero, there is only one real solution.

(c) Graph $Y_1 = X^2 - 4X + 4$ in [0, 4, 1] by [–3, 3, 1]. See Figure 39. There is one x-intercept.

41. (a) $\left(\dfrac{3}{2}\right)^2 - 4\left(\dfrac{1}{2}\right)(2) = -\dfrac{7}{4}$

(b) Since the discriminant is negative, there are no real solutions.

(c) Graph $Y_1 = (1/2)X^2 + (3/2)X + 2$ in [–10, 10, 1] by [0, 10, 1]. See Figure 41. There are no x-intercepts.

[–3, 3, 1] by [–3, 3, 1]   [0, 4, 1] by [–3, 3, 1]   [–10, 10, 1] by [0, 10, 1]

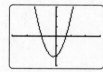   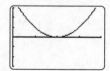   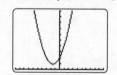

Figure 37   Figure 39   Figure 41

43. (a) $(3)^2 - 4(1)(-3) = 21$

(b) Since the discriminant is positive, there are two real solutions.

(c) Graph $Y_1 = X^2 + 3X - 3$ in [–5, 5, 1] by [–8, 5, 1]. See Figure 43. There are two x-intercepts.

[–5, 5, 1] by [–8, 5, 1]

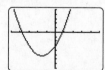

Figure 43

45. $x = \dfrac{-(-2) \pm \sqrt{(-2)^2 - 4(1)(-1)}}{2(1)} = \dfrac{2 \pm \sqrt{8}}{2} = \dfrac{2 \pm 2\sqrt{2}}{2} = \dfrac{2(1 \pm \sqrt{2})}{2} = 1 \pm \sqrt{2}$

47. $x = \dfrac{-(-1) \pm \sqrt{(-1)^2 - 4(-2)(3)}}{2(-2)} = \dfrac{1 \pm \sqrt{25}}{-4} = \dfrac{1 \pm 5}{-4} = -\dfrac{3}{2}$ or 1

49. $x = \dfrac{-1 \pm \sqrt{(1)^2 - 4(1)(5)}}{2(1)} = \dfrac{-1 \pm \sqrt{-19}}{2} \Rightarrow$ No real solutions. No x-intercepts.

51. $x = \dfrac{-(0) \pm \sqrt{(0)^2 - 4(1)(9)}}{2(1)} = \dfrac{0 \pm \sqrt{-36}}{2} \Rightarrow$ No real solutions. No x-intercepts.

53. $x = \dfrac{-4 \pm \sqrt{(4)^2 - 4(3)(-2)}}{2(3)} = \dfrac{-4 \pm \sqrt{40}}{6} = \dfrac{-4 \pm 2\sqrt{10}}{6} = \dfrac{2(-2 \pm \sqrt{10})}{6} = \dfrac{-2 \pm \sqrt{10}}{3}$

Complex Solutions

55. $x^2 + 9 = 0 \Rightarrow x^2 = -9 \Rightarrow x = \pm\sqrt{-9} \Rightarrow x = \pm 3i$

57. $x^2 + 80 = 0 \Rightarrow x^2 = -80 \Rightarrow x = \pm\sqrt{-80} \Rightarrow x = \pm\sqrt{-16 \cdot 5} \Rightarrow x = \pm 4i\sqrt{5}$

59. $x^2 + \dfrac{1}{4} = 0 \Rightarrow x^2 = -\dfrac{1}{4} \Rightarrow x = \pm\sqrt{-\dfrac{1}{4}} \Rightarrow x = \pm\dfrac{1}{2}i$

61. $16x^2 + 9 = 0 \Rightarrow 16x^2 = -9 \Rightarrow x^2 = -\dfrac{9}{16} \Rightarrow x = \pm\sqrt{-\dfrac{9}{16}} \Rightarrow x = \pm\dfrac{3}{4}i$

63. $x^2 = -6 \Rightarrow x = \pm\sqrt{-6} \Rightarrow x = \pm i\sqrt{6}$

65. $x^2 - 3 = 0 \Rightarrow x^2 = 3 \Rightarrow x = \pm\sqrt{3}$

67. $x^2 + 2 = 0 \Rightarrow x^2 = -2 \Rightarrow x = \pm\sqrt{-2} \Rightarrow x = \pm i\sqrt{2}$

69. $x = \dfrac{-(-1) \pm \sqrt{(-1)^2 - 4(1)(2)}}{2(1)} = \dfrac{1 \pm \sqrt{-7}}{2} = \dfrac{1 \pm i\sqrt{7}}{2} = \dfrac{1}{2} \pm i\dfrac{\sqrt{7}}{2}$

71. $x = \dfrac{-3 \pm \sqrt{3^2 - 4(2)(4)}}{2(2)} = \dfrac{-3 \pm \sqrt{-23}}{4} = \dfrac{-3 \pm i\sqrt{23}}{4} = -\dfrac{3}{4} \pm i\dfrac{\sqrt{23}}{4}$

73. $x = \dfrac{-(-4) \pm \sqrt{(-4)^2 - 4(1)(1)}}{2(1)} = \dfrac{4 \pm \sqrt{12}}{2} = \dfrac{4 \pm 2\sqrt{3}}{2} = 2 \pm \sqrt{3}$

75. $x = \dfrac{-1 \pm \sqrt{1^2 - 4(1)(2)}}{2(1)} = \dfrac{-1 \pm \sqrt{-7}}{2} = \dfrac{-1 \pm i\sqrt{7}}{2} = -\dfrac{1}{2} \pm i\dfrac{\sqrt{7}}{2}$

77. $x = \dfrac{-2 \pm \sqrt{(2)^2 - 4(5)(4)}}{2(5)} = \dfrac{-2 \pm \sqrt{-76}}{10} = \dfrac{-2 \pm 2i\sqrt{19}}{10} = -\dfrac{1}{5} \pm i\dfrac{\sqrt{19}}{5}$

79. $x = \dfrac{-\frac{3}{4} \pm \sqrt{\left(\frac{3}{4}\right)^2 - 4\left(\frac{1}{2}\right)(1)}}{2\left(\frac{1}{2}\right)} = \dfrac{-\frac{3}{4} \pm \sqrt{-\frac{23}{16}}}{1} = -\dfrac{3}{4} \pm \sqrt{\dfrac{-23}{16}} = -\dfrac{3}{4} \pm \dfrac{\sqrt{-23}}{4} = -\dfrac{3}{4} \pm i\dfrac{\sqrt{23}}{4}$

81. $x = \dfrac{-1 \pm \sqrt{(1)^2 - 4(1)(4)}}{2(1)} = \dfrac{-1 \pm \sqrt{-15}}{2} = -\dfrac{1}{2} \pm i\dfrac{\sqrt{15}}{2}$

83. $x = \dfrac{-(-2) \pm \sqrt{(-2)^2 - 4(2)(-1)}}{2(2)} = \dfrac{2 \pm \sqrt{12}}{4} = \dfrac{2 \pm 2\sqrt{3}}{4} = \dfrac{2(1 \pm \sqrt{3})}{4} = \dfrac{1 \pm \sqrt{3}}{2}$

85. $x = \dfrac{-(-1) \pm \sqrt{(-1)^2 - 4(2)(2)}}{2(2)} = \dfrac{1 \pm \sqrt{-15}}{4} = \dfrac{1 \pm i\sqrt{15}}{4} = \dfrac{1}{4} \pm i\dfrac{\sqrt{15}}{4}$

You Decide the Method

87. $x^2 - 3x + 2 = 0 \Rightarrow (x-1)(x-2) = 0 \Rightarrow x = 1 \text{ or } x = 2$

89. Multiply both sides of the equation by 4 to obtain the following result.

$$0.5x^2 - 1.75x - 1 = 0 \Rightarrow 2x^2 - 7x - 4 = 0 \Rightarrow (2x+1)(x-4) = 0 \Rightarrow x = -\frac{1}{2} \text{ or } x = 4$$

91. $x = \dfrac{-(-5) \pm \sqrt{(-5)^2 - 4(1)(2)}}{2(1)} = \dfrac{5 \pm \sqrt{17}}{2}$; Quadratic formula or completing the square.

93. $x = \dfrac{-1 \pm \sqrt{(1)^2 - 4(2)(8)}}{2(2)} = \dfrac{-1 \pm \sqrt{-63}}{4} = \dfrac{-1 \pm 3i\sqrt{7}}{4} = -\dfrac{1}{4} \pm \dfrac{3}{4}i\sqrt{7}$; Quadratic formula.

95. $4x^2 - 1 = 0 \Rightarrow (2x-1)(2x+1) = 0 \Rightarrow \text{Either } 2x-1 = 0 \Rightarrow x = \dfrac{1}{2} \text{ or } 2x+1 = 0 \Rightarrow x = -\dfrac{1}{2}$

97. $3x^2 + 6 = 0 \Rightarrow 3(x^2 + 2) = 0 \Rightarrow x^2 + 2 = 0 \Rightarrow x^2 = -2 \Rightarrow x = \pm\sqrt{-2} \Rightarrow x = \pm i\sqrt{2}$

99. $9x^2 + 1 = 6x \Rightarrow 9x^2 - 6x + 1 = 0 \Rightarrow (3x-1)(3x-1) = 0 \Rightarrow 3x - 1 = 0 \Rightarrow x = \dfrac{1}{3}$

Applications

101. $\dfrac{1}{9}x^2 + \dfrac{11}{3}x - 42 = 0 \Rightarrow x^2 + 33x - 378 = 0 \Rightarrow x = (x-9)(x+42) = 0 \Rightarrow x = 9$ mph

   The value $x = -42$ has no physical meaning.

103. $\dfrac{1}{9}x^2 + \dfrac{11}{3}x - 390 = 0 \Rightarrow x^2 + 33x - 3510 = 0 \Rightarrow x = (x-45)(x+78) = 0 \Rightarrow x = 45$ mph

   The value $x = -78$ has no physical meaning.

105. $2.39x^2 + 5.04x + 5.1 = 200 \Rightarrow 2.39x^2 + 5.04x - 194.9 = 0$

   $$x = \dfrac{-5.04 \pm \sqrt{(5.04)^2 - 4(2.39)(-194.9)}}{2(2.39)} = \dfrac{-5.04 \pm \sqrt{1888.6456}}{4.78} \Rightarrow x \approx 8.04 \text{ ; about 1992}$$

   The value $x \approx -10.15$ has no meaning in this problem. Our answer agrees with the graph.

107. Use the Pythagorean Theorem to write the equation: $x^2 + (x+10)^2 = 50^2 \Rightarrow x^2 + x^2 + 20x + 100 = 2500 \Rightarrow$

   $2x^2 + 20x - 2400 = 0 \Rightarrow x^2 + 10x - 1200 = 0 \Rightarrow (x+40)(x-30) = 0 \Rightarrow x = 30$ ($x = -40$ is not

   reasonable). The cars traveled 30 miles and 40 miles.

109. Let x represent the height of the screen. Then $x+3$ represents the width. The equation is $x(x+3)=154$.

(a) Graph $Y_1 = X^2 + 3X - 154$ in [–20, 20, 2] by [–200, 100, 50]. See Figure 109a. $x = 11$

(b) Table $Y_1 = X^2 + 3X - 154$ with TblStart = 7 and ΔTbl = 1. See Figure 109b. $x = 11$

(c) $x^2 + 3x - 154 = 0 \Rightarrow (x-11)(x+14) = 0 \Rightarrow x = 11$; The value $x = -14$ has no physical meaning.

The screen is 11 inches by 14 inches.

[–20, 20, 2] by [–200, 100, 50]

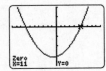   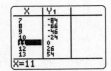

Figure 109a   Figure 109b

111. (a) The rate of change is not constant.

(b) From the table, the height was 7 centimeters after about 75 seconds. *Answers may vary.*

(c) $x = \dfrac{-(-0.15) \pm \sqrt{(0.15)^2 - 4(0.0004)(9)}}{2(0.0004)} = \dfrac{0.15 \pm \sqrt{0.0081}}{0.0008} = \dfrac{0.15 \pm 0.09}{0.0008} \Rightarrow x = 75$ seconds

The value $x = 300$ is not possible for this problem.

Checking Basic Concepts for Sections 8.3 & 8.4

1. Symbolic:

$2x^2 - 7x + 3 = 0 \Rightarrow (2x-1)(x-3) = 0 \Rightarrow$ Either $2x - 1 = 0 \Rightarrow x = \dfrac{1}{2}$ or $x - 3 = 0 \Rightarrow x = 3$

Graphical: Graph $Y_1 = 2X^2 - 7X + 3$ in [–5, 5, 1] by [–5, 5, 1]. See Figures 1a & 1b. $x = \dfrac{1}{2}$ or $x = 3$

[–5, 5, 1] by [–5, 5, 1]   [–5, 5, 1] by [–5, 5, 1]

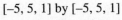   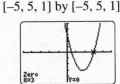

Figure 1a   Figure 1b

2. $x^2 = 5 \Rightarrow x = \pm\sqrt{5}$

3. $x^2 - 4x + 1 = 0 \Rightarrow x^2 - 4x + 4 = -1 + 4 \Rightarrow (x-2)^2 = 3 \Rightarrow x - 2 = \pm\sqrt{3} \Rightarrow x = 2 \pm \sqrt{3}$

4. $x^2 + y^2 = 1 \Rightarrow y^2 = 1 - x^2 \Rightarrow y = \pm\sqrt{1 - x^2}$

5. (a) $2x^2 = 3x + 1 \Rightarrow 2x^2 - 3x - 1 = 0 \Rightarrow x = \dfrac{-(-3) \pm \sqrt{(-3)^2 - 4(2)(-1)}}{2(2)} = \dfrac{3 \pm \sqrt{17}}{4}$

(b) $9x^2 - 24x + 16 = 0 \Rightarrow x = \dfrac{-(-24) \pm \sqrt{(-24)^2 - 4(9)(16)}}{2(9)} = \dfrac{24 \pm \sqrt{0}}{18} = \dfrac{24}{18} = \dfrac{4}{3}$

(c) $x = \dfrac{-1 \pm \sqrt{1^2 - 4(1)(2)}}{2(1)} = \dfrac{-1 \pm \sqrt{-7}}{2} = \dfrac{-1 \pm i\sqrt{7}}{2} = -\dfrac{1}{2} \pm i\dfrac{\sqrt{7}}{2}$

6.   (a) $(-5)^2 - 4(1)(5) = 5$; Since the discriminant is positive, there are two real solutions.

(b) $(-5)^2 - 4(2)(4) = -7$; Since the discriminant is negative, there are no real solutions.

(c) $(-56)^2 - 4(49)(16) = 0$; Since the discriminant is zero, there is one real solution.

8.5: Quadratic Inequalities

Concepts

1.   An inequality has an inequality sign rather than an equals sign.

2.   No, they often have an infinite number of solutions.

3.   No, since $9 \not< 7$.

4.   Yes, since $25 \geq 25$.

5. $-2 < x < 4$

6. $x < -3$ or $x > 1$

7.   Yes

8.   No, the degree of the polynomial on the right side is 3.

9.   Yes

10.   Yes

11.   No, this inequality is linear.

12.   No, the degree of the polynomial on the left side is 3.

13.   Yes, since $2(3)^2 + (3) - 1 = 20$ and $20 > 0$.

14.   Yes, since $(2)^2 - 3(2) + 2 = 0$ and $0 \leq 0$.

15.   No, since $(0)^2 + 2 = 2$ and $2 \not\leq 0$.

16.   No, since $2(1)(1-3) = -4$ and $-4 \not\geq 0$.

17.   No, since $(-3)^2 - 3(-3) = 18$ and $18 \not\leq 1$.

18.   No, since $4(-2)^2 - 5(-2) + 1 = 27$ and $27 \not> 30$.

Solving Quadratic Inequalities

19.   (a)   The solutions are the *x*-intercepts, $x = -3$ or $x = 2$.

(b) $-3 < x < 2$

(c) $x < -3$ or $x > 2$

21.   (a)   The solutions are the *x*-intercepts, $x = -2$ or $x = 2$.

(b) $-2 < x < 2$

(c) $x < -2$ or $x > 2$

23.   (a)   The solutions are the *x*-intercepts, $x = -10$ or $x = 5$.

(b) $x < -10$ or $x > 5$

(c) $-10 < x < 5$

25.   (a) The solutions are $x = -2$ or $x = 2$.

(b) $-2 < x < 2$

(c) $x < -2$ or $x > 2$

27.   (a) The solutions are $x = -4$ or $x = 0$.

(b) $-4 < x < 0$

(c) $x < -4$ or $x > 0$

29.   First replace the inequality symbol with an equals sign and solve the resulting equation.

$x^2 + 4x + 3 = 0 \Rightarrow (x+3)(x+1) = 0 \Rightarrow x = -3$ or -1. The parabola given by $y = x^2 + 4x + 3$ lies below the x-axis when $-3 < x < -1$. The interval is $(-3, -1)$.

31.   First replace the inequality symbol with an equals sign and solve the resulting equation.

$2x^2 - x - 15 = 0 \Rightarrow (2x+5)(x-3) = 0 \Rightarrow x = -2.5$ or 3. The parabola given by $y = 2x^2 - x - 15$ lies above the x-axis when $x \le -2.5$ or $x \ge 3$. The interval is $(-\infty, -2.5] \cup [3, \infty)$.

33.   First replace the inequality symbol with an equals sign and solve the resulting equation.

$2x^2 = 8 \Rightarrow x^2 = 4 \Rightarrow x^2 - 4 = 0 \Rightarrow (x+2)(x-2) = 0 \Rightarrow x = -2$ or 2. The parabola given by $y = 2x^2 - 8$ lies below the x-axis when $-2 \le x \le 2$. The interval is $[-2, 2]$.

35.   The value of x^2 is greater than -5 for all values of x. The parabola given by $y = x^2 + 5$ lies entirely above the x-axis. The interval is $(-\infty, \infty)$.

37.   First replace the inequality symbol with an equals sign and solve the resulting equation.

$-x^2 + 3x = 0 \Rightarrow -x(x-3) = 0 \Rightarrow x = 0$ or 3. The parabola given by $y = -x^2 + 3x$ lies above the x-axis when $0 < x < 3$. The interval is $(0, 3)$.

39. $x^2 + 2 \le 0$ has no solutions; the left side of the inequality is always greater than or equal to 2.

41. $(x-2)^2 \le 0 \Rightarrow (x-2)^2 = 0$ because $(x-2)^2$ can never be less than 0. $(x-2)^2 = 0 \Rightarrow x - 2 = 0 \Rightarrow x = 2$.

43.   First replace the inequality symbol with an equals sign and solve the resulting equation.

$(x+1)^2 = 0 \Rightarrow x + 1 = 0 \Rightarrow x = -1$. The parabola given by $y = (x+1)^2$ lies above the x-axis when $x < -1$ or $x > -1$. The interval is $(-\infty, -1) \cup (-1, \infty)$.

45. $x(1-x) \ge -2 \Rightarrow x(1-x) + 2 \ge 0$. Replace the inequality symbol with an equals sign and solve the resulting equation. $x(1-x) + 2 = 0 \Rightarrow x - x^2 + 2 = 0 \Rightarrow -x^2 + x + 2 = 0 \Rightarrow (-x+2)(x+1) = 0 \Rightarrow x = 2$ or $x = -1$. The parabola given by $y = -x^2 + x + 2$ lies above or on the x-axis when $-1 \le x \le 2$. The interval is $[-1, 2]$.

47.   (a) $x^2 - 4 = 0 \Rightarrow x^2 = 4 \Rightarrow x = -2$ or $x = 2$

(b) Since the parabola opens upward, the solution is $-2 < x < 2$.

(c) Since the parabola opens upward, the solution is $x < -2$ or $x > 2$.

49. (a) $x^2 + x - 1 = 0 \Rightarrow x = \dfrac{-1 \pm \sqrt{(1)^2 - 4(1)(-1)}}{2(1)} \Rightarrow x = \dfrac{-1 \pm \sqrt{5}}{2}$

   (b) Since the parabola opens upward, the solution is $\dfrac{-1 - \sqrt{5}}{2} < x < \dfrac{-1 + \sqrt{5}}{2}$.

   (c) Since the parabola opens upward, the solution is $x < \dfrac{-1 - \sqrt{5}}{2}$ or $x > \dfrac{-1 + \sqrt{5}}{2}$.

51. $x^2 + 10x + 21 = 0 \Rightarrow (x+7)(x+3) = 0 \Rightarrow x = -7$ or $x = -3$

   Since the parabola opens upward, the solution is $[-7, \ -3]$.

53. $3x^2 - 9x + 6 = 0 \Rightarrow 3(x-1)(x-2) = 0 \Rightarrow x = 1$ or $x = 2$

   Since the parabola opens upward, the solution is $(-\infty, \ 1) \cup (2, \ \infty)$.

55. $x^2 = 10 \Rightarrow x = -\sqrt{10}$ or $x = \sqrt{10}$

   Since the parabola opens upward, the solution is $(-\sqrt{10}, \ \sqrt{10})$.

57. $x(x-6) = 0 \Rightarrow x = 0$ or $x = 6$

   Since the parabola opens upward, the solution is $(-\infty, \ 2) \cup (6, \ \infty)$.

59. $x(4 - x) = 2 \Rightarrow -x^2 + 4x - 2 = 0 \Rightarrow x = \dfrac{-4 \pm \sqrt{(4)^2 - 4(-1)(-2)}}{2(-1)} = 2 \pm \sqrt{2}$

   Since the parabola opens downward, the solution is $(-\infty, \ 2 - \sqrt{2}] \cup [2 + \sqrt{2}, \ \infty)$.

61. Graph $Y_1 = \pi X \wedge 2 - \sqrt{(3)} X$ and $Y_2 = 3/11$ in $[-4.7, 4.7, 1]$ by $[-3.1, 3.1, 1]$. The solution is $[-0.128, 0.679]$.

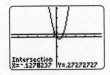   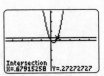

Figure 61a   Figure 61b

Applications

63. (a) Graph $Y_1 = 0.0000375X^2 - 0.175X + 1000$ and $Y_2 = 850$ in $[0, 4000, 1000]$ by $[500, 1200, 100]$.

   See Figures 63a & 63b. The elevation is 850 feet or less from 1131 feet to 3535 feet (approximately).

   (b) The elevation is 850 feet or more before 1131 feet or after 3535 feet (approximately).

   $[0, 4000, 1000]$ by $[500, 1200, 100]$ $[0, 4000, 1000]$ by $[500, 1200, 100]$

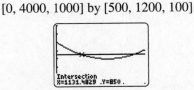   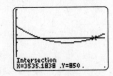

Figure 63a   Figure 63b

65.   (a)   From the formula, $f(1985) = -0.05107(1985)^2 + 194.74(1985) - 184,949 \approx 383$.

           From the graph $f(1985) \approx 383$.

      (b)   The death rate was 500 or less about 1969 and after.

      (c) $-0.05107x^2 + 194.74x - 184,949 = 500 \Rightarrow -0.05107x^2 + 194.74x - 185,449 = 0 \Rightarrow$

$$x = \frac{-194.74 \pm \sqrt{(194.74)^2 - 4(-0.05107)(-185,449)}}{2(-0.05107)} \approx 1969 \text{ or after}.$$

           The only valid solution is 1969. The death rate was 250 or less about 1969 or after.

67.   Let w represent the width of the pen. Then $w + 5$ represents the length and the area is given by

      $w(w+5) = w^2 + 5w$. We want to find w so that $w^2 + 5w \geq 176$ and $w^2 + 5w \leq 500$.

      $w^2 + 5w - 176 = 0 \Rightarrow (w+16)(w-16) = 0 \Rightarrow w = -16$ or $w = 11$, so $w \geq 11$. The values $w \leq -16$ have no

      meaning. $w^2 + 5w - 500 = 0 \Rightarrow (w+25)(w-20) = 0 \Rightarrow w = -25$ or $w = 20$, so $w \leq 20$. The values $w \leq -25$

      have no meaning. The width must be from 11 feet to 20 feet.

8.6: Equations in Quadratic Form

Concepts

1. $u^2 - 7u + 6 = 0 \Rightarrow (u-1)(u-6) = 0 \Rightarrow u = 1$ or 6

      When $u = 1$, $x^2 = 1 \Rightarrow x = \pm 1$. When $u = 6$, $x^2 = 6 \Rightarrow x = \pm\sqrt{6}$.

2. $2u^2 - 7u + 6 = 0 \Rightarrow (2u-3)(u-2) = 0 \Rightarrow u = \dfrac{3}{2}$ or 2

      When $u = \dfrac{3}{2}$, $k^2 = \dfrac{3}{2} \Rightarrow k = \pm\dfrac{\sqrt{3}}{\sqrt{2}} = \pm\dfrac{\sqrt{6}}{2}$. When $u = 2$, $k^2 = 2 \Rightarrow k = \pm\sqrt{2}$.

3. $3u^2 + u - 10 = 0 \Rightarrow (u+2)(3u-5) = 0 \Rightarrow u = -2$ or $\dfrac{5}{3}$

      When $u = -2$, $z^3 = -2 \Rightarrow z = -\sqrt[3]{2}$. When $u = \dfrac{5}{3}$, $z^3 = \dfrac{5}{3} \Rightarrow z = \sqrt[3]{\dfrac{5}{3}}$.

4. $2u^2 + 17u + 8 = 0 \Rightarrow (u+8)(2u+1) = 0 \Rightarrow u = -8$ or $-\dfrac{1}{2}$

      When $u = -8$, $x^3 = -8 \Rightarrow x = -\sqrt[3]{8} = -2$. When $u = -\dfrac{1}{2}$, $x^3 = -\dfrac{1}{2} \Rightarrow x = -\sqrt[3]{\dfrac{1}{2}}$.

5. $4u^2 + 17u + 15 = 0 \Rightarrow (u+3)(4u+5) = 0 \Rightarrow u = -3$ or $-\dfrac{5}{4}$

      When $u = -3$, $n^{-1} = -3 \Rightarrow n = -\dfrac{1}{3}$. When $u = -\dfrac{5}{4}$, $n^{-1} = -\dfrac{5}{4} \Rightarrow n = -\dfrac{4}{5}$.

6. $u^2 + 24 = 10u \Rightarrow u^2 - 10u + 24 = 0 \Rightarrow (u-6)(u-4) = 0 \Rightarrow u = 6$ or 4

When $u = 6$, $m^{-1} = 6 \Rightarrow m = \dfrac{1}{6}$. When $u = 4$, $m^{-1} = 4 \Rightarrow m = \dfrac{1}{4}$.

7. Let $u = x^2$. Then $u^2 = 8u + 9 \Rightarrow u^2 - 8u - 9 = 0 \Rightarrow (u+1)(u-9) = 0 \Rightarrow u = -1$ or 9.

When $u = -1$, $x^2 = -1 \Rightarrow$ no solutions. When $u = 9$, $x^2 = 9 \Rightarrow x = \pm\sqrt{9} = -3$ or 3.

8. Let $u = x^2$. Then $3u^2 = 10u + 8 \Rightarrow 3u^2 - 10u - 8 = 0 \Rightarrow (3u+2)(u-4) = 0 \Rightarrow u = -\dfrac{2}{3}$ or 4.

When $u = -\dfrac{2}{3}$, $x^2 = -\dfrac{2}{3} \Rightarrow$ no solutions. When $u = 4$, $x^2 = 4 \Rightarrow x = \pm\sqrt{4} = -2$ or 2.

9. Let $u = x^3$. Then $3u^2 - 5u - 2 = 0 \Rightarrow (3u+1)(u-2) = 0 \Rightarrow u = -\dfrac{1}{3}$ or 2.

When $u = -\dfrac{1}{3}$, $x^3 = -\dfrac{1}{3} \Rightarrow x = -\sqrt[3]{\dfrac{1}{3}}$. When $u = 2$, $x^3 = 2 \Rightarrow x = \sqrt[3]{2}$.

10. Let $u = x^3$. Then $6u^2 + 11u + 4 = 0 \Rightarrow (3u+4)(2u+1) = 0 \Rightarrow u = -\dfrac{4}{3}$ or $-\dfrac{1}{2}$.

When $u = -\dfrac{4}{3}$, $x^3 = -\dfrac{4}{3} \Rightarrow x = -\sqrt[3]{\dfrac{4}{3}}$. When $u = -\dfrac{1}{2}$, $x^3 = -\dfrac{1}{2} \Rightarrow x = -\sqrt[3]{\dfrac{1}{2}}$.

11. Let $u = z^{-1}$. Then $2u^2 + 11u = 40 \Rightarrow 2u^2 + 11u - 40 = 0 \Rightarrow (u+8)(2u-5) = 0 \Rightarrow u = -8$ or $\dfrac{5}{2}$.

When $u = -8$, $z^{-1} = -8 \Rightarrow z = -\dfrac{1}{8}$. When $u = \dfrac{5}{2}$, $z^{-1} = \dfrac{5}{2} \Rightarrow z = \dfrac{2}{5}$.

12. Let $u = z^{-1}$. Then $u^2 - 10u + 25 = 0 \Rightarrow (u-5)^2 = 0 \Rightarrow u = 5$. When $u = 5$, $z^{-1} = 5 \Rightarrow z = \dfrac{1}{5}$.

13. Let $u = x^{1/3}$. Then $u^2 - 2u + 1 = 0 \Rightarrow (u-1)^2 = 0 \Rightarrow u = 1$. When $u = 1$, $x^{1/3} = 1 \Rightarrow x = 1^3 = 1$.

14. Let $u = x^{1/3}$. Then $3u^2 + 18u = 48 \Rightarrow 3u^2 + 18u - 48 = 0 \Rightarrow 3(u+8)(u-2) = 0 \Rightarrow u = -8$ or 2.

When $u = -8$, $x^{1/3} = -8 \Rightarrow x = (-8)^3 = -512$. When $u = 2$, $x^{1/3} = 2 \Rightarrow x = 2^3 = 8$.

15. Let $u = x^{1/5}$. Then $u^2 - 33u + 32 = 0 \Rightarrow (u-1)(u-32) = 0 \Rightarrow u = 1$ or 32.

When $u = 1$, $x^{1/5} = 1 \Rightarrow x = 1^5 = 1$. When $u = 32$, $x^{1/5} = 32 \Rightarrow x = 32^5 = 33,554,432$.

16. Let $u = x^{1/5}$. Then $u^2 - 80u - 81 = 0 \Rightarrow (u+1)(u-81) = 0 \Rightarrow u = -1$ or 81.

When $u = -1$, $x^{1/5} = -1 \Rightarrow x = (-1)^5 = -1$. When $u = 81$, $x^{1/5} = 81 \Rightarrow x = 81^5 = 3,486,784,401$.

17. Let $u = x^{1/2}$. Then $u^2 - 13u + 36 = 0 \Rightarrow (u-4)(u-9) = 0 \Rightarrow u = 4$ or 9.

When $u = 4$, $x^{1/2} = 4 \Rightarrow x = 4^2 = 16$. When $u = 9$, $x^{1/2} = 9 \Rightarrow x = 9^2 = 81$.

18. Let $u = x^{1/2}$. Then $u^2 - 17u + 16 = 0 \Rightarrow (u-1)(u-16) = 0 \Rightarrow u = 1$ or 16.

When $u = 1$, $x^{1/2} = 1 \Rightarrow x = 1^2 = 1$. When $u = 16$, $x^{1/2} = 16 \Rightarrow x = 16^2 = 256$.

19. Let $u = z^{1/4}$. Then $u^2 - 2u + 1 = 0 \Rightarrow (u-1)^2 = 0 \Rightarrow u = 1$. When $u = 1$, $z^{1/4} = 1 \Rightarrow x = 1^4 = 1$.

20. Let $u = z^{1/4}$. Then $u^2 - 4u + 4 = 0 \Rightarrow (u-2)^2 = 0 \Rightarrow u = 2$. When $u = 2$, $z^{1/4} = 2 \Rightarrow x = 2^4 = 16$.

21. Let $u = x + 1$. Then $u^2 - 5u - 14 = 0 \Rightarrow (u+2)(u-7) = 0 \Rightarrow u = -2$ or 7.

    When $u = -2$, $x + 1 = -2 \Rightarrow x = -3$. When $u = 7$, $x + 1 = 7 \Rightarrow x = 6$.

22. Let $u = x - 5$. Then $2u^2 + 5u + 3 = 0 \Rightarrow (2u+3)(u+1) = 0 \Rightarrow u = -\dfrac{3}{2}$ or -1.

    When $u = -\dfrac{3}{2}$, $x - 5 = -\dfrac{3}{2} \Rightarrow x = \dfrac{7}{2}$. When $u = -1$, $x - 5 = -1 \Rightarrow x = 4$.

23. Let $u = x^2 - 1$. Then $u^2 - 4 = 0 \Rightarrow (u+2)(u-2) = 0 \Rightarrow u = -2$ or 2.

    When $u = -2$, $x^2 - 1 = -2 \Rightarrow x^2 = -1 \Rightarrow$ no solutions.

    When $u = 2$, $x^2 - 1 = 2 \Rightarrow x^2 = 3 \Rightarrow x = -\sqrt{3}$ or $\sqrt{3}$.

24. Let $u = x^2 - 9$. Then $u^2 - 8u + 16 = 0 \Rightarrow (u-4)^2 = 0 \Rightarrow u = 4$.

    When $u = 4$, $x^2 - 9 = 4 \Rightarrow x^2 = 13 \Rightarrow x = -\sqrt{13}$ or $\sqrt{13}$.

Equations Having Complex Solutions

25. $x^4 - 16 = 0 \Rightarrow (x^2 - 4)(x^2 + 4) = 0 \Rightarrow (x-2)(x+2)(x^2 + 4) = 0 \Rightarrow x = 2$ or $x = -2$ or

    $x^2 = -4 \Rightarrow x = \pm\sqrt{-4} \Rightarrow x = \pm 2i$

27. $x^3 + x = 0 \Rightarrow x(x^2 + 1) = 0 \Rightarrow x = 0$ or $x^2 = -1 \Rightarrow x = \pm\sqrt{-1} \Rightarrow x = \pm i$

29. $x^4 + 2 = x^2 \Rightarrow x^4 - x^2 + 2 = 0 \Rightarrow (x^2 - 2)(x^2 + 1) = 0 \Rightarrow x^2 = 2$ or $x^2 = -1 \Rightarrow x = \pm\sqrt{2}$ or $x = \pm i$

31. $\dfrac{1}{x} + \dfrac{1}{x^2} = -\dfrac{1}{2} \Rightarrow 2x^2\left(\dfrac{1}{x} + \dfrac{1}{x^2}\right) = 2x^2\left(-\dfrac{1}{2}\right) \Rightarrow 2x + 2 = -x^2 \Rightarrow x^2 + 2x + 2 = 0 \Rightarrow x^2 + 2x = -2$

    $\Rightarrow x^2 + 2x + 1 = -2 + 1 \Rightarrow (x+1)^2 = -1 \Rightarrow x + 1 = \pm\sqrt{-1} \Rightarrow x = -1 \pm i$

33. $\dfrac{2}{x-2} - \dfrac{1}{x} = -\dfrac{1}{2} \Rightarrow 2x(x-2)\left(\dfrac{2}{x-2} - \dfrac{1}{x}\right) = 2x(x-2)\left(-\dfrac{1}{2}\right) \Rightarrow 4x - 2(x-2) = -x(x-2)$

    $\Rightarrow 4x - 2x + 4 = -x^2 + 2x \Rightarrow x^2 + 4 = 0 \Rightarrow x^2 = -4 \Rightarrow x = \pm\sqrt{-4} \Rightarrow x = \pm 2i$

Checking Basic Concepts for Sections 8.5 & 8.6

1. $x^2 - x - 6 = 0 \Rightarrow (x+2)(x-3) = 0 \Rightarrow x = -2$ or $x = 3$

   Since the parabola opens upward, the solution is $(-\infty, -2) \cup (3, \infty)$.

2. $3x^2 + 5x + 2 = 0 \Rightarrow (x+1)(3x+2) = 0 \Rightarrow x = -1$ or $x = -\dfrac{2}{3}$

   Since the parabola opens upward, the solution is $\left[-1, -\dfrac{2}{3}\right]$.

3. Let $u = x^3$. Then $u^2 + 6u - 16 = 0 \Rightarrow (u+8)(u-2) = 0 \Rightarrow u = -8$ or 2.

When $u = -8$, $x^3 = -8 \Rightarrow x = \sqrt[3]{-8} = -2$. When $u = 2$, $x^3 = 2 \Rightarrow x = \sqrt[3]{2}$.

4.   Let $u = x^{1/3}$. Then $u^2 - 7u - 8 = 0 \Rightarrow (u+1)(u-8) = 0 \Rightarrow u = -1$ or 8.

When $u = -1$, $x^{1/3} = -1 \Rightarrow x = (-1)^3 = -1$. When $u = 8$, $x^{1/3} = 8 \Rightarrow x = 8^3 = 512$.

5. $x^4 + 2x^2 + 1 = 0 \Rightarrow (x^2 + 1)^2 = 0 \Rightarrow x^2 + 1 = 0 \Rightarrow x^2 = -1 \Rightarrow x = \pm\sqrt{-1} \Rightarrow x = \pm i$

Chapter 8 Review Exercises

Section 8.1

1.   Vertex: $(-3, 4)$; Axis of symmetry: $x = -3$; Opens downward; Increasing: $x \le -3$; Decreasing: $x \ge -3$

2.   Vertex: $(1, 0)$; Axis of symmetry: $x = 1$; Opens upward; Increasing: $x \ge 1$; Decreasing: $x \le 1$

3.   (a) See Figure 3.

   (b) The vertex is $(0, -2)$. The axis of symmetry is $x = 0$.

   (c) $f(-1) = (-1)^2 - 2 = 1 - 2 = -1$

4.   (a) See Figure 4.

   (b) The vertex is $(2, 1)$. The axis of symmetry is $x = 2$.

   (c) $f(3) = -(3)^2 + 4(3) - 3 = -9 + 12 - 3 = 0$

5.   (a) See Figure 5.

   (b) The vertex is $(1, 2)$. The axis of symmetry is $x = 1$.

   (c) $f(-2) = -\dfrac{1}{2}(-2)^2 + (-2) + \dfrac{3}{2} = -2 - 2 + \dfrac{3}{2} = -2.5$

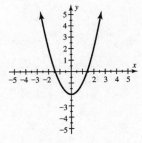   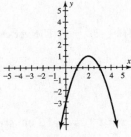   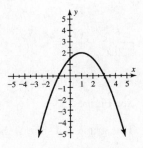

Figure 3   Figure 4   Figure 5

6.   (a) See Figure 6.

   (b) The vertex is $(-2, -3)$. The axis of symmetry is $x = -2$.

   (c) $f(-3) = 2(-3)^2 + 8(-3) + 5 = 18 - 24 + 5 = -1$

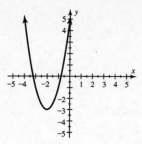

Figure 6

7.   Because $-\dfrac{b}{2a} = -\dfrac{-6}{2(2)} = \dfrac{3}{2}$ and $f\left(\dfrac{3}{2}\right) = 2\left(\dfrac{3}{2}\right)^2 - 6\left(\dfrac{3}{2}\right) + 1 = -\dfrac{7}{2}$, the vertex is $\left(\dfrac{3}{2},\ -\dfrac{7}{2}\right)$.

The minimum y-value on the graph is $-\dfrac{7}{2}$.

8.   Because $-\dfrac{b}{2a} = -\dfrac{2}{2(-3)} = \dfrac{1}{3}$ and $f\left(\dfrac{1}{3}\right) = -3\left(\dfrac{1}{3}\right)^2 + 2\left(\dfrac{1}{3}\right) - 5 = -\dfrac{14}{3}$, the vertex is $\left(\dfrac{1}{3},\ -\dfrac{14}{3}\right)$.

The maximum y-value on the graph is $-\dfrac{14}{3}$.

9. $x = -\dfrac{b}{2a} = -\dfrac{(-4)}{2(1)} = 2$, $f(2) = (2)^2 - 4(2) - 2 = -6$; The vertex is $(2, -6)$.

10. $x = -\dfrac{b}{2a} = -\dfrac{(0)}{2(-1)} = 0$, $f(0) = 5 - (0)^2 = 5$; The vertex is $(0, 5)$.

11. $x = -\dfrac{b}{2a} = -\dfrac{(1)}{2(-\frac{1}{4})} = 2$, $f(2) = -\dfrac{1}{4}(2)^2 + (2) + 1 = 2$; The vertex is $(2, 2)$.

12. $x = -\dfrac{b}{2a} = -\dfrac{(2)}{2(1)} = -1$, $f(-1) = 2 + 2(-1) + (-1)^2 = 1$; The vertex is $(-1, 1)$.

Section 8.2

13.   (a) See Figure 13.

(b) This graph is a shift of the graph of $y = x^2$ upward 2 units.

14.   (a) See Figure 14.

(b) This graph is more narrow than the graph of $y = x^2$.

15.   (a) See Figure 15.

(b) This graph is a shift of the graph of $y = x^2$ right 2 units.

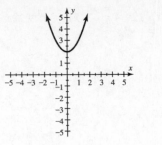

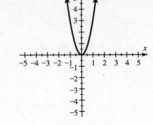

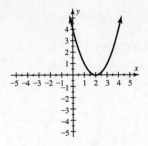

Figure 13   Figure 14   Figure 15

16.   (a) See Figure 16.

       (b) This graph is a shift of the graph of $y = x^2$ left 1 unit and downward 3 units.

17.   (a) See Figure 17.

       (b) This graph is wider than the graph of $y = x^2$ and is shifted left 1 unit and upward 2 units.

18.   (a) See Figure 18.

       (b) This graph is more narrow than the graph of $y = x^2$ and is shifted right 1 unit and downward 3 units.

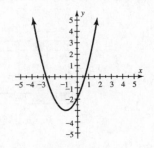

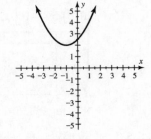

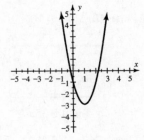

Figure 16   Figure 17   Figure 18

19. $y = a(x-h)^2 + k$ where $(h,\ k)$ is the vertex, gives $y = -4(x-2)^2 - 5$.

20.   Opening downward means $a = -1$. $y = a(x-h)^2 + k$, gives $y = -1(x+4)^2 + 6$.

21. $y = x^2 + 4x - 7 \Rightarrow y = (x^2 + 4x + 4) - 7 - 4 \Rightarrow y = (x+2)^2 - 11$. The vertex is $(-2,\ -11)$.

22. $y = x^2 - 7x + 1 \Rightarrow y = \left(x^2 - 7x + \dfrac{49}{4}\right) + 1 - \dfrac{49}{4} \Rightarrow y = \left(x - \dfrac{7}{2}\right)^2 - \dfrac{45}{4}$. The vertex is $\left(\dfrac{7}{2},\ -\dfrac{45}{2}\right)$.

23. $y = 2x^2 - 3x - 8 \Rightarrow y = 2\left(x^2 - \dfrac{3}{2}x + \dfrac{9}{16}\right) - 8 - \dfrac{9}{8} \Rightarrow y = 2\left(x - \dfrac{3}{4}\right)^2 - \dfrac{73}{8}$. The vertex is $\left(\dfrac{3}{4},\ -\dfrac{73}{8}\right)$.

24. $y = 3x^2 + 6x - 2 \Rightarrow y = 3(x^2 + 2x + 1) - 2 - 3 \Rightarrow y = 3(x+1)^2 - 5$. The vertex is $(-1, -5)$.

25. $a(-1)^2 - 1 = 2 \Rightarrow a = 3$

26. $a(-1)^2 - 1 = -\dfrac{3}{4} \Rightarrow a = \dfrac{1}{4}$

27. $f(x) = -5(x-3)^2 + 4 \Rightarrow f(x) = -5(x^2 - 6x + 9) + 4$

       $\Rightarrow f(x) = -5x^2 + 30x - 45 + 4 \Rightarrow f(x) = -5x^2 + 30x - 41$; the y-intercept is $c = -41$.

28. $f(x) = 3(x+2)^2 - 4 \Rightarrow f(x) = 3(x^2 + 4x + 4) - 4 \Rightarrow f(x) = 3x^2 + 12x + 12 - 4 \Rightarrow f(x) = 3x^2 + 12x + 8$; the y-

intercept is $c = 8$.

Section 8.3

29. $-2, 3$

30. -1

31.   No real solutions.

32. $-4, 6$

33. $-10, 5$

34. $-0.5, 0.25$

35.   (a) Graph $Y_1 = X^2 - 5X - 50$ in $[-10, 20, 5]$ by $[-100, 20, 10]$. See Figures 35a & 35b.

The solutions are the x-intercepts, $x = -5$ and $x = 10$.

(b) Table $Y_1 = X^2 - 5X - 50$ with TblStart $= -10$ and ΔTbl $= 5$. See Figure 35c.

Since $Y_1 = 0$ when $x = -5$ and when $x = 10$, the solutions are $x = -5$ or $x = 10$.

$[-10, 20, 5]$ by $[-100, 20, 10]$ $[-10, 20, 5]$ by $[-100, 20, 10]$

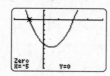   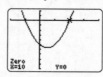   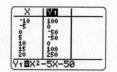

Figure 35a   Figure 35b   Figure 35c

36.   (a) Graph $Y_1 = (1/2)X^2 + X - (3/2)$ in $[-5, 5, 1]$ by $[-5, 5, 1]$. See Figures 36a & 36b.

The solutions are the x-intercepts, $x = -3$ or $x = 1$.

(b) Table $Y_1 = (1/2)X^2 + X - (3/2)$ with TblStart $= -4$ and ΔTbl $= 1$. See Figure 36c.

Since $Y_1 = 0$ when $x = -3$ and when $x = 1$, the solutions are $x = -3$ and $x = 1$.

$[-5, 5, 1]$ by $[-5, 5, 1]$ $[-5, 5, 1]$ by $[-5, 5, 1]$

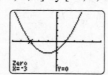   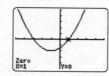

Figure 36a   Figure 36b   Figure 36c

37.   (a) Graph $Y_1 = (1/4)X^2 + (1/2)X - 2$ in $[-5, 5, 1]$ by $[-3, 3, 1]$. See Figures 37a & 37b.

The solutions are the x-intercepts, $x = -4$ and $x = 2$.

(b) Table $Y_1 = (1/4)X^2 + (1/2)X - 2$ with TblStart $= -8$ and ΔTbl $= 2$. See Figure 37c.

Since $Y_1 = 0$ when $x = -4$ and when $x = 2$, the solutions are $x = -4$, and $x = 2$.

[–5, 5, 1] by [–3, 3, 1]   [–5, 5, 1] by [–3, 3, 1]

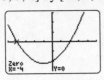   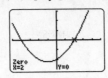

Figure 37a   Figure 37b   Figure 37c

38.   (a) Graph $Y_1 = (1/4)X^2 - (1/2)X - (3/4)$ in [–5, 5, 1] by [–3, 3, 1]. See Figures 38a & 38b.

The solutions are the *x*-intercepts, $x = -1$ and $x = 3$.

(b) Table $Y_1 = (1/4)X^2 - (1/2)X - (3/4)$ with TblStart = –2 and ΔTbl = 1. See Figure 38c.

Since $Y_1 = 0$ when $x = -1$ and when $x = 3$, the solutions are $x = -1$ and $x = 3$.

[–5, 5, 1] by [–3, 3, 1]   [–5, 5, 1] by [–3, 3, 1]

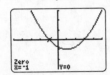   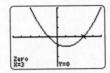

Figure 38a   Figure 38b   Figure 38c

39. $x^2 + x - 20 = 0 \Rightarrow (x+5)(x-4) = 0 \Rightarrow x = -5$ or $x = 4$

40. $x^2 + 11x + 24 = 0 \Rightarrow (x+8)(x+3) = 0 \Rightarrow x = -8$ or $x = -3$

41. $15x^2 - 4x - 4 = 0 \Rightarrow (5x+2)(3x-2) = 0 \Rightarrow x = -\dfrac{2}{5}$ or $x = \dfrac{2}{3}$

42. $7x^2 - 25x + 12 = 0 \Rightarrow (7x-4)(x-3) = 0 \Rightarrow x = \dfrac{4}{7}$ or $x = 3$

43. $x^2 = 100 \Rightarrow x = \pm\sqrt{100} \Rightarrow x = \pm 10$

44. $3x^2 = \dfrac{1}{3} \Rightarrow x^2 = \dfrac{1}{9} \Rightarrow x = \pm\sqrt{\dfrac{1}{9}} \Rightarrow x = \pm\dfrac{1}{3}$

45. $4x^2 - 6 = 0 \Rightarrow x^2 = \dfrac{6}{4} \Rightarrow x = \pm\sqrt{\dfrac{6}{4}} \Rightarrow x = \pm\dfrac{\sqrt{6}}{2}$

46. $5x^2 = x^2 - 4 \Rightarrow 4x^2 = -4 \Rightarrow x^2 = -1 \Rightarrow x = \pm\sqrt{-1} \Rightarrow$ No real solutions.

47. $x^2 + 6x = -2 \Rightarrow x^2 + 6x + 9 = -2 + 9 \Rightarrow (x+3)^2 = 7 \Rightarrow x+3 = \pm\sqrt{7} \Rightarrow x = -3 \pm\sqrt{7}$

48. $x^2 - 4x = 6 \Rightarrow x^2 - 4x + 4 = 6 + 4 \Rightarrow (x-2)^2 = 10 \Rightarrow x-2 = \pm\sqrt{10} \Rightarrow x = 2 \pm\sqrt{10}$

49. $x^2 - 2x - 5 = 0 \Rightarrow x^2 - 2x + 1 = 5 + 1 \Rightarrow (x-1)^2 = 6 \Rightarrow x-1 = \pm\sqrt{6} \Rightarrow x = 1 \pm\sqrt{6}$

50. $2x^2 + 6x - 1 = 0 \Rightarrow x^2 + 3x + \dfrac{9}{4} = \dfrac{1}{2} + \dfrac{9}{4} \Rightarrow \left(x+\dfrac{3}{2}\right)^2 = \dfrac{11}{4} \Rightarrow x+\dfrac{3}{2} = \pm\sqrt{\dfrac{11}{4}} \Rightarrow x = \dfrac{-3 \pm\sqrt{11}}{2}$

51. $F = \dfrac{k}{(R+r)^2} \Rightarrow F(R+r)^2 = k \Rightarrow (R+r)^2 = \dfrac{k}{F} \Rightarrow R+r = \pm\sqrt{\dfrac{k}{F}} \Rightarrow R = -r \pm\sqrt{\dfrac{k}{F}}$

52. $2x^2 + 3y^2 = 12 \Rightarrow 3y^2 = 12 - 2x^2 \Rightarrow y^2 = \dfrac{12 - 2x^2}{3} \Rightarrow y = \pm\sqrt{\dfrac{12 - 2x^2}{3}}$

Section 8.4

53. $x = \dfrac{-(-9) \pm \sqrt{(-9)^2 - 4(1)(18)}}{2(1)} = \dfrac{9 \pm \sqrt{9}}{2} = \dfrac{9 \pm 3}{2} \Rightarrow x = 3 \text{ or } x = 6$

54. $x = \dfrac{-(-24) \pm \sqrt{(-24)^2 - 4(1)(143)}}{2(1)} = \dfrac{24 \pm \sqrt{4}}{2} = \dfrac{24 \pm 2}{2} \Rightarrow x = 11 \text{ or } x = 13$

55. $x = \dfrac{-1 \pm \sqrt{(1)^2 - 4(6)(-1)}}{2(6)} = \dfrac{-1 \pm \sqrt{25}}{12} = \dfrac{-1 \pm 5}{12} \Rightarrow x = -\dfrac{1}{2} \text{ or } x = \dfrac{1}{3}$

56. $x = \dfrac{-(-5) \pm \sqrt{(-5)^2 - 4(5)(1)}}{2(5)} = \dfrac{5 \pm \sqrt{5}}{10}$

57. $x = \dfrac{-(-8) \pm \sqrt{(-8)^2 - 4(1)(-5)}}{2(1)} = \dfrac{8 \pm \sqrt{84}}{2} = \dfrac{8 \pm 2\sqrt{21}}{2} = \dfrac{2(4 \pm \sqrt{21})}{2} = 4 \pm \sqrt{21}$

58. $x = \dfrac{-(-6) \pm \sqrt{(-6)^2 - 4(2)(3)}}{2(2)} = \dfrac{6 \pm \sqrt{12}}{4} = \dfrac{6 \pm 2\sqrt{3}}{4} = \dfrac{2(3 \pm \sqrt{3})}{2(2)} = \dfrac{3 \pm \sqrt{3}}{2}$

59. $x^2 - 4 = 0 \Rightarrow (x - 2)(x + 2) = 0 \Rightarrow x = 2 \text{ or } x = -2$

60. $4x^2 - 1 = 0 \Rightarrow (2x - 1)(2x + 1) = 0 \Rightarrow x = \dfrac{1}{2} \text{ or } x = -\dfrac{1}{2}$

61. $2x^2 + 15 = 11x \Rightarrow 2x^2 - 11x + 15 = 0 \Rightarrow (2x - 5)(x - 3) = 0 \Rightarrow x = \dfrac{5}{2} \text{ or } x = 3$

62. $2x^2 + 15 = 13x \Rightarrow 2x^2 - 13x + 15 = 0 \Rightarrow (2x - 3)(x - 5) = 0 \Rightarrow x = \dfrac{3}{2} \text{ or } x = 5$

63. $x(5 - x) = 2x + 1 \Rightarrow 5x - x^2 - 2x - 1 = 0 \Rightarrow -x^2 + 3x - 1 = 0 \,;\; a = -1, b = 3, c = -1$

$x = \dfrac{-b \pm \sqrt{b^2 - 4(ac)}}{2a} \Rightarrow x = \dfrac{-3 \pm \sqrt{3^2 - 4(-1)(-1)}}{2(-1)} \Rightarrow x = \dfrac{-3 \pm \sqrt{5}}{-2} \Rightarrow x = \dfrac{3 \pm \sqrt{5}}{2}$

64. $-2x(x - 1) = x - \dfrac{1}{2} \Rightarrow 2(-2x)(x - 1) = 2\left(x - \dfrac{1}{2}\right) \Rightarrow -4x^2 + 4x = 2x - 1 \Rightarrow -4x^2 + 2x + 1 = 0 \,;$

$a = -4, b = 2, c = 1 \quad x = \dfrac{-b \pm \sqrt{b^2 - 4ac}}{2a} \Rightarrow x = \dfrac{-2 \pm \sqrt{2^2 - 4(-4)(1)}}{2(-4)}$

$\Rightarrow x = \dfrac{-2 \pm \sqrt{20}}{-8} \Rightarrow x = \dfrac{-2 \pm 2\sqrt{5}}{-8} \Rightarrow x = \dfrac{1 \pm \sqrt{5}}{4}$

65.   (a) Since the parabola opens upward, $a > 0$.

(b) The solutions are the x-intercepts, $x = -2$ or $x = 3$.

(c) Since there are two unique solutions, the discriminant is positive.

66.   (a) Since the parabola opens upward, $a > 0$.

       (b) The solution is the x-intercept, $x = 2$.

       (c) Since there is one solution, the discriminant is zero.

67.   (a) Since the parabola opens downward, $a < 0$.

       (b) There are no x-intercepts. No real solutions.

       (c) Since there are no real solutions, the discriminant is negative.

68.   (a) Since the parabola opens downward, $a < 0$.

       (b) The solutions are the x-intercepts, $x = -4$ or $x = 2$.

       (c) Since there are two unique solutions, the discriminant is positive.

69.   (a) $(-3)^2 - 4(2)(1) = 1$

       (b) Since the discriminant is positive, there are two real solutions.

       (c) Graph $Y_1 = 2X^2 - 3X + 1$ in [0, 2, 1] by [−1, 1, 1]. See Figure 69. There are two x-intercepts.

70.   (a) $(2)^2 - 4(7)(-5) = 144$

       (b) Since the discriminant is positive, there are two real solutions.

       (c) Graph $Y_1 = 7X^2 + 2X - 5$ in [−2, 2, 1] by [−10, 5, 1]. See Figure 70. There are two x-intercepts.

      [0, 2, 1] by [−1, 1, 1]   [−2, 2, 1] by [−10, 5, 1]

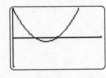   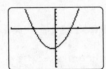

         Figure 69   Figure 70

71.   (a) $(1)^2 - 4(3)(2) = -23$

       (b) Since the discriminant is negative, there are no real solutions.

       (c) Graph $Y_1 = 3X^2 + X + 2$ in [−3, 3, 1] by [−5, 10, 1]. See Figure 71. There are no x-intercepts.

72.   (a) $(-12.6)^2 - 4(4.41)(9) = 0$

       (b) Since the discriminant is zero, there is one real solution.

       (c) Graph $4.41X^2 - 12.6X + 9$ in [0, 3, 1] by [−5, 10, 1]. See Figure 72. There is one x-intercept.

      [−3, 3, 1] by [−5, 10, 1]   [0, 3, 1] by [−5, 10, 1]

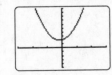   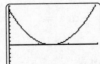

         Figure 71   Figure 72

73. $x = \dfrac{-1 \pm \sqrt{(1)^2 - 4(1)(5)}}{2(1)} = \dfrac{-1 \pm \sqrt{-19}}{2} = -\dfrac{1}{2} \pm i\dfrac{\sqrt{19}}{2}$

74. $2x^2 + 8 = 0 \Rightarrow 2x^2 = -8 \Rightarrow x^2 = -4 \Rightarrow x = \pm\sqrt{-4} \Rightarrow x = \pm 2i$

75. $x = \dfrac{-(-1) \pm \sqrt{(-1)^2 - 4(2)(1)}}{2(2)} = \dfrac{1 \pm \sqrt{-7}}{4} = \dfrac{1}{4} \pm i\dfrac{\sqrt{7}}{4}$

76. $x = \dfrac{-(-2) \pm \sqrt{(-2)^2 - 4(7)(5)}}{2(7)} = \dfrac{2 \pm \sqrt{-136}}{14} = \dfrac{2 \pm 2\sqrt{-34}}{14} = \dfrac{2(1 \pm \sqrt{-34})}{14} = \dfrac{1}{7} \pm i\dfrac{\sqrt{34}}{7}$

Section 8.5

77.   (a) The solutions are the *x*-intercepts, $x = -2$ or $x = 6$.

(b) $-2 < x < 6$

(c) $x < -2$ or $x > 6$

78.   (a) The solutions are the *x*-intercepts, $x = -2$ or $x = 0$.

(b) $x < -2$ or $x > 0$

(c) $-2 < x < 0$

79.   (a) The solutions are $x = -4$ or $x = 4$.

(b) $-4 < x < 4$

(c) $x < -4$ or $x > 4$

80.   (a) The solutions are $x = -2$ or $x = 1$.

(b) $-2 < x < 1$

(c) $x < -2$ or $x > 1$

81.   (a) $x^2 - 2x - 3 = 0 \Rightarrow (x+1)(x-3) = 0 \Rightarrow x = -1$ or $x = 3$

(b) Since the parabola opens upward, the solution is $-1 < x < 3$.

(c) Since the parabola opens upward, the solution is $x < -1$ or $x > 3$.

82.   (a) $2x^2 - 7x - 15 = 0 \Rightarrow (2x+3)(x-5) = 0 \Rightarrow x = -\dfrac{3}{2}$ or $x = 5$

(b) Since the parabola opens upward, the solution is $-\dfrac{3}{2} \leq x \leq 5$.

(c) Since the parabola opens upward, the solution is $x \leq -\dfrac{3}{2}$ or $x \geq 5$.

83. $x^2 + 4x + 3 = 0 \Rightarrow (x+3)(x+1) = 0 \Rightarrow x = -3$ or $x = -1$

Since the parabola opens upward, the solution is $[-3,\ -1]$.

84. $5x^2 - 16x + 3 = 0 \Rightarrow (5x-1)(x-3) = 0 \Rightarrow x = \dfrac{1}{5}$ or $x = 3$

Since the parabola opens upward, the solution is $\left(\dfrac{1}{5},\ 3\right)$.

85. $6x^2 - 13x + 2 = 0 \Rightarrow (6x-1)(x-2) = 0 \Rightarrow x = \dfrac{1}{6}$ or $x = 2$

Since the parabola opens upward, the solution is $\left(-\infty, \dfrac{1}{6}\right) \cup (2, \infty)$.

86. $x^2 = 5 \Rightarrow \Rightarrow x = -\sqrt{5}$ or $x = \sqrt{5}$

Since the parabola opens upward, the solution is $(-\infty, \sqrt{5}] \cup [\sqrt{5}, \infty)$.

87.   The graph of the parabola $y = (x-1)^2$ lies on or above the x-axis for all real numbers x. The interval is

$(-\infty, \infty)$.

88. $x^2 + 3 < 2 \Rightarrow x^2 < -1$, which is not true for any real number x. The graph of the parabola $y = x^2 + 1$ lies

above the x-axis for all real numbers, so there are no solutions.

Section 8.6

89.   Let $u = x^2$. Then $u^2 - 14u + 45 = 0 \Rightarrow (u-5)(u-9) = 0 \Rightarrow u = 5$ or 9.

When $u = 5$, $x^2 = 5 \Rightarrow x = \pm\sqrt{5}$. When $u = 9$, $x^2 = 9 \Rightarrow x = \pm\sqrt{9} = \pm 3$.

90.   Let $u = z^{-1}$. Then $2u^2 + u - 28 = 0 \Rightarrow (u+4)(2u-7) = 0 \Rightarrow u = -4$ or $\dfrac{7}{2}$.

When $u = -4$, $z^{-1} = -4 \Rightarrow z = -\dfrac{1}{4}$. When $u = \dfrac{7}{2}$, $z^{-1} = \dfrac{7}{2} \Rightarrow z = \dfrac{2}{7}$.

91.   Let $u = x^{1/3}$. Then $u^2 - 9u + 8 = 0 \Rightarrow (u-1)(u-8) = 0 \Rightarrow u = 1$ or 8.

When $u = 1$, $x^{1/3} = 1 \Rightarrow x = 1^3 = 1$. When $u = 8$, $x^{1/3} = 8 \Rightarrow x = 8^3 = 512$.

92.   Let $u = x - 1$. Then $u^2 + 2u + 1 = 0 \Rightarrow (u+1)^2 = 0 \Rightarrow u = -1$. When $u = -1$, $x - 1 = -1 \Rightarrow x = 0$.

93. $4x^4 + 4x^2 + 1 = 0 \Rightarrow (2x^2 + 1)^2 = 0 \Rightarrow 2x^2 + 1 = 0 \Rightarrow x^2 = -\dfrac{1}{2} \Rightarrow x = \pm\sqrt{-\dfrac{1}{2}} \Rightarrow x = \pm i\dfrac{\sqrt{1}}{\sqrt{2}}$

$\Rightarrow x = \pm i\dfrac{1}{\sqrt{2}} \Rightarrow x = \pm i\dfrac{\sqrt{2}}{2}$

94. $\dfrac{1}{x-2} - \dfrac{3}{x} = -1 \Rightarrow x(x-2)\left(\dfrac{1}{x-2} - \dfrac{3}{x}\right) = x(x-2)(-1) \Rightarrow x - 3(x-2) = -x^2 + 2x \Rightarrow x^2 - 4x + 6 = 0$

$\Rightarrow x^2 - 4x = -6 \Rightarrow x^2 - 4x + 4 = -6 + 4 \Rightarrow (x-2)^2 = -2 \Rightarrow x - 2 = \pm\sqrt{-2} \Rightarrow x = 2 \pm i\sqrt{2}$

Applications

95.   (a) $f(x) = x(12 - 2x)$

(b)   Note that $f(x) = x(12 - 2x) \Rightarrow f(x) = -2x^2 + 12 + 0$. Then $-\dfrac{b}{2a} = -\dfrac{12}{2(-2)} = 3$.

The maximum area occurs when $x = 3$. The dimensions should be 6 inches by 3 inches.

96.   (a) $-16t^2 + 44t + 4 = 32 \Rightarrow -16t^2 + 44t - 28 = 0 \Rightarrow 4x^2 - 11x + 7 = 0 \Rightarrow (x-1)(4x-7) = 0 \Rightarrow$

$x = 1$ or $x = \dfrac{7}{4}$. The height of the stone is 32 feet at 1 second and 1.75 seconds.

(b) Find the vertex. $-\dfrac{b}{2a} = -\dfrac{44}{2(-16)} = 1.375$; $f(1.375) = -16(1.375)^2 + 44(1.375) + 4 = 34.25$

After 1.375 seconds, the stone is at a height of 34.25 feet.

97. (a) $f(x) = x(90 - 3x)$

(b) Graph $Y_1 = X(90 - 3X)$ in $[0, 30, 5]$ by $[0, 800, 100]$. See Figure 97a.

(c) Graph $Y_1 = X(90 - 3X)$ and $Y_2 = 600$ in $[0, 30, 5]$ by $[0, 800, 100]$. See Figures 97b & 97c. 10 or 20 rooms.

(d) Graph $Y_1 = X(90 - 3X)$ in $[0, 30, 5]$ by $[0, 800, 100]$. See Figure 97d. Rent 15 rooms.

$[0, 30, 5]$ by $[0, 800, 100]$ $\qquad$ $[0, 30, 5]$ by $[0, 800, 100]$

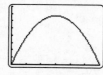   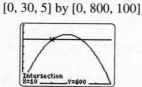

Figure 97a $\qquad\qquad$ Figure 97b

$[0, 30, 5]$ by $[0, 800, 100]$ $\qquad$ $[0, 30, 5]$ by $[0, 800, 100]$

    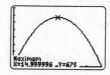

Figure 97c $\qquad\qquad$ Figure 97d

98. (a) $f(1999) = 0.4(1999 - 1997)^2 + 0.8 = 2.4$; In 1999 there were 2.4 complaints per 100,000 passengers.

(b) Graph $Y_1 = 0.4(X - 1997)^2 + 0.8$ in $[1997, 1999, 1]$ by $[0.5, 3, 0.5]$. See Figure 98. They have increased.

$[1997, 1999, 1]$ by $[0.5, 3, 0.5]$

Figure 98

99. (a) $\dfrac{x^2}{12} = 144 \Rightarrow x^2 = 1728 \Rightarrow x = \sqrt{1728} \approx 41.6$ mph $(-\sqrt{1728}$ mph has no meaning).

(b) $\dfrac{x^2}{12} = 300 \Rightarrow x^2 = 3600 \Rightarrow x = \sqrt{3600} = 60$ mph $(-60$ mph has no meaning).

100. (a) $x(x + 2) = 143$ or $x^2 + 2x - 143 = 0$

(b) $x^2 + 2x - 143 = 0 \Rightarrow (x + 13)(x - 11) = 0 \Rightarrow x = -13$ or 11. There are two possible number pairs.

Either $x = -13$, and the other number is -11, or $x = 11$, and the other number is 13.

101. (a) Plot the data in $[1935, 1995, 10]$ by $[0, 100, 10]$. See Figure 101.

(b) Yes, the data is nearly linear.

(c) Since $m = \dfrac{78-25}{1991-1940} = \dfrac{53}{51} \approx 1.04$.

The function could be $f(x) = 1.04(x-1940) + 25$. *Answers may vary.*

(d) $f(1995) = 1.04(1995-1940) + 25 = 1.04(55) + 25 = 82.2$,

$f(2005) = 1.04(2005-1940) + 25 = 1.04(65) + 25 = 92.6$. *Answers may vary.*

102. (a) The vertex is (1950, 220). In 1950 the per-capita energy consumption was at a low of 220 million Btu.

(b) Graph $Y_1 = (1/4)(X-1950)^2 + 220$ in [1950, 1970, 5] by [200, 350, 25]. See Figure 102. It Increased.

(c) $f(1996) = \dfrac{1}{4}(1996-1950)^2 + 220 = 749$. The function is not a good model for 1996 because the trend

represented by the model did not continue after 1970.

[1935, 1995, 10] by [0, 100, 10]   [1950, 1970, 5] by [200, 350, 25]

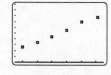   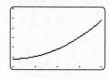

Figure 101   Figure 102

103. $\sqrt{123} \approx 11.1$; the screen is about 11.1 inches by 11.1 inches.

104. $x^2 + (x+70)^2 = 130^2 \Rightarrow x^2 + x^2 + 140x + 4900 = 16,900 \Rightarrow 2x^2 + 140x - 12,000 = 0 \Rightarrow$

$2(x-50)(x+120) = 0 \Rightarrow x = 50$ or $x = -120$. The solution is $x = 50$ feet ($x = -120$ has no meaning).

105. $(30+2x)(50+2x) - 30(50) = 250 \Rightarrow 1500 + 160x + 4x^2 - 1500 = 250 \Rightarrow 4x^2 + 160x - 250 = 0 \Rightarrow$

$x = \dfrac{-160 \pm \sqrt{(160)^2 - 4(4)(-250)}}{2(4)} = \dfrac{-160 \pm \sqrt{29,600}}{8} \Rightarrow x \approx -41.5$ or $x \approx 1.5$

The width of the strip of grass is about 1.5 feet. The value $x \approx -41.5$ has no physical meaning.

106. We must find r so that $750 \le \dfrac{1}{3}\pi r^2 (20) \le 1700$.

$\dfrac{1}{3}\pi r^2 (20) \ge 750 \Rightarrow r^2 \ge \dfrac{2250}{20\pi} \Rightarrow r \ge \sqrt{\dfrac{2250}{20\pi}} \approx 5.98$

$\dfrac{1}{3}\pi r^2 (20) \le 1700 \Rightarrow r^2 \ge \dfrac{5100}{20\pi} \Rightarrow r \ge \sqrt{\dfrac{5100}{20\pi}} \approx 9.01$

The values or r can range from about 6 inches to about 9 inches.

Chapter 8 Test

1. $x = -\dfrac{b}{2a} = -\dfrac{1}{2(-\frac{1}{2})} = \dfrac{1}{1} = 1$, $f(1) = -\dfrac{1}{2}(1)^2 + (1) + 1 = \dfrac{3}{2}$; Vertex: $\left(1,\ \dfrac{3}{2}\right)$; Axis of symmetry: $x = 1$.

    $f(-2) = -\frac{1}{2}(-2)^2 + (-2) + 1 = -\frac{1}{2}(4) + (-2) + 1 = -3$

2. Because $-\dfrac{b}{2a} = -\dfrac{3}{2(1)} = -\dfrac{3}{2}$ and $f\left(-\dfrac{3}{2}\right) = \left(-\dfrac{3}{2}\right)^2 + 3\left(-\dfrac{3}{2}\right) - 5 = -\dfrac{29}{4}$, the vertex is $\left(-\dfrac{3}{2},\ -\dfrac{29}{4}\right)$.

    The minimum y-value on the graph is $-\dfrac{29}{4}$.

3. Since $f(x) = 0$ when $x = -2$, $a(-2)^2 + 2 = 0 \Rightarrow 4a = -2 \Rightarrow a = -\dfrac{1}{2}$.

4. See Figure 4.

    This graph is wider than the graph of $y = x^2$ and is shifted right 3 units and upward 2 units.

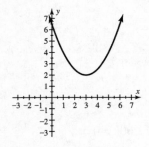

    Figure 4

5. $y = x^2 - 6x + 2 \Rightarrow y = (x^2 - 6x + 9) + 2 - 9 \Rightarrow y = (x-3)^2 - 7$. The vertex is $(3,\ -7)$. The axis of symmetry is $x = 3$.

6. The solutions are the x-intercepts, $x = -1$ or $x = 2$. $f(1) = 2$.

7. $3x^2 + 11x - 4 = 0 \Rightarrow (x+4)(3x-1) = 0 \Rightarrow x = -4$ or $x = \dfrac{1}{3}$

8. $2x^2 = 2 - 6x^2 \Rightarrow 8x^2 = 2 \Rightarrow x^2 = \dfrac{1}{4} \Rightarrow x = \pm\sqrt{\dfrac{1}{4}} \Rightarrow x = -\dfrac{1}{2}$ or $x = \dfrac{1}{2}$

9. $x^2 - 8x = 1 \Rightarrow x^2 - 8x + 16 = 1 + 16 \Rightarrow (x-4)^2 = 17 \Rightarrow x - 4 = \pm\sqrt{17} \Rightarrow x = 4 \pm \sqrt{17}$

10. $x = \dfrac{-3 \pm \sqrt{(3)^2 - 4(-2)(1)}}{2(-2)} = \dfrac{-3 \pm \sqrt{17}}{-4} = \dfrac{3 \pm \sqrt{17}}{4}$

11. $9x^2 - 16 = 0 \Rightarrow (3x+4)(3x-4) = 0 \Rightarrow x = \pm\dfrac{4}{3}$

12. $F = \dfrac{Gm^2}{r^2} \Rightarrow Fr^2 = Gm^2 \Rightarrow m^2 = \dfrac{Fr^2}{G} \Rightarrow m = \pm\sqrt{\dfrac{Fr^2}{G}}$

13. (a) Since the parabola opens downward, $a < 0$.

(b) The solutions are the x-intercepts, $x = -3$ or $x = 1$.

(c) Since there are two real solutions, the discriminant is positive.

14.   (a) $(4)^2 - 4(-3)(-5) = -44$

(b) Since the discriminant is negative, there are no real solutions.

(c) Graph $Y_1 = -3X^2 + 4X - 5$ in $[-5, 5, 1]$ by $[-20, 10, 5]$. See Figure 14. It does not intersect the x-axis.

$[-5, 5, 1]$ by $[-20, 10, 5]$

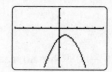

Figure 14

15.   (a) The solutions are the x-intercepts, $x = -1$ or $x = 1$.

(b) $-1 < x < 1$

(c) $x < -1$ or $x > 1$

16.   (a) The solutions are the x-intercepts, $x = -10$ or $x = 20$.

(b) $x < -10$ or $x > 20$

(c) $-10 < x < 20$

17.   (a) $8x^2 - 2x - 3 = 0 \Rightarrow (2x+1)(4x-3) = 0 \Rightarrow x = -\dfrac{1}{2}$ or $x = \dfrac{3}{4}$

(b) Since the parabola opens upward, the solution is $\left[-\dfrac{1}{2}, \dfrac{3}{4}\right]$.

(c) Since the parabola opens upward, the solution is $\left(-\infty, -\dfrac{1}{2}\right] \cup \left[\dfrac{3}{4}, \infty\right)$.

18. $x^2 + 2x = 0 \Rightarrow x(x+2) = 0 \Rightarrow x = -2$ or $x = 0$. Solutions to $x^2 + 2x \le 0$ lie between and include these two

values. The solution set is $[-2, 0]$.

19.   Let $u = x^3$. Then $u^2 - 3u + 2 = 0 \Rightarrow (u-1)(u-2) = 0 \Rightarrow u = 1$ or 2.

When $u = 1$, $x^3 = 1 \Rightarrow x = \sqrt[3]{1} = 1$. When $u = 2$, $x^3 = 2 \Rightarrow x = \sqrt[3]{2}$.

20. $2x^2 + 4x + 3 = 0 \Rightarrow x = \dfrac{-4 \pm \sqrt{(4)^2 - 4(2)(3)}}{2(2)} = \dfrac{-4 \pm \sqrt{-8}}{4} = -1 \pm i\dfrac{\sqrt{2}}{2}$

21.   Graph $Y_1 = \sqrt{2} - \pi X^2$ and $Y_2 = 2.12X - 0.5\pi$ in $[-5, 5, 1]$ by $[-7, 3, 1]$.

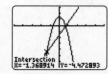   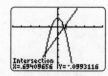

Figure 21a   Figure 21b

The solutions are approximately $x = -1.37$ and $x = 0.69$.

22.　　$\dfrac{x^2}{9} = 250 \Rightarrow x^2 = 2250 \Rightarrow x = \sqrt{2250} \approx 47.4$ mph (the value $x \approx -47.4$ mph has no physical meaning).

23.　　(a)　$2z + 2x + 20 = 200 \Rightarrow 2x + 2z = 180 \Rightarrow z = 90 - x$, so the formula is $f(x) = (x+20)(90-x)$.

　　　(b)　Note that $f(x) = (x+20)(90-x) \Rightarrow f(x) = -x^2 + 70x + 1800$. The area is greatest for the value of x at

　　　　　the vertex of the parabola, $x = -\dfrac{b}{2a} = -\dfrac{70}{2(-1)} = 35$. The enclosed area is greatest when $x = 35$.

24.　　(a)　Graph $Y_1 = -16X^2 + 88X + 8$ in [0, 6, 1] by [0, 150, 50]. See Figure 24.

　　　(b)　$t = \dfrac{-88 \pm \sqrt{(88)^2 - 4(-16)(8)}}{2(-16)} = \dfrac{-88 \pm \sqrt{8256}}{-32} \approx 5.6$ seconds (the value $t \approx -0.089$ has no meaning).

　　　(c)　The stone reaches a maximum height of 129 feet after 2.75 seconds. See Figure 24.

　　　　　[0, 6, 1] by [0, 150, 50]

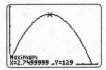

Figure 24

Chapter 8 Extended and Discovery Exercises

1.　　(a)　For the first 3 years of life, the likelihood of survival increases with age. After 3 years of life, it
　　　　　decreases with age.

　　　(b)　Plot the data in [0, 10, 1] by [0, 75, 5]. See Figure 1b. A quadratic function could model these data since
　　　　　the data points form a parabola.

　　　(c)　Graph $Y_1 = -3.57X + 71.1$ in [0, 10, 1] by [0, 75, 5]. See Figure 1c.

　　　　　Graph $Y_1 = -2.07X^2 + 17.1X + 33$ in [0, 10, 1] by [0, 75, 5]. See Figure 1d.

　　　　　The function f_2 models the data better.

　　　(d)　Evaluate f at $x = 6.5$ to find the likelihood of a 5.5-year-old sparrowhawk living 1 more year.

　　　　　$f_2(6.5) = -2.07(6.5)^2 + 17.1(6.5) + 33 \approx 56.7\ \%$

　　　[0, 10, 1] by [0, 75, 5]　　　　　　[0, 10, 1] by [0, 75, 5]　　　　　　[0, 10, 1] by [0, 75, 5]

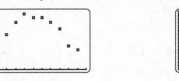　　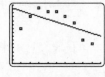　　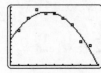

　　　　Figure 1b　　　　　　　　　　Figure 1c　　　　　　　　　　Figure 1d

2.　　(a)　Plot the data in [−5, 35, 5] by [0, 100, 10]. See Figure 2.

　　　(b)　A quadratic function could model these data since the data points form a parabola.

　　　(c)　Using the data point (12, 95) as the vertex, the function has the form $f(x) = a(x-12)^2 + 95$.

　　　　　By trial and error we find that $a = -0.25$ gives a fairly good fit. The function is given by

$f(x) = -0.25(x-12)^2 + 95$. *Answers may vary.*

(d) The x-coordinate of the vertex, $12°C$, represents the the temperature that photosynthesis is most efficient.

[–5, 35, 5] by [0, 100, 10]   [–4, 4, 1] by [0, 6, 1]   [–4, 4, 1] by [0, 6, 1]

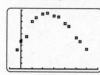   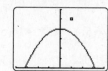   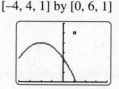

Figure 2   Figure 3a   Figure 3b

3. (a) Plot (1, 5) and graph $Y_1 = -0.4X^2 + 4$ in [–4, 4, 1] by [0, 6, 1]. See Figure 3a.

   (b) After 10 seconds the plane would have moved 2 kilometers. We will shift the parabola 2 units left.

      Plot (1, 5) and graph $Y_1 = -0.4(X+2)^2 + 4$ in [–4, 4, 1] by [0, 6, 1]. See Figure 3b.

4. We would need to translate the mountain both to the right and downward so that the airplane would appear to move to the left and upward. One example could be $f(x) = -0.4(x-2)^2 + 4 - 2$.

5. The discriminant is $(-1)^2 - 4(10)(-3) = 121 = 11^2$. The trinomial factors as $(2x+1)(5x-3)$.

6. The discriminant is $(-3)^2 - 4(4)(-6) = 105$. The trinomial will not factor.

7. The discriminant is $(2)^2 - 4(3)(-2) = 28$. The trinomial will not factor.

8. The discriminant is $(1)^2 - 4(2)(3) = -23$. The trinomial will not factor.

9. $x^3 - x^2 - 6x = 0 \Rightarrow x(x^2 - x - 6) = 0 \Rightarrow x(x-3)(x+2) = 0 \Rightarrow x = 0,\ 3,\ \text{or}\ -2$

In the interval $(-\infty,\ -2)$, we test $x = -3$. $\ f(-3) = (-3)^3 - (-3)^2 - 6(-3) = -18 < 0$

In the interval $(-2,\ 0)$, we test $x = -1$. $\ f(-1) = (-1)^3 - (-1)^2 - 6(-1) = 4 > 0$

In the interval $(0,\ 3)$, we test $x = 1$. $\ f(1) = (1)^3 - (1)^2 - 6(1) = -6 < 0$

In the interval $(3,\ \infty)$, we test $x = 4$. $\ f(4) = (4)^3 - (4)^2 - 6(4) = 24 > 0$

The solution to the given inequality is $(-2,\ 0) \cup (3,\ \infty)$.

10. $x^3 - 3x^2 + 2x = 0 \Rightarrow x(x^2 - 3x + 2) = 0 \Rightarrow x(x-1)(x-2) = 0 \Rightarrow x = 0,\ 1,\ \text{or}\ 2$

In the interval $(-\infty,\ 0)$, we test $x = -1$. $\ f(-1) = (-1)^3 - 3(-1)^2 + 2(-1) = -6 < 0$

In the interval $(0,\ 1)$, we test $x = 0.5$. $\ f(0.5) = (0.5)^3 - 3(0.5)^2 + 2(0.5) = 0.375 > 0$

In the interval $(1,\ 2)$, we test $x = 1.5$. $\ f(1.5) = (1.5)^3 - 3(1.5)^2 + 2(1.5) = -0.375 < 0$

In the interval $(2,\ \infty)$, we test $x = 3$. $\ f(3) = (3)^3 - 3(3)^2 + 2(3) = 6 > 0$

The solution to the given inequality is $(-\infty,\ 0) \cup (1,\ 2)$.

11. $x^3 - 7x^2 + 14x - 8 = 0 \Rightarrow (x^3 - 8) - 7x^2 + 14x = 0 \Rightarrow (x-2)(x^2 + 2x + 4) - 7x(x-2) = 0 \Rightarrow$

$(x-2)((x^2 + 2x + 4) - 7x) = 0 \Rightarrow (x-2)(x^2 - 5x + 4) = 0 \Rightarrow (x-2)(x-1)(x-4) = 0 \Rightarrow x = 1,\ 2,\ \text{or}\ 4$

In the interval $(-\infty,\ 1)$, we test $x=0$. $f(0)=(0)^3-7(0)^2+14(0)-8=-8\le 0$

In the interval $(1,\ 2)$, we test $x=1.5$. $f(1.5)=(1.5)^3-7(1.5)^2+14(1.5)-8=0.625\ge 0$

In the interval $(2,\ 4)$, we test $x=3$. $f(3)=(3)^3-7(3)^2+14(3)-8=-2\le 0$

In the interval $(4,\ \infty)$, we test $x=5$. $f(5)=(5)^3-7(5)^2+14(5)-8=12\ge 0$

The solution to the given inequality is $(-\infty,\ 1]\cup[2,\ 4]$.

12. $9x-x^3=0\Rightarrow x(9-x^2)=0\Rightarrow x(3-x)(3+x)=0\Rightarrow x=0,\ 3,\ \text{or}\ -3$

In the interval $(-\infty,\ -3)$, we test $x=-4$. $f(-4)=9(-4)-(-4)^3=28\ge 0$

In the interval $(-3,\ 0)$, we test $x=-1$. $f(-1)=9(-1)-(-1)^3=-8\le 0$

In the interval $(0,\ 3)$, we test $x=1$. $f(1)=9(1)-(1)^3=8\ge 0$

In the interval $(3,\ \infty)$, we test $x=4$. $f(4)=9(4)-(4)^3=-28\le 0$

The solution to the given inequality is $(-\infty,\ -3]\cup[0,\ 3]$.

13. $x^4-5x^2+4=0\Rightarrow (x^2-1)(x^2-4)=0\Rightarrow (x-1)(x+1)(x-2)(x+2)=0\Rightarrow x=-2,\ -1,\ 1,\ \text{or}\ 2$

In the interval $(-\infty,\ -2)$, we test $x=-3$. $f(-3)=(-3)^4-5(-3)^2+4=40>0$

In the interval $(-2,\ -1)$, we test $x=-1.5$. $f(-1.5)=(-1.5)^4-5(-1.5)^2+4=-2.1875<0$

In the interval $(-1,\ 1)$, we test $x=0$. $f(0)=(0)^4-5(0)^2+4=4>0$

In the interval $(1,\ 2)$, we test $x=1.5$. $f(1.5)=(1.5)^4-5(1.5)^2+4=-2.1875<0$

In the interval $(2,\ \infty)$, we test $x=3$. $f(3)=(3)^4-5(3)^2+4=40>0$

The solution to the given inequality is $(-\infty,\ -2)\cup(-1,\ 1)\cup(2,\ \infty)$.

14. $1-x^4=0\Rightarrow (1-x^2)(1+x^2)=0\Rightarrow (1-x)(1+x)(1+x^2)=0\Rightarrow x=-1\ \text{or}\ 1$

In the interval $(-\infty,\ -1)$, we test $x=-2$. $f(-2)=1-(-2)^4=-15<0$

In the interval $(-1,\ 1)$, we test $x=0$. $f(0)=1-(0)^4=1>0$

In the interval $(1,\ \infty)$, we test $x=2$. $f(2)=1-(2)^4=-15<0$

The solution to the given inequality is $(-\infty,\ -1)\cup(1,\ \infty)$.

15. $\dfrac{3-x}{3x}=0\Rightarrow 3-x=0\Rightarrow x=3$. The expression is undefined when $3x=0\Rightarrow x=0$.

In the interval $(-\infty,\ 0)$, we test $x=-1$. $f(-1)=\dfrac{3-(-1)}{3(-1)}=-\dfrac{4}{3}\le 0$

In the interval $(0,\ 3)$, we test $x=1$. $f(1)=\dfrac{3-(1)}{3(1)}=\dfrac{2}{3}\ge 0$

In the interval $(3, \infty)$, we test $x = 4$. $f(4) = \dfrac{3-(4)}{3(4)} = -\dfrac{1}{12} \le 0$

The solution to the given inequality is $(0, 3]$.

16. $\dfrac{x-2}{x+2} = 0 \Rightarrow x-2 = 0 \Rightarrow x = 2$. The expression is undefined when $x+2 = 0 \Rightarrow x = -2$.

In the interval $(-\infty, -2)$, we test $x = -3$. $f(-3) = \dfrac{(-3)-2}{(-3)+2} = 5 > 0$

In the interval $(-2, 2)$, we test $x = 0$. $f(0) = \dfrac{(0)-2}{(0)+2} = -1 < 0$

In the interval $(2, \infty)$, we test $x = 3$. $f(3) = \dfrac{(3)-2}{(3)+2} = \dfrac{1}{5} > 0$

The solution to the given inequality is $(-\infty, -2) \cup (2, \infty)$.

17. $\dfrac{3-2x}{1+x} = 3 \Rightarrow 3-2x = 3+3x \Rightarrow 5x = 0 \Rightarrow x = 0$.

The expression is undefined when $1+x = 0 \Rightarrow x = -1$.

In the interval $(-\infty, -1)$, we test $x = -2$. $f(-2) = \dfrac{3-2(-2)}{1+(-2)} - 3 = -10 < 0$

In the interval $(-1, 0)$, we test $x = -0.5$. $f(-0.5) = \dfrac{3-2(-0.5)}{1+(-0.5)} - 3 = 5 > 0$

In the interval $(0, \infty)$, we test $x = 1$. $f(1) = \dfrac{3-2(1)}{1+(1)} - 3 = -\dfrac{5}{2} < 0$

The solution to the given inequality is $(-\infty, -1) \cup (0, \infty)$.

18. $\dfrac{x+1}{4-2x} = 1 \Rightarrow x+1 = 4-2x \Rightarrow 3x = 3 \Rightarrow x = 1$.

The expression is undefined when $4-2x = 0 \Rightarrow 2x = 4 \Rightarrow x = 2$.

In the interval $(-\infty, 1)$, we test $x = 0$. $f(0) = \dfrac{(0)+1}{4-2(0)} - 1 = -\dfrac{3}{4} \le 0$

In the interval $(1, 2)$, we test $x = 1.5$. $f(1.5) = \dfrac{(1.5)+1}{4-2(1.5)} - 1 = \dfrac{3}{2} \ge 0$

In the interval $(2, \infty)$, we test $x = 3$. $f(3) = \dfrac{(3)+1}{4-2(3)} - 1 = -3 \le 0$

The solution to the given inequality is $[1, 2)$.

19. The expression is undefined when $x^2 - 4 = 0 \Rightarrow (x+2)(x-2) = 0 \Rightarrow x = -2$ or 2.

In the interval $(-\infty, -2)$, we test $x = -3$. $f(-3) = \dfrac{5}{(-3)^2 - 4} = 1 > 0$

In the interval $(-2, 2)$, we test $x = 0$. $f(0) = \dfrac{5}{(0)^2 - 4} = -\dfrac{5}{4} < 0$

In the interval $(2, \infty)$, we test $x = 3$. $f(3) = \dfrac{5}{(3)^2 - 4} = 1 > 0$

The solution to the given inequality is $(-2, 2)$.

20. $\dfrac{x}{x^2 - 1} = 0 \Rightarrow x = 0$.   The expression is undefined when $x^2 - 1 = 0 \Rightarrow (x+1)(x-1) = 0 \Rightarrow x = -1$ or 1.

In the interval $(-\infty, -1)$, we test $x = -2$. $f(-2) = \dfrac{(-2)}{(-2)^2 - 1} = -\dfrac{2}{3} \le 0$

In the interval $(-1, 0)$, we test $x = -0.5$. $f(-0.5) = \dfrac{(-0.5)}{(-0.5)^2 - 1} = \dfrac{2}{3} \ge 0$

In the interval $(0, 1)$, we test $x = 0.5$. $f(0.5) = \dfrac{(0.5)}{(0.5)^2 - 1} = -\dfrac{2}{3} \le 0$

In the interval $(1, \infty)$, we test $x = 2$. $f(2) = \dfrac{(2)}{(2)^2 - 1} = \dfrac{2}{3} \ge 0$

The solution to the given inequality is $(-1, 0] \cup (1, \infty)$.

Chapters 1–8 Cumulative Review Exercises

1. $F = \dfrac{5}{(-2)^2 + 1} = \dfrac{5}{4 + 1} = \dfrac{5}{5} = 1$

2.   Natural number: $\sqrt[3]{8} = 2$; whole number: $0, \sqrt[3]{8}$; integer: $0, -5, \sqrt[3]{8}$; rational number: $0.\overline{4}, 0, -5, \sqrt[3]{8}, -\dfrac{4}{3}$;

   irrational number: $\sqrt{7}$

3.   (a) $\left(\dfrac{x^2 y^6}{x^{-3}}\right)^2 = \dfrac{x^4 y^{12}}{x^{-6}} = x^{4-(-6)} y^{12} = x^{10} y^{12}$

   (b) $\dfrac{(xy^{-3})^2}{x(y^{-2})^{-1}} = \dfrac{x^2 y^{-6}}{xy^2} = x^{2-1} y^{-6-2} = x^1 y^{-8} = \dfrac{x}{y^8}$

   (c) $(a^2 b)^2 (ab^3)^{-4} = a^4 b^2 a^{-4} b^{-12} = a^{4-4} b^{2-12} = a^0 b^{-10} = \dfrac{1}{b^{10}}$

4. $9{,}290{,}000 = 9.29 \times 10^6$

5. $f(-2) = \sqrt{2 - (-2)} = \sqrt{4} = 2$; the radicand must be greater than or equal to zero, so $2 - x \ge 0 \Rightarrow x \le 2$.

6. $(2, 5)$ lies on the graph of f.

7.   See Figure 7.

8.   See Figure 8.

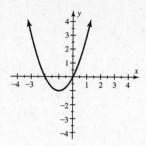

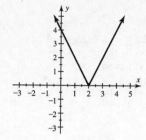

Figure 7   Figure 8

9.   Parallel lines have the same slope, so $m = \dfrac{4-1}{-2-0} = \dfrac{3}{-2} = -\dfrac{3}{2}$.

$$y - (-1) = -\frac{3}{2}(x-4) \Rightarrow y = -\frac{3}{2}x + 6 - 1 \Rightarrow y = -\frac{3}{2}x + 5$$

10. $x = -3$

11. $2x - 3(x+2) = 6 \Rightarrow 2x - 3x - 6 = 6 \Rightarrow -x = 12 \Rightarrow x = -12$

12. $7 - x > 3x \Rightarrow 7 > 4x \Rightarrow \dfrac{7}{4} > x$; $\left(-\infty, \dfrac{7}{4}\right)$

13. $|3x-2| \le 1 \Rightarrow 3x - 2 \le 1$ and $3x - 2 \ge -1 \Rightarrow 3x \le 3$ and $3x \ge 1 \Rightarrow x \le 1$ and $x \ge \dfrac{1}{3}$; $\left[\dfrac{1}{3}, 1\right]$

14. $-4 \le 1 - x < 2 \Rightarrow -5 \le -x < 1 \Rightarrow 5 \ge x > -1 \Rightarrow -1 < x \le 5$; $(-1, 5]$

15.   Multiply the first equation by 5 and add to the second equation.

$$\begin{array}{r} -5x - 20y = -15 \\ \underline{5x + y = -4} \\ -19y = -19 \end{array} \Rightarrow y = 1$$

Substitute $y = 1$ into the first equation: $-x - 4(1) = -3 \Rightarrow -x = -3 + 4 \Rightarrow -x = 1 \Rightarrow x = -1$

The solution is $(-1, 1)$.

16.   Note that $3x + y \le 3 \Rightarrow y \le -3x + 3$ and $x - 3y \le 3 \Rightarrow -3y \le -x + 3 \Rightarrow y \ge \dfrac{1}{3}x - 1$. See Figure 16.

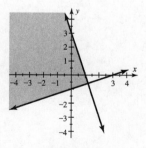

Figure 16

17.   Add the first two equations.

$$\begin{array}{r} x + y - z = 3 \\ \underline{x - y + z = 1} \\ 2x = 4 \end{array} \Rightarrow x = 2$$

Add the second and third equations.

$$x - \ y + z = 1$$
$$2x - \ y - z = 1$$
$$\overline{3x - 2y \quad = 2} \ \Rightarrow 3(2) - 2y = 2 \Rightarrow 6 - 2y = 2 \Rightarrow -2y = -4 \Rightarrow y = 2$$

Substitute $x = 2$ and $y = 2$ into the first equation.

$$2 + 2 - z = 3 \Rightarrow 4 - z = 3 \Rightarrow -z = -1 \Rightarrow z = 1$$

The solution is $(2, 2, 1)$.

18. $(3x - 2)(2x + 7) = 6x^2 + 21x - 4x - 14 = 6x^2 + 17x - 14$

19. $3xy(x^2 + y^2) = 3x^3 y + 3xy^3$

20. $(\sqrt{x} + 3)(\sqrt{x} - 3) = (\sqrt{x})^2 - (3)^2 = x - 9$

21. $x^3 - x^2 - 2x = x(x^2 - x - 2) = x(x - 2)(x + 1)$

22. $4x^2 - 25 = (2x - 5)(2x + 5)$

23. $x^2 - 3 = 0 \Rightarrow x^2 = 3 \Rightarrow x = \pm\sqrt{3}$

24. $x^2 + 1 = 2x \Rightarrow x^2 - 2x + 1 = 0 \Rightarrow (x - 1)^2 = 0 \Rightarrow x - 1 = 0 \Rightarrow x = 1$

25. $\dfrac{(x+3)^2}{x+2} \cdot \dfrac{x+2}{2x+6} = \dfrac{(x+3)(x+3)(x+2)}{(x+2)(2)(x+3)} = \dfrac{x+3}{2} \ \text{or} \ \dfrac{1}{2}(x+3)$

26. $\dfrac{1}{x+2} - \dfrac{1}{x} = \dfrac{1}{x+2} \cdot \dfrac{x}{x} - \dfrac{1}{x} \cdot \dfrac{x+2}{x+2} = \dfrac{x}{x(x+2)} - \dfrac{x+2}{x(x+2)} = \dfrac{x-x-2}{x(x+2)} = \dfrac{-2}{x(x+2)}$

27. $\sqrt{16x^6} = \sqrt{16} \cdot \sqrt{x^6} = 4x^3$

28. $16^{-3/2} = (16^{1/2})^{-3} = 4^{-3} = \dfrac{1}{64}$

29. $\dfrac{\sqrt[3]{81x}}{\sqrt[3]{3x}} = \sqrt[3]{\dfrac{81x}{3x}} = \sqrt[3]{27} = 3$

30. $\sqrt{8x} + \sqrt{2x} = \sqrt{4 \cdot 2x} + \sqrt{2x} = 2\sqrt{2x} + \sqrt{2x} = 3\sqrt{2x}$

31.   See Figure 31.

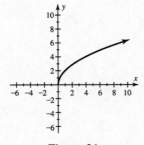

Figure 31

32. $d = \sqrt{(x_2 - x_1)^2 + (y_2 - y_1)^2} = \sqrt{(4 - (-1))^2 + (3 - 2)^2} = \sqrt{(5)^2 + (1)^2} = \sqrt{25 + 1} = \sqrt{26}$

33. $\dfrac{3-i}{2+i} = \dfrac{3-i}{2+i} \cdot \dfrac{2-i}{2-i} = \dfrac{6 - 3i - 2i + i^2}{4 - i^2} = \dfrac{6 - 5i - 1}{4 - (-1)} = \dfrac{5 - 5i}{5} = 1 - i$

34. $3\sqrt{x+1} = 2x \Rightarrow 9(x+1) = 4x^2 \Rightarrow 9x+9 = 4x^2 \Rightarrow 4x^2 - 9x - 9 = 0 \Rightarrow (x-3)(4x+3) = 0 \Rightarrow x = 3$ ($x = -\dfrac{3}{4}$

    does not check).

35. Graph $Y_1 = 2X$ and $Y_2 = \sqrt{(2.1-X)} + (0.1X) \wedge (1/3)$ in $[-4.7, 4.7, 1]$ by $[-3.1, 3.1, 1]$. The solution is

    $x \approx 0.79$.

36. See Figure 36.

    (a) $(1, 2)$

    (b) $f(-1) = (-1)^2 - 2(-1) + 3 = 1 + 2 + 3 = 6$

    (c) $x = 1$

    (d) f is increasing when $x \geq 1$

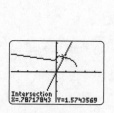

Figure 35

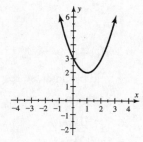

Figure 36

37. $f(x) = 2x^2 - 4x - 1 \Rightarrow f(x) = 2(x^2 - 2x) - 1 \Rightarrow f(x) = 2(x^2 - 2x + 1) - 1 - 2 \Rightarrow f(x) = 2(x-1)^2 - 3$

38. The graph is shifted 1 unit left and 2 units downward. The graph of $f(x)$ is narrower.

39. $x^2 + 6x = 2 \Rightarrow x^2 + 6x + 9 = 2 + 9 \Rightarrow (x+3)^2 = 11 \Rightarrow x + 3 = \pm\sqrt{11} \Rightarrow x = -3 \pm \sqrt{11}$

40. $2x^2 - 3x = 1 \Rightarrow 2x^2 - 3x - 1 = 0$; $a = 2, b = -3, c = -1$

    $$x = \frac{-b \pm \sqrt{b^2 - 4ac}}{2a} = \frac{-(-3) \pm \sqrt{(-3)^2 - 4(2)(-1)}}{2(2)} = \frac{3 \pm \sqrt{17}}{4}$$

41. $x(4-x) = 3 \Rightarrow 4x - x^2 = 3 \Rightarrow x^2 - 4x + 3 = 0 \Rightarrow (x-3)(x-1) = 0 \Rightarrow x = 3$ or $x = 1$

42. (a) $x = -2$ or $x = 1$

    (b) $-2 \leq x \leq 1$

43. $x^2 - 3x + 2 > 0 \Rightarrow (x-2)(x-1) > 0 \Rightarrow x < 1$ or $x > 2$

44. $x^4 - 256 = 0 \Rightarrow (x^2 - 16)(x^2 + 16) = 0 \Rightarrow (x-4)(x+4)(x^2 + 16) = 0 \Rightarrow x = 4$ or $x = -4$ or

    $x^2 = -16 \Rightarrow x = \pm\sqrt{-16} \Rightarrow x = \pm 4i$

45. c; $y = ax - b$ is a linear equation and $a > 0$.

46. f; $y = b$ is a horizontal line.

47. g; $y = -ax^2 + c$ is a quadratic equation whose graph is a parabola opening downward (since $a > 0$).

48. e; $y = \dfrac{a}{x}$ is a rational equation, undefined when $x = 0$ and $y = 0$.

49. d; $y = ax^3$ is a cubic equation.

50. a; $y = |ax + b|$ has only nonnegative values.

51. b; $y = a\sqrt{x}$ is a radical function defined only for $x \geq 0$.

52. h; $y = a\sqrt[3]{x}$ is a cube root equation.

53. (a) $G(0) = 300$ gallons ; initially, the tank holds 300 gallons.

    (b) The t-intercept is 6; after 6 minutes the tank is empty.

    (c) (0, 300) and (6, 0) are on the graph so $m = \dfrac{0 - 300}{6 - 0} = \dfrac{-300}{6} = -50$; water is pumped out at 50 gallons per

    minute.

    (d) Since (0, 300) is on the graph and $m = -50$, $G(t) = -50t + 300$ or $G(t) = 300 - 50t$

54. Let x, y, and z be the amounts invested at 4% ,5%, and 6% interest, respectively. The system needed is

    $$x + y + z = 4000 \qquad\qquad x + y + z = 4000$$
    $$z = y + 1000 \qquad \Rightarrow \qquad y - z = -1000$$
    $$0.04x + 0.05y + 0.06z = 216 \qquad 4x + 5y + 6z = 21,600$$

    Multiply the first equation by -4 and add to the third equation

    $$\begin{array}{r} -4x - 4y - 4z = -16,000 \\ 4x + 5y + 6z = 21,600 \\ \hline y + 2z = 5600 \end{array}$$

    Multiply the second equation by 2 and add to this new equation.

    $$\begin{array}{r} 2y - 2z = -2000 \\ y + 2z = 5600 \\ \hline 3y = 3600 \end{array} \qquad \Rightarrow y = 1200$$

    Substitute $y = 1200$ into the second equation.

    $$1200 - z = -1000 \Rightarrow -z = -2200 \Rightarrow z = 2200$$

    Substitute $y = 1200$ and $z = 2200$ into the first equation.

    $$x + 1200 + 2200 = 4000 \Rightarrow x = 600$$

    \$600 is invested at 4%, \$1200 is invested at 5%, and \$2200 is invested at 6%.

55. Let x represent the length of one of the two equal sides of the garden without the gate. Then the other sides of

    the garden have lengths $\dfrac{1}{2}(490 - 2x + 10) = \dfrac{1}{2}(500 - 2x) = 250 - x$. $(250 - x)x = 250x - x^2$; The graph of

    $y = -x^2 + 250x$ is a parabola opening downward; its vertex is its maximum point. The x-value of the vertex is

    on the axis of symmetry, so $x = -\dfrac{b}{2a} = -\dfrac{250}{2(-1)} = 125$. The dimensions of the garden of largest area are 125

    feet by 125 feet.

56.   Let h represent the height of the tree.

$$\frac{6}{10} = \frac{h}{55} \quad \Rightarrow \quad 10h = 6(55) \Rightarrow 10h = 330 \Rightarrow h = 33 \text{ feet}$$

Critical Thinking Solutions for Chapter 8

Section 8.2

- The productof $(4)(4) = 16$ goes in the remaining square.

Section 8.3

- The product of $(3)(3) = 9$ goes in the remaining square.

- If we write the equation in $x^2 + kx = d$ form, it would become $x^2 + 0x = 7$. To complete the square we would add zero to both sides of the equation. Completing the square will not work in this case. The square root property should be used for this equation.

Section 8.4

- Both expressions equal zero because these values of x are the solutions to $2x^2 - 3x - 1 = 0$.

- Both expressions equal zero because these values of x are the solutions to $2x^2 + x + 3 = 0$.

Section 8.5

- i.   The parabola is above the x-axis for values of x between the intercepts: $(-3, 4)$.

  ii.   The parabola is below the x-axis for values of x outside of the intercepts: $(-\infty, -3) \cup (4, \infty)$.

Chapter 9: Exponential and Logarithmic Functions

9.1: Composite and Inverse Functions

Concepts

1. $g\left(f\left(7\right)\right)$

2. $f\left(g\left(x\right)\right)$

3. No

4. inputs; outputs

5. No

6. one-to-one

7. adding 10

8. 7

9. 8; 6

10. x; y

11. one-to-one

12. reflection; $y = x$

Composite Functions

13. (a) $f\left(-2\right) = \left(-2\right)^2 = 4$, then $\left(g \circ f\right)\left(-2\right) = g\left(f\left(-2\right)\right) = g\left(4\right) = 4 + 3 = 7$

(b) $g\left(4\right) = 4 + 3 = 7$, then $\left(f \circ g\right)\left(4\right) = f\left(g\left(4\right)\right) = f\left(7\right) = 7^2 = 49$

(c) $\left(g \circ f\right)\left(x\right) = g\left(f\left(x\right)\right) = g\left(x^2\right) = x^2 + 3$

(d) $\left(f \circ g\right)\left(x\right) = f\left(g\left(x\right)\right) = f\left(x + 3\right) = \left(x + 3\right)^2$

15. (a) $f\left(-2\right) = 2\left(-2\right) = -4$, then $\left(g \circ f\right)\left(-2\right) = g\left(f\left(-2\right)\right) = g\left(-4\right) = \left(-4\right)^3 - 1 = -65$

(b) $g\left(4\right) = 4^3 - 1 = 63$, then $\left(f \circ g\right)\left(4\right) = f\left(g\left(4\right)\right) = f\left(63\right) = 2\left(63\right) = 126$

(c) $\left(g \circ f\right)\left(x\right) = g\left(f\left(x\right)\right) = g\left(2x\right) = \left(2x\right)^3 - 1 = 8x^3 - 1$

(d) $\left(f \circ g\right)\left(x\right) = f\left(g\left(x\right)\right) = f\left(x^3 - 1\right) = 2\left(x^3 - 1\right) = 2x^3 - 2$

17. (a) $f\left(-2\right) = \frac{1}{2}\left(-2\right) = -1$, then $\left(g \circ f\right)\left(-2\right) = g\left(f\left(-2\right)\right) = g\left(-1\right) = \left|\left(-1\right) - 2\right| = 3$

(b) $g\left(4\right) = \left|4 - 2\right| = 2$, then $\left(f \circ g\right)\left(4\right) = f\left(g\left(4\right)\right) = f\left(2\right) = \frac{1}{2}\left(2\right) = 1$

(c) $\left(g \circ f\right)\left(x\right) = g\left(f\left(x\right)\right) = g\left(\frac{1}{2}x\right) = \left|\frac{1}{2}x - 2\right|$

(d) $(f \circ g)(x) = f(g(x)) = f(|x-2|) = \frac{1}{2}|x-2|$

19.   (a) $f(-2) = \frac{1}{-2} = -\frac{1}{2}$, then $(g \circ f)(-2) = g(f(-2)) = g\left(-\frac{1}{2}\right) = 3 - 5\left(-\frac{1}{2}\right) = \frac{11}{2}$

(b) $g(4) = 3 - 5(4) = -17$, then $(f \circ g)(4) = f(g(4)) = f(-17) = \frac{1}{-17} = -\frac{1}{17}$

(c) $(g \circ f)(x) = g(f(x)) = g\left(\frac{1}{x}\right) = 3 - 5\left(\frac{1}{x}\right) = 3 - \frac{5}{x}$

(d) $(f \circ g)(x) = f(g(x)) = f(3 - 5x) = \frac{1}{3 - 5x}$

21.   (a) $f(-2) = 2(-2) = -4$, then $(g \circ f)(-2) = g(f(-2)) = g(-4) = 4(-4)^2 - 2(-4) + 5 = 77$

(b) $g(4) = 4(4)^2 - 2(4) + 5 = 61$, then $(f \circ g)(4) = f(g(4)) = f(61) = 2(61) = 122$

(c) $(g \circ f)(x) = g(f(x)) = g(2x) = 4(2x)^2 - 2(2x) + 5 = 16x^2 - 4x + 5$

(d) $(f \circ g)(x) = f(g(x)) = f(4x^2 - 2x + 5) = 2(4x^2 - 2x + 5) = 8x^2 - 4x + 10$

23.   (a) $(f \circ g)(0) = f(g(0)) = f(-1) = 1$

(b) $(g \circ f)(-1) = g(f(-1)) = g(1) = 2$

25.   (a) $(f \circ f)(-1) = f(f(-1)) = f(1) = -1$

(b) $(g \circ g)(0) = g(g(0)) = g(-1) = 1$

27.   (a) $\left(f^{-1} \circ g\right)(-2) = f^{-1}(g(-2)) = f^{-1}(0) = 0$

(b) $\left(g^{-1} \circ f\right)(2) = g^{-1}(f(2)) = g^{-1}(-2) = 2$

29.   (a) $(f \circ g)(0) = f(g(0)) = f(-1) = 2$

(b) $(g \circ f)(1) = g(f(1)) = g(2) = -3$

(c) $(f \circ f)(-1) = f(f(-1)) = f(2) = -1$

31. $f(1) = f(-1) = 5$; *Answers may vary.*

33. $f(1) = f(-1) = 101$; *Answers may vary.*

35. $f(2) = f(-2) = 4$; *Answers may vary.*

37.   This graph passes the horizontal line test. The function is one-to-one.

39.   This graph does not pass the horizontal line test. The function is not one-to-one.

41.   This graph passes the horizontal line test. The function is one-to-one.

43. Divide x by 7. $f(x) = 7x$; $g(x) = \dfrac{x}{7}$

45. Multiply x by 2 and then subtract 5. $f(x) = \dfrac{x+5}{2}$; $g(x) = 2x - 5$

47. Add 3 to x and then multiply the result by 2. $f(x) = \dfrac{1}{2}x - 3$; $g(x) = 2(x+3)$

49. Take the cube root of x and then subtract 5. $f(x) = (x+5)^3$; $g(x) = \sqrt[3]{x} - 5$

51. $\left(f \circ f^{-1}\right)(x) = f\left(f^{-1}(x)\right) = f\left(\dfrac{x}{4}\right) = 4\left(\dfrac{x}{4}\right) = x$ and $\left(f^{-1} \circ f\right)(x) = f^{-1}(f(x)) = f^{-1}(4x) = \dfrac{4x}{4} = x$

53. $\left(f \circ f^{-1}\right)(x) = f\left(f^{-1}(x)\right) = f\left(\dfrac{x-5}{3}\right) = 3\left(\dfrac{x-5}{3}\right) + 5 = x - 5 + 5 = x$

    $\left(f^{-1} \circ f\right)(x) = f^{-1}(f(x)) = f^{-1}(3x+5) = \dfrac{(3x+5)-5}{3} = \dfrac{3x}{3} = x$

55. $\left(f \circ f^{-1}\right)(x) = f\left(f^{-1}(x)\right) = f\left(\sqrt[3]{x}\right) = \left(\sqrt[3]{x}\right)^3 = x$ and $\left(f^{-1} \circ f\right)(x) = f^{-1}(f(x)) = f^{-1}\left(x^3\right) = \sqrt[3]{x^3} = x$

57. $\left(f \circ f^{-1}\right)(x) = f\left(f^{-1}(x)\right) = f\left(\dfrac{1}{x}\right) = \dfrac{1}{\frac{1}{x}} = x$ and $\left(f^{-1} \circ f\right)(x) = f^{-1}(f(x)) = f^{-1}\left(\dfrac{1}{x}\right) = \dfrac{1}{\frac{1}{x}} = x$

59. $f(x) = 12x \Rightarrow y = 12x$, interchange x and y and solve for y. $x = 12y \Rightarrow y = \dfrac{x}{12} \Rightarrow f^{-1}(x) = \dfrac{x}{12}$

61. $f(x) = x + 8 \Rightarrow y = x + 8$, interchange x and y and solve for y.

    $x = y + 8 \Rightarrow y = x - 8 \Rightarrow f^{-1}(x) = x - 8$

63. $f(x) = 5x - 2 \Rightarrow y = 5x - 2$, interchange x and y and solve for y.

    $x = 5y - 2 \Rightarrow 5y = x + 2 \Rightarrow y = \dfrac{x+2}{5} \Rightarrow f^{-1}(x) = \dfrac{x+2}{5}$

65. $f(x) = -\dfrac{1}{2}x + 1 \Rightarrow y = -\dfrac{1}{2}x + 1$, interchange x and y and solve for y.

    $x = -\dfrac{1}{2}y + 1 \Rightarrow -\dfrac{1}{2}y = x - 1 \Rightarrow y = -2(x-1) \Rightarrow f^{-1}(x) = -2(x-1)$

67. $f(x) = 8 - x \Rightarrow y = 8 - x$, interchange x and y and solve for y.

    $x = 8 - y \Rightarrow y = 8 - x \Rightarrow f^{-1}(x) = 8 - x$

69. $f(x) = \dfrac{x+1}{2} \Rightarrow y = \dfrac{x+1}{2}$, interchange x and y and solve for y.

    $x = \dfrac{y+1}{2} \Rightarrow y + 1 = 2x \Rightarrow y = 2x - 1 \Rightarrow f^{-1}(x) = 2x - 1$

71. $f(x) = \sqrt[3]{2x} \Rightarrow y = \sqrt[3]{2x}$, interchange x and y and solve for y.

$$x = \sqrt[3]{2y} \Rightarrow 2y = x^3 \Rightarrow y = \frac{x^3}{2} \Rightarrow f^{-1}(x) = \frac{x^3}{2}$$

73. $f(x) = x^3 - 8 \Rightarrow y = x^3 - 8,$ interchange x and y and solve for y.

$$x = y^3 - 8 \Rightarrow y^3 = x + 8 \Rightarrow y = \sqrt[3]{x+8} \Rightarrow f^{-1}(x) = \sqrt[3]{x+8}$$

75.   See Figure 75. The domain of f = the range of $f^{-1} = \{0, 1, 2, 3, 4\}$.

The range of f = the domain of $f^{-1} = \{0, 5, 10, 15, 20\}$.

77.   See Figure 77. The domain of f = the range of $f^{-1} = \{-5, 0, 5, 10, 15\}$.

The range of f = the domain of $f^{-1} = \{-4, -2, 0, 2, 4\}$.

x	0	5	10	15	20
$f^{-1}(x)$	0	1	2	3	4

x	4	2	0	-2	-4
$f^{-1}(x)$	-5	0	5	10	15

Figure 75   Figure 77

79.   See Figure 79.

81.   See Figure 81.

83.   The graph of f^{-1} is a reflection of the graph of f across the line $y = x$. See Figure 83.

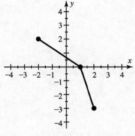
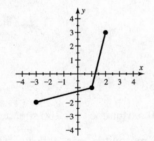
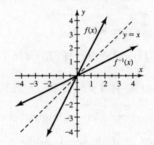

Figure 79   Figure 81   Figure 83

85.   The graph of f^{-1} is a reflection of the graph of f across the line $y = x$. See Figure 85.

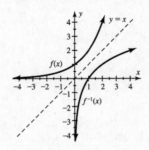

Figure 85

87.   (a) $(fg)(2) = (2^2 - 2)(2^2 + 2) = (4 - 2)(4 + 2) = (2)(6) = 12$

(b) $(f - g)(x) = (x^2 - 2) - (x^2 + 2) = x^2 - x^2 - 2 - 2 = -4$

(c) $(f \circ g)(x) = f(g(x)) = f(x^2 + 2) = (x^2 + 2)^2 - 2 = x^4 + 4x^2 + 4 - 2 = x^4 + 4x^2 + 2$

89.   (a) $(fg)(2) = \left(\frac{1}{2}\right)\left(\frac{2}{2}\right) = \frac{2}{4} = \frac{1}{2}$

(b) $(f - g)(x) = \frac{1}{x} - \frac{2}{x} = -\frac{1}{x}$

(c) $(f \circ g)(x) = f(g(x)) = f\left(\frac{2}{x}\right) = \frac{1}{\frac{2}{x}} = \frac{x}{2}$

Applications

91.   (a) $(C \circ r)(5) = C(r(5)) = C(2 \cdot 5) = C(10) = 2\pi \cdot 10 = 20\pi$

After 5 seconds, the wave has a circumference of $20\pi \approx 62.8$ feet.

(b) $(C \circ r)(t) = C(r(t)) = C(2t) = 2\pi \cdot 2t = 4\pi t$

93.   (a) $P(1980) = 16$; In 1980, 16% of people 25 or older completed 4 or more years of college.

(b) See Figure 93.

(c) $P^{-1}(16) = 1980$

x	8	16	27
$P^{-1}(x)$	1960	1980	2000

Figure 93

95.   (a) $T(1) = 75°$ and $M(75) = 150$

(b) $(M \circ T)(1) = M(T(1)) = M(75) = 150$

One hour after midnight there are 150 mosquitoes per 100 square feet.

(c) $(M \circ T)(h)$ calculates the number of mosquitoes per 100 square feet, h hours after midnight.

(d) For $T(h), m = \frac{50 - 80}{6 - 0} = \frac{-30}{6} = -5.$ Since the y-intercept is 80, the equation is $T(h) = -5h + 80.$

For $M(T), m = \frac{150 - 100}{75 - 50} = \frac{50}{25} = 2.$ Since the line passes through (50, 100), the equation is

$M(T) = 2T.$

(e) $(M \circ T)(h) = M(T(h)) = M(-5h + 80) = 2(-5h + 80) = -10h + 160$

97.   (a) Yes, this is a one-to-one function because different inputs result in different outputs.

(b) $f(x) = \frac{5}{9}x + 32 \Rightarrow y = \frac{5}{9}x + 32,$ interchange x and y and solve for y.

$x = \frac{5}{9}y + 32 \Rightarrow \frac{5}{9}y = x - 32 \Rightarrow y = \frac{9}{5}(x - 32) \Rightarrow f^{-1}(x) = \frac{9}{5}(x - 32)$

This formula converts x degrees Fahrenheit to an equivalent temperature in degrees Celsius.

99.   Since there are 4 quarts in 1 gallon, the function $f(x) = 4x$ converts x gallons to quarts.

$f(x) = 4x \Rightarrow y = 4x,$ interchange x and y and solve for $y.$ $x = 4y \Rightarrow y = \dfrac{x}{4} \Rightarrow f^{-1}(x) = \dfrac{x}{4}$

This formula converts x quarts to gallons.

9.2: Exponential Functions

Concepts

1. $f(x) = Ca^x$

2.   See Figure 2. *Answers may vary.*

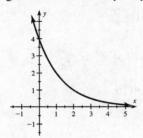

Figure 2

3.   The graph illustrates growth, since $a > 1$.

4. $2^5 = 32; 5^2 = 25$

5. $e \approx 2.718$

6. $e^2 \approx 7.389; \pi^2 \approx 9.870$

7.   factor

8.   The growth factor is 1.5.

Evaluating and Graphing Exponential Functions

9. $f(-2) = 3^{-2} = \dfrac{1}{3^2} = \dfrac{1}{9}$ and $f(2) = 3^2 = 9$

11. $f(0) = 5(2^0) = 5(1) = 5$ and $f(5) = 5(2^5) = 5(32) = 160$

13. $f(-2) = \left(\dfrac{1}{2}\right)^{-2} = 2^2 = 4$ and $f(3) = \left(\dfrac{1}{2}\right)^3 = \dfrac{1}{2^3} = \dfrac{1}{8}$

15. $f(-1) = 5(3)^{-(-1)} = 5(3)^1 = 15$ and $f(2) = 5(3)^{-2} = 5\left(\dfrac{1}{3^2}\right) = \dfrac{5}{9}$

17. $f(-3) = 1.8^{-3} \approx 0.17$ and $f(1.5) = 1.8^{1.5} \approx 2.41$

19. $f(-1) = 3(0.6)^{-1} = 5$ and $f(2) = 3(0.6)^2 = 1.08$

21. $f(0) = a^0 = 1; f(-1) = a^{-1} = \dfrac{1}{a}$

23.   (a) Exponential decay. For each unit increase in x, $f(x)$ decreases by a factor of $\dfrac{1}{4}$.

(b) Since $f(x) = 64$ when $x = 0$, $f(x) = 64\left(\dfrac{1}{4}\right)^x$

25.   (a) Linear growth. For each unit increase in x, $f(x)$ increases by 3 units.

(b) Since $f(x) = 8$ when $x = 0$, $f(x) = 3x + 8$

27.   (a) Exponential growth. For each unit increase in x, $f(x)$ increases by a factor of 1.25.

(b) Since $f(x) = 4$ when $x = 0$, $f(x) = 4(1.25)^x$

29.   Since $y = 1$ when $x = 0$, $1 = Ca^0 \Rightarrow C = 1$. Since $y = 2$ when $x = 1$, $2 = 1(a)^1 \Rightarrow a = 2$.

31.   Since $y = 4$ when $x = 0$, $4 = Ca^0 \Rightarrow C = 4$. Since $y = 1$ when $x = 1$, $1 = 4(a)^1 \Rightarrow a = \dfrac{1}{4}$.

33.   c. This function models exponential growth and passes through the point (0, 1).

35.   d. This function models exponential decay and passes through the point (2, 1).

37.   See Figure 37. The graph illustrates exponential growth.

39.   See Figure 39. The graph illustrates exponential decay.

41.   See Figure 41. The graph illustrates exponential decay.

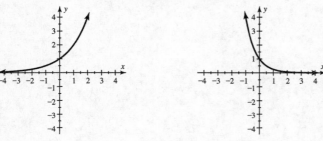

Figure 37   Figure 39   Figure 41

43.   See Figure 43. The graph illustrates exponential growth.

45.   See Figure 45. The graph illustrates exponential growth.

47.   See Figure 47. The graph illustrates exponential decay.

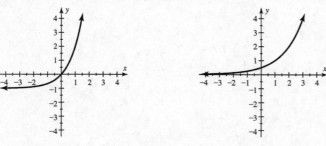

Figure 43   Figure 45   Figure 47

49. $e^2 e^5 = e^{2+5} = e^7$

51. $\dfrac{a^4 a^{-2}}{a^3} = \dfrac{a^{4+(-2)}}{a^3} = \dfrac{a^2}{a^3} = a^{2-3} = a^{-1} = \dfrac{1}{a}$

53. $e^x e^{y-x} = e^{x+(y-x)} = e^y$

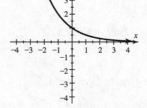

55. $\dfrac{2^{x+3}}{2^x} = 2^{x+3-x} = 2^3 = 8$

57. $5^{-y} \cdot 5^{3y} = 5^{-y+3y} = 5^{2y} = 25^y$

Compound Interest

59. $1500(1+0.09)^{10} = \$3551.05$

61. $200(1+0.20)^{50} = \$1,820,087.63$

63. $560(1+0.014)^{25} = \$792.75$

65. Yes. This is equivalent to having two accounts, each containing $1000 initially.

67. $A = 300(1+0.0495)^{30} \approx \$1278.2$ billion or about $1.28 trillion

The Natural Exponential Function

69. $f(1.2) = e^{1.2} \approx 3.32$

71. $f(-2) = 1 - e^{-2} \approx 0.86$

73. Graph $Y_1 = e\wedge(0.5X)$ in $[-4, 4, 1]$ by $[0, 8, 1]$. See Figure 73. The graph illustrates exponential growth.

75. Graph $Y_1 = 1.5e\wedge(-0.32X)$ in $[-4, 4, 1]$ by $[0, 8, 1]$. See Figure 75. The graph illustrates exponential decay.

[-4, 4, 1] by [0, 8, 1]   [-4, 4, 1] by [0, 8, 1]

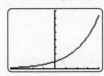   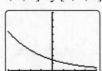

Figure 73   Figure 75

Applications

77. (a) $f(x) = 34e^{0.013x}$

   (b) $f(10) = 34e^{0.013(10)} \approx 38.7$; In 2010, the population of California will be about 38.7 million.

79. (a) $f(x) = 8e^{0.019x}$

   (b) $f(10) = 8e^{0.019(10)} \approx 9.7$; In 2010, the population of North Carolina will be about 9.7 million.

81. (a) $f(x) = 5e^{0.031x}$

   (b) Graph $Y_1 = 5e\wedge(0.031X)$ in $[0,10,1]$ by $[4,7,1]$.

   [0, 10, 1] by [4, 7, 1]

   (c) Since 2010 is 10 years after 2000, evaluate $f(10) = 5e^{0.031(10)} \approx 6.82$ million

83. (a) Since $f(t) = 500$ when $t = 0$, $500 = Ca^0 \Rightarrow C = 500$. Since $f(t) = 1000$ when $t = 50$,

$$1000 = 500a^{50/50} \Rightarrow a = \frac{1000}{500} = 2$$

(b) $f(170) = 500(2)^{170/50} \approx 5278$ thousand bacteria per milliliter or 5.278 million bacteria per milliliter

(c) The growth in the number of bacteria is exponential.

85. (a) $f(1995) = 0.0272(1.495)^{1995-1980} = 0.0272(1.495)^{15} \approx 11.3$ million.

In 1995 there were about 11.3 million cellular phone subscribers.

(b) The growth factor is 1.495. This means that each year from 1985 to 2000 the number of subscribers increases by a factor of 1.495 or by 49.5%.

87. Table $Y_1 = (0.905)^\wedge X$ with TblStart = 0 and ΔTbl = 10. See Figure 87.

(a) $f(0) = (0.905)^0 = 1$.

The probability that no vehicle will enter the intersection during a period of zero seconds is 1.

(b) Since $f(30) = (0.905)^{30} = 0.05006$ and $f(31) = (0.905)^{31} = 0.04530$, this occurs after about 30 seconds.

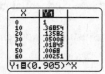

Figure 87

Checking Basic Concepts for Sections 9.1 & 9.2

1. (a) $f(1) = 2(1)^2 + 5(1) - 1 = 6$, and so $(g \circ f)(1) = g(f(1)) = g(6) = 6 + 1 = 7$

(b) $(f \circ g)(x) = f(g(x)) = f(x+1) = 2(x+1)^2 + 5(x+1) - 1 = 2x^2 + 9x + 6$

2. See Figure 2.

(a) No, this is not a one-to-one function because it does not pass the horizontal line test.

(b) No, this function does not have an inverse because it is not one-to-one.

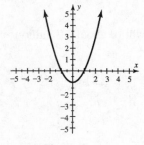

Figure 2

3. $f(x) = 4x - 3 \Rightarrow y = 4x - 3$, interchange x and y and solve for y.

$$x = 4y - 3 \Rightarrow 4y = x + 3 \Rightarrow y = \frac{x+3}{4} \Rightarrow f^{-1}(x) = \frac{x+3}{4}$$

4. $f(-2) = 3(2^{-2}) = 3 \cdot \dfrac{1}{2^2} = 3 \cdot \dfrac{1}{4} = \dfrac{3}{4}$

5.   See Figure 5.

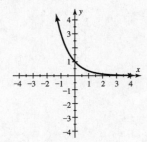

Figure 5

6.   Since $y = 2$ when $x = 0$, $2 = Ca^0 \Rightarrow C = 2$. Since $y = 1$ when $x = 1$, $1 = 2(a)^1 \Rightarrow a = \dfrac{1}{2}$.

9.3: Logarithmic Functions

Concepts

1.   10

2. e

3.   D: $\{x \mid x > 0\}$; R: all real numbers

4.   D: $\{x \mid x > 0\}$; R: all real numbers

5. k

6. k

7. x

8. x

9. $\log 5$

10. $\log_2 5$

11. $\log_a 1 = 0$ because $a^0 = 1$.

12. $\log_a -1$ is undefined because there is no value x such that $a^x = -1$.

Evaluating and Graphing Logarithmic Functions

13. 5

15. -4

17. $\log 1 = \log 10^0 = 0$

19. $\log \dfrac{1}{100} = \log 10^{-2} = -2$

21. $\log 10^{4.7} = 4.7$

23. $\log_5 5^{6x} = 6x$

25. $\log \sqrt{\dfrac{1}{1000}} = \log \sqrt{10^{-3}} = \log 10^{-3/2} = -\dfrac{3}{2}$

27. $\log_2 2^8 = 8$

29. $\log_2 \sqrt{8} = \log_2 \left(2^3\right)^{1/2} = \log_2 2^{3/2} = \dfrac{3}{2}$

31. $\log_2 \sqrt[3]{\dfrac{1}{4}} = \log_2 \left(2^{-2}\right)^{1/3} = \log_2 2^{-2/3} = -\dfrac{2}{3}$

33. $\log_2 -8$ is undefined.

35. $\log_2 4 = \log_2 2^2 = 2$

37. $\log_2 \dfrac{1}{16} = \log_2 2^{-4} = -4$

39. $\log_3 \dfrac{1}{9} = \log_3 3^{-2} = -2$

41. $\ln 1 = \ln e^0 = 0$

43. $\log 0.001 = \log 10^{-3} = -3$

45. $\log_5 \dfrac{1}{25} = \log_5 5^{-2} = -2$

47. $10^{\log 2} = 2$

49. $10^{\log x^2} = x^2$

51. $5^{\log_5 17} = 17$

53. $4^{\log_4 (2x)^2} = (2x)^2$

55. $10^{\log 5} = 5$

57. $\ln e^{-5x} = -5x$

59. $\log 10^{(2x-7)} = 2x - 7$

61. $5^{\log_5 0.6z} = 0.6z$

63. $\log 25 \approx 1.398$

65. $\log 1.45 \approx 0.161$

67. $\ln 7 \approx 1.946$

69. $\ln \dfrac{4}{7} \approx -0.560$

71. See Figure 71.

73. See Figure 73.

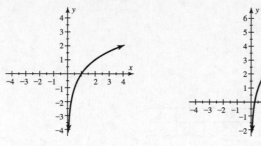

Figure 71   Figure 73

75.   Graph $Y_1 = \ln(\text{abs}(X))$ in [–4, 4, 1] by [–4, 4, 1]. See Figure 75. The graph is a reflection across the y-axis together with the graph of $y = \ln x$. The domain is $\{x \mid x \neq 0\}$.

77.   Graph $Y_1 = \ln(X+2)$ in [–4, 4, 1] by [–4, 4, 1]. See Figure 77. The graph is shifted 2 units to the left. The domain is $\{x \mid x > -2\}$.

[–4, 4, 1] by [–4, 4, 1]   [–4, 4, 1] by [–4, 4, 1]

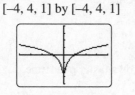   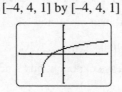

Figure 75   Figure 77

79.   d. The graph of this function passes through the point (1, 0) but does not pass through (3, 1).

81.   a. The graph of this function is shifted 2 units upward.

83.   See Figure 83. Compared to the graph of $y = \log x$, this graph is shifted 1 unit downward.

85.   See Figure 85. Compared to the graph of $y = \log x$, this graph is shifted 1 unit to the left.

87.   See Figure 87. Compared to the graph of $y = \log x$, this graph is shifted 1 unit to the right.

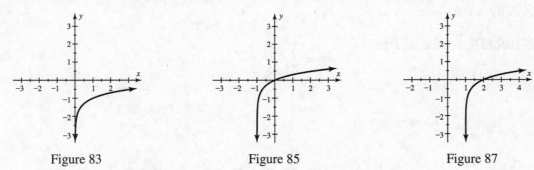

Figure 83   Figure 85   Figure 87

89.   See Figure 89. Compared to the graph of $y = \log x$, this graph increases faster.

91.   See Figure 91.

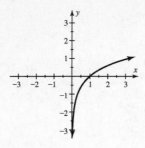

Figure 89

x	$\frac{1}{4}$	$\frac{1}{2}$	1	$\sqrt{2}$	64
$\log_2 x$	-2	-1	0	$\frac{1}{2}$	6

Figure 91

Applications

93. $f\left(10^{-4}\right)=160+10\log\left(10^{-4}\right)=160+10(-4)=160+(-40)=120$ db.   Yes, this could cause pain.

95.   (a) $P(0)=0.48\ln(0+1)+27=0.48\ln(1)+27=0.48(0)+27=27.$ The air pressure at the eye of the

hurricane is 27 inches of mercury. $P(50)=0.48\ln(50+1)+27=0.48\ln(51)+27\approx28.9.$ The air

pressure 50 miles from the eye of the hurricane is about 28.9 inches of mercury.

(b)   The pressure increases rapidly at first and then more slowly.

(c)   The eye of a hurricane is a low pressure area.

97.   Most people would want their salaries to increase exponentially since it grows faster over time.

99.   (a) $M=6-2.5\log\dfrac{10}{1}=6-2.5\log(10)=6-2.5(1)=3.5$

(b) $M=6-2.5\log\dfrac{100}{1}=6-2.5\log(100)=6-2.5(2)=1$

(c)   The intensity of the star decreases by 2.5.

101.   (a) $\log\dfrac{x}{1}=6.0\Rightarrow\log x=6.0\Rightarrow x=10^6$ and $\log\dfrac{x}{1}=8.0\Rightarrow\log x=8.0\Rightarrow x=10^8$

(b) $10^8\div10^6=100$ times

9.4: Properties of Logarithms

Concepts

1.   4

2.   4

3.   3

4. $\log m+\log n$

5. $\log m-\log n$

6. $r\log m$

7.   No

8.   Yes

9.   No; $\log(xy)=\log x+\log y$

10.   No; $\log\left(\dfrac{x}{y}\right) = \log x - \log y$

11. $\log_a x = \dfrac{\log x}{\log a}$ or $\log_a x = \dfrac{\ln x}{\ln a}$

12.   0; 1

Basic Properties of Logarithms

13. $\ln\left(3 \cdot 5\right) = \ln 3 + \ln 5$

15. $\log_3 xy = \log_3 x + \log_3 y$

17. $\ln 10z = \ln\left(2 \cdot 5 \cdot z\right) = \ln 2 + \ln 5 + \ln z$

19. $\log\dfrac{7}{3} = \log 7 - \log 3$

21. $\ln\dfrac{x}{y} = \ln x - \ln y$

23. $\log_2 \dfrac{45}{x} = \log_2 45 - \log_2 x$

25. $\log 45 + \log 5 = \log\left(45 \cdot 5\right) = \log 225$

27. $\ln x + \ln y = \ln xy$

29. $\ln 7x^2 + \ln 2x = \ln\left(7x^2 \cdot 2x\right) = \ln 14x^3$

31. $\ln x + \ln y^2 - \ln y = \ln xy^2 - \ln y = \ln\dfrac{xy^2}{y} = \ln xy$

33. $\log 20 - \log 4 = \log\left(\dfrac{20}{4}\right) = \log 5$

35. $\ln x^4 - \ln x^2 = \ln\left(\dfrac{x^4}{x^2}\right) = \ln x^2$

37. $\log_2 12x - \log_2 3x = \log_2\left(\dfrac{12x}{3x}\right) = \log_2 4 = 2$

39. $\log 3^6 = 6\log 3$

41. $\ln 2^x = x\ln 2$

43. $\log_2 5^{1/4} = \dfrac{1}{4}\log_2 5$

45. $\log_4 \sqrt[3]{z} = \log_4 z^{1/3} = \dfrac{1}{3}\log_4 z$

47. $\log x^{y-1} = \left(y-1\right)\log x$

49. $4 \log z - \log z^3 = \log z^4 - \log z^3 = \log \dfrac{z^4}{z^3} = \log z$

51. $\log x + 2 \log x + 2 \log y = \log x + \log x^2 + \log y^2 = \log \left(x \cdot x^2 \cdot y^2 \right) = \log x^3 y^2$

53. $\log x - 2 \log \sqrt{x} = \log x - \log \left(\sqrt{x} \right)^2 = \log x - \log x = 0$

55. $\ln 2^{x+1} - \ln 2 = \ln \dfrac{2^{x+1}}{2} = \ln 2^x$

57. $2 \log_3 \sqrt{x} - 3 \log_3 x = \log_3 \left(\sqrt{x} \right)^2 - \log_3 x^3 = \log_3 x - \log_3 x^3 = \log_3 \dfrac{x}{x^3} = \log_3 \dfrac{1}{x^2}$

59. $2 \log_a (x+1) - \log_a \left(x^2 - 1 \right) = \log_a (x+1)^2 - \log_a \left(x^2 - 1 \right) = \log_a \dfrac{(x+1)(x+1)}{(x+1)(x-1)} = \log_a \dfrac{x+1}{x-1}$

61. $\log xy^2 = \log x + \log y^2 = \log x + 2 \log y$

63. $\ln \dfrac{x^4 y}{z} = \ln x^4 y - \ln z = \ln x^4 + \ln y - \ln z = 4 \ln x + \ln y - \ln z$

65. $\log_4 \dfrac{\sqrt[3]{z}}{\sqrt{y}} = \log_4 \dfrac{z^{1/3}}{y^{1/2}} = \log_4 z^{1/3} - \log_4 y^{1/2} = \dfrac{1}{3} \log_4 z - \dfrac{1}{2} \log_4 y$

67. $\log \left(x^4 y^3 \right) = \log x^4 + \log y^3 = 4 \log x + 3 \log y$

69. $\ln \dfrac{1}{y} - \ln \dfrac{1}{x} = \ln y^{-1} - \ln x^{-1} = -1 \ln y - (-1) \ln x = \ln x - \ln y$

71. $\log_4 \sqrt{\dfrac{x^3 y}{z^2}} = \log_4 \left(\dfrac{x^3 y}{z^2} \right)^{1/2} = \dfrac{1}{2} \log_4 \dfrac{x^3 y}{z^2} = \dfrac{1}{2} \left(\log_4 x^3 + \log_4 y - \log_4 z^2 \right) = \dfrac{3}{2} \log_4 x + \dfrac{1}{2} \log_4 y - \log_4 z$

73.   Graph $Y_1 = \log \left(X^3 \right)$ and $Y_2 = 3 \log (X)$ in [–6, 6, 1] by [–4, 4, 1]. See Figures 73a & 73b.

By the power rule $\log x^3 = 3 \log x$.

[–6, 6, 1] by [–4, 4, 1]   [–6, 6, 1] by [–4, 4, 1]

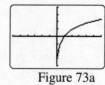   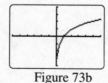

Figure 73a   Figure 73b

75.   Graph $Y_1 = \ln (X+5)$ and $Y_2 = \ln (X) + \ln (5)$ in [–6, 6, 1] by [–4, 4, 1]. See Figure 75a & 75b. Not the same.

[–6, 6, 1] by [–4, 4, 1]   [–6, 6, 1] by [–4, 4, 1]

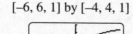

Figure 75a   Figure 75b

77. $\log 16 = \log 2^4 = 4 \log 2 = 4(0.3) = 1.2$

79. $\log 65 = \log(5 \cdot 13) = \log 5 + \log 13 = 0.7 + 1.1 = 1.8$

81. $\log 130 = \log(2 \cdot 5 \cdot 13) = \log 2 + \log 5 + \log 13 = 0.3 + 0.7 + 1.1 = 2.1$

83. $\log \dfrac{5}{2} = \log 5 - \log 2 = 0.7 - 0.3 = 0.4$

85. $\log \dfrac{1}{13} = \log 13^{-1} = -\log 13 = -1.1$

87. $\log_3 5 = \dfrac{\log 5}{\log 3} \approx 1.46$

90. $\log_7 8 = \dfrac{\log 8}{\log 7} \approx 1.07$

91. $\log_9 102 = \dfrac{\log 102}{\log 9} \approx 2.10$

Applications

93. $f(x) = 10 \log(10^{16} x) = 10(\log 10^{16} + \log x) = 10(16 + \log x) = 160 + 10 \log x$

Checking Basic Concepts for Sections 9.3 & 9.4

1.   (a) $\log 10^4 = 4$

      (b) $\ln e^x = x$

      (c) $\log_2 \dfrac{1}{8} = \log_2 2^{-3} = -3$

      (d) $\log_5 \sqrt{5} = \log_5 5^{1/2} = \dfrac{1}{2}$

2.   See Figure 2.

      (a) $D : \{x \mid x > 0\};\ R :$ all real numbers

      (b) $f(1) = 0$

      (c)   Yes, for example $\log \dfrac{1}{10} = -1.$

      (d)   No, since negative numbers are not in the domain of $f(x) = \log x.$

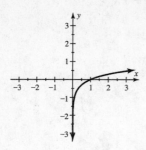

Figure 2

3. (a) $\log xy = \log x + \log y$

    (b) $\ln \dfrac{x}{yz} = \ln x - \ln yz = \ln x - \left(\ln y + \ln z \right) = \ln x - \ln y - \ln z$

    (c) $\ln x^2 = 2 \ln x$

    (d) $\log \dfrac{x^2 y^3}{\sqrt{z}} = \log x^2 y^3 - \log z^{1/2} = \log x^2 + \log y^3 - \log z^{1/2} = 2 \log x + 3 \log y - \dfrac{1}{2} \log z$

4. (a) $\log x + \log y = \log xy$

    (b) $\ln 2x - 3 \ln y = \ln 2x - \ln y^3 = \ln \dfrac{2x}{y^3}$

    (c) $2 \log_2 x + 3 \log_2 y - \log_2 z = \log_2 x^2 + \log_2 y^3 - \log_2 z = \log_2 x^2 y^3 - \log_2 z = \log_2 \dfrac{x^2 y^3}{z}$

9.5: Exponential and Logarithmic Equations

Concepts

1. Add 5 to both sides.

2. Divide both sides by 5.

3. Take the common logarithm of both sides.

4. Exponentiate both sides using base 10.

5. x

6. x

7. $2x$

8. $x + 7$

9. No, $\log \dfrac{5}{4} = \log 5 - \log 4$

10. No, $\log 5 - \log 4 = \log \dfrac{5}{4}$

11. One

12. One

Exponential Equations

13. $10^x = 1000 \Rightarrow 10^x = 10^3 \Rightarrow x = 3$

15. $2^x = 64 \Rightarrow 2^x = 2^6 \Rightarrow x = 6$

17. $2^{x-3} = 8 \Rightarrow 2^{x-3} = 2^3 \Rightarrow x - 3 = 3 \Rightarrow x = 6$

19. $4^x + 3 = 259 \Rightarrow 4^x = 256 \Rightarrow 4^x = 4^4 \Rightarrow x = 4$

21. $10^{0.4x} = 124 \Rightarrow \log 10^{0.4x} = \log 124 \Rightarrow 0.4x = \log 124 \Rightarrow x = \dfrac{\log 124}{0.4} \approx 5.23$

23. $e^{-x} = 1 \Rightarrow e^{-x} = e^0 \Rightarrow -x = 0 \Rightarrow x = 0$

25. $e^x = 25 \Rightarrow \ln e^x = \ln 25 \Rightarrow x = \ln 25 \Rightarrow x \approx 3.22$

27. $0.4^x = 2 \Rightarrow \ln 0.4^x = \ln 2 \Rightarrow x \ln 0.4 = \ln 2 \Rightarrow x = \dfrac{\ln 2}{\ln 0.4} \Rightarrow x \approx -0.76$

29. $e^x - 1 = 6 \Rightarrow e^x = 7 \Rightarrow \ln e^x = \ln 7 \Rightarrow x = \ln 7 \approx 1.95$

31. $2(10)^{x+2} = 35 \Rightarrow 10^{x+2} = \dfrac{35}{2} \Rightarrow \log 10^{x+2} = \log \dfrac{35}{2} \Rightarrow x + 2 = \log \dfrac{35}{2} \Rightarrow x = \log \dfrac{35}{2} - 2 \approx -0.76$

33. $3.1^{2x} - 4 = 16 \Rightarrow 3.1^{2x} = 20 \Rightarrow \log_{3.1} 3.1^{2x} = \log_{3.1} 20 \Rightarrow 2x = \log_{3.1} 20 \Rightarrow x = \dfrac{\log 20}{2 \log 3.1} \approx 1.32$

35. $e^{3x} = e^{2x-1} \Rightarrow 3x = 2x - 1 \Rightarrow x = -1$

37. $5^{4x} = 5^{x^2 - 5} \Rightarrow 4x = x^2 - 5 \Rightarrow x^2 - 4x - 5 = 0 \Rightarrow (x+1)(x-5) = 0 \Rightarrow x = -1 \text{ or } 5$

39. $e^{2x} \cdot e^x = 10 \Rightarrow e^{2x+x} = 10 \Rightarrow e^{3x} = 10 \Rightarrow \ln e^{3x} = \ln 10 \Rightarrow 3x = \ln 10 \Rightarrow x = \dfrac{\ln 10}{3} \approx 0.77$

41. $e^x = 2^{x+2} \Rightarrow \ln e^x = \ln 2^{x+2} \Rightarrow x = (x+2)\ln 2 \Rightarrow x = x \ln 2 + 2 \ln 2 \Rightarrow x - x \ln 2 = 2 \ln 2 \Rightarrow$

   $x(1 - \ln 2) = 2 \ln 2 \Rightarrow x = \dfrac{2 \ln 2}{1 - \ln 2} \approx 4.52$

43. $4^{0.5x} = 5^{x+2} \Rightarrow \log 4^{0.5x} = \log 5^{x+2} \Rightarrow 0.5x \log 4 = (x+2)\log 5 \Rightarrow 0.5x \log 4 = x \log 5 + 2 \log 5 \Rightarrow$

   $0.5x \log 4 - x \log 5 = 2 \log 5 \Rightarrow x(0.5 \log 4 - \log 5) = 2 \log 5 \Rightarrow x = \dfrac{2 \log 5}{0.5 \log 4 - \log 5} \approx -3.51$

45. (a) The solution is the *x*-coordinate of the intersection point, $x = 1$.

   (b) $0.2(10^x) = 2 \Rightarrow 10^x = 10 \Rightarrow x = 1$

47. (a) The solution is the *x*-coordinate of the intersection point, $x = -2$.

   (b) $2^{-x} = 4 \Rightarrow 2^{-x} = 2^2 \Rightarrow -x = 2 \Rightarrow x = -2$

49. $10^x = 0.1 \Rightarrow 10^x = 10^{-1} \Rightarrow x = -1$

   For numerical support, table $Y_1 = 10 \wedge X$ and $Y_2 = 0.1$ with TblStart $= -3$ and ΔTbl $= 1$. See Figure 49.

51. $4e^x + 5 = 9 \Rightarrow 4e^x = 4 \Rightarrow e^x = 1 \Rightarrow e^x = e^0 \Rightarrow x = 0$

For numerical support, table $Y_1 = 4e \wedge X + 5$ and $Y_2 = 9$ with TblStart $= -3$ and ΔTbl $= 1$. See Figure 51.

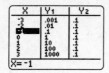   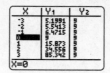

Figure 49   Figure 51

53. $4^x = 1024 \Rightarrow 4^x = 4^5 \Rightarrow x = 5$

For numerical support, table $Y_1 = 4 \wedge X$ and $Y_2 = 1024$ with TblStart $= 0$ and ΔTbl $= 1$. See Figure 53.

55. $(0.55)^x + 0.55 = 2 \Rightarrow 0.55^x = 1.45 \Rightarrow \log_{0.55} 0.55^x = \log_{0.55} 1.45 \Rightarrow x = \dfrac{\log 1.45}{\log 0.55} \approx -0.62$

For graphical support, graph $Y_1 = 0.55 \wedge X + 0.55$ and $Y_2 = 2$ in $[-1, 1, 1]$ by $[0, 3, 1]$. See Figure 55.

$$[-1, 1, 1] \text{ by } [0, 3, 1]$$

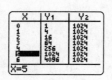   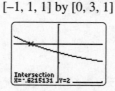

Figure 53   Figure 55

57.   Graph $Y_1 = e \wedge X - X$ and $Y_2 = 2$ in $[-5, 5, 1]$ by $[-5, 5, 1]$. See Figures 57a & 57b.

The solutions are the x-coordinates of the intersection points, $x \approx -1.84$ and $x \approx 1.15$.

$$[-5, 5, 1] \text{ by } [-5, 5, 1] \qquad [-5, 5, 1] \text{ by } [-5, 5, 1]$$

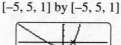

Figure 57a   Figure 57b

59.   Graph $Y_1 = \ln(X)$ and $Y_2 = e \wedge (-X)$ in $[-5, 5, 1]$ by $[-5, 5, 1]$. See Figure 59.

The solution is the x-coordinate of the intersection point, $x \approx 1.31$.

$$[-5, 5, 1] \text{ by } [-5, 5, 1]$$

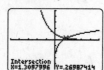

Figure 59

Logarithmic Equations

61. $\log x = 2 \Rightarrow 10^{\log x} = 10^2 \Rightarrow x = 100$

63. $\ln x = 5 \Rightarrow e^{\ln x} = e^5 \Rightarrow x = e^5 \approx 148.41$

65. $\log 2x = 7 \Rightarrow 10^{\log 2x} = 10^7 \Rightarrow 2x = 10,000,000 \Rightarrow x = 5,000,000$

67. $\log_2 x = 4 \Rightarrow 2^{\log_2 x} = 2^4 \Rightarrow x = 16$

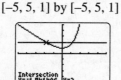

69. $\log_2 5x = 2.3 \Rightarrow 2^{\log_2 5x} = 2^{2.3} \Rightarrow 5x = 2^{2.3} \Rightarrow x = \dfrac{2^{2.3}}{5} \approx 0.98$

71. $2\log x + 5 = 7.8 \Rightarrow 2\log x = 2.8 \Rightarrow \log x = 1.4 \Rightarrow 10^{\log x} = 10^{1.4} \Rightarrow x = 10^{1.4} \approx 25.12$

73. $5\ln(2x+1) = 55 \Rightarrow \ln(2x+1) = 11 \Rightarrow e^{\ln(2x+1)} = e^{11} \Rightarrow 2x+1 = e^{11} \Rightarrow x = \dfrac{e^{11} - 1}{2} \approx 29{,}936.57$

75. $\log x^2 = \log x \Rightarrow x^2 = x \Rightarrow x^2 - x = 0 \Rightarrow x(x-1) = 0 \Rightarrow x = 0 \text{ or } 1$

    The solution $x = 0$ causes an undefined expression in the original equation. The only solution is 1.

77. $\ln x + \ln(x+1) = \ln 30 \Rightarrow \ln x(x+1) = \ln 30 \Rightarrow x(x+1) = 30 \Rightarrow x^2 + x - 30 = 0 \Rightarrow$

    $(x+6)(x-5) = 0 \Rightarrow x = -6 \text{ or } 5$

    The solution $x = -6$ causes an undefined expression in the original equation. The only solution is 5.

79. $\log_3 3x - \log_3(x+2) = \log_3 2 \Rightarrow \log_3 \dfrac{3x}{x+2} = \log_3 2 \Rightarrow \dfrac{3x}{x+2} = 2 \Rightarrow 3x = 2x+4 \Rightarrow x = 4$

81. $\log_2(x-1) + \log_2(x+1) = 3 \Rightarrow \log_2(x^2-1) = 3 \Rightarrow 2^{\log_2(x^2-1)} = 2^3 \Rightarrow x^2 - 1 = 8 \Rightarrow$

    $x^2 - 9 = 0 \Rightarrow (x+3)(x-3) = 0 \Rightarrow x = -3 \text{ or } 3$

    The solution $x = -3$ causes an undefined expression in the original equation. The only solution is 3.

83. (a) The solution is the x-coordinate of the intersection point, $x = 2$.

    (b) $\ln x = 0.7 \Rightarrow e^{\ln x} = e^{0.7} \Rightarrow x = e^{0.7} \approx 2.01$

85. (a) The solution is the x-coordinate of the intersection point, $x = 2$.

    (b) $5\log 2x = 3 \Rightarrow \log 2x = 0.6 \Rightarrow 10^{\log 2x} = 10^{0.6} \Rightarrow 2x = 10^{0.6} \Rightarrow x = \dfrac{10^{0.6}}{2} \approx 1.99$

87. $\log x = 1.6 \Rightarrow 10^{\log x} = 10^{1.6} \Rightarrow x = 10^{1.6} \approx 39.81$

    For graphical support, graph $Y_1 = \log(X)$ and $Y_2 = 1.6$ in $[0, 50, 10]$ by $[-2, 2, 1]$. See Figure 87.

89. $\ln(x+1) = 1 \Rightarrow e^{\ln(x+1)} = e^1 \Rightarrow x+1 = e \Rightarrow x = e - 1 \approx 1.72$

    For graphical support, graph $Y_1 = \ln(X+1)$ and $Y_2 = 1$ in $[-1, 2, 1]$ by $[-2, 2, 1]$. See Figure 89.

91. $17 - 6\log_3 x = 5 \Rightarrow 6\log_3 x = 12 \Rightarrow \log_3 x = 2 \Rightarrow 3^{\log_3 x} = 3^2 \Rightarrow x = 3^2 = 9$

    For graphical support, graph $Y_1 = 17 - 6\left(\ln(X)/\ln(3)\right)$ and $Y_2 = 5$ in $[0, 10, 1]$ by $[-3, 3, 1]$. See Figure 91.

$[0, 50, 10]$ by $[-2, 2, 1]$ $[-1, 2, 1]$ by $[-2, 2, 1]$ $[0, 10, 1]$ by $[-3, 3, 1]$

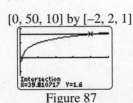
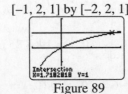
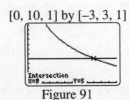

Figure 87   Figure 89   Figure 91

Applications

93. $2000(1+0.15)^t = 6000 \Rightarrow 1.15^t = 3 \Rightarrow \log_{1.15} 1.15^t = \log_{1.15} 3 \Rightarrow t = \dfrac{\log 3}{\log 1.15} \approx 7.86 \approx 8$ years

95. (a) $f(1994) = 2339(1.24)^{(1994-1988)} \approx 8503$; In 1994 there were about 8503 people waiting for liver

   transplants.

   (b) $2339(1.24)^{(x-1988)} = 20{,}000 \Rightarrow (1.24)^{(x-1988)} = \dfrac{20{,}000}{2339} \Rightarrow \log_{1.24}(1.24)^{(x-1988)} = \log_{1.24}\left(\dfrac{20{,}000}{2339}\right) \Rightarrow$

   $x - 1988 = \log_{1.24}\left(\dfrac{20{,}000}{2339}\right) \Rightarrow x = \log_{1.24}\left(\dfrac{20{,}000}{2339}\right) + 1988 = \dfrac{\log\left(\frac{20{,}000}{2339}\right)}{\log 1.24} + 1988 \approx 1998$

97. $3\log x = 3.960 \Rightarrow \log x = 1.320 \Rightarrow 10^{\log x} = 10^{1.320} \Rightarrow x = 10^{1.320} \approx 20.893 = 20{,}893$ lb

99. (a) $f(1) = 230\left(10^{-0.055 \cdot 1}\right) \approx 203$; In 1975 there were about 203 thousand bluefin tuna.

   (b) About 1979.

   (c) $230\left(10^{-0.055x}\right) = 115 \Rightarrow 10^{-0.055x} = 0.5 \Rightarrow \log 10^{-0.055x} = \log 0.5 \Rightarrow -0.055x = \log 0.5 \Rightarrow$

   $x = \dfrac{\log 0.5}{-0.055} \approx 5.47$ or about 1979.

101. From the data point $(1, 25)$, $25 = a + b\log 1 \Rightarrow 25 = a + b(0) \Rightarrow a = 25$.

   Using the data point $(10, 28)$ and the fact that $a = 25$, $28 = 25 + b\log 10 \Rightarrow 28 = 25 + b \Rightarrow b = 3$

103. (a) Graph $Y_1 = 645\log(X+1) + 1925$ and $Y_2 = 2200$ in $[0, 4, 1]$ by $[0, 3000, 1000]$. See Figure 103.

   A person consuming 2200 calories would typically own about 1.67 acres.

   (b) $645\log(x+1) + 1925 = 2200 \Rightarrow 645\log(x+1) = 275 \Rightarrow \log(x+1) = \dfrac{275}{645} \Rightarrow$

   $10^{\log(x+1)} = 10^{275/645} \Rightarrow x+1 = 10^{275/645} \Rightarrow x = 10^{275/645} - 1 \approx 1.67$ acres

   $[0, 4, 1]$ by $[0, 3000, 1000]$

   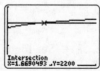

   Figure 103

105. (a) The data is nonlinear. It does not increase at a constant rate.

   (b) Each year the amount of fertilizer increases by a factor of 1.06 or 6%.

   (c) $5(1.06)^{(x-1950)} = 15 \Rightarrow 1.06^{(x-1950)} = 3 \Rightarrow \log_{1.06} 1.06^{(x-1950)} = \log_{1.06} 3 \Rightarrow$

   $x - 1950 = \log_{1.06} 3 \Rightarrow x = \log_{1.06} 3 + 1950 = \dfrac{\log 3}{\log 1.06} + 1950 \approx 1968.85$ or in 1968

107. $160 + 10\log x = 100 \Rightarrow 10\log x = -60 \Rightarrow \log x = -6 \Rightarrow 10^{\log x} = 10^{-6} \Rightarrow x = 10^{-6}$ w/cm^2

109. $0.48 \ln(x+1) + 27 = 28 \Rightarrow 0.48 \ln(x+1) = 1 \Rightarrow \ln(x+1) = \dfrac{1}{0.48} \Rightarrow e^{\ln(x+1)} = e^{1/0.48} \Rightarrow$

$x+1 = e^{1/0.48} \Rightarrow x = e^{1/0.48} - 1 \approx 7.03$ or about 7 miles

Checking Basic Concepts for Section 9.5

1.   (a) $2(10^x) = 40 \Rightarrow 10^x = 20 \Rightarrow \log 10^x = \log 20 \Rightarrow x = \log 20 \approx 1.30$

(b) $2^{3x} + 3 = 150 \Rightarrow 2^{3x} = 147 \Rightarrow \log_2 2^{3x} = \log_2 147 \Rightarrow 3x = \dfrac{\log 147}{\log 2} \Rightarrow x = \dfrac{\log 147}{3 \log 2} \approx 2.40$

(c) $\ln x = 4.1 \Rightarrow e^{\ln x} = e^{4.1} \Rightarrow x = e^{4.1} \approx 60.34$

(d) $4 \log 2x = 12 \Rightarrow \log 2x = 3 \Rightarrow 10^{\log 2x} = 10^3 \Rightarrow 2x = 1000 \Rightarrow x = 500$

2. $\log(x+4) + \log(x-4) = \log 48 \Rightarrow \log(x^2 - 16) = \log 48 \Rightarrow x^2 - 16 = 48 \Rightarrow$

$x^2 - 64 = 0 \Rightarrow (x+8)(x-8) = 0 \Rightarrow x = -8$ or 8

The solution $x = -8$ causes an undefined expression in the original equation. The only solution is 8.

3. $500(1.03)^x = 900 \Rightarrow 1.03^x = \dfrac{9}{5} \Rightarrow \log_{1.03} 1.03^x = \log_{1.03}\left(\dfrac{9}{5}\right) \Rightarrow x = \dfrac{\log\left(\frac{9}{5}\right)}{\log 1.03} \approx 19.88$ or about 20 years

Chapter 9 Review Exercises

Section 9.1

1.   (a) $f(-2) = 2(-2)^2 - 4(-2) = 16$, then $(g \circ f)(-2) = g(f(-2)) = g(16) = 5(16) + 1 = 81$

(b) $(f \circ g)(x) = f(g(x)) = f(5x+1) = 2(5x+1)^2 - 4(5x+1) = 50x^2 - 2$

2.   (a) $f(-2) = \sqrt[3]{-2-6} = -2$, then $(g \circ f)(-2) = g(f(-2)) = g(-2) = 4(-2)^3 = -32$

(b) $(f \circ g)(x) = f(g(x)) = f(4x^3) = \sqrt[3]{4x^3 - 6}$

3.   (a) $(f \circ g)(2) = f(g(2)) = f(3) = 0$

(b) $(g \circ f)(1) = g(f(1)) = g(2) = 3$

4.   (a) $(f \circ g)(-1) = f(g(-1)) = f(2) = 3$

(b) $(g \circ f)(2) = g(f(2)) = g(3) = -2$

(c) $(f \circ f)(1) = f(f(1)) = f(0) = -1$

5. $f(1) = f(-1) = 2$

6. $f(0) = f(2) = 1$

7.   This graph does not pass the horizontal line test. The function is not one-to-one.

8.   This graph passes the horizontal line test. The function is one-to-one.

9. $\left(f \circ f^{-1}\right)(x) = f\left(f^{-1}(x)\right) = f\left(\dfrac{x+9}{2}\right) = 2\left(\dfrac{x+9}{2}\right) - 9 = x + 9 - 9 = x$

$\left(f^{-1} \circ f\right)(x) = f^{-1}\left(f(x)\right) = f^{-1}(2x-9) = \dfrac{(2x-9)+9}{2} = \dfrac{2x}{2} = x$

10. $\left(f \circ f^{-1}\right)(x) = f\left(f^{-1}(x)\right) = f\left(\sqrt[3]{x-1}\right) = \left(\sqrt[3]{x-1}\right)^3 + 1 = x - 1 + 1 = x$

$\left(f^{-1} \circ f\right)(x) = f^{-1}\left(f(x)\right) = f^{-1}\left(x^3 + 1\right) = \sqrt[3]{\left(x^3 + 1\right) - 1} = \sqrt[3]{x^3} = x$

11. $f(x) = 5x \Rightarrow y = 5x$, interchange x and y and solve for y. $x = 5y \Rightarrow y = \dfrac{x}{5} \Rightarrow f^{-1}(x) = \dfrac{x}{5}$

12. $f(x) = x - 11 \Rightarrow y = x - 11$, interchange x and y and solve for y.

$x = y - 11 \Rightarrow y = x + 11 \Rightarrow f^{-1}(x) = x + 11$

13. $f(x) = 2x + 7 \Rightarrow y = 2x + 7$, interchange x and y and solve for y.

$x = 2y + 7 \Rightarrow 2y = x - 7 \Rightarrow y = \dfrac{x-7}{2} \Rightarrow f^{-1}(x) = \dfrac{x-7}{2}$

14. $f(x) = \dfrac{4}{x} \Rightarrow y = \dfrac{4}{x}$, interchange x and y and solve for y. $x = \dfrac{4}{y} \Rightarrow xy = 4 \Rightarrow y = \dfrac{4}{x} \Rightarrow f^{-1}(x) = \dfrac{4}{x}$

15. See Figure 15. $D = \{3, 7, 8, 10\};\ R = \{0, 1, 2, 3\}$

16. The graph of f^{-1} is a reflection of the graph of f across the line $y = x$. See Figure 16.

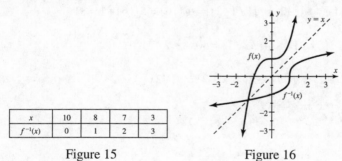

x	10	8	7	3
$f^{-1}(x)$	0	1	2	3

Figure 15   Figure 16

Sections 9.2 and 9.3

17. $f(-1) = 6^{-1} = \dfrac{1}{6}$ and $f(2) = 6^2 = 36$

18. $f(0) = 5\left(2^0\right) = 5(1) = 5$ and $f(3) = 5\left(2^{-3}\right) = 5\left(\dfrac{1}{8}\right) = \dfrac{5}{8}$

19. $f(-1) = \left(\dfrac{1}{3}\right)^{-1} = 3$ and $f(4) = \left(\dfrac{1}{3}\right)^4 = \dfrac{1}{3^4} = \dfrac{1}{81}$

20. $f(0) = 3\left(\dfrac{1}{6}\right)^0 = 3(1) = 3$ and $f(1) = 3\left(\dfrac{1}{6}\right)^1 = 3\left(\dfrac{1}{6}\right) = \dfrac{3}{6} = \dfrac{1}{2}$

21. See Figure 21. The graph illustrates exponential growth.

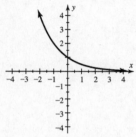

Figure 21

22.   See Figure 22. The graph illustrates exponential decay.

23.   See Figure 23. The graph illustrates logarithmic growth.

24.   See Figure 24. The graph illustrates exponential decay.

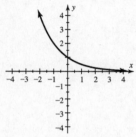

Figure 22

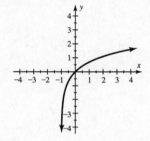

Figure 23

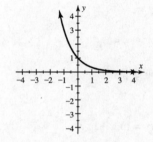

Figure 24

25.   (a) Exponential growth. For each unit increase in x, $f(x)$ increases by a factor of 2.

(b) Since $f(x) = 5$ when $x = 0$, $f(x) = 5(2)^x$

26.   (a) Linear growth. For each unit increase in x, $f(x)$ increases by 5 units.

(b) Since $f(x) = 5$ when $x = 0$, $f(x) = 5x + 5$

27.   Since $y = \dfrac{1}{2}$ when $x = 0$, $\dfrac{1}{2} = Ca^0 \Rightarrow C = \dfrac{1}{2}$. Since $y = 1$ when $x = 1$, $1 = \dfrac{1}{2}(a)^1 \Rightarrow a = 2$.

28.   Since $y = 2$ when $x = 2$, $2 = k\log_2 2 \Rightarrow 2 = k(1) \Rightarrow k = 2$.

29. $1200(1 + 0.10)^9 = \$2829.54$

30. $900(1 + 0.18)^{40} = \$675,340.51$

31. $f(5.3) = 2e^{5.3} - 1 \approx 399.67$

32. $f(2.1) = 0.85^{2.1} \approx 0.71$

33. $f(55) = 2\log 55 \approx 3.48$

34. $f(23) = \ln(2 \cdot 23 + 3) \approx 3.89$

35. $\log 0.001 = \log 10^{-3} = -3$

36. $\log \sqrt{10,000} = \log 100 = 2$

37. $\ln e^{-4} = -4$

38. $\log_4 16 = \log_4 4^2 = 2$

39. $\log 65 \approx 1.813$

40. $\ln 0.85 \approx -0.163$

41. $\ln 120 \approx 4.787$

42. $\log_2 \dfrac{2}{5} \approx -1.322$

43. $10^{\log 7} = 7$

44. $\log_2 2^{5/9} = \dfrac{5}{9}$

45. $\ln e^{6-x} = 6 - x$

46. $e^{2\ln x} = e^{\ln x^2} = x^2$

Section 9.4

47. $\ln xy = \ln x + \ln y$

48. $\log \dfrac{x}{y} = \log x - \log y$

49. $\ln x^2 y^3 = \ln x^2 + \ln y^3 = 2\ln x + 3\ln y$

50. $\log \dfrac{\sqrt{x}}{z^3} = \log \dfrac{x^{1/2}}{z^3} = \log x^{1/2} - \log z^3 = \dfrac{1}{2}\log x - 3\log z$

51. $\log_2 \dfrac{x^2 y}{z} = \log_2 x^2 y - \log_2 z = \log_2 x^2 + \log_2 y - \log_2 z = 2\log_2 x + \log_2 y - \log_2 z$

52. $\log_3 \sqrt[3]{\dfrac{x}{y}} = \log_3 \left(\dfrac{x}{y}\right)^{1/3} = \dfrac{1}{3}\log_3 \left(\dfrac{x}{y}\right) = \dfrac{1}{3}\left(\log_3 x - \log_3 y\right) = \dfrac{1}{3}\log_3 x - \dfrac{1}{3}\log_3 y$

53. $\log 45 + \log 5 - \log 3 = \log (45 \cdot 5) - \log 3 = \log 225 - \log 3 = \log \dfrac{225}{3} = \log 75$

54. $\log_4 2x + \log_4 5x = \log_4 (2x \cdot 5x) = \log_4 \left(10x^2\right)$

55. $2\ln x - 3\ln y = \ln x^2 - \ln y^3 = \ln \dfrac{x^2}{y^3}$

56. $\log x^4 - \log x^3 + \log y = \log \dfrac{x^4}{x^3} + \log y = \log x + \log y = \log xy$

57. $\log 6^3 = 3\log 6$

58. $\ln x^2 = 2\ln x$

59. $\log_2 5^{2x} = (2x)\log_2 5$

60. $\log_4 (0.6)^{x+1} = (x+1)\log_4 0.6$

Section 9.5

61. $10^x = 100 \Rightarrow 10^x = 10^2 \Rightarrow x = 2$

62. $2^{2x} = 256 \Rightarrow 2^{2x} = 2^8 \Rightarrow 2x = 8 \Rightarrow x = 4$

63. $3e^x + 1 = 28 \Rightarrow 3e^x = 27 \Rightarrow e^x = 9 \Rightarrow \ln e^x = \ln 9 \Rightarrow x = \ln 9 \approx 2.20$

64. $0.85^x = 0.2 \Rightarrow \log_{0.85} 0.85^x = \log_{0.85} 0.2 \Rightarrow x = \log_{0.85} 0.2 = \dfrac{\log 0.2}{\log 0.85} \approx 9.90$

65. $5\ln x = 4 \Rightarrow \ln x = 0.8 \Rightarrow e^{\ln x} = e^{0.8} \Rightarrow x = e^{0.8} \approx 2.23$

66. $\ln 2x = 5 \Rightarrow e^{\ln 2x} = e^5 \Rightarrow 2x = e^5 \Rightarrow x = \dfrac{e^5}{2} \approx 74.21$

67. $2\log x = 80 \Rightarrow \log x = 40 \Rightarrow 10^{\log x} = 10^{40} \Rightarrow x = 10^{40}$

68. $3\log x - 5 = 1 \Rightarrow 3\log x = 6 \Rightarrow \log x = 2 \Rightarrow 10^{\log x} = 10^2 \Rightarrow x = 100$

69. $2^{x+4} = 3^x \Rightarrow \log 2^{x+4} = \log 3^x \Rightarrow (x+4)\log 2 = x\log 3 \Rightarrow x\log 2 + 4\log 2 = x\log 3 \Rightarrow$

       $x\log 3 - x\log 2 = 4\log 2 \Rightarrow x(\log 3 - \log 2) = 4\log 2 \Rightarrow x = \dfrac{4\log 2}{\log 3 - \log 2} \approx 6.84$

70. $\ln(2x+1) + \ln(x-5) = \ln 13 \Rightarrow \ln(2x+1)(x-5) = \ln 13 \Rightarrow (2x+1)(x-5) = 13 \Rightarrow$

       $2x^2 - 9x - 5 = 13 \Rightarrow 2x^2 - 9x - 18 = 0 \Rightarrow (2x+3)(x-6) = 0 \Rightarrow x = -\dfrac{3}{2}$ or 6

       The solution $x = -\dfrac{3}{2}$ causes an undefined expression in the original equation. The only solution is 6.

71.   (a) The solution is the *x*-coordinate of the intersection point, $x = 3$.

       (b) $\dfrac{1}{2}(2^x) = 4 \Rightarrow 2^x = 8 \Rightarrow 2^x = 2^3 \Rightarrow x = 3$

72.   (a) The solution is the *x*-coordinate of the intersection point, $x = 4$.

       (b) $\log_2 2x = 3 \Rightarrow 2^{\log_2 2x} = 2^3 \Rightarrow 2x = 8 \Rightarrow x = 4$

Applications

73.   (a) $(S \circ r)(8) = S(r(8)) = S(\sqrt{2 \cdot 8}) = S(4) = 4\pi(4)^2 = 64\pi$

       After 8 seconds, the balloon has a surface area of $64\pi \approx 201$ square inches.

       (b) $(S \circ r)(t) = S(r(t)) = S(\sqrt{2t}) = 4\pi(\sqrt{2t})^2 = 4\pi \cdot 2t = 8\pi t$

74.   (a) Yes, this is a one-to-one function because different inputs result in different outputs.

       (b) $f(x) = 0.08x \Rightarrow y = 0.08x,$ interchange *x* and *y* and solve for *y*.

$$x = 0.08y \Rightarrow y = \frac{x}{0.08} \Rightarrow f^{-1}(x) = \frac{x}{0.08}$$

This formula calculates the cost of an item whose sales tax is x dollars.

75. $1500(1+0.12)^t = 3000 \Rightarrow 1.12^t = 2 \Rightarrow \log_{1.12} 1.12^t = \log_{1.12} 2 \Rightarrow t = \dfrac{\log 2}{\log 1.12} \approx 6.12 \approx 7$ years

76. $100 = a + b \log 1 \Rightarrow 100 = a + b(0) \Rightarrow a = 100$

$150 = 100 + b \log 10 \Rightarrow 150 = 100 + b(1) \Rightarrow b = 50$

77. $3 = Ca^0 \Rightarrow C = 3$ and $6 = 3a^1 \Rightarrow a = 2$

78. $\log \dfrac{x}{1} = 7 \Rightarrow 10^{\log x} = 10^7 \Rightarrow x = 10^7$

79.   (a) Graph $Y_1 = 2e \wedge (0.051X)$ in $[0, 10, 2]$ by $[0, 4, 1]$.

[0, 10, 2] by [0, 4, 1]

(b) In 2010, $x = 10$ thus $f(10) = 2e^{0.051(10)} \approx 3.3$ million

(c) $2e^{0.051x} = 3 \Rightarrow e^{0.051x} = \dfrac{3}{2} \Rightarrow \ln e^{0.051x} = \ln\left(\dfrac{3}{2}\right)$

$\Rightarrow 0.051x = \ln(1.5) \Rightarrow x = \dfrac{\ln(1.5)}{0.051} \approx 7.95$

That is about 8 years after 2000, which is 2008.

80.   (a) $N(0) = 1000e^{0.0014(0)} = 1000$; There were initially 1000 bacteria.

(b) $1000e^{0.0014x} = 2000 \Rightarrow e^{0.0014x} = 2 \Rightarrow \ln e^{0.0014x} = \ln 2 \Rightarrow 0.0014x = \ln 2 \Rightarrow x = \dfrac{\ln 2}{0.0014} \approx 495.11$ min.

81.   (a) $f(5) = 1.2 \ln 5 + 5 \approx 6.93$ m/sec

(b) $1.2 \ln x + 5 = 8 \Rightarrow 1.2 \ln x = 3 \Rightarrow \ln x = 2.5 \Rightarrow e^{\ln x} = e^{2.5} \Rightarrow x = e^{2.5} \approx 12.18$ meters

Chapter 9 Test

1. $f(1) = 4(1)^3 - 5(1) = -1$, then $(g \circ f)(1) = g(f(1)) = g(-1) = (-1) + 7 = 6$

$(f \circ g)(x) = f(g(x)) = f(x+7) = 4(x+7)^3 - 5(x+7)$

2.   (a) $(f \circ g)(-1) = f(g(-1)) = f(-1) = 3$

(b) $(g \circ f)(1) = g(f(1)) = g(1) = 3$

3.   Two different inputs result in the same output. For example $-5 \neq 5$, but $f(-5) = f(5) = 0$.

4. $f(x) = 5 - 2x \Rightarrow y = 5 - 2x$, interchange x and y and solve for y.

$$x = 5 - 2y \Rightarrow 2y = 5 - x \Rightarrow y = \frac{5-x}{2} \Rightarrow f^{-1}(x) = \frac{5-x}{2}$$

5.   The graph of f^{-1} is a reflection of the graph of f across the line $y = x$. See Figure 5.

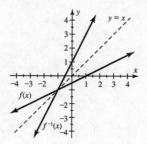

Figure 5

6.   See Figure 6. $D = \{2, 4, 6, 8\};\ \ R = \{1, 2, 3, 4\}$

x	8	6	4	2
$f^{-1}(x)$	1	2	3	4

Figure 6

7. $f(2) = 3\left(\dfrac{1}{4}\right)^2 = 3\left(\dfrac{1}{16}\right) = \dfrac{3}{16}$

8.   See Figure 8. This graph represents exponential decay.

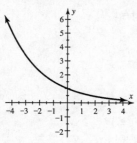

Figure 8

9.   (a) Exponential growth. For each unit increase in x, $f(x)$ increases by a factor of 2.

(b) Since $f(x) = 3$ when $x = 0$, $f(x) = 3(2)^x$

10.   (a) Linear growth. For each unit increase in x, $f(x)$ increases by 1.5 units.

(b) Since $f(x) = -1$ when $x = 0$, $f(x) = 1.5x - 1$

11. $1 = Ca^0 \Rightarrow C = 1$ and $2 = 1a^{-1} \Rightarrow a = \dfrac{1}{2}$

12. $750(1 + 0.07)^5 = \$1051.91$

13. $f(21) = 1.5\ln(21 - 5) = 1.5\ln 16 \approx 4.16$

14. $\log \sqrt{10} = \log 10^{1/2} = \dfrac{1}{2}$

15.　　$\log_2 43 = \dfrac{\log 43}{\log 2} \approx 5.426$

16.　　See Figure 16. The graph is shifted 2 units to the right.

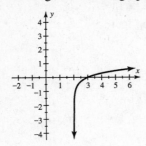

Figure 16

17.　　$\log \dfrac{x^3 y^2}{\sqrt{x}} = \log \dfrac{x^3 y^2}{x^{1/2}} = \log x^3 + \log y^2 - \log x^{1/2} = 3 \log x + 2 \log y - \dfrac{1}{2} \log z$

18.　　$4 \ln x - 5 \ln y + \ln z = \ln x^4 - \ln y^5 + \ln z = \ln \dfrac{x^4}{y^5} + \ln z = \ln \dfrac{x^4 z}{y^5}$

19.　　$\log 7^{2x} = 2x \log 7$

20.　　$\ln e^{1-3x} = 1 - 3x$

21.　　$2e^x = 50 \Rightarrow e^x = 25 \Rightarrow \ln e^x = \ln 25 \Rightarrow x = \ln 25 \approx 3.22$

22.　　$3(10)^x - 7 = 143 \Rightarrow 3(10)^x = 150 \Rightarrow 10^x = 50 \Rightarrow \log 10^x = \log 50 \Rightarrow x = \log 50 \approx 1.70$

23.　　$5 \log x = 9 \Rightarrow \log x = 1.8 \Rightarrow 10^{\log x} = 10^{1.8} \Rightarrow x = 10^{1.8} \approx 63.10$

24.　　$3 \ln 5x = 27 \Rightarrow \ln 5x = 9 \Rightarrow e^{\ln 5x} = e^9 \Rightarrow 5x = e^9 \Rightarrow x = \dfrac{e^9}{5} \approx 1620.62$

25.　　$5 = a + b \log 1 \Rightarrow 5 = a + b(0) \Rightarrow a = 5$ and $8 = 5 + b \log 10 \Rightarrow 8 = 5 + b(1) \Rightarrow b = 3$

26.　　(a)　$f(0) = 4e^{0.09(0)} = 4e^0 = 4(1) = 4$ million

　　　(b)　$f(5) = 4e^{0.09(5)} \approx 6.27$; After 5 hours there were about 6.27 million bacteria.

　　　(c)　This represents exponential growth.

　　　(d)　$4e^{0.09x} = 6 \Rightarrow e^{0.09x} = 1.5 \Rightarrow \ln e^{0.09x} = \ln 1.5 \Rightarrow 0.09x = \ln 1.5 \Rightarrow x = \dfrac{\ln 1.5}{0.09} \approx 4.51$

　　　　　There were 6 million bacteria after about 4.51 hours.

Chapter 9 Extended and Discovery Exercises

1.　　$a^{5730} = 0.5 \Rightarrow \left(a^{5730}\right)^{1/5730} = 0.5^{1/5730} \Rightarrow a \approx 0.9998790392$

2.　　$P(10,000) = 0.9998790392^{10,000} \approx 0.298$ or 29.8%

3. $0.9998790392^x = 0.9 \Rightarrow \log_{0.9998790392} 0.9998790392^x = \log_{0.9998790392} 0.9 \Rightarrow x = \dfrac{\log 0.9}{\log 0.9998790392} \Rightarrow$

    $x \approx 871$ years

4. $0.9998790392^x = 0.01 \Rightarrow \log_{0.9998790392} 0.999879^x = \log_{0.9998790392} 0.01 \Rightarrow x = \dfrac{\log 0.01}{\log 0.9998790392} \Rightarrow$

    $x \approx 38,069$ years   (38,100 rounded to the nearest 100 years)

5. Plot the data and graph $Y_1 = 0.133\big(0.878(0.73\wedge X) + 0.122(0.92\wedge X)\big)$ in [0, 25, 5] by [0, 0.11, 0.01].

    See Figure 5. The fit is quite good.

    [0, 25, 5] by [0, 0.11, 0.01]

    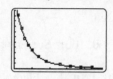

    Figure 5

6. $f(0) = 0.133\big(0.878(0.73^0) + 0.122(0.92^0)\big) = 0.133(0.878 + 0.122) = 0.133(1) = 0.133$

    The initial concentration is 0.133 mg/mL.

7. The concentration decreases to 0 as the body eliminates the dye from the blood stream.

8. Note that 40% of 0.133 is 0.0532.

    Graph $Y_1 = 0.133\big(0.878(0.73\wedge X) + 0.122(0.92\wedge X)\big)$ and $Y_2 = 0.0532$ in [0, 25, 5] by [0, 0.11, 0.01].

    See Figure 8. This happens after about 3.33 minutes. Solving this problem symbolically would be very difficult.

    [0, 25, 5] by [0, 0.11, 0.01]

    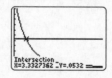

    Figure 8

9. $f\big(10^{-4.7}\big) = -\log 10^{-4.7} = -(-4.7) = 4.7$;   This rain could cause the pH to drop below 5.6.

10. The ion concentration in seawater is $-\log x = 8.2 \Rightarrow \log x = -8.2 \Rightarrow 10^{\log x} = 10^{-8.2} \Rightarrow x = 10^{-8.2}$.

    This is $\dfrac{10^{-4.7}}{10^{-8.2}} \approx 3162$ times greater.

11. $A = 100\left[\dfrac{\left(1 + \frac{0.09}{26}\right)^{260} - 1}{\frac{0.09}{26}}\right] \approx \$42,055.97$

12. A 19-year-old student would have 46 years to deposit money before age 65. *Answers may vary.*

$$x \left[\frac{\left(1 + \frac{0.12}{26}\right)^{1196} - 1}{\frac{0.12}{26}} \right] = 1,000,000 \Rightarrow x = 1,000,000 \left[\frac{\frac{0.12}{26}}{\left(1 + \frac{0.12}{26}\right)^{1196} - 1} \right] = \$18.80$$

13.   Plot the data in [0, 20, 2] by [0, 700, 100]. See Figure 13.

[0, 20, 2] by [0, 700, 100]

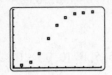

Figure 13

14.   The yeast grew slowly at first and then, as more yeast was present, the growth rate increased. Finally, as the yeast began to use up the fixed amount of nourishment, the growth rate slowed.

15.   Plot the data and graph $Y_1 = 663/\left(1 + 71.6(0.579)^{\wedge} X\right)$ in [0, 20, 2] by [0, 700, 100]. See Figure 15.

16.   Graph $Y_1 = 663/\left(1 + 71.6(0.579)^{\wedge} X\right)$ and $Y_2 = 400$ in [0, 20, 2] by [0, 700, 100]. See Figure 16.

The amount of yeast was equal to 400 units after about 8.6 hours.

[0, 20, 2] by [0, 700, 100]   [0, 20, 2] by [0, 700, 100]

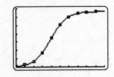   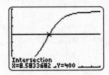

Figure 15   Figure 16

17. $400 = \dfrac{663}{1 + 71.6(0.579)^x} \Rightarrow 400\left(1 + 71.6(0.579)^x\right) = 663 \Rightarrow 400 + 28,640(0.579)^x = 663 \Rightarrow$

$28,640(0.579)^x = 263 \Rightarrow 0.579^x = \dfrac{263}{28,640} \Rightarrow \log_{0.579} 0.579^x = \log_{0.579} \dfrac{263}{28,640} \Rightarrow$

$x = \log_{0.579} \dfrac{263}{28,640} = \dfrac{\log\left(\frac{263}{28,640}\right)}{\log 0.579} \approx 8.6 \text{ hours}$

Chapters 1–9 Cumulative Review Exercises

1.   Move the decimal point 4 places to the right: $0.000429 = 4.29 \times 10^{-4}$.

2.   Natural: none; Whole: 0; Integer: $-3, 0$; Rational: $-\dfrac{11}{7}, -3, 0, 5.\overline{18}$; Irrational: $\sqrt{6}, \pi$

3.   By substitution of values, the formula that best fits the data is (iii).

4.   This equation illustrates the commutative property for addition.

5. $\left(\dfrac{1}{d^2}\right)^{-2} = \left(d^2\right)^2 = d^4$

6. $\left(\dfrac{8a^2}{2b^3}\right)^{-3} = \left(\dfrac{4a^2}{b^3}\right)^{-3} = \left(\dfrac{b^3}{4a^2}\right)^3 = \dfrac{\left(b^3\right)^3}{\left(4a^2\right)^3} = \dfrac{b^9}{4^3\left(a^2\right)^3} = \dfrac{b^9}{64a^6}$

7. $\dfrac{\left(2x^{-2}y^3\right)^2}{xy^{-2}} = \dfrac{2^2\left(x^{-2}\right)^2\left(y^3\right)^2}{xy^{-2}} = \dfrac{4x^{-4}y^6}{xy^{-2}} = 4x^{-4-1}y^{6-(-2)} = 4x^{-5}y^8 = \dfrac{4y^8}{x^5}$

8. $\dfrac{x^{-3}y}{4x^2y^{-3}} = \dfrac{1}{4}x^{-3-2}y^{1-(-3)} = \dfrac{1}{4}x^{-5}y^4 = \dfrac{y^4}{4x^5}$

9. The graph rises 5 units for each 4 units of run. The slope is $\dfrac{5}{4}$. The y-intercept is 1. So $y = \dfrac{5}{4}x + 1$.

10. The function is defined for all values of the variable except -3. The domain is $\{x \mid x \neq -3\}$.

11. The slope is $m = \dfrac{5-1}{2-1} = \dfrac{4}{1} = 4$. Since $f(x) = -3$ when $x = 0$ the y-intercept is -3. Here $y = 4x - 3$.

12. Vertical lines have equations of the form $x = k$. The equation of the vertical line passing through $(4, 7)$ is $x = 4$.

13. $m = \dfrac{-3-(-1)}{2-4} = \dfrac{-2}{-2} = 1$

14. See Figure 14.

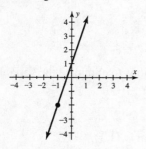

Figure 14

15. Since the line is perpendicular to $y = -\dfrac{1}{7}x - 8$, the slope is $m = 7$. Using the point-slope form gives

$y = 7(x-1) + 1 \Rightarrow y = 7x - 7 + 1 \Rightarrow y = 7x - 6$

16. Since the line is parallel to $y = 3x - 1$, the slope is $m = 3$. The y-intercept is given, so $b = 5$.

The equation is $y = 3x + 5$.

17. The graphs intersect at the point $(-1, 1)$, so $y_1 = y_2$ when $x = -1$.

18. Here $y < -4$ when $x < 0$.

19. $\dfrac{2}{3}(x-3) + 8 = -6 \Rightarrow \dfrac{2}{3}(x-3) = -14 \Rightarrow x - 3 = -21 \Rightarrow x = -18$

20. $\dfrac{1}{3}z + 6 < \dfrac{1}{4}z - (5z-6) \Rightarrow 4z + 72 < 3z - 12(5z-6) \Rightarrow 4z + 72 < 3z - 60z + 72 \Rightarrow$

$4z + 72 < -57z + 72 \Rightarrow 61z < 0 \Rightarrow z < 0$. The interval is $(-\infty, 0)$.

21. $\left(\dfrac{t+2}{3}\right) - 10 = \dfrac{1}{3}t - (5t + 8) \Rightarrow t + 2 - 30 = t - 3(5t + 8) \Rightarrow 15t = 4 \Rightarrow t = \dfrac{4}{15}$

22. $-10 \le -\dfrac{3}{5}x - 4 < -1 \Rightarrow -6 \le -\dfrac{3}{5}x < 3 \Rightarrow 10 \ge x > -5 \Rightarrow -5 < x \le 10$. The interval is $(-5, 10]$.

23. First divide each side of $-2|t - 4| \ge -12$ by -2 to obtain $|t - 4| \le 6$.

    The solutions to $|t - 4| \le 6$ satisfy $c \le t \le d$ where c and d are the solutions to $|t - 4| = 6$.

    $|t - 4| = 6$ is equivalent to $t - 4 = -6 \Rightarrow t = -2$ and $t - 4 = 6 \Rightarrow t = 10$. The interval is $[-2, 10]$.

24. $\left|\dfrac{1}{2}x - 5\right| = 3 \Rightarrow \dfrac{1}{2}x - 5 = -3 \Rightarrow \dfrac{1}{2}x = 2 \Rightarrow x = 4$ or $\dfrac{1}{2}x - 5 = 3 \Rightarrow \dfrac{1}{2}x = 8 \Rightarrow x = 16$

25. See Figure 25.

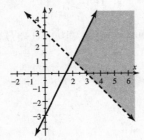

    Figure 25

26. $\det A = -1(4) - 3(-2) = -4 + 6 = 2$

27. Multiply the first equation by 2 and the second equation by 3. Add the equations to eliminate the variable y.

    $8x - 6y = 2$
    $\underline{15x + 6y = 21}$
    $23x = 23$ Thus, $x = 1$. And so $4(1) - 3y = 1 \Rightarrow y = 1$. The solution is $(1, 1)$.

28. Multiplying the first equation by 3 and adding the equations will eliminate both variables.

    $6x - 9y = -6$
    $\underline{-6x + 9y = 5}$
    $0 = -1$ Thus, the system has no solutions.

29. Multiply the first equation by 5 and add the first and second equations to eliminate the variable y.

    $10x - 5y + 15z = -10$
    $\underline{x + 5y - 2z = -8}$
    $11x + 13z = -18$

    Multiply the third equation by 5 and add the second and third equations to eliminate the variable y.

    $x + 5y - 2z = -8$
    $\underline{-15x - 5y - 15z = 30}$
    $-14x - 17z = 22$

Multiply the first *new* equation by 14 and the second *new* equation by 11. Add the equations to eliminate x.

$$154x + 182z = -252$$
$$\underline{-154x - 187z = 242}$$
$$-5z = -10$$

And so $z = 2$. Substitute $z = 2$ into the first *new* equation: $11x + 13(2) = -18 \Rightarrow x = -4$

Substitute $x = -4$ and $z = 2$ into the *original* first equation: $2(-4) - y + 3(2) = -2 \Rightarrow y = 0$

The solution is $(-4, 0, 2)$.

30. $\begin{bmatrix} 1 & 1 & -1 & | & -1 \\ -1 & -1 & -1 & | & -1 \\ 1 & -2 & 1 & | & 1 \end{bmatrix} \begin{matrix} \\ R_2 + R_1 \to \\ R_3 - R_1 \to \end{matrix} \begin{bmatrix} 1 & 1 & -1 & | & -1 \\ 0 & 0 & -2 & | & -2 \\ 0 & -3 & 2 & | & 2 \end{bmatrix} \begin{matrix} Exchange \\ \\ R_2 \leftrightarrow R_3 \end{matrix} \begin{bmatrix} 1 & 1 & -1 & | & -1 \\ 0 & -3 & 2 & | & 2 \\ 0 & 0 & -2 & | & -2 \end{bmatrix}$

$\begin{matrix} \\ \\ (-1/2)R_3 \to \end{matrix} \begin{bmatrix} 1 & 1 & -1 & | & -1 \\ 0 & -3 & 2 & | & 2 \\ 0 & 0 & 1 & | & 1 \end{bmatrix} \begin{matrix} R_1 + R_3 \to \\ R_2 - 2R_3 \to \\ \end{matrix} \begin{bmatrix} 1 & 1 & 0 & | & 0 \\ 0 & -3 & 0 & | & 0 \\ 0 & 0 & 1 & | & 1 \end{bmatrix} \begin{matrix} R_1 + (1/3)R_2 \to \\ (-1/3)R_2 \to \\ \end{matrix} \begin{bmatrix} 1 & 0 & 0 & | & 0 \\ 0 & 1 & 0 & | & 0 \\ 0 & 0 & 1 & | & 1 \end{bmatrix}$

The solution is $(0, 0, 1)$.

31. From the graph of the region of feasible solutions (not shown), the vertices are $(0, 0)$, $(0, 4)$, $(3, 3)$, and $(4, 0)$.

The maximum value of R occurs at one of the vertices. For $(0, 0)$, $R = 2(0) + 5(0) = 0$.

For $(0, 4)$, $R = 2(0) + 5(4) = 20$. For $(3, 3)$, $R = 2(3) + 5(3) = 21$. For $(4, 0)$, $R = 2(4) + 5(0) = 8$.

The maximum value is $R = 21$.

32. The triangle has vertices $(-2, -3), (-1, 2)$ and $(2, 1)$. The matrix needed is $A = \begin{bmatrix} -2 & -1 & 2 \\ -3 & 2 & 1 \\ 1 & 1 & 1 \end{bmatrix}$.

The area is $D = \left| \dfrac{1}{2} \det([A]) \right| = 8 \text{ in}^2$.

33. $2x^3 - 4x^2 + 2x = 2x(x^2 - 2x + 1) = 2x(x-1)^2$

34. $4a^2 - 25b^2 = (2a)^2 - (5b)^2 = (2a - 5b)(2a + 5b)$

35. $8t^3 - 27 = (2t)^3 - 3^3 = (2t - 3)(4t^2 + 6t + 9)$

36. $4a^3 - 2a^2 + 10a - 5 = 2a^2(2a - 1) + 5(2a - 1) = (2a^2 + 5)(2a - 1)$

37. $6x^2 - 7x - 10 = 0 \Rightarrow (6x + 5)(x - 2) = 0 \Rightarrow x = -\dfrac{5}{6} \text{ or } x = 2$

38. $9x^2 = 4 \Rightarrow 9x^2 - 4 = 0 \Rightarrow (3x + 2)(3x - 2) = 0 \Rightarrow x = -\dfrac{2}{3} \text{ or } x = \dfrac{2}{3}$

39. $x^4 - 2x^3 = 15x^2 \Rightarrow x^4 - 2x^3 - 15x^2 = 0 \Rightarrow x^2(x + 3)(x - 5) = 0 \Rightarrow x = -3, 0, \text{ or } 5$

40. $5x - 10x^2 = 0 \Rightarrow 5x(1-2x) = 0 \Rightarrow x = 0$ or $x = \dfrac{1}{2}$

41. $\dfrac{x^2 + 5x + 6}{x^2 - 9} \cdot \dfrac{x-3}{x+2} = \dfrac{(x+2)(x+3)(x-3)}{(x-3)(x+3)(x+2)} = 1$

42. $\dfrac{x^2 - 2x - 8}{x^2 + x - 12} \div \dfrac{(x-4)^2}{x^2 - 16} = \dfrac{x^2 - 2x - 8}{x^2 + x - 12} \cdot \dfrac{x^2 - 16}{(x-4)^2} = \dfrac{(x+2)(x-4)(x-4)(x+4)}{(x+4)(x-3)(x-4)(x-4)} = \dfrac{x+2}{x-3}$

43. $\dfrac{2}{x+2} - \dfrac{1}{x-2} = \dfrac{-3}{x^2 - 4} \Rightarrow 2(x-2) - 1(x+2) = -3 \Rightarrow x - 6 = -3 \Rightarrow x = 3$

44. $\dfrac{3y}{y^2 + y - 2} = \dfrac{1}{y-1} - 2 \Rightarrow 3y = 1(y+2) - 2(y+2)(y-1) \Rightarrow 3y = y + 2 - 2y^2 - 2y + 4 \Rightarrow$

   $2y^2 + 4y - 6 = 0 \Rightarrow 2(y+3)(y-1) = 0 \Rightarrow y = -3$ or $y = 1.$ The value $y = 1$ is not valid, thus $y = -3.$

45. $P = \dfrac{J + 2z}{J} \Rightarrow JP = J + 2z \Rightarrow JP - J = 2z \Rightarrow J(P-1) = 2z \Rightarrow J = \dfrac{2z}{P-1}$

46. $\dfrac{\dfrac{3}{x^2} + x}{x - \dfrac{3}{x^2}} = \dfrac{\dfrac{3}{x^2} + x}{x - \dfrac{3}{x^2}} \cdot \dfrac{x^2}{x^2} = \dfrac{3 + x^3}{x^3 - 3} = \dfrac{x^3 + 3}{x^3 - 3}$

47. $y = kx \Rightarrow 15 = 3k \Rightarrow k = 5 \Rightarrow y = 5x,$ so when $x = 8,$ $y = 5(8) = 40$

48.
$$
\begin{array}{r}
3x^2 + 6x + 10 \\
x-2\overline{\smash{\big)}\,3x^3 + 0x^2 - 2x - 15} \\
\underline{3x^2 - 6x^2} \\
6x^2 - 2x \\
\underline{6x^2 - 12x} \\
10x - 15 \\
\underline{10x - 20} \\
5
\end{array}
$$

The solution is: $3x^2 + 6x + 10 + \dfrac{5}{x-2}$

49. $\left(\dfrac{x^6}{y^9}\right)^{2/3} = \dfrac{\left(x^6\right)^{2/3}}{\left(y^9\right)^{2/3}} = \dfrac{x^{6 \cdot (2/3)}}{y^{9 \cdot (2/3)}} = \dfrac{x^4}{y^6}$

50. $\sqrt[3]{-x^4} \cdot \sqrt[3]{-x^5} = \sqrt[3]{\left(-x^4\right)\left(-x^5\right)} = \sqrt[3]{x^9} = x^{9/3} = x^3$

51. $\sqrt{5ab} \cdot \sqrt{20ab} = \sqrt{5 \cdot 20 \cdot ab \cdot ab} = \sqrt{100(ab)^2} = \sqrt{(10ab)^2} = 10ab$

52. $2\sqrt{24} - \sqrt{54} = 2\sqrt{4 \cdot 6} - \sqrt{9 \cdot 6} = 2\sqrt{4} \cdot \sqrt{6} - \sqrt{9} \cdot \sqrt{6} = 4\sqrt{6} - 3\sqrt{6} = \sqrt{6}$

53. $\sqrt[3]{a^5 b^4} + 3\sqrt[3]{a^5 b} = \sqrt[3]{(ab)^3 \cdot a^2 b} + 3\sqrt[3]{a^3 \cdot a^2 b} = ab\sqrt[3]{a^2 b} + 3a\sqrt[3]{a^2 b} = (b+3)a\sqrt[3]{a^2 b}$

54. $\left(5+\sqrt{5}\right)\left(5-\sqrt{5}\right)=5^2-\left(\sqrt{5}\right)^2=25-5=20$

55. $P=\dfrac{2}{5-\sqrt{3}}=\dfrac{2}{5-\sqrt{3}}\cdot\dfrac{5+\sqrt{3}}{5+\sqrt{3}}=\dfrac{10+2\sqrt{3}}{25-3}=\dfrac{10+2\sqrt{3}}{22}=\dfrac{2\left(5+\sqrt{3}\right)}{2\cdot11}=\dfrac{5+\sqrt{3}}{11}$

56. For the function to be defined, $x-4>0\Rightarrow x>4$. The interval is $(4,\infty)$.

57. $2(x+1)^2=50\Rightarrow(x+1)^2=25\Rightarrow x+1=\pm\sqrt{25}\Rightarrow x+1=\pm5\Rightarrow x=-6$ or 4

58. $\sqrt{x+6}=x\Rightarrow x+6=x^2\Rightarrow x^2-x-6=0\Rightarrow(x+2)(x-3)=0\Rightarrow x=-2$ or 3

    The value $x=-2$ does not check. The only solution is 3.

59. $(-2+3i)-(-5-2i)=-2-(-5)+3i+2i=3+5i$

60. $\dfrac{3-i}{1+3i}=\dfrac{3-i}{1+3i}\cdot\dfrac{1-3i}{1-3i}=\dfrac{3-9i-i+3i^2}{1-9i^2}=\dfrac{3-10i-3}{1+9}=\dfrac{-10i}{10}=-i$

61. $-\dfrac{b}{2a}=-\dfrac{-12}{2(3)}=\dfrac{12}{6}=2;\ f(2)=3(2)^2-12(2)+13=1.$ The vertex is $(2,1)$.

62. $-\dfrac{b}{2a}=-\dfrac{6}{2(-2)}=\dfrac{6}{4}=\dfrac{3}{2};\ f\left(\dfrac{3}{2}\right)=-2\left(\dfrac{3}{2}\right)^2+6\left(\dfrac{3}{2}\right)-1=\dfrac{7}{2}.$ The vertex is $\left(\dfrac{3}{2},\dfrac{7}{2}\right)$. The maximum value is $\dfrac{7}{2}$.

63. Compared to $y=x^2$, the graph of $f(x)$ is shifted right 3 units and up 2 units.

64. $y=x^2+6x-2\Rightarrow y=\left(x^2+6x+9\right)-2-9\Rightarrow y=(x+3)^2-11.$ The vertex is $(-3,-11)$.

65. $x^2-13x+40=0\Rightarrow(x-5)(x-8)=0\Rightarrow x=5$ or $x=8$

66. $2d^2-5=d\Rightarrow2d^2-d-5=0.$ Let $a=2,b=-1$ and $c=-5$ in the quadratic formula.

    $$d=\dfrac{-(-1)\pm\sqrt{(-1)^2-4(2)(-5)}}{2(2)}=\dfrac{1\pm\sqrt{41}}{4}$$

67. $z^2-4z=-2\Rightarrow z^2-4z+4=-2+4\Rightarrow(z-2)^2=2\Rightarrow z-2=\pm\sqrt{2}\Rightarrow z=2\pm\sqrt{2}$

68. $x^4-10x^2+24=0\Rightarrow\left(x^2-6\right)\left(x^2-4\right)=0\Rightarrow x^2-6=0$ or $x^2-4=0\Rightarrow x=\pm\sqrt{6}$ or $x=\pm2$

69. (a) The graph intersects the x-axis at -1 and 3.

    (b) Because the parabola opens downward, $a<0$.

    (c) Because there are two real solutions, the discriminant is positive.

70. $x^2+5x-14=0\Rightarrow(x+7)(x-2)=0\Rightarrow x=-7$ or $x=2$

    Since the parabola opens upward, the solution is $(-\infty,-7]\cup[2,\infty)$.

71. (a) $g(1)=2(1)+1=3$, then $(f\circ g)(1)=f\left(g(1)\right)=f(3)=(3)^2-2=7$

    (b) $(g\circ f)(x)=g\left(f(x)\right)=g\left(x^2-2\right)=2\left(x^2-2\right)+1=2x^2-4+1=2x^2-3$

72. $f(-4) = f(3) = 6$

73. $f(x) = \dfrac{3}{x} \Rightarrow y = \dfrac{3}{x}$, interchange x and y and solve for y. $x = \dfrac{3}{y} \Rightarrow xy = 3 \Rightarrow y = \dfrac{3}{x} \Rightarrow f^{-1}(x) = \dfrac{3}{x}$

74. $A = 800(1+0.075)^{15} \approx \$2367.10$

75. $\log_3 81 = \log_3 3^4 = 4$

76. $e^{\ln(2x)} = 2x$

77. $\log \dfrac{\sqrt{x}}{y^2} = \log \dfrac{x^{1/2}}{y^2} = \log x^{1/2} - \log y^2 = \dfrac{1}{2} \log x - 2 \log y$

78. $2 \ln x + \ln 5x = \ln x^2 + \ln 5x = \ln \left(x^2 \cdot 5x\right) = \ln \left(5x^3\right)$

79. $6 \log x - 2 = 9 \Rightarrow 6 \log x = 11 \Rightarrow \log x = \dfrac{11}{6} \Rightarrow 10^{\log x} = 10^{11/6} \Rightarrow x \approx 68.13$

80. $2^{3x} = 17 \Rightarrow \log_2 2^{3x} = \log_2 17 \Rightarrow 3x = \dfrac{\log 17}{\log 2} \Rightarrow x = \dfrac{\log 17}{3 \log 2} \approx 1.36$

81. $12{,}000(1+r)^5 = 14{,}600 \Rightarrow (1+r)^5 = \dfrac{73}{60} \Rightarrow 1+r = \sqrt[5]{\dfrac{73}{60}} \Rightarrow r = \sqrt[5]{\dfrac{73}{60}} - 1 \approx 0.04$ or 4%

82. $27.4 \sqrt[3]{W} = 36 \Rightarrow \sqrt[3]{W} = \dfrac{36}{27.4} \Rightarrow W = \left(\dfrac{36}{27.4}\right)^3 \approx 2.27$ pounds

83.   (a) $-\dfrac{b}{2a} = -\dfrac{-975}{2(0.25)} = 1950$

      (b) $f(1950) = 0.25(1950)^2 - 975(1950) + 950{,}845 = 220$ million Btu

84. $\dfrac{x^2}{12} = 350 \Rightarrow x^2 = 4200 \Rightarrow x = \sqrt{4200} \approx 64.8$ miles per hour

85. $8000(1+r)^{45} = 1{,}000{,}000 \Rightarrow (1+r)^{45} = 125 \Rightarrow 1+r = 125^{1/45} \Rightarrow r = 125^{1/45} + 1 \approx 0.113$ or 11.3%

86.   (a) $f(8) = 1.4 \ln 8 + 7 \approx 9.91$ meters per second

      (b) $1.4 \ln x + 7 = 10 \Rightarrow 1.4 \ln x = 3 \Rightarrow \ln x = \dfrac{3}{1.4} \Rightarrow e^{\ln x} = e^{3/1.4} \Rightarrow x \approx 8.52$ meters

Critical Thinking Solutions for Chapter 9

Section 9.2

- Every graph of $y = a^x$ passes through the point $(0, 1)$ because any nonzero base raised to the power 0 equals 1.

- The graph of $y = e^x$ will be between the graphs of $y = 2^x$ and $y = 3^x$ since $2 < e < 3$.

    It is closer to the graph of $y = 3^x$.

Section 9.3

- If the sound level increases by 10 db the intensity increases by a factor of 10.

- $\log_a 1 = 0$ because $a^0 = 1$ for any positive base a.

Section 9.5

- Yes, they are approximately the same since $3 \log x = 3\left(\dfrac{\ln x}{\ln 10}\right) = \dfrac{3}{\ln 10} \cdot \ln x \approx 1.3 \ln x$.

Chapter 10: Conic Sections

10.1: Parabolas and Circles

Concepts

1.   Parabola, ellipse and hyperbola

2.   The first parabola has a vertical axis and the second has a horizontal axis.

3.   No, it does not pass the vertical line test for functions.

4.   See Figure 4.

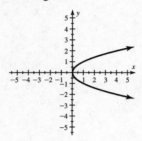

Figure 4

5.   No, it does not pass the vertical line test for functions.

6. (h, k)

7.   left

8.   right

9.   circle; (h, k)

10. $(0, 0)$; r

Parabolas

11.   Since $x = (y-0)^2 + 0,$ the vertex is $(0, 0)$ and the axis of symmetry is $y = 0.$ See Figure 11.

13.   Since $x = (y-0)^2 + 1,$ the vertex is $(1, 0)$ and the axis of symmetry is $y = 0.$ See Figure 13.

15.   Since $x = 2(y-0)^2 + 0,$ the vertex is $(0, 0)$ and the axis of symmetry is $y = 0.$ See Figure 15.

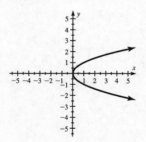   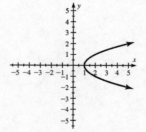   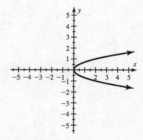

Figure 11   Figure 13   Figure 15

17.   Since $x = (y-1)^2 + 2,$ the vertex is $(2, 1)$ and the axis of symmetry is $y = 1.$ See Figure 17.

19.   Since $y = (x+2)^2 + 1,$ the vertex is $(-2, 1)$ and the axis of symmetry is $x = -2.$ See Figure 19.

21. Since $x = \frac{1}{2}(y+1)^2 - 3$, the vertex is $(-3, -1)$ and the axis of symmetry is $y = -1$. See Figure 21.

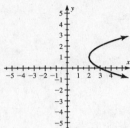

Figure 17

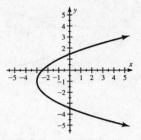

Figure 19

Figure 21

23. Since $x = -3(y-1)^2 + 0$, the vertex is $(0, 1)$ and the axis of symmetry is $y = 1$. See Figure 23.

25. See Figure 25.

$x = -\frac{b}{2a} = -\frac{-1}{2(2)} = \frac{1}{4}$ and $y = 2\left(\frac{1}{4}\right)^2 - \left(\frac{1}{4}\right) + 1 = \frac{7}{8}$. Vertex: $\left(\frac{1}{4}, \frac{7}{8}\right)$. Axis of symmetry: $x = \frac{1}{4}$.

27. See Figure 27.

$y = -\frac{b}{2a} = -\frac{3}{2(-2)} = \frac{3}{4}$ and $x = -2\left(\frac{3}{4}\right)^2 + 3\left(\frac{3}{4}\right) + 2 = \frac{25}{8}$. Vertex: $\left(\frac{25}{8}, \frac{3}{4}\right)$. Axis of symmetry: $y = \frac{3}{4}$.

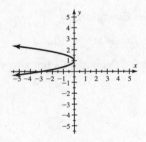

Figure 23

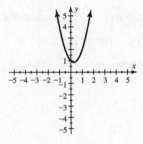

Figure 25

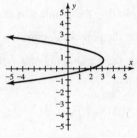

Figure 27

29. See Figure 29.

$y = -\frac{b}{2a} = -\frac{1}{2(3)} = -\frac{1}{6}$ and $x = 3\left(-\frac{1}{6}\right)^2 + \left(-\frac{1}{6}\right) = -\frac{1}{12}$. Vertex: $\left(-\frac{1}{12}, -\frac{1}{6}\right)$. Axis of symmetry: $y = -\frac{1}{6}$.

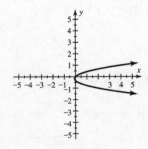

Figure 29

31. See Figure 31.

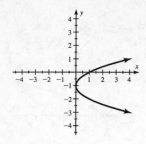

Figure 31

$$y = -\frac{b}{2a} = -\frac{2}{2(1)} = -\frac{2}{2} = -1 \text{ and } x = (-1)^2 + 2(-1) + 1 = 0.$$

Vertex: $(0, -1)$. Axis of symmetry: $y = -1$

33. Since the parabola opens upward and the vertex is $(0, 0)$, the equation has the form $y = a(x-0)^2 + 0.$

   Since the parabola passes through $(1, 1)$, $1 = a(1-0)^2 + 0 \Rightarrow a = 1.$ The equation is $y = x^2.$

35. Since the parabola opens to the right and the vertex is $(-2, -1)$, the equation has the form $x = a(y+1)^2 - 2.$

   Since the parabola passes through $(-1, 0)$, $-1 = a(0+1)^2 - 2 \Rightarrow a = 1.$ The equation is $x = (y+1)^2 - 2.$

37. By plotting the points by hand, we see that the parabola must open upward.

39. By plotting the points and axis by hand, we see that the parabola must open downward.

41. Since the parabola opens to the left and the vertex is $(0, 0)$, the possible x-values are $x \geq 0.$

43. Since the parabola opens to the right and the vertex is to the left of the y-axis, the parabola has two y-intercepts.

45. $x = 3(0)^2 - (0) + 1 \Rightarrow x = 1$

Circles

47. $(x-0)^2 + (y-0)^2 = 1^2 \Rightarrow x^2 + y^2 = 1$

49. $(x-(-1))^2 + (y-5)^2 = 3^2 \Rightarrow (x+1)^2 + (y-5)^2 = 9$

51. $(x-(-4))^2 + (y-(-6))^2 = (\sqrt{2})^2 \Rightarrow (x+4)^2 + (y+6)^2 = 2$

53. Since the center is $(0, 0)$ and the radius is 4, the equation is $x^2 + y^2 = 16.$

55. Since the center is $(-3, 2)$ and the radius is 1, the equation is $(x+3)^2 + (y-2)^2 = 1.$

57. The radius is 3 and the center is $(0, 0)$. Solving the equation for y results in $y = \pm\sqrt{9-x^2}.$ See Figure 57.

59. The radius is 3 and the center is $(1, 3)$. Solving the equation for y results in $y = 3 \pm \sqrt{9-(x-1)^2}.$

   See Figure 59.

61.   The radius is 5 and the center is (–5, 5). Solving the equation for y results in $y = 5 \pm \sqrt{25 - (x+5)^2}$.

See Figure 61.

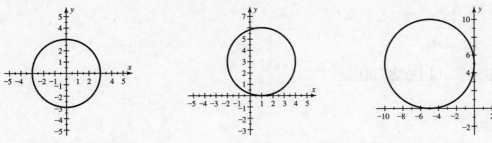

Figure 57   Figure 59   Figure 61

63. $x^2 + 6x + y^2 - 2y = -1 \Rightarrow x^2 + 6x + 9 + y^2 - 2y + 1 = -1 + 9 + 1 \Rightarrow (x+3)^2 + (y-1)^2 = 9$

The radius is 3 and the center is (–3, 1). Solving the equation for y results in $y = 1 \pm \sqrt{9 - (x+3)^2}$.

See Figure 63.

65. $x^2 + 6x + y^2 - 2y + 3 = 0 \Rightarrow x^2 + 6x + 9 + y^2 - 2y + 1 = -3 + 9 + 1 \Rightarrow (x+3)^2 + (y-1)^2 = 7$

The radius is $\sqrt{7}$ and the center is (–3, 1). Solving the equation for y results in $y = 1 \pm \sqrt{7 - (x+3)^2}$.

See Figure 65.

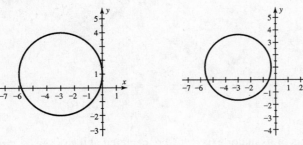

Figure 63   Figure 65

Applications

67.   (a) Graph $Y_1 = (32/11025)X^2$ using the DrawInv feature in [–40, 40, 10] by [–120, 120, 20]. See Figure 67.

(b) When $y = 105$, $x = \dfrac{32}{11,025}(105)^2 = 32$ feet.

69.   (a) Plot (–0.1, 0) and graph $Y_1 = -2.5X^2$ using the DrawInv feature in [–1.5, 1.5, 0.5] by [–1, 1, 0.5]. See Figure 69.

(b) $d = \sqrt{(-2.5 - (-0.1))^2 + (1-0)^2} = \sqrt{(-2.4)^2 + 1^2} = \sqrt{6.76} = 2.6$ A.U. or 241,800,000 miles.

[–40, 40, 10] by [–120, 120, 20]   [–1.5, 1.5, 0.5] by [–1, 1, 0.5]

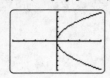

Figure 67

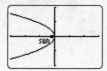

Figure 69

10.2: Ellipses and Hyperbolas

Concepts

1.   See Figure 1.

2.   See Figure 2.

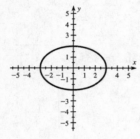

Figure 1

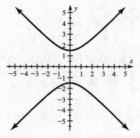

Figure 2

3.   horizontal

4.   vertical

5.   2

6.   4

7.   left and right

8.   lower and upper

9.   They are the diagonals extended.

10.   No, these points could not be the endpoints of its major axis if it were centered at the origin.

Ellipses

11.   The ellipse has a vertical major axis with vertices $(0, \pm 5)$ and minor axis endpoints $(\pm 3, 0)$. See Figure 11.

13.   The ellipse has a horizontal major axis with vertices $(\pm 3, 0)$ and minor axis endpoints $(0, \pm 2)$. See Figure 13.

15.   The ellipse has a vertical major axis with vertices $(0, \pm 2)$ and minor axis endpoints $(\pm 1, 0)$. See Figure 15.

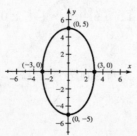

Figure 11

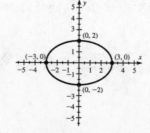

Figure 13

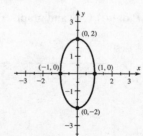

Figure 15

17.   The ellipse has a horizontal major axis with vertices $\left(\pm\sqrt{7}, 0\right)$ and minor axis endpoints $\left(0, \pm\sqrt{5}\right)$. See Figure 17.

19. $36x^2 + 4y^2 = 144 \Rightarrow \dfrac{36x^2}{144} + \dfrac{4y^2}{144} = 1 \Rightarrow \dfrac{x^2}{4} + \dfrac{y^2}{36} = 1$

The ellipse has a vertical major axis with vertices $\left(0, \pm 6\right)$ and minor axis endpoints $\left(\pm 2, 0\right)$. See Figure 19.

21. $6y^2 + 7x^2 = 42 \Rightarrow \dfrac{6y^2}{42} + \dfrac{7x^2}{42} = 1 \Rightarrow \dfrac{y^2}{7} + \dfrac{x^2}{6} = 1$

The ellipse has a vertical major axis with vertices $\left(0, \pm\sqrt{7}\right)$ and minor axis endpoints $\left(\pm\sqrt{6}, 0\right)$. See Figure 21.

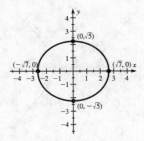
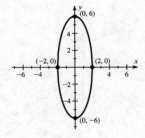
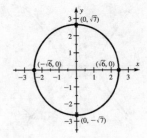

| Figure 17 | Figure 19 | Figure 21 |

23.   Horizontal major axis with vertices $\left(\pm 3, 0\right)$ and minor axis endpoints $\left(0, \pm 2\right)$ $\Rightarrow \dfrac{x^2}{9} + \dfrac{y^2}{4} = 1$

25.   Vertical major axis with vertices $\left(0, \pm 5\right)$ and minor axis endpoints $\left(\pm 4, 0\right)$ $\Rightarrow \dfrac{y^2}{25} + \dfrac{x^2}{16} = 1$

Hyperbolas

27.   The hyperbola has a horizontal transverse axis with vertices $\left(\pm 2, 0\right)$ and asymptotes $y = \pm\dfrac{3}{2}x$. See Figure 27.

29.   The hyperbola has a horizontal transverse axis with vertices $\left(\pm 5, 0\right)$ and asymptotes $y = \pm\dfrac{4}{5}x$. See Figure 29.

31.   The hyperbola has a horizontal transverse axis with vertices $\left(\pm 1, 0\right)$ and asymptotes $y = \pm x$. See Figure 31.

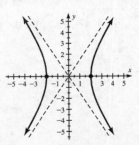
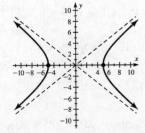
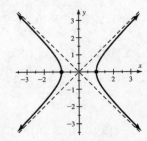

| Figure 27 | Figure 29 | Figure 31 |

33.   The hyperbola has a horizontal transverse axis with vertices $\left(\pm\sqrt{3}, 0\right)$ and asymptotes $y = \pm\dfrac{2}{\sqrt{3}}x$. See Figure 33.

35. $9y^2 - 4x^2 = 36 \Rightarrow \dfrac{9y^2}{36} - \dfrac{4x^2}{36} = 1 \Rightarrow \dfrac{y^2}{4} - \dfrac{x^2}{9} = 1$

The hyperbola has a vertical transverse axis with vertices $(0, \pm 2)$ and asymptotes $y = \pm\dfrac{2}{3}x$. See Figure 35.

37. $16x^2 - 4y^2 = 64 \Rightarrow \dfrac{16x^2}{64} - \dfrac{4y^2}{64} = 1 \Rightarrow \dfrac{x^2}{4} - \dfrac{y^2}{16} = 1$

The hyperbola has a horizontal transverse axis with vertices $(\pm 2, 0)$ and asymptotes $y = \pm 2x$. See Figure 37.

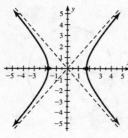
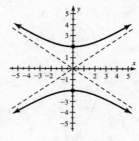
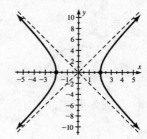

Figure 33   Figure 35   Figure 37

39.   Horizontal transverse axis with vertices $(\pm 1, 0)$ and asymptotes $y = \pm\dfrac{1}{1}x \Rightarrow x^2 - y^2 = 1$

41.   Vertical transverse axis with vertices $(0, \pm 2)$ and asymptotes $y = \pm\dfrac{2}{3}x \Rightarrow \dfrac{y^2}{4} - \dfrac{x^2}{9} = 1$

Applications

43.   (a) $A = \pi(5)(4) \approx 62.83; \ P = 2\pi\sqrt{\dfrac{5^2 + 4^2}{2}} \approx 28.45$

(b) $A = \pi\left(\sqrt{7}\right)\left(\sqrt{2}\right) \approx 11.75; \ P = 2\pi\sqrt{\dfrac{\left(\sqrt{7}\right)^2 + \left(\sqrt{2}\right)^2}{2}} \approx 13.33$

45.   (a) $\dfrac{x^2}{39.44^2} + \dfrac{y^2}{38.20^2} = 1 \Rightarrow \dfrac{y^2}{38.20^2} = 1 - \dfrac{x^2}{39.44^2} \Rightarrow y = \pm\sqrt{38.20^2\left(1 - \dfrac{x^2}{39.44^2}\right)}$

Graph $Y_1 = \sqrt{\left(38.20^2\left(1 - \left(X^2/39.44^2\right)\right)\right)}$ and $Y_2 = -\sqrt{\left(38.20^2\left(1 - \left(X^2/39.44^2\right)\right)\right)}$ and plot the point

(9.82,0) in [–60, 60, 10] by [–40, 40, 10]. See Figure 45.

(b) $P = 2\pi\sqrt{\dfrac{39.44^2 + 38.20^2}{2}} \approx 243.9$ A.U. or about 2.27×10^{10} miles

$A = \pi(39.44)(38.20) \approx 4733$ A.U. or about 4.09×10^{19} square miles

[–60, 60, 10] by [–40, 40, 10]

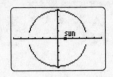

Figure 45

47.   The maximum and minimum heights occur when $y = 0$. These values are calculated below.

Earth: $\dfrac{(x-164)^2}{3960^2} + 0 = 1 \Rightarrow (x-164)^2 = 3960^2 \Rightarrow x - 164 = \pm 3960 \Rightarrow x = 4124$ or $x = -3796$.

Explorer VII: $\dfrac{x^2}{4464^2} + 0 = 1 \Rightarrow x^2 = 4464^2 \Rightarrow x = \pm 4464 \Rightarrow x = 4464$ or $x = -4464$.

The maximum height is $-3796 - (-4464) = 668$ miles. The minimum height is $4464 - 4124 = 340$ miles.

49.   The height is half the length of the minor axis and the width is the full length of the major axis.

$$400x^2 + 10,000y^2 = 4,000,000 \Rightarrow \frac{400x^2}{4,000,000} + \frac{10,000y^2}{4,000,000} = 1 \Rightarrow \frac{x^2}{10,000} + \frac{y^2}{400} = 1$$

The height is $\sqrt{400} = 20$ feet and the width is $2\sqrt{10,000} = 2(100) = 200$ feet.

Checking Basic Concepts for Sections 10.1 & 10.2

1.   See Figure 1. Vertex: $(1, 2)$. Axis of symmetry: $y = 2$.

2.   The equation is $(x-1)^2 + (y+2)^2 = 4$.

$(x-1)^2 + (y+2)^2 = 4 \Rightarrow (y+2)^2 = 4 - (x-1)^2 \Rightarrow y = -2 \pm \sqrt{4 - (x-1)^2}$. See Figure 2.

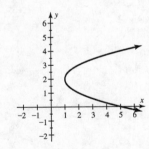

Figure 1   Figure 2

3. x-intercepts: $\dfrac{x^2}{4} + \dfrac{0^2}{9} = 1 \Rightarrow \dfrac{x^2}{4} = 1 \Rightarrow x^2 = 4 \Rightarrow x = \pm 2$

       y-intercepts: $\dfrac{0^2}{4} + \dfrac{y^2}{9} = 1 \Rightarrow \dfrac{y^2}{9} = 1 \Rightarrow y^2 = 9 \Rightarrow y = \pm 3$

4.   (a) This parabola has vertex $(0, 0)$ and axis of symmetry $y = 0$. See Figure 4a.

       (b) This ellipse has a vertical major axis with vertices $(0, \pm 5)$ and minor axis endpoints $(\pm 4, 0)$. See Figure

           4b.

(c) This hyperbola has a horizontal transverse axis with vertices $(\pm 2, 0)$ and asymptotes $y = \pm\dfrac{3}{2}x.$ See Figure 4c.

(d) This circle is centered at $(1, -2)$ and has radius 3. See Figure 4d.

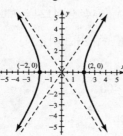

Figure 4a

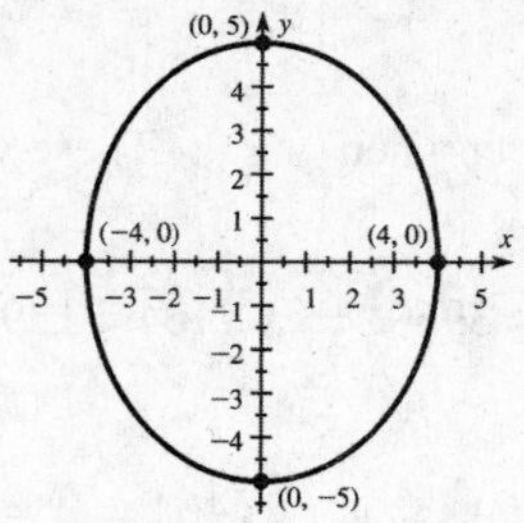

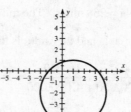

Figure 4b

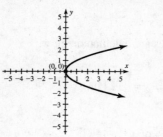

Figure 4c

Figure 4d

10.3: Nonlinear Systems of Equations and Inequalities

Concepts

1.   Any number.

2.   2

3.   Two, the line intersects the circle twice.

4.   All points inside and including a circle of radius 1 centered at the origin.

5.   No. $5(-2)^2 - 2(-1)^2 = 5(4) - 2(1) = 18 \not> 18$

6.   No. $(3)^2 - 2(4) = 9 - 8 = 1 \not\geq 4$

7.   See Figure 7. *Answers may vary.*

8.   See Figure 8. *Answers may vary.*

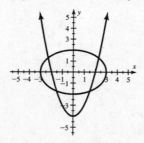

Figure 7

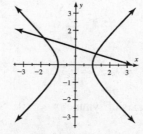

Figure 8

Nonlinear Systems of Equations

9.   The solutions are the intersection points (1, 3) and (–1, –3). Both solutions check.

11. The solutions are the intersection points (0, –1) and (0, 1). Both solutions check.

13. Substitute $y = 2x$ into the second equation.

$$x^2 + (2x)^2 = 45 \Rightarrow x^2 + 4x^2 = 45 \Rightarrow 5x^2 = 45 \Rightarrow x^2 = 9 \Rightarrow x = \pm 3$$

When $x = -3$, $y = 2(-3) = -6$. When $x = 3$, $y = 2(3) = 6$. The solutions are $(-3, -6)$ and $(3, 6)$.

15. From the first equation, $y = 1 - x$. Substitute $y = 1 - x$ into the second equation.

$$x^2 - (1-x)^2 = 3 \Rightarrow x^2 - 1 + 2x - x^2 = 3 \Rightarrow 2x - 1 = 3 \Rightarrow 2x = 4 \Rightarrow x = 2$$

When $x = 2$, $y = 1 - (2) = -1$ The solution is $(2, -1)$.

17. From the first equation, $x^2 = y$. Substitute $x^2 = y$ into the second equation.

$$y + y^2 = 6 \Rightarrow y^2 + y - 6 = 0 \Rightarrow (y+3)(y-2) = 0 \Rightarrow \text{Either } y = -3 \text{ or } y = 2$$

When $y = -3$, $x = \pm\sqrt{-3}$, (not real numbers). When $y = 2$, $x = \pm\sqrt{2}$. The solutions are

$\left(-\sqrt{2}, 2\right)$ and $\left(\sqrt{2}, 2\right)$.

19. Solve the second equation for x : $x = y - 2$. Substitute $x = y - 2$ into the first equation:

$$3(y-2)^2 + 2y^2 = 5 \Rightarrow 3\left(y^2 - 4y + 4\right) + 2y^2 = 5$$

$$\Rightarrow 3y^2 - 12y + 12 + 2y^2 = 5 \Rightarrow 5y^2 - 12y + 7 = 0 \Rightarrow (5y-7)(y-1) = 0$$

$$\Rightarrow y = \frac{7}{5} \text{ or } y = 1 \Rightarrow x = -\frac{3}{5} \text{ or } x = -1. \text{ The solutions are } \left(-\frac{3}{5}, \frac{7}{5}\right) \text{ and } (-1, 1).$$

21. Multiply the second equation by -1 and add to the first equation.

$$\begin{aligned} x^2 + y^2 &= 4 \\ -x^2 + 9y^2 &= -9 \\ \hline 10y^2 &= -5 \end{aligned}$$

$y^2 = -\dfrac{1}{2} \Rightarrow y = \pm\sqrt{-\dfrac{1}{2}}$, which is not a real number. There are no real solutions.

23. Add the equations.

$$\begin{aligned} x^2 + y^2 &= 10 \\ 2x^2 - y^2 &= 17 \\ \hline 3x^2 &= 27 \end{aligned} \qquad \Rightarrow x^2 = 9 \Rightarrow x = \pm 3$$

$$(3)^2 + y^2 = 10 \Rightarrow y^2 = 1 \Rightarrow y = \pm 1$$

$$(-3)^2 + y^2 = 10 \Rightarrow y^2 = 1 \Rightarrow y = \pm 1$$

The solutions are $(\pm 3, \pm 1)$

25. Solve the second equation for y. $2x^2 - y = 1 - 3x \Rightarrow y = 2x^2 + 3x - 1$

Graph $Y_1 = X^2 - 3$ and $Y_2 = 2X^2 + 3X - 1$ in $[-5, 5, 1]$ by $[-5, 5, 1]$. See Figures 25a & 25b.

The solutions are the intersection points $(-2, 1)$ and $(-1, -2)$.

$[-5, 5, 1]$ by $[-5, 5, 1]$

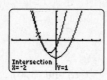

$[-5, 5, 1]$ by $[-5, 5, 1]$

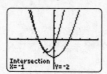

Figure 25a   Figure 25b

27.   Solve both equations for y. $y - x = -4 \Rightarrow y = x - 4$ and $x - y^2 = -2 \Rightarrow y^2 = x + 2 \Rightarrow y = \pm\sqrt{x + 2}$

Graph $Y_1 = X - 4$, $Y_2 = \sqrt{(X + 2)}$ and $Y_3 = -\sqrt{(X + 2)}$ in $[-9.4, 9.4, 1]$ by $[-6.2, 6.2, 1]$. See Figures 27a & 27b.

The solutions are the intersection points $(7, 3)$ and $(2, -2)$.

$[-9.4, 9.4, 1]$ by $[-6.2, 6.2, 1]$

$[-9.4, 9.4, 1]$ by $[-6.2, 6.2, 1]$

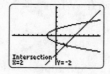

Figure 27a   Figure 27b

29.   (a) Substitute $y = -2x$ into the second equation.

$x^2 + (-2x) = 3 \Rightarrow x^2 - 2x - 3 = 0 \Rightarrow (x + 1)(x - 3) = 0 \Rightarrow$ Either $x = -1$ or $x = 3$

When $x = -1$, $y = -2(-1) = 2$. When $x = 3$, $y = -2(3) = -6$. The solutions are $(-1, 2)$ and $(3, -6)$.

(b) Solve the second equation for y. $x^2 + y = 3 \Rightarrow y = 3 - x^2$

Graph $Y_1 = -2X$ and $Y_2 = 3 - X^2$ in $[-10, 10, 1]$ by $[-10, 10, 1]$. See Figures 29a & 29b.

(c) Table $Y_1 = -2X$ and $Y_2 = 3 - X^2$ with TblStart $= -3$ and ΔTbl $= 1$. See Figure 29c.

$[-10, 10, 1]$ by $[-10, 10, 1]$

$[-10, 10, 1]$ by $[-10, 10, 1]$

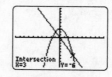

Figure 29a   Figure 29b   Figure 29c

31.   (a) From the second equation, $y = x$. Substitute $y = x$ into the first equation.

$x \cdot x = 1 \Rightarrow x^2 = 1 \Rightarrow x = \pm 1$

When $x = -1$, $y = -1$. When $x = 1$, $y = 1$. The solutions are $(-1, -1)$ and $(1, 1)$.

(b) Solve both equations for y. $xy = 1 \Rightarrow y = \dfrac{1}{x}$ and $x - y = 0 \Rightarrow y = x$

Graph $Y_1 = 1/X$ and $Y_2 = X$ in $[-4.7, 4.7, 1]$ by $[-3.1, 3.1, 1]$. See Figures 31a & 31b.

(c)   Table $Y_1 = 1/X$ and $Y_2 = X$ with TblStart $= -3$ and ΔTbl $= 1$. See Figure 31c.

$[-4.7, 4.7, 1]$ by $[-3.1, 3.1, 1]$ $[-4.7, 4.7, 1]$ by $[-3.1, 3.1, 1]$

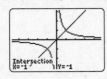   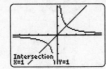   

Figure 31a   Figure 31b   Figure 31c

Nonlinear Systems of Inequalities

33.   Sketch the parabola given by $y = x^2$ using a solid line. Try test point $(0, 2)$ to determine shading.

Since $2 \geq 0^2$, we shade the portion of the *xy*-plane containing the point $(0, 2)$. See Figure 33.

35.   Sketch the ellipse given by $\dfrac{x^2}{4} + \dfrac{y^2}{9} = 1$ using a dashed line. Try test point $(4, 0)$ to determine shading.

Since $\dfrac{4^2}{4} + \dfrac{0^2}{9} = 4 > 1$, we shade the portion of the *xy*-plane containing the point $(4, 0)$. See Figure 35.

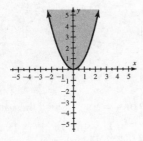   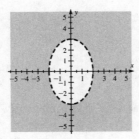

Figure 33   Figure 35

37.   Sketch the parabola given by $y = x^2 + 1$ using a dashed line and the line given by $y = 3$ using a dashed line.

Since the point $(0, 2)$ satisfies both inequalities, we shade the portion of the *xy*-plane containing the point $(0, 2)$. See Figure 37. One solution is $(0, 2)$. *Answers may vary.*

39.   Sketch the circle given by $x^2 + y^2 = 1$ using a solid line and the line given by $y = x$ using a dashed line.

Since the point $\left(\dfrac{1}{2}, -\dfrac{1}{2}\right)$ satisfies both inequalities, we shade the portion of the *xy*-plane containing the point $\left(\dfrac{1}{2}, -\dfrac{1}{2}\right)$. See Figure 39. One solution is $(0.5, -0.5)$. *Answers may vary.*

41.   Sketch the circle given by $x^2 + y^2 = 1$ using a solid line and the circle given by $(x-2)^2 + y^2 = 1$ using a solid line.

Since the point $(1, 0)$ is the only point that satisfies both inequalities, we mark only the point $(1, 0)$ on the graph. See Figure 41. The only solution is $(1, 0)$.

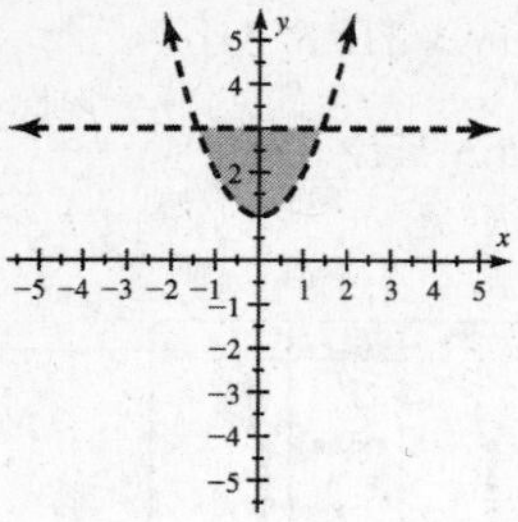

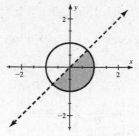

Figure 37

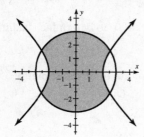

Figure 39

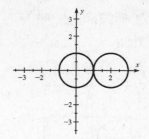

Figure 41

43.   Sketch the hyperbola given by $x^2 - y^2 = 4$ using a solid line and the circle given by $x^2 + y^2 = 9$ using a solid line.

Since the point (0, 0) satisfies both inequalities, we shade the portion of the xy-plane containing the point (0, 0). See Figure 43. One solution is (0, 0). *Answers may vary.*

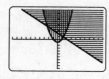

Figure 43

45.   a

47.   A parabola with vertex (0, 0) has an equation of the form $y = a(x-0)^2 + 0$ or $y = ax^2$. Since the parabola also passes through the point (1, 1), $1 = a \cdot 1^2 \Rightarrow a = 1$. The equation of the parabola is $y = x^2$.

Since the line passes through (0, 4) and (4, 0), its slope is $m = \dfrac{0-4}{4-0} = \dfrac{-4}{4} = -1$. The y-intercept is (0, 4).

The equation of the line is $y = -x + 4$ or $y = 4 - x$. Since the line is dashed and the parabola is solid, the system of inequalities is $y \ge x^2$ and $y < 4 - x$.

49.   See Figure 49.

[−10, 10, 1] by [−10, 10, 1]

Figure 49

Applications

51.   The necessary equations are $\pi r^2 h = 40$ and $2\pi r h = 50$. Start by solving each of these equations for h.

$$\pi r^2 h = 40 \Rightarrow h = \frac{40}{\pi r^2} \text{ and } 2\pi r h = 50 \Rightarrow h = \frac{50}{2\pi r}$$

(a) Graph $Y_1 = 40/(\pi X^2)$ and $Y_2 = 50/(2\pi X)$ in [0, 5, 1] by [0, 10, 1]. See Figure 51.

The graphs intersect near the point (1.6, 4.97). The dimensions are $r = 1.6$ inches and $h \approx 4.97$ inches.

(b) $\dfrac{40}{\pi r^2} = \dfrac{50}{2\pi r} \Rightarrow 80\pi r = 50\pi r^2 \Rightarrow 8r = 5r^2 \Rightarrow 5r^2 - 8r = 0 \Rightarrow r(5r - 8) = 0 \Rightarrow r = 0$ or $r = \dfrac{8}{5}$

Since $r = 0$ has no physical meaning, $r = \dfrac{8}{5} = 1.6$. And so $h = \dfrac{50}{2\pi\left(\frac{8}{5}\right)} = \dfrac{125}{8\pi} \approx 4.97$.

The dimensions are $r = 1.6$ inches and $h \approx 4.97$ inches.

53.   (a) $V = \dfrac{1}{3}\pi r^2 h \Rightarrow h = \dfrac{3V}{\pi r^2}$ and $S = \pi r\sqrt{r^2 + h^2} \Rightarrow r^2 + h^2 = \left(\dfrac{S}{\pi r}\right)^2 \Rightarrow h = \sqrt{\left(\dfrac{S}{\pi r}\right)^2 - r^2}$

       (b) Graph $Y_1 = 102/(\pi X^2)$ and $Y_2 = \sqrt{\left((52/(\pi X))^2 - X^2\right)}$ in [0, 5, 1] by [0, 10, 1]. See Figures 53a & 53b.

There are two possibilities: $r \approx 2.02$ ft, $h \approx 7.92$ ft or $r \approx 3.76$ ft and $h \approx 2.30$ ft.

[0, 5, 1] by [0, 10, 1]   [0, 5, 1] by [0, 10, 1]   [0, 5, 1] by [0, 10, 1]

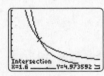   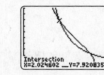   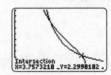

Figure 51   Figure 53a   Figure 53b

Checking Basic Concepts for Section 10.3

1.   Symbolically: From the second equation $y = 2x - 3$. Substitute $y = 2x - 3$ into the first equation.

$x^2 - (2x - 3) = 2x \Rightarrow x^2 - 4x + 3 = 0 \Rightarrow (x - 1)(x - 3) = 0 \Rightarrow x = 1$ or $x = 3$

When $x = 1$, $y = 2(1) - 3 = -1$. When $x = 3$, $y = 2(3) - 3 = 3$. The solutions are (1, –1) and (3, 3).

Graphically: Solve each equation for y. $x^2 - y = 2x \Rightarrow y = x^2 - 2x$ and $2x - y = 3 \Rightarrow y = 2x - 3$

Graph $Y_1 = X^2 - 2X$ and $Y_2 = 2X - 3$ in [–5, 5, 1] by [–5, 5, 1]. See Figures 1a & 1b.

[–5, 5, 1] by [–5, 5, 1]   [–5, 5, 1] by [–5, 5, 1]

Figure 1a   Figure 1b

2.   Since graphing these equations would result in a parabola intersected twice by a line, there are two solutions.

3.   (a) The point (0, 3) is in the shaded region and is a solution. The point (4, 4) is not in the shaded region and is not a solution. *Answers may vary.*

       (b) A parabola with vertex (0, 4) has an equation of the form $y = a(x - 0)^2 + 4$ or $y = ax^2 + 4$. Since the parabola also passes through the point (2, 0), $0 = a \cdot 2^2 + 4 \Rightarrow 4a = -4 \Rightarrow a = -1$. The equation of the parabola is $y = -x^2 + 4$ or $y = 4 - x^2$. Since the line passes through (0, 2) and (2, 0), its slope is

$m = \dfrac{0-2}{2-0} = \dfrac{-2}{2} = -1$. The y-intercept is (0, 2). The equation of the line is $y = -x + 2$ or $y = 2 - x$.

Since both the line and the parabola are solid, the system of inequalities is $y \le 4 - x^2$ and $y \ge 2 - x$.

4.   Sketch the circle given by $x^2 + y^2 = 4$ using a solid line and the line given by $y = 1$ using a dashed line.

The point (0, 0) satisfies both inequalities. Shade the portion of the xy-plane containing (0, 0). See Figure 4.

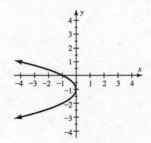

Figure 4

Chapter 10 Review Exercises

Section 10.1

1.   Since $x = 2(y+0)^2 + 0$, the vertex is (0, 0) and the axis of symmetry is $y = 0$. See Figure 1.

2.   Since $x = -(y+1)^2 + 0$, the vertex is (0, –1) and the axis of symmetry is $y = -1$. See Figure 2.

3.   Since $x = -2(y-2)^2 + 0$, the vertex is (0, 2) and the axis of symmetry is $y = 2$. See Figure 3.

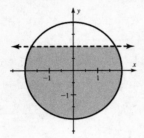 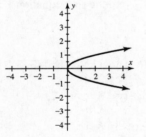 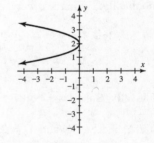

Figure 1   Figure 2   Figure 3

4.   Since $x = (y+2)^2 - 1$, the vertex is (–1, –2) and the axis of symmetry is $y = -2$. See Figure 4.

5.   Since $x = -3(y-0)^2 + 1$, the vertex is (1, 0) and the axis of symmetry is $y = 0$. See Figure 5.

6.   Since $x = \dfrac{1}{2}(y+1)^2 - \dfrac{7}{2}$, the vertex is $\left(-\dfrac{7}{2}, -1\right)$ and the axis of symmetry is $y = -1$. See Figure 6.

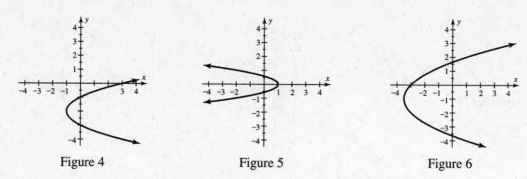

Figure 4   Figure 5   Figure 6

7.   Since the vertex is $(0, 0)$, the parabola has an equation of the form $x = a(x-0)^2 + 0$ or $x = ay^2$.

Since the parabola passes through the point $(1, 1)$, $1 = a(1) \Rightarrow a = 1$. The equation is $x = y^2$.

8.   This is a circle of radius 4 centered at $(-2, 2)$. The equation is $(x+2)^2 + (y-2)^2 = 16$.

9. $(x-0)^2 + (y-0)^2 = 1^2 \Rightarrow x^2 + y^2 = 1$

10. $(x-2)^2 + (y-(-3))^2 = 4^2 \Rightarrow (x-2)^2 + (y+3)^2 = 16$

11.   The radius is 5 and the center is $(0, 0)$. Solving the equation for y results in $y = \pm\sqrt{25 - x^2}$. See Figure 11.

12.   The radius is 3 and the center is $(2, 0)$. Solving the equation for y results in $y = \pm\sqrt{9 - (x-2)^2}$.

See Figure 12.

13.   The radius is $\sqrt{5}$ and the center is $(-3, 1)$. Solving the equation for y results in $y = 1 \pm \sqrt{5 - (x+3)^2}$.

See Figure 13.

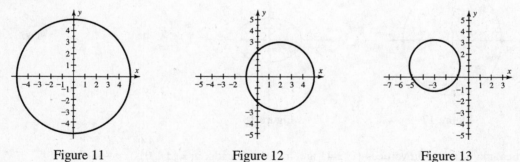

Figure 11   Figure 12   Figure 13

14. $x^2 - 2x + y^2 + 2y = 7 \Rightarrow x^2 - 2x + 1 + y^2 + 2y + 1 = 7 + 1 + 1 \Rightarrow (x-1)^2 + (y+1)^2 = 9$

The radius is 3 and the center is $(1, -1)$. Solving the equation for y results in $y = -1 \pm \sqrt{9 - (x-1)^2}$.

See Figure 14.

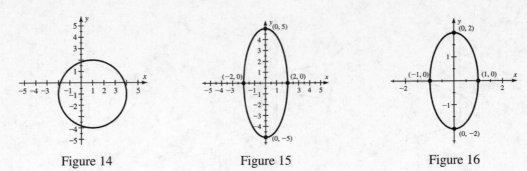

Figure 14	Figure 15	Figure 16

Section 10.2

15. The ellipse has a vertical major axis with vertices $(0, \pm 5)$ and minor axis endpoints $(\pm 2, 0)$. See Figure 15.

16. The ellipse has a vertical major axis with vertices $(0, \pm 2)$ and minor axis endpoints $(\pm 1, 0)$. See Figure 16.

17. $25x^2 + 20y^2 = 500 \Rightarrow \dfrac{25x^2}{500} + \dfrac{20y^2}{500} = 1 \Rightarrow \dfrac{x^2}{20} + \dfrac{y^2}{25} = 1$

   The ellipse has a vertical major axis with vertices $(0, \pm 5)$ and minor axis endpoints $\left(\pm \sqrt{20}, 0\right)$. See Figure 17.

18. $4x^2 + 9y^2 = 36 \Rightarrow \dfrac{4x^2}{36} + \dfrac{9y^2}{36} = 1 \Rightarrow \dfrac{x^2}{9} + \dfrac{y^2}{4} = 1$

   The ellipse has a horizontal major axis with vertices $(\pm 3, 0)$ and minor axis endpoints $(0, \pm 2)$. See Figure 18.

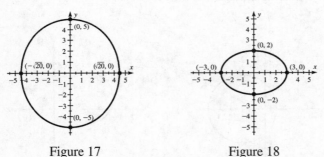

Figure 17	Figure 18

19. Vertical major axis with vertices $(0, \pm 4)$ and minor axis endpoints $(\pm 2, 0)$ $\Rightarrow \dfrac{y^2}{16} + \dfrac{x^2}{4} = 1$

20. Horizontal transverse axis with vertices $(\pm 1, 0)$ and asymptotes $y = \pm 2x$ $\Rightarrow x^2 - \dfrac{y^2}{4} = 1$

21. The hyperbola has a horizontal transverse axis with vertices $(\pm 3, 0)$ and asymptotes $y = \pm \dfrac{2}{3} x$. See Figure 21.

22. The hyperbola has a vertical transverse axis with vertices $(0, \pm 5)$ and asymptotes $y = \pm \dfrac{5}{4} x$. See Figure 22.

23. The hyperbola has a vertical transverse axis with vertices $(0, \pm 1)$ and asymptotes $y = \pm x$. See Figure 23.

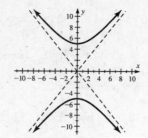

Figure 21

Figure 22

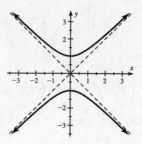

Figure 23

24. $25x^2 - 16y^2 = 400 \Rightarrow \dfrac{25x^2}{400} - \dfrac{16y^2}{400} = 1 \Rightarrow \dfrac{x^2}{16} - \dfrac{y^2}{25} = 1$

The hyperbola has a horizontal transverse axis with vertices $(\pm 4, 0)$ and asymptotes $y = \pm\dfrac{5}{4}x$. See Figure 24.

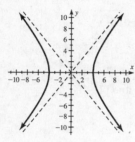

Figure 24

Section 10.3

25. The solutions are the intersection points (0, 3) and (3, 0). Both solutions check.

26. The solutions are the intersection points (–1, –2) and (1, 2). Both solutions check.

27. The solutions are the intersection points (0, 0) and (2, 2). Both solutions check.

28. The solutions are the intersection points (–2, –1), (–2, 1), (2, –1) and (2, 1). All four solutions check.

29. Substitute $y = x$ into the second equation.

$x^2 + (x)^2 = 32 \Rightarrow 2x^2 = 32 \Rightarrow x^2 = 16 \Rightarrow x = \pm 4$

When $x = -4$, $y = -4$. When $x = 4$, $y = 4$. The solutions are (–4, –4) and (4, 4).

30. From the first equation, $y = x - 4$. Substitute $y = x - 4$ into the second equation.

$x^2 + (x-4)^2 = 16 \Rightarrow x^2 + x^2 - 8x + 16 = 16 \Rightarrow 2x^2 - 8x = 0 \Rightarrow 2x(x-4) = 0 \Rightarrow x = 0 \text{ or } x = 4$

When $x = 0$, $y = (0) - 4 = -4$. When $x = 4$, $y = (4) - 4 = 0$. The solutions are (0, –4) and (4, 0).

31. Substitute $y = x^2$ into the second equation.

$2x^2 + (x^2) = 3 \Rightarrow 3x^2 = 3 \Rightarrow x^2 = 1 \Rightarrow x = \pm 1$

When $x = -1$, $y = (-1)^2 = 1$. When $x = 1$, $y = (1)^2 = 1$. The solutions are (–1, 1) and (1, 1).

32. Substitute $y = x^2 + 1$ into the second equation.

$$2x^2 - \left(x^2 + 1\right) = 3x - 3 \Rightarrow x^2 - 3x + 2 = 0 \Rightarrow (x-1)(x-2) = 0 \Rightarrow x = 1 \text{ or } x = 2$$

When $x = 1$, $y = (1)^2 + 1 = 2$. When $x = 2$, $y = (2)^2 + 1 = 5$. The solutions are $(1, 2)$ and $(2, 5)$.

33.   Solve both equations for y. $2x - y = 4 \Rightarrow y = 2x - 4$ and $x^2 + y = 4 \Rightarrow y = 4 - x^2$

Graph $Y_1 = 2X - 4$ and $Y_2 = 4 - X^2$ in $[-8, 8, 1]$ by $[-25, 10, 5]$. See Figures 33a & 33b.

The solutions are the intersection points $(-4, -12)$ and $(2, 0)$.

[-8, 8, 1] by [-25, 10, 5]   [-8, 8, 1] by [-25, 10, 5]

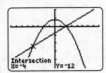   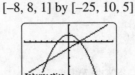

Figure 33a   Figure 33b

34.   Solve both equations for y. $x^2 + y = 0 \Rightarrow y = -x^2$ and $x^2 + y^2 = 2 \Rightarrow y = \pm\sqrt{2 - x^2}$

Graph $Y_1 = -X^2$, $Y_2 = \sqrt{\left(2 - X^2\right)}$ and $Y_3 = -\sqrt{\left(2 - X^2\right)}$ in $[-4.7, 4.7, 1]$ by $[-4.1, 2.1, 1]$.

See Figures 34a and 34b. The solutions are the intersection points $(-1, -1)$ and $(1, -1)$.

[-4.7, 4.7, 1] by [-4.1, 2.1, 1]   [-4.7, 4.7, 1] by [-4.1, 2.1, 1]

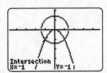   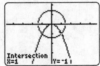

Figure 34a   Figure 34b

35.   (a)   Substitute $y = x$ into the second equation.

$$x^2 + 2(x) = 8 \Rightarrow x^2 + 2x - 8 = 0 \Rightarrow (x+4)(x-2) = 0 \Rightarrow x = -4 \text{ or } x = 2$$

When $x = -4$, $y = -4$. When $x = 2$, $y = 2$. The solutions are $(-4, -4)$ and $(2, 2)$.

(b)   Solve the second equation for y. $x^2 + 2y = 8 \Rightarrow y = 4 - \dfrac{x^2}{2}$

Graph $Y_1 = X$ and $Y_2 = 4 - \left(X^2/2\right)$ in $[-10, 10, 1]$ by $[-10, 10, 1]$. See Figures 35a & 35b.

(c)   Table $Y_1 = X$ and $Y_2 = 4 - \left(X^2/2\right)$ with TblStart $= -2$ and ΔTbl $= 1$. See Figure 35c.

[-10, 10, 1] by [-10, 10, 1]   [-10, 10, 1] by [-10, 10, 1]

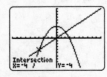      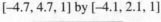

Figure 35a   Figure 35b   Figure 35c

36.   (a)   Substitute $y = x^3$ into the second equation.

$$x^2 - x^3 = 0 \Rightarrow x^2(1-x) = 0 \Rightarrow x = 0 \text{ or } x = 1$$

When $x = 0$, $y = 0^3 = 0$. When $x = 1$, $y = 1^3 = 1$. The solutions are $(0, 0)$ and $(1, 1)$.

(b) Solve the second equation for y. $x^2 - y = 0 \Rightarrow y = x^2$

Graph $Y_1 = X^{\wedge}3$ and $Y_2 = X^2$ in $[-1.5, 1.5, 1]$ by $[-1.5, 1.5, 1]$. See Figures 36a & 36b.

(c) Table $Y_1 = X^{\wedge}3$ and $Y_2 = X^2$ with TblStart = -1 and ΔTbl = 0.5. See Figure 36c.

$[-1.5, 1.5, 1]$ by $[-1.5, 1.5, 1]$ $[-1.5, 1.5, 1]$ by $[-1.5, 1.5, 1]$

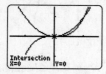   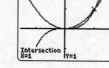   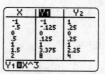

Figure 36a   Figure 36b   Figure 36c

37.   Sketch the parabola given by $y = 2x^2$ using a solid line. Try test point $(0, 2)$ to determine shading.

Since $2 \geq 2(0^2)$, we shade the portion of the xy-plane containing the point $(0, 2)$. See Figure 37.

38.   Sketch the line given by $y = 2x - 3$ using a dashed line. Try test point $(3, 0)$ to determine shading.

Since $0 < 2(3) - 3$, we shade the portion of the xy-plane containing the point $(3, 0)$. See Figure 38.

39.   Sketch the parabola given by $y = -x^2$ using a dashed line. Try test point $(0, -1)$ to determine shading.

Since $-1 < -(0)^2$, we shade the portion of the xy-plane containing the point $(0, -1)$. See Figure 39.

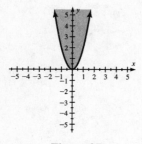   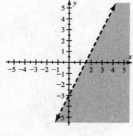   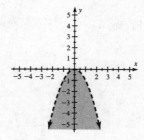

Figure 37   Figure 38   Figure 39

40.   Sketch the ellipse given by $\dfrac{x^2}{9} + \dfrac{y^2}{16} = 1$ using a solid line. Try test point $(0, 0)$ to determine shading.

Since $0 + 0 \leq 1$, we shade the portion of the xy-plane containing the point $(0, 0)$. See Figure 40.

41.   Sketch the parabola given by $y = x^2 + 1$ using a solid line and the line given by $y = 2$ using a solid line.

Since the point $(0, 1.5)$ satisfies both inequalities, we shade the portion of the xy-plane containing the point $(0, 1.5)$.

See Figure 41.

42.   Sketch the parabola given by $y = 4 - x^2$ using a solid line and the line given by $3x + 2y = 6$ using a solid line.

Since the point (1, 2) satisfies both inequalities, we shade the portion of the xy-plane containing the point (1, 2).

See Figure 42.

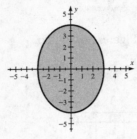

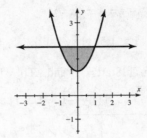

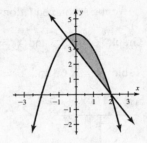

| Figure 40 | Figure 41 | Figure 42 |

43. Sketch the parabola given by $y = x^2$ using a dashed line and the parabola given by $y = 4 - x^2$ using a dashed line.

Since the point (0, 2) satisfies both inequalities, we shade the portion of the xy-plane containing the point (0, 2).

See Figure 43.

44. Sketch the ellipse given by $\dfrac{x^2}{4} + \dfrac{y^2}{9} = 1$ using a dashed line and the circle given by $x^2 + y^2 = 16$ using a dashed line.

Since the point (3, 0) satisfies both inequalities, we shade the portion of the xy-plane containing the point (3, 0).

See Figure 44.

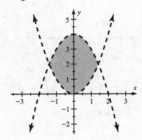

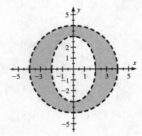

| Figure 43 | Figure 44 |

45. A parabola with vertex (0, –2) has an equation of the form $y = a(x - 0)^2 - 2$ or $y = ax^2 - 2$. Since the parabola also passes through the point (1, –1), $-1 = a \cdot 1^2 - 2 \Rightarrow a = 1$. The equation of the parabola is

$y = x^2 - 2$. Since the line passes through (0, 2) and (2, 0), its slope is $m = \dfrac{0 - 2}{2 - 0} = \dfrac{-2}{2} = -1$.

The y-intercept is (0, 2). The equation of the line is $y = -x + 2$ or $y = 2 - x$. Since both the line and the parabola are solid, the system of inequalities is $y \ge x^2 - 2$ and $y \le 2 - x$.

46. A circle of radius 2 centered at (0, 0) has the equation $x^2 + y^2 = 4$. Since the line passes through the points

$(0, 0)$ and $(1, 1)$, its slope is $m = \dfrac{1-0}{1-0} = \dfrac{1}{1} = 1$. The y-intercept is $(0, 0)$. The equation of the line is $y = x$.

Since both the circle and the line are solid, the system of inequalities is $y \geq x$ and $x^2 + y^2 \leq 4$.

Applications

47. (a) $xy = 1000$ and $2x + 2y = 130$

   (b) Solve each equation for y. $xy = 1000 \Rightarrow y = \dfrac{1000}{x}$ and $2x + 2y = 130 \Rightarrow y = 65 - x$

   Graph $Y_1 = 1000/X$ and $Y_2 = 65 - X$ in $[0, 50, 10]$ by $[0, 50, 10]$. See Figure 47.

   The table has dimensions $x = 25$ inches and $y = 40$ inches.

   (c) From the second equation, $y = 65 - x$. Substitute $y = 65 - x$ into the first equation.

   $x(65 - x) = 1000 \Rightarrow x^2 - 65x + 1000 = 0 \Rightarrow (x - 25)(x - 40) = 0 \Rightarrow x = 25$ or $x = 40$

   When $x = 25$, $y = 65 - 25 = 40$. When $x = 40$, $y = 65 - 40 = 25$. These answers are equivalent.

48. (a) $xy = 60$ and $y - x = 7$

   (b) Solve each equation for y. $xy = 60 \Rightarrow y = \dfrac{60}{x}$ and $y - x = 7 \Rightarrow y = x + 7$

   Graph $Y_1 = 60/X$ and $Y_2 = X + 7$ in $[0, 16, 4]$ by $[0, 16, 4]$. See Figure 48.

   The numbers are $x = 5$ and $y = 12$.

   (c) From the second equation, $y = x + 7$. Substitute $y = x + 7$ into the first equation.

   $x(x + 7) = 60 \Rightarrow x^2 + 7x - 60 = 0 \Rightarrow (x + 12)(x - 5) = 0 \Rightarrow x = -12$ or $x = 5$

   When $x = -12$, $y = -12 + 7 = -5$. This does not fit the question. When $x = 5$, $y = 5 + 7 = 12$.

49. Solve each equation for h. $V = \pi r^2 h \Rightarrow h = \dfrac{V}{\pi r^2}$ and $A = 2\pi rh \Rightarrow h = \dfrac{A}{2\pi r}$

   Graph $Y_1 = 50/\!\left(\pi X^2\right)$ and $Y_2 = 100/(2\pi X)$ in $[0, 5, 1]$ by $[0, 50, 10]$. See Figure 49.

   The unique answer is $r = 1$ foot and $h \approx 15.92$ feet.

$[0, 50, 10]$ by $[0, 50, 10]$ $[0, 16, 4]$ by $[0, 16, 4]$ $[0, 5, 1]$ by $[0, 50, 10]$

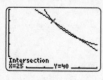
Figure 47

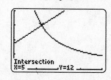
Figure 48

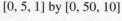
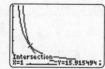
Figure 49

50. Solve each equation for h. $V = \pi r^2 h \Rightarrow h = \dfrac{V}{\pi r^2}$ and $A = 2\pi rh + 2\pi r^2 \Rightarrow h = \dfrac{A - 2\pi r^2}{2\pi r}$

   Graph $Y_1 = 35/\!\left(\pi X^2\right)$ and $Y_2 = \left(80 - 2\pi X^2\right)/(2\pi X)$ in $[0, 4, 1]$ by $[-5, 25, 5]$. See Figures 50a & 50b.

   Two solutions are possible. Either

$r \approx 0.94$ inches and $h \approx 12.60$ inches or $r \approx 3.00$ inches and $h \approx 1.23$ inches. The answer is not unique.

51.   (a) $\dfrac{x^2}{5} + \dfrac{y^2}{12} = 1 \Rightarrow \dfrac{y^2}{12} = 1 - \dfrac{x^2}{5} \Rightarrow y^2 = 12\left(1 - \dfrac{x^2}{5}\right) \Rightarrow y = \pm\sqrt{12\left(1 - \dfrac{x^2}{5}\right)}$

Graph $Y_1 = \sqrt{\left(12\left(1 - X^2/5\right)\right)}$ and $Y_2 = -\sqrt{\left(12\left(1 - X^2/5\right)\right)}$ in $[-7.5, 7.5, 1]$ by $[-5, 5, 1]$. See Figure 51.

(b) $A = \pi\left(\sqrt{12}\right)\left(\sqrt{5}\right) \approx 24.33$ square units and $P = 2\pi\sqrt{\dfrac{5+12}{2}} \approx 18.32$ units

52.   (a) $\dfrac{x^2}{1.524^2} + \dfrac{y^2}{1.517^2} = 1 \Rightarrow \dfrac{y^2}{1.517^2} = 1 - \dfrac{x^2}{1.524^2} \Rightarrow y = \pm\sqrt{1.517^2\left(1 - \dfrac{x^2}{1.524^2}\right)}$

Plot the point $(0.15, 0)$ and Graph $Y_1 = \sqrt{\left(1.517^2\left(1 - X^2/1.524^2\right)\right)}$ and $Y_2 = -\sqrt{\left(1.517^2\left(1 - X^2/1.524^2\right)\right)}$

in $[-3, 3, 1]$ by $[-2, 2, 1]$. See Figure 52.

(b) $P = 2\pi\sqrt{\dfrac{1.524^2 + 1.517^2}{2}} \approx 9.55$ A.U. or about 8.9×10^8 miles

$A = \pi(1.524)(1.517) \approx 7.26$ square A.U. or about 6.3×10^{16} square miles

| $[0, 4, 1]$ by $[-5, 25, 5]$ | $[0, 4, 1]$ by $[-5, 25, 5]$ | $[-7.5, 7.5, 1]$ by $[-5, 5, 1]$ | $[-3, 3, 1]$ by $[-2, 2, 1]$ |

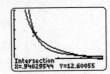 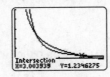 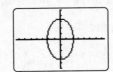 

| Figure 50a | Figure 50b | Figure 51 | Figure 52 |

Chapter 10 Test

1.   Since $x = (y-4)^2 - 2$, the vertex is $(-2, 4)$ and the axis of symmetry is $y = 4$. See Figure 1.

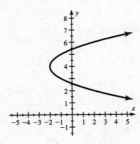

Figure 1

2.   Since the parabola opens to the left and the vertex is $(1, 0)$, the equation has the form $x = a(y-0)^2 + 1$.

Since the parabola passes through $(0, 1)$, $0 = a(1-0)^2 + 1 \Rightarrow a = -1$. The equation is $x = -y^2 + 1$.

3.   Since the center is $(2, -4)$ and the radius is 2, the equation is $(x-2)^2 + (y+4)^2 = 4$.

4. $\left(x - (-5)\right)^2 + (y-2)^2 = 10^2 \Rightarrow (x+5)^2 + (y-2)^2 = 100$

5. $x^2 + 4x + y^2 - 6y = 3 \Rightarrow x^2 + 4x + 4 + y^2 - 6y + 9 = 3 + 4 + 9 \Rightarrow (x+2)^2 + (y-3)^2 = 16$

The radius is 4 and the center is (–2, 3). Solving the equation for y results in $y = 3 \pm \sqrt{16 - (x+2)^2}$.

See Figure 5.

6. The ellipse has a vertical major axis with vertices $(0, \pm 7)$ and minor axis endpoints $(\pm 4, 0)$. See Figure 6.

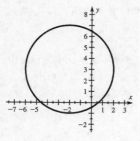

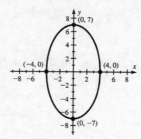

Figure 5   Figure 6

7. Horizontal major axis with vertices $(\pm 10, 0)$ and minor axis endpoints $(0, \pm 8)$ $\Rightarrow \dfrac{x^2}{100} + \dfrac{y^2}{64} = 1$

8. $4x^2 - 9y^2 = 36 \Rightarrow \dfrac{4x^2}{36} - \dfrac{9y^2}{36} = 1 \Rightarrow \dfrac{x^2}{9} - \dfrac{y^2}{4} = 1$

The hyperbola has a horizontal transverse axis with vertices $(\pm 3, 0)$ and asymptotes $y = \pm \dfrac{2}{3}x$. See Figure 8.

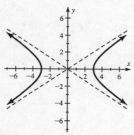

Figure 8

9. The solutions are the intersection points (0, –4) and (4, 0). Both solutions check.

10. From the first equation, $y = x - 3$. Substitute $y = x - 3$ into the second equation.

$x^2 + (x-3)^2 = 17 \Rightarrow x^2 + x^2 - 6x + 9 = 17 \Rightarrow 2(x-4)(x+1) = 0 \Rightarrow x = -1 \text{ or } x = 4$

When $x = -1$, $y = (-1) - 3 = -4$. When $x = 4$, $y = (4) - 3 = 1$. The solutions are (–1, –4) and (4, 1).

11. Solve both equations for y. $2x^2 - y = 4 \Rightarrow y = 2x^2 - 4$ and $x^2 + y = 8 \Rightarrow y = 8 - x^2$

Graph $Y_1 = 2X^2 - 4$, $Y_2 = 8 - X^2$ in [–10, 10, 1] by [–10, 10, 1]. See Figures 11a & 11b.

The solutions are the intersection points $(-2, 4)$ and $(2, 4)$.

[-10, 10, 1] by [-10, 10, 1]   [-10, 10, 1] by [-10, 10, 1]

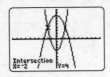 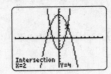

Figure 11a   Figure 11b

12. Sketch the line given by $3x + y = 6$ using a dashed line and the circle given by $x^2 + y^2 = 25$ using a dashed line.

Since the point (4, 0) satisfies both inequalities, we shade the portion of the xy-plane containing the point (4, 0).

See Figure 12.

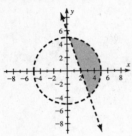

Figure 12

13. A parabola with vertex (0, –4) has an equation of the form $y = a(x - 0)^2 - 4$ or $y = ax^2 - 4$. Since the parabola also passes through the point (2, 0), $0 = a \cdot 2^2 - 4 \Rightarrow a = 1$. The equation of the parabola is $y = x^2 - 4$. The other parabola is a reflection of this parabola across the x-axis. Its equation is $y = 4 - x^2$. Since both the parabolas are solid, the system of inequalities is $y \le 4 - x^2$ and $y \ge x^2 - 4$.

14. (a) $xy = 5000$ and $2x + 2y = 300$

    (b) $\dfrac{5000}{x} = 150 - x \Rightarrow 5000 = 150x - x^2 \Rightarrow x^2 - 150x + 5000 = 0 \Rightarrow (x - 50)(x - 100) = 0 \Rightarrow$

    Either $x = 50$ or $x = 100$. When $x = 50$, $y = 150 - 50 = 100$. When $x = 100$, $y = 150 - 100 = 50$.

    Since the width is shorter than the length, $x = 50$ and $y = 100$. The solution is 50 by 100 feet.

15. From the hint, the two equations to graph are $y = \dfrac{1183}{x^2}$ and $y = \dfrac{702 - x^2}{4x}$. Figures not shown. There are two possible solutions: Either $x \approx 22.08$ and $y \approx 2.43$ or $x \approx 7.29$ and $y \approx 22.24$. The answer is not unique.

16. (a) $\dfrac{x^2}{19.18^2} + \dfrac{y^2}{19.16^2} = 1 \Rightarrow \dfrac{y^2}{19.16^2} = 1 - \dfrac{x^2}{19.18^2} \Rightarrow y = \pm\sqrt{19.16^2\left(1 - \dfrac{x^2}{19.18^2}\right)}$

    Plot the point (0.9, 0) and graph $Y_1 = \sqrt{\left(19.16^2\left(1 - X^2/19.18^2\right)\right)}$ and $Y_2 = -\sqrt{\left(19.16^2\left(1 - X^2/19.18^2\right)\right)}$ in [–30, 30, 10] by [–20, 20, 10]. See Figure 16.

(b) The minimum distance occurs when $x = 19.18$ A.U. The distance is $19.18 - 0.9 = 18.28$ A.U.

This is about 1,700,040,000 miles.

$[-30, 30, 10]$ by $[-20, 20, 10]$

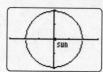

Figure 16

Chapter 10 Extended and Discovery Exercises

1. (a) $x^2 = 4y \Rightarrow x^2 = 4(1)y$. Since $p = 1$, the focus is (0, 1). See Figure 1a.

(b) $y^2 = -8x \Rightarrow y^2 = 4(-2)x$. Since $p = -2$, the focus is (-2, 0). See Figure 1b.

(c) $x = 2y^2 \Rightarrow y^2 = \frac{1}{2}x \Rightarrow y^2 = 4\left(\frac{1}{8}\right)x$. Since $p = \frac{1}{8}$, the focus is $\left(\frac{1}{8}, 0\right)$. See Figure 1c.

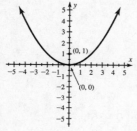   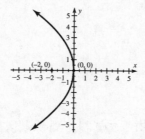   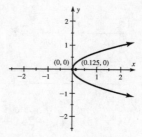

Figure 1a   Figure 1b   Figure 1c

2. (a) Since the vertex of the parabola is (0, 0), the cross section of the dish has an equation of the form

$y = ax^2$.

Since the parabola passes through (150, 44), $44 = a(150)^2 \Rightarrow 44 = 22,500a \Rightarrow a = \dfrac{44}{22,500} = \dfrac{11}{5625}$.

The equation is $y = \dfrac{11}{5625}x^2$.

(b) Writing this equation in the form $x^2 = 4py$ yields $x^2 = 4\left(\dfrac{5625}{44}\right)y$. The focus is located at $\left(0, \dfrac{5625}{44}\right)$.

That is, the focus is $\dfrac{5625}{44} \approx 127.8$ feet from the vertex.

3. (a) This is an ellipse with horizontal major axis, centered at (3, 1). See Figure 3a.

(b) This is an ellipse with vertical major axis, centered at (-1, -2). See Figure 3b.

(c) This is a hyperbola centered at (-1, 3). The asymptotes are $y = \pm\dfrac{3}{2}(x+1) + 3$. See Figure 3c.

(d) This is a hyperbola centered at (-1, 4). The asymptotes are $y = \pm 2(x+1) + 4$. See Figure 3d.

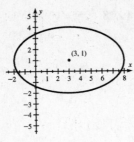

Figure 3a

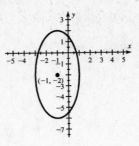

Figure 3b

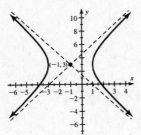

Figure 3c

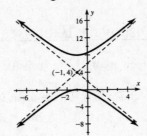

Figure 3d

4.   (a) $\dfrac{(x+3)^2}{16}+\dfrac{(y-5)^2}{4}=1$

   (b) $\dfrac{(y+3)^2}{25}+\dfrac{(x-2)^2}{9}=1$

5.   (a) $9x^2-18x+4y^2+24y+9=0 \Rightarrow 9\left(x^2-2x+1\right)+4\left(y^2+6y+9\right)=-9+9+36\Rightarrow$

   $9(x-1)^2+4(y+3)^2=36 \Rightarrow \dfrac{(x-1)^2}{4}+\dfrac{(y+3)^2}{9}=1 \Rightarrow$ Center: $(1,-3)$

   (b) $25x^2+150x-16y^2+32y-191=0 \Rightarrow 25\left(x^2+6x+9\right)-16\left(y^2-2y+1\right)=191+225-16\Rightarrow$

   $25(x+3)^2-16(y-1)^2=400 \Rightarrow \dfrac{(x+3)^2}{16}-\dfrac{(y-1)^2}{25}=1 \Rightarrow$ Center: $(-3,1)$

Chapters 1-10 Cumulative Review Exercises

1. $K=(4)^2+(-3)^2=16+9=25$

2. $\dfrac{\left(a^{-2}b\right)^2}{a^{-1}\left(b^3\right)^{-2}}=\dfrac{a^{-4}b^2}{a^{-1}b^{-6}}=a^{-4-(-1)}b^{2-(-6)}=a^{-3}b^8=\dfrac{b^8}{a^3}$

3. $7.345\times10^{-3}=0.007345$

4. $f(-4)=\dfrac{-4}{-4-4}=\dfrac{-4}{-8}=\dfrac{1}{2}$; the denominator cannot equal zero, so $x-4\neq0\Rightarrow x\neq4$.

5. $f(3)=4$

6.   See Figure 6.

7. See Figure 7.

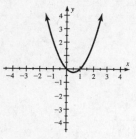

Figure 6   Figure 7

8. A line perpendicular to $y = -\dfrac{2}{3}x + 1$ has slope $m = \dfrac{3}{2}$.

$$y - (-2) = \frac{3}{2}(x-2) \Rightarrow y = \frac{3}{2}x - 3 - 2 \Rightarrow y = \frac{3}{2}x - 5$$

9. $2(1-x) - 4x = x \Rightarrow 2 - 2x - 4x - x = 0 \Rightarrow -7x = -2 \Rightarrow x = \dfrac{2}{7}$

10. $-5 \le 1 - 2x < 3 \Rightarrow -6 \le -2x < 2 \Rightarrow 3 \ge x > -1 \Rightarrow -1 < x \le 3; (-1, 3]$

11. $x^2 - 4 \le 0$; replace the inequality symbol with an equals sign and solve the resulting equation.

$x^2 - 4 = 0 \Rightarrow (x-2)(x+2) = 0 \Rightarrow x = 2$ or $x = -2$.

The graph of $y = x^2 - 4$ is a parabola that lies on or below the x-axis when $x \ge -2$ and $x \le 2$.

The solution is $[-2, 2]$.

12. $|1-x| \ge 2 \Rightarrow 1 - x \ge 2$ or $1 - x \le -2 \Rightarrow -x \ge 1$ or $-x \le -3 \Rightarrow x \le -1$ or $x \ge 3$; The solution is $(-\infty, -1] \cup [3, \infty)$.

13. Add the equations.

$$\begin{aligned} -2x + y &= 1 \\ 5x - y &= 2 \\ \hline 3x \phantom{{}+y} &= 3 \end{aligned} \Rightarrow x = 1.$$ Substitute $x = 1$ into the first equation.

$-2(1) + y = 1 \Rightarrow -2 + y = 1 \Rightarrow y = 3$. The solution is $(1, 3)$.

14. Note that $x + y \le 4 \Rightarrow y \le -x + 4$ and $x - y \ge 2 \Rightarrow y \le x - 2$. See Figure 14.

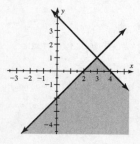

Figure 14

15. $(2x-1)(x+5) = 2x^2 + 10x - x - 5 = 2x^2 + 9x - 5$

16. $xy(2x-3y^2+1) = 2x^2y - 3xy^3 + xy$

17. $6x^2 - 13x - 5 = (3x+1)(2x-5)$

18. $x^3 - 4x = x(x^2-4) = x(x-2)(x+2)$

19. $x^2 + 3x + 2 = 0 \Rightarrow (x+2)(x+1) = 0 \Rightarrow x = -2$ or $x = -1$

20. $x^2 + 1 = -3x \Rightarrow x^2 + 3x + 1 = 0; a = 1, b = 3, c = 1 \quad x = \dfrac{-b \pm \sqrt{b^2-4ac}}{2a} = \dfrac{-3 \pm \sqrt{3^2-4(1)(1)}}{2(1)} = \dfrac{-3 \pm \sqrt{5}}{2}$

21. $\dfrac{x-2}{x+2} \div \dfrac{2x-4}{3x+6} = \dfrac{x-2}{x+2} \cdot \dfrac{3(x+2)}{2(x-2)} = \dfrac{3}{2}$

22. $\dfrac{1}{x+1} + \dfrac{1}{x-1} = \dfrac{1}{x+1} \cdot \dfrac{x-1}{x-1} + \dfrac{1}{x-1} \cdot \dfrac{x+1}{x+1} = \dfrac{x-1}{x^2-1} + \dfrac{x+1}{x^2-1} = \dfrac{x-1+x+1}{x^2-1} = \dfrac{2x}{x^2-1}$

23. $\sqrt{8x^2} = \sqrt{4x^2 \cdot 2} = 2x\sqrt{2}$

24. $8^{2/3} = \left(\sqrt[3]{8}\right)^2 = 2^2 = 4$

25. $\sqrt[3]{2x} \cdot \sqrt[3]{32x^2} = \sqrt[3]{2x \cdot 32x^2} = \sqrt[3]{64x^3} = 4x$

26. $3\sqrt{3x} + \sqrt{12x} = 3\sqrt{3x} + 2\sqrt{3x} = 5\sqrt{3x}$

27. See Figure 27.

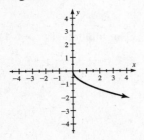

Figure 27

28. $d = \sqrt{(x_2-x_1)^2 + (y_2-y_1)^2} = \sqrt{(-2-2)^2 + (0-(-3))^2} = \sqrt{(-4)^2 + (3)^2} = \sqrt{16+9} = \sqrt{25} = 5$

29. $(2+3i)(2-3i) = (2)^2 - (3i)^2 = 4 - 9i^2 = 4 - 9(-1) = 4 + 9 = 13$

30. $\sqrt{x+2} = x \Rightarrow x+2 = x^2 \Rightarrow x^2 - x - 2 = 0 \Rightarrow (x-2)(x+1) = 0 \Rightarrow x = 2 \ (x = -1 \text{ does not check.})$

31. $x = -\dfrac{b}{2a} = -\dfrac{(-6)}{2(1)} = \dfrac{-6}{-2} = 3; \ y = (3)^2 - 6(3) + 3 \Rightarrow y = 9 - 18 + 3 \Rightarrow y = -6.$ The vertex is $(3, -6)$.

32. $f(x) = x^2 - 2x + 3 \Rightarrow f(x) = (x^2 - 2x + 1) + 3 - 1 \Rightarrow f(x) = (x-1)^2 + 2$

33. The graph of $f(x)$ is shifted 4 units right.

34. $x(3-x) = 2 \Rightarrow 3x - x^2 - 2 = 0 \Rightarrow x^2 - 3x + 2 = 0 \Rightarrow (x-2)(x-1) = 0 \Rightarrow x = 2$ or $x = 1$

35. $x^3 + x = 0 \Rightarrow x(x^2 + 1) = 0 \Rightarrow x = 0$ or $x^2 = -1 \Rightarrow x = \pm\sqrt{-1} \Rightarrow x = \pm i$

36.   (a) $\log 10,000 = 4$ because $10^4 = 10,000$

       (b) $\log_2 8 = 3$

       (c) $\log_3 3^x = x$

       (d) $e^{\ln 6} = 6$

       (e) $\log 2 + \log 50 = \log(2 \cdot 50) = \log(100) = 2$ because $10^2 = 100$

       (f) $\log_2 24 - \log_2(3) = \log_2\left(\dfrac{24}{3}\right) = \log_2(8) = 3$

37.   (a) $(f \circ g)(2) = f(g(2)) = f(2 \cdot 2) = f(4) = 4^2 + 1 = 16 + 1 = 17$

       (b) $(g \circ f)(x) = g(f(x)) = g(x^2 + 1) = 2(x^2 + 1) = 2x^2 + 2$

38.   Let $x = 2 - 3y$. Then $x - 2 = -3y \Rightarrow y = \dfrac{x-2}{-3}$ or $f^{-1}(x) = -\dfrac{1}{3}x + \dfrac{2}{3}$

39. $A = 1000(1 + 0.05)^6 = 1000(1.05)^6 \approx \$1340.10$

40. $\log \dfrac{x^2\sqrt{y}}{z^3} = \log\left(x^2\sqrt{y}\right) - \log\left(z^3\right)$

       $= \log\left(x^2\right) + \log\left(y^{1/2}\right) - \log\left(z^3\right)$

       $= 2\log x + \dfrac{1}{2}\log y - 3\log z$

41. $2e^x - 1 = 17 \Rightarrow 2e^x = 18 \Rightarrow e^x = 9 \Rightarrow \ln e^x = \ln 9 \Rightarrow x = \ln 9$

42. $3 + \log 4x = 5 \Rightarrow \log 4x = 2 \Rightarrow 10^{\log 4x} = 10^2 \Rightarrow 4x = 100 \Rightarrow x = 25$

43. (a)   (b)   (c)   (d)

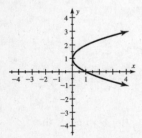   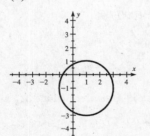   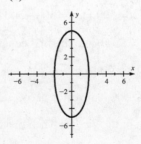   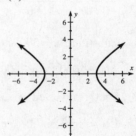

44.   Multiply the first equation by -1 and add to the second equation.

       $-x^2 - y^2 = -1$

       $\underline{x^2 + 9y^2 = 9}$

       $\qquad 8y^2 = 8 \qquad \Rightarrow y^2 = 1 \Rightarrow y = \pm 1$

       $x^2 + (\pm 1)^2 = 1 \Rightarrow x^2 + 1 = 1 \Rightarrow x^2 = 0 \Rightarrow x = 0$

The solutions are $(0, 1)$ and $(0, -1)$.

45.   See Figure 45.

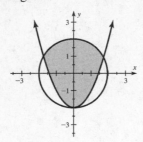

Figure 45

Applications

46.   (a) $D(0) = 400 - 50(0) = 400$; initially, the driver is 400 miles from home.

    (b) $400 - 50x = 0 \Rightarrow 400 = 50x \Rightarrow x = 8$; after 8 hours the driver arrives at home.

    (c) -50; the driver is traveling 50 miles per hour toward home.

47.   Let x, y, z represent the amounts invested at 5%, 6%, and 7%, respectively. The system needed is

$$x + y + z = 2000 \qquad\qquad\qquad x + y + z = 2000$$

$$y = x + 500 \qquad\qquad \Rightarrow \qquad -x + y \qquad = 500$$

$$0.05x + 0.06y + 0.07z = 120: \qquad\qquad 5x + 6y = 7z = 12,000$$

Multiply the first equation by -7 and add to the third equation.

$$-7x - 7y - 7z = -14,000$$
$$\underline{5x + 6y + 7z = 12,000}$$
$$-2x - y = -2000$$

Add this new equation to the second equation.

$$-2x - y = -2000$$
$$\underline{-x + y = 500}$$
$$-3x = -1500 \qquad \Rightarrow x = 500 \Rightarrow y = 500 + 500 = 1000$$

Substitute $x = 500$ and $y = 1000$ into the first equation.

$$500 + 1000 + z = 2000 \Rightarrow z = 500,$$

$500 is invested at 5%, $1000 is invested at 6%, and $500 is invested at 7%.

48.   Let x represent the length of the garden. Then the area is $x(600 - x) = 600x - x^2$. Find the x-value of the

vertex of $y = -x^2 + 600x.$

$$x = -\frac{b}{2a} = -\frac{600}{2(-1)} = \frac{-600}{-2} = 300$$

The dimensions should be 300 feet by 300 feet.

49. $2e^{0.02x} = 4 \Rightarrow e^{0.02x} = 2 \Rightarrow \ln e^{0.02x} = \ln 2 \Rightarrow 0.02x = \ln 2 \Rightarrow x = \dfrac{\ln 2}{0.02} \Rightarrow 50 \ln 2 \approx 34.7$ years

50. $V = \pi r^2 h = 60$ and $S = 2\pi r h = 50$, solve for r

$$\Rightarrow h = \frac{50}{2\pi r}, \text{ substitute into } \pi r^2 h = 60$$

$$\Rightarrow \pi r^2 \left(\frac{50}{2\pi r} \right) = 60 \Rightarrow 25r = 60 \Rightarrow r = \frac{60}{25} = 2.4 \text{ inches}$$

Critical Thinking Solutions for Chapter 10

Section 10.1.

*   The circle is undefined since the radius is $\sqrt{-7}$.

Section 10.2

*   When the nails are moved farther apart the ellipse becomes more flat (ultimately a straight line).

    When the nails are moved closer together the ellipse becomes more circular.

    When the nails come together, a circle would be formed.

*   It is a circle with radius a. We can find the equation of the circle by multiplying both sides of the equation by

    a^2.

*   Minimum distance: about 0.307 A.U. or about 28,551,000 miles

    Maximum distance: about 0.467 A.U. or about 43,431,000 miles

Chapter 11: Sequences and Series

11.1: Sequences

Concepts

1. 1, 2, 3, 4; *Answers may vary.*

2. 1, 3, 5, 7, …; *Answers may vary.*

3. function; natural numbers

4. sequence

5. 6

6. scatterplot

7. $f(2)$

8. a_4

Evaluating and Representing Sequences

9. $f(1) = 1^2 = 1, f(2) = 2^2 = 4, f(3) = 3^2 = 9, f(4) = 4^2 = 16 \Rightarrow 1, 4, 9, 16$

11. $f(1) = \dfrac{1}{1+5} = \dfrac{1}{6}, f(2) = \dfrac{1}{2+5} = \dfrac{1}{7}, f(3) = \dfrac{1}{3+5} = \dfrac{1}{8}, f(4) = \dfrac{1}{4+5} = \dfrac{1}{9} \Rightarrow \dfrac{1}{6}, \dfrac{1}{7}, \dfrac{1}{8}, \dfrac{1}{9}$

13. $f(1) = 5\left(\dfrac{1}{2}\right)^1 = \dfrac{5}{2}, f(2) = 5\left(\dfrac{1}{2}\right)^2 = \dfrac{5}{4}, f(3) = 5\left(\dfrac{1}{2}\right)^3 = \dfrac{5}{8}, f(4) = 5\left(\dfrac{1}{2}\right)^4 = \dfrac{5}{16} \Rightarrow \dfrac{5}{2}, \dfrac{5}{4}, \dfrac{5}{8}, \dfrac{5}{16}$

15. $f(1) = 9, f(2) = 9, f(3) = 9, f(4) = 9 \Rightarrow 9, 9, 9, 9$

17. $a_1 = 1^3 = 1, a_2 = 2^3 = 8, a_3 = 3^3 = 27 \Rightarrow 1, 8, 27$

19. $a_1 = \dfrac{4(1)}{3+1} = 1, a_2 = \dfrac{4(2)}{3+2} = \dfrac{8}{5}, a_3 = \dfrac{4(3)}{3+3} = 2 \Rightarrow 1, \dfrac{8}{5}, 2$

21. $a_1 = 2(1)^2 + 1 - 1 = 2, a_2 = 2(2)^2 + 2 - 1 = 9, a_3 = 2(3)^2 + 3 - 1 = 20 \Rightarrow 2, 9, 20$

23. $a_1 = -2, a_2 = -2, a_3 = -2 \Rightarrow -2, -2, -2$

25. $a_1 = b(1) + c = b + c; a_2 = b(2) + c = 2b + c$

27. $\dfrac{1}{2}(a_1 + a_4) = \dfrac{1}{2}(10 + 4) = \dfrac{1}{2}(14) = 7$

29. The points shown are (1, 3), (2, 4), (3, 5), (4, 3) and (5, 1). The sequence is 3, 4, 5, 3, 1.

31. The points shown are (1, 6), (2, 5), (3, 4), (4, 3), (5, 2) and (6, 1). The sequence is 6, 5, 4, 3, 2, 1.

33. Numerical: See Figure 33a. Graphical: See Figure 33b.

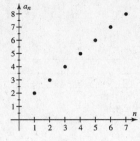

n	1	2	3	4	5	6	7
a_n	2	3	4	5	6	7	8

Figure 33a   Figure 33b

35.   Numerical: See Figure 35a. Graphical: See Figure 35b.

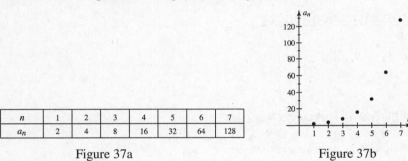

n	1	2	3	4	5	6	7
a_n	0	2	6	12	20	30	42

Figure 35a   Figure 35b

37.   Numerical: See Figure 37a. Graphical: See Figure 37b.

n	1	2	3	4	5	6	7
a_n	2	4	8	16	32	64	128

Figure 37a   Figure 37b

Applications

39.   Symbolic: $a_n = 30n$ for $n = 1, 2, 3, \ldots, 7$

      Graphical: See Figure 39a. Numerical: See Figure 39b.

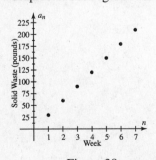

n	1	2	3	4	5	6	7
a_n	30	60	90	120	150	180	210

Figure 39a   Figure 39b

41.   (a) $a_1 = 1^2 = 1$, $a_2 = 2^2 = 4$, $a_3 = 3^2 = 9$, $a_4 = 4^2 = 16 \Rightarrow 1, 4, 9, 16$

      (b) $a_1 = 4(1) = 4$, $a_2 = 4(2) = 8$, $a_3 = 4(3) = 12$, $a_4 = 4(4) = 16 \Rightarrow 4, 8, 12, 16$

43.   (a) After 1 year it is worth $25,000(0.80) = \$20,000$.   After 2 years it is worth $20,000(0.80) = \$16,000$.

(b) $a_n = 25,000(0.8)^n$

(c) See Figure 43.

n	1	2	3	4	5	6	7
a_n	20,000	16,000	12,800	10,240	8192	6553.6	5242.9

Figure 43

45.   (a) $a_n = 2048(0.5)^{n-1}$ for $n = 1, 2, 3, \ldots, 7$

(b) See Figure 45b.

(c) See Figure 45c.

n	1	2	3	4	5	6	7
a_n	2048	1024	512	256	128	64	32

Figure 45b

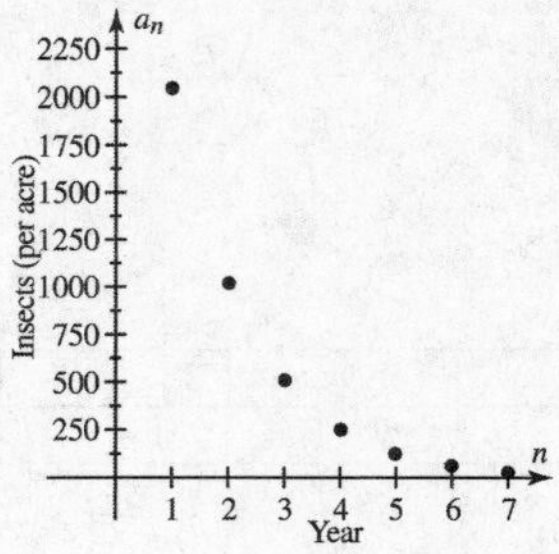

Figure 45c

11.2: Arithmetic and Geometric Sequences

Concepts

1.   linear

2.   exponential

3. $a_n = 3n + 1$; the common difference is 3. *Answers may vary.*

4. $a_n = 5(2)^{n-1}$; the common ratio is 2. *Answers may vary.*

5.   add; previous

6.   multiply; common ratio

7.   19; 4

8.   32; –2

9. $a_n = a_1(r)^{n-1}$

10. $a_n = a_1 + (n-1)d$

Arithmetic Sequences

11.   Yes, the common difference is 10.

13.   Yes, the common difference is –1.

15.   No, there is no common difference.

17.   Yes, the common difference is 3.

19.   Yes, the common difference is –3.

21.   No, there is no common difference.

23.   Yes, the common difference is 1.

25.   No, there is no common difference.

27.   Yes, the common difference is 2.

29. $a_n = 7 + (n-1)(-2) \Rightarrow a_n = 7 - 2n + 2 \Rightarrow a_n = -2n + 9$

31.   Note: $d = \dfrac{6-(-2)}{2} = 4$, thus $a_n = -2 + (n-1)(4) \Rightarrow a_n = -2 + 4n - 4 \Rightarrow a_n = 4n - 6$

33.   Note: $d = \dfrac{8-16}{4} = -2$ and $a_1 = 16 - 7(-2) = 30$, thus

      $a_n = 30 + (n-1)(-2) \Rightarrow a_n = 30 - 2n + 2 \Rightarrow a_n = -2n + 32$

35. $a_{32} = -3 + (32-1)(2) \Rightarrow -3 + (31)(2) = -3 + 62 = 59$

37.   Note: $d = 0 - (-3) = 3$, thus $a_9 = -3 + (9-1)(3) \Rightarrow -3 + (8)(3) = -3 + 24 = 21$

Geometric Sequences

39.   Yes, the common ratio is 3.

41.   Yes, the common ratio is 0.8.

43.   No, there is no common ratio.

45.   Yes, the common ratio is 2.

47.   No, there is no common ratio.

49.   Yes, the common ratio is 4.

51.   Yes, the common ratio is 2.

53.   No, there is no common ratio.

55. $a_n = 1.5(4)^{n-1}$

57.   Note: $r = \dfrac{6}{-3} = -2$, thus $a_n = -3(-2)^{n-1}$

59.   Note: $16 = 1 \cdot r^2 \Rightarrow r^2 = 16 \Rightarrow r = 4 \,(\text{since } r > 0)$, thus $a_n = 1(4)^{n-1}$

61. $a_8 = 2(3)^{8-1} = 2(3)^7 = 4374$

63.   Note: $r = \dfrac{3}{-1} = -3$, thus $a_6 = -1(-3)^{6-1} = -1(-3)^5 = 243$

Applications

65.   (a)   3000, 6000, 9000, 12,000, 15,000; This sequence is arithmetic.

      (b) $a_n = 3000n$

      (c) $a_{20} = 3000(20) = 60,000$; When there are 20 people, the ventilation should be 60,000 cubic feet per

            hour.

      (d)   See Figure 65. The points are collinear.

67.   (a)   Since 20% of the chlorine dissipates, 80% will remain. The function is $a_n = 3(0.8)^{n-1}$.

(b) See Figure 67. The points are not collinear. The sequence is geometric.

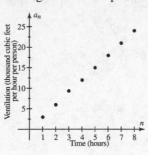

Figure 65   Figure 67

69.   (a) $a_n = 5(0.85)^{n-1}$

      (b) The sequence is geometric and the common ratio is 0.85.

      (c) $a_8 = 5(0.85)^{8-1} = 5(0.85)^7 \approx 1.6$; On the 8th bounce, the ball reaches a maximum height of about 1.6 feet.

71.   (a) The number of seats can be modeled by an arithmetic sequence whose common difference is 2.

      (b) Since $a_1 = 40$ and $d = 2$, $a_n = 40 + 2(n-1)$ or $a_n = 2n + 38$

      (c) $a_{20} = 40 + 2(20-1) = 40 + 2(19) = 40 + 38 = 78$ seats

Checking Basic Concepts for Sections 11.1 & 11.2

1. $a_1 = \dfrac{1}{1+4} = \dfrac{1}{5}, a_2 = \dfrac{2}{2+4} = \dfrac{1}{3}, a_3 = \dfrac{3}{3+4} = \dfrac{3}{7}, a_4 = \dfrac{4}{4+4} = \dfrac{1}{2} \Rightarrow \dfrac{1}{5}, \dfrac{1}{3}, \dfrac{3}{7}, \dfrac{1}{2}$

2.   Graphical: See Figure 2a. Numerical: See Figure 2b.

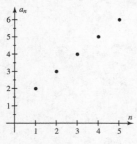

n	1	2	3	4	5
a_n	2	3	4	5	6

Figure 2a   Figure 2b

3.   (a) Arithmetic. Here $d = 1 - (-2) = 3$ and $a_1 = -2$, thus $a_n = -2 + (n-1)(3) \Rightarrow a_n = 3n - 5$

      (b) Geometric. Here $r = \dfrac{-6}{3} = -2$ and $a_1 = 3$, thus $a_n = 3(-2)^{n-1}$

4. $a_n = 5 + (n-1)(2) \Rightarrow a_n = 5 + 2n - 2 \Rightarrow a_n = 2n + 3$

5. $a_n = 5(2)^{n-1}$

11.3: Series

Concepts

1. series

2. $1+2+3+4 = 10$

3. arithmetic (the common difference is 2)

4. geometric (the common ratio is 3)

5. $n\left(\dfrac{a_1 + a_n}{2}\right)$ or $\dfrac{n}{2}\left(2a_1 + (n-1)d\right)$

6. $a_1\left(\dfrac{1 - r^n}{1 - r}\right)$

7. sum

8. $a_1 + a_2 + a_3 + a_4$

9. arithmetic

10. geometric

Sums of Series

11. $6\left(\dfrac{3+13}{2}\right) = 6(8) = 48$

13. $40\left(\dfrac{1+40}{2}\right) = 40(20.5) = 820$

15. $5\left(\dfrac{-7+5}{2}\right) = 5(-1) = -5$

17. Here $r = 3$ and $n = 7$ thus $S_7 = 3\left(\dfrac{1-3^7}{1-3}\right) = 3\left(\dfrac{-2186}{-2}\right) = 3(1093) = 3279$

19. Here $r = -2$ and $n = 8$ thus $S_8 = 1\left(\dfrac{1-(-2)^8}{1-(-2)}\right) = 1\left(\dfrac{-255}{3}\right) = -85$

21. Here $r = 3$ and $n = 6$ thus $S_6 = 0.5\left(\dfrac{1-3^6}{1-3}\right) = 0.5\left(\dfrac{-728}{-2}\right) = 0.5(364) = 182$

23. $S_{20} = 2000\left(\dfrac{(1+0.08)^{20} - 1}{0.08}\right) \approx \$91,523.93$

25. $S_5 = 10,000\left(\dfrac{(1+0.11)^5 - 1}{0.11}\right) \approx \$65,278.01$

Summation Notation

27. $2(1)+2(2)+2(3)+2(4) \Rightarrow 2+4+6+8 = 20$

29. $4+4+4+4+4+4+4+4 = 32$

31. $1^2 + 2^2 + 3^2 + 4^2 + 5^2 + 6^2 + 7^2 \Rightarrow 1+4+9+16+25+36+49 = 140$

33. $\left(4^2-4\right)+\left(5^2-5\right) \Rightarrow 12+20 = 32$

35. $\displaystyle\sum_{k=1}^{6} k^4$

37. $\displaystyle\sum_{k=1}^{5} \frac{1}{k^2}$

39. $\displaystyle\sum_{k=1}^{n} k = n\left(\frac{a_1+a_n}{2}\right) = n\left(\frac{1+n}{2}\right) = \frac{n(n+1)}{2}$

Applications

41.   (a) $8518+9921+10,706+14,035+14,307+12,249$

      (b) $8518+9921+10,706+14,035+14,307+12,249 = 69,736$

43.   (a)   Each air filter removes 80% or 0.8 of the impurities, so 20% or 0.2 passes through it. If we let 100% or 1 represent the amount of impurities entering the first air filter, the amount removed by *n* filters equals

         $(0.8)(1)+(0.8)(0.2)+(0.8)(0.04)+(0.8)(0.008)\ +\cdots+(0.8)(0.2)^{n-1}$. In summation notation,

         $\displaystyle\sum_{k=1}^{n} 0.8(0.2)^{k-1}$

      (b)   To remove 96% or 0.96 of the impurities requires 2 filters because

         $\displaystyle\sum_{k=1}^{2} 0.8(0.2)^{k-1} = (0.8)(1)+(0.8)(0.2) = 0.8+0.16 = 0.96$

45.   (a)   Each successive square has half the area of the square before it. $1, \dfrac{1}{2}, \dfrac{1}{4}, \dfrac{1}{8}, \dfrac{1}{16}$

      (b)   Here $r = \dfrac{1}{2}$ and $n = 10$ thus $S_5 = 1\left(\dfrac{1-\left(\frac{1}{2}\right)^{10}}{1-\left(\frac{1}{2}\right)}\right) = \left(\dfrac{\frac{1023}{1024}}{\frac{1}{2}}\right) = \dfrac{1023}{512}$.

47.   The sum is $14+13+12+11+10+9+8+7+6$. This is an arithmetic series with $a_1 = 14$ and $a_9 = 6$.

      The sum is $S_9 = 9\left(\dfrac{14+6}{2}\right) = 9(10) = 90$ logs.

49.   This is an arithmetic series with $a_1 = 35,000$, $n = 20$ and $d = 2000$

      The sum is $S_{20} = \dfrac{20}{2}\left(2(35,000)+(20-1)(2000)\right) = 10(70,000+38,000) = 10(108,000) = \$1,080,000$.

51.   This is an geometric sequence given by $a_n = 10(0.75)^n$. The distance it falls is $a_4 = 10(0.75)^4 \approx 3.16$ feet.

11.4: The Binomial Theorem

Concepts

1.   5

2. $n+1$

3.   Row 4

4.
$$1$$
$$1 \quad 1$$
$$1 \quad 2 \quad 1$$
$$1 \quad 3 \quad 3 \quad 1$$
$$1 \quad 4 \quad 6 \quad 4 \quad 1$$

5. $4! = 1 \cdot 2 \cdot 3 \cdot 4 = 24$

6. $1 \cdot 2 \cdot 3 \cdot 4 \cdot 5 \cdot 6 = 6! = 720$

7. $\dfrac{n!}{(n-r)!r!}$

8. $a^2 + 2ab + b^2$

9. $\dfrac{n!}{(n-1)!} = \dfrac{n \cdot (n-1)!}{(n-1)!} = n$

10. ${}_nC_n = \dfrac{n!}{(n-n)!n!} = \dfrac{n!}{0!n!} = 1$

Using Pascal's Triangle

11. Row 4 of Pascal's triangle is 1, 3, 3, 1. $(x+y)^3 = x^3 + 3x^2y + 3xy^2 + y^3$

13. Row 5 of Pascal's triangle is 1, 4, 6, 4, 1. $(2x+1)^4 = (2x)^4 + 4(2x)^3(1) + 6(2x)^2(1)^2 + 4(2x)(1)^3 + (1)^4 \Rightarrow$

$(2x+1)^4 = 16x^4 + 32x^3 + 24x^2 + 8x + 1$

15. Row 6 of Pascal's triangle is 1, 5, 10, 10, 5, 1. $(a-b)^5 = a^5 - 5a^4b + 10a^3b^2 - 10a^2b^3 + 5ab^4 - b^5$

17. Row 4 of Pascal's triangle is 1, 3, 3, 1. $(x^2+1)^3 = (x^2)^3 + 3(x^2)^2(1) + 3(x^2)(1)^2 + (1)^3 \Rightarrow$

$(x^2+1)^3 = x^6 + 3x^4 + 3x^2 + 1$

Factorials and Binomial Coefficients

19. $3! = 1 \cdot 2 \cdot 3 = 6$

21. $\dfrac{4!}{3!} = \dfrac{1 \cdot 2 \cdot 3 \cdot 4}{1 \cdot 2 \cdot 3} = 4$

23. $\dfrac{2!}{0!} = \dfrac{1 \cdot 2}{1} = 2$

25. $\dfrac{5!}{2!3!} = \dfrac{1 \cdot 2 \cdot 3 \cdot 4 \cdot 5}{(1 \cdot 2)(1 \cdot 2 \cdot 3)} = 2 \cdot 5 = 10$

27. $_5C_4 = \dfrac{5!}{4!1!} = \dfrac{1 \cdot 2 \cdot 3 \cdot 4 \cdot 5}{(1 \cdot 2 \cdot 3 \cdot 4)(1)} = 5$

29. $_6C_5 = \dfrac{6!}{5!1!} = \dfrac{1 \cdot 2 \cdot 3 \cdot 4 \cdot 5 \cdot 6}{(1 \cdot 2 \cdot 3 \cdot 4 \cdot 5)(1)} = 6$

31. $_4C_0 = \dfrac{4!}{0!4!} = \dfrac{1 \cdot 2 \cdot 3 \cdot 4}{(1)(1 \cdot 2 \cdot 3 \cdot 4)} = 1$

33. $_{12}C_7 = 792$

35. $_9C_5 = 126$

37. $_{19}C_{11} = 75,582$

The Binomial Theorem

39. $(m+n)^3 = \left(_3C_0\right)m^3 + \left(_3C_1\right)m^2 n + \left(_3C_2\right)mn^2 + \left(_3C_3\right)n^3 \Rightarrow (m+n)^3 = m^3 + 3m^2 n + 3mn^2 + n^3$

41. $(x-y)^4 = \left(_4C_0\right)x^4 - \left(_4C_1\right)x^3 y + \left(_4C_2\right)x^2 y^2 - \left(_4C_3\right)xy^3 + \left(_4C_4\right)y^4 \Rightarrow$

    $(x-y)^4 = x^4 - 4x^3 y + 6x^2 y^2 - 4xy^3 + y^4$

43. $(2a+1)^3 = \left(_3C_0\right)(2a)^3 + \left(_3C_1\right)(2a)^2 (1) + \left(_3C_2\right)(2a)(1)^2 + \left(_3C_3\right)(1)^3 \Rightarrow (2a+1)^3 = 8a^3 + 12a^2 + 6a + 1$

45. $(x+2)^5 = \left(_5C_0\right)x^5 + \left(_5C_1\right)x^4 (2) + \left(_5C_2\right)x^3 (2)^2 + \left(_5C_3\right)x^2 (2)^3 + \left(_5C_4\right)x(2)^4 + \left(_5C_5\right)(2)^5 \Rightarrow$

    $(x+2)^5 = x^5 + 10x^4 + 40x^3 + 80x^2 + 80x + 32$

47. $(3+2m)^4 = \left(_4C_0\right)(3)^4 + \left(_4C_1\right)(3)^3 (2m) + \left(_4C_2\right)(3)^2 (2m)^2 + \left(_4C_3\right)(3)(2m)^3 + \left(_4C_4\right)(2m)^4 \Rightarrow$

    $(3+2m)^4 = 81 + 216m + 216m^2 + 96m^3 + 16m^4$

49. $(2x-y)^3 = \left(_3C_0\right)(2x)^3 - \left(_3C_1\right)(2x)^2 y + \left(_3C_2\right)(2x) y^2 - \left(_3C_3\right)y^3 \Rightarrow (2x-y)^3 = 8x^3 - 12x^2 y + 6xy^2 - y^3$

51. Here $r = 0$ and $n = 8$. The first term is $\left(_8C_0\right)a^{8-0}b^0 = a^8$.

53. Here $r = 3$ and $n = 7$. The fourth term is $\left(_7C_3\right)x^{7-3}y^3 = 35x^4 y^3$.

55. Here $r = 0$ and $n = 9$. The first term is $\left(_9C_0\right)(2m)^{9-0} n^0 = 512m^9$.

Checking Basic Concepts for Sections 11.3 & 11.4

1. (a) Geometric. The common ratio is $\dfrac{1}{2}$.

   (b) Arithmetic. The common difference is 2.

2. $12\left(\dfrac{4+48}{2}\right) = 12(26) = 312$

3. Here $r = -2$ and $n = 10$ thus $S_6 = 1\left(\dfrac{1-(-2)^{10}}{1-(-2)}\right) = 1\left(\dfrac{-1023}{3}\right) = -341$

4. Row 5 of Pascal's triangle is 1, 4, 6, 4, 1. $(x-y)^4 = x^4 - 4x^3 y + 6x^2 y^2 - 4xy^3 + y^4$

5. $(x+2)^3 = \left(_3 C_0\right)x^3 + \left(_3 C_1\right)x^2(2) + \left(_3 C_2\right)x(2)^2 + \left(_3 C_3\right)(2)^3 \Rightarrow \left(x^2 - 1\right)^3 = x^3 + 6x^2 + 12x + 8$

Chapter 11 Review Exercises

Section 11.1

1. $f(1) = 1^3 = 1,\ f(2) = 2^3 = 8,\ f(3) = 3^3 = 27,\ f(4) = 4^3 = 64 \Rightarrow 1, 8, 27, 64$

2. $f(1) = 5 - 2(1) = 3,\ f(2) = 5 - 2(2) = 1,\ f(3) = 5 - 2(3) = -1,\ f(4) = 5 - 2(4) = -3 \Rightarrow 3, 1, -1, -3$

3. $f(1) = \dfrac{2(1)}{1^2 + 1} = 1,\ f(2) = \dfrac{2(2)}{2^2 + 1} = \dfrac{4}{5},\ f(3) = \dfrac{2(3)}{3^2 + 1} = \dfrac{3}{5},\ f(4) = \dfrac{2(4)}{4^2 + 1} = \dfrac{8}{17} \Rightarrow 1, \dfrac{4}{5}, \dfrac{3}{5}, \dfrac{8}{17}$

4. $f(1) = (-2)^1 = -2,\ f(2) = (-2)^2 = 4,\ f(3) = (-2)^3 = -8,\ f(4) = (-2)^4 = 16 \Rightarrow -2, 4, -8, 16$

5. The points shown are (1, –2), (2, 0), (3, 4), and (4, 2). The sequence is –2, 0, 4, 2.

6. The points shown are (1, 5), (2, 3), (3, 2), and (4, 1). The sequence is 5, 3, 2, 1.

7. Numerical: See Figure 7a. Graphical: See Figure 7b.

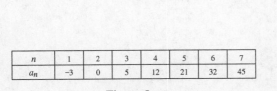

n	1	2	3	4	5	6	7
a_n	2	4	6	8	10	12	14

Figure 7a

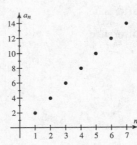

Figure 7b

8. Numerical: See Figure 8a. Graphical: See Figure 8b.

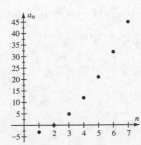

n	1	2	3	4	5	6	7
a_n	–3	0	5	12	21	32	45

Figure 8a

Figure 8b

9. Numerical: See Figure 9a. Graphical: See Figure 9b.

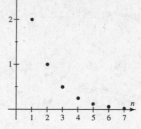

n	1	2	3	4	5	6	7
a_n	2	1	0.5	0.25	0.125	0.0625	0.0313

Figure 9a   Figure 9b

10.   Numerical: See Figure 10a. Graphical: See Figure 10b.

n	1	2	3	4	5	6	7
a_n	1	1.4142	1.7321	2	2.2361	2.4495	2.6458

Figure 10a   Figure 10b

Section 11.2

11.   Yes, the common difference is 5.

12.   No, there is no common difference.

13.   No, there is no common difference.

14.   Yes, the common difference is $-\dfrac{1}{3}$.

15.   Yes, the common difference is –3.

16.   No, there is no common difference.

17.   Yes, the common difference is –1.

18.   No, there is no common difference.

19. $a_n = -3 + (n-1)(4) \Rightarrow a_n = -3 + 4n - 4 \Rightarrow a_n = 4n - 7$

20.   Note: $d = -3 - 2 = -5,$ thus $a_n = 2 + (n-1)(-5) \Rightarrow a_n = 2 - 5n + 5 \Rightarrow a_n = -5n + 7$

21.   Yes, the common ratio is 4.

22.   No, there is no common ratio.

23.   No, there is no common ratio.

24.   Yes, the common ratio is 0.7.

25.   No, there is no common ratio.

26.   Yes, the common ratio is $-\dfrac{1}{3}$.

27.   No, there is no common ratio.

28.   Yes, the common ratio is 2.

29. $a_n = 5(0.9)^{n-1}$

30.   Note: $r = \dfrac{8}{2} = 4$, thus $a_n = 2(4)^{n-1}$

Section 11.3

31. $9\left(\dfrac{4+44}{2}\right) = 9(24) = 216$

32. $5\left(\dfrac{4.5+(-1.5)}{2}\right) = 5(1.5) = 7.5$

33.   Here $r = -4$ and $n = 7$ thus $S_7 = 1\left(\dfrac{1-(-4)^7}{1-(-4)}\right) = \left(\dfrac{16,385}{5}\right) = 3277$

34.   Here $r = \dfrac{1}{2}$ and $n = 9$ thus $S_9 = 1\left(\dfrac{1-\left(\frac{1}{2}\right)^9}{1-\left(\frac{1}{2}\right)}\right) = \left(\dfrac{\frac{511}{512}}{\frac{1}{2}}\right) = \dfrac{511}{256}$

35. $(2(1)+1)+(2(2)+1)+(2(3)+1)+(2(4)+1)+(2(5)+1) \Rightarrow 3+5+7+9+11$

36. $\dfrac{1}{1+1}+\dfrac{1}{2+1}+\dfrac{1}{3+1}+\dfrac{1}{4+1} \Rightarrow \dfrac{1}{2}+\dfrac{1}{3}+\dfrac{1}{4}+\dfrac{1}{5}$

37. $1^3 + 2^3 + 3^3 + 4^3 \Rightarrow 1+8+27+64$

38. $(1-2)+(1-3)+(1-4)+(1-5)+(1-6)+(1-7) \Rightarrow -1+(-2)+(-3)+(-4)+(-5)+(-6)$

39. $\displaystyle\sum_{k=1}^{20} k$

40. $\displaystyle\sum_{k=1}^{20} \dfrac{1}{k}$

41. $\displaystyle\sum_{k=1}^{9} \dfrac{k}{k+1}$

42. $\displaystyle\sum_{k=1}^{7} k^2$

Section 11.4

43.   Row 4 of Pascal's triangle is 1, 3, 3, 1. $(x+4)^3 = x^3 + 3x^2(4) + 3x(4)^2 + (4)^3 \Rightarrow$

        $(x+4)^3 = x^3 + 12x^2 + 48x + 64$

44.   Row 5 of Pascal's triangle is 1, 4, 6, 4, 1. $(2x+1)^4 = (2x)^4 + 4(2x)^3(1) + 6(2x)^2(1)^2 + 4(2x)(1)^3 + (1)^4 \Rightarrow$

        $(2x+1)^4 = 16x^4 + 32x^3 + 24x^2 + 8x + 1$

45.   Row 6 of Pascal's triangle is 1, 5, 10, 10, 5, 1. $(x-y)^5 = x^5 - 5x^4 y + 10x^3 y^2 - 10x^2 y^3 + 5xy^4 - y^5$

46.   Row 7 of Pascal's triangle is 1, 6, 15, 20, 15, 6, 1.

$$(a-1)^6 = a^6 - 6a^5(1) + 15a^4(1)^2 - 20a^3(1)^3 + 15a^2(1)^4 - 6a(1)^5 + (1)^6 \Rightarrow$$

$$(a-1)^6 = a^6 - 6a^5 + 15a^4 - 20a^3 + 15a^2 - 6a + 1$$

47. $3! = 1 \cdot 2 \cdot 3 = 6$

48. $\dfrac{5!}{3!2!} = \dfrac{1 \cdot 2 \cdot 3 \cdot 4 \cdot 5}{(1 \cdot 2 \cdot 3)(1 \cdot 2)} = 2 \cdot 5 = 10$

49. $_6C_3 = \dfrac{6!}{3!3!} = \dfrac{1 \cdot 2 \cdot 3 \cdot 4 \cdot 5 \cdot 6}{(1 \cdot 2 \cdot 3)(1 \cdot 2 \cdot 3)} = 2 \cdot 5 \cdot 2 = 20$

50. $_4C_3 = \dfrac{4!}{3!1!} = \dfrac{1 \cdot 2 \cdot 3 \cdot 4}{(1 \cdot 2 \cdot 3)(1)} = 4$

51. $(m+2)^4 = (_4C_0)m^4 + (_4C_1)m^3(2) + (_4C_2)m^2(2)^2 + (_4C_3)m(2)^3 + (_4C_4)(2)^4 \Rightarrow$

$$(m+2)^4 = m^4 + 8m^3 + 24m^2 + 32m + 16$$

52. $(a+b)^5 = (_5C_0)a^5 + (_5C_1)a^4b + (_5C_2)a^3b^2 + (_5C_3)a^2b^3 + (_5C_4)ab^4 + (_5C_5)b^5 \Rightarrow$

$$(a+b)^5 = a^5 + 5a^4b + 10a^3b^2 + 10a^2b^3 + 5ab^4 + b^5$$

53. $(x-3y)^4 = (_4C_0)x^4 - (_4C_1)x^3(3y) + (_4C_2)x^2(3y)^2 - (_4C_3)x(3y)^3 + (_4C_4)(3y)^4 \Rightarrow$

$$(x-3y)^4 = x^4 - 12x^3y + 54x^2y^2 - 108xy^3 + 81y^4$$

54. $(3x-2)^3 = (_3C_0)(3x)^3 - (_3C_1)(3x)^2(2) + (_3C_2)(3x)(2)^2 - (_3C_3)(2)^3 \Rightarrow (3x-2)^3 = 27x^3 - 54x^2 + 36x - 8$

Applications

55.   Symbolic: $a_n = 45,000(1.10)^{\wedge}(n-1)$ for $n = 1, 2, 3, \ldots, 7$. This is a geometric sequence.

Numerical: See Figure 55a. Graphical: See Figure 55b.

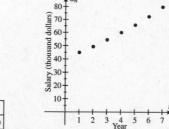

n	1	2	3	4	5	6	7
a_n	45,000	49,500	54,450	59,895	65,885	72,473	79,720

Figure 55a   Figure 55b

56.   Symbolic: $a_n = 45,000 + 5000(n-1)$ for $n = 1, 2, 3, \ldots, 7$. This is an arithmetic sequence.

Numerical: See Figure 56a. Graphical: See Figure 56b.

n	1	2	3	4	5	6	7
a_n	45,000	50,000	55,000	60,000	65,000	70,000	75,000

Figure 56a

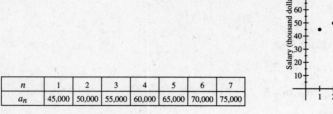

Figure 56b

57.   Symbolic: $a_n = 49n$ for $n = 1, 2, 3, \ldots, 7$

Graphical: See Figure 57a. Numerical: See Figure 57b.

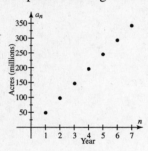

Figure 57a

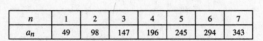

n	1	2	3	4	5	6	7
a_n	49	98	147	196	245	294	343

Figure 57b

58.   (a) $a_n = 1087(1.025)^{n-1}$

(b)   The sequence is geometric. The common ratio is 1.025.

(c) $a_5 = 1087(1.025)^{5-1} \approx 1200$; The average mortgage payment in 2000 was about \$1200.

(d)   See Figure 58.

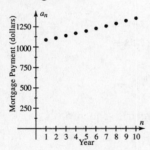

Figure 58

Chapter 11 Test

1. $f(1) = \dfrac{1^2}{1+1} = \dfrac{1}{2}, f(2) = \dfrac{2^2}{2+1} = \dfrac{4}{3}, f(3) = \dfrac{3^2}{3+1} = \dfrac{9}{4}, f(4) = \dfrac{4^2}{4+1} = \dfrac{16}{5} \Rightarrow \dfrac{1}{2}, \dfrac{4}{3}, \dfrac{9}{4}, \dfrac{16}{5}$

2.   The points shown are $(1, -3), (2, 2), (3, 1), (4, -2)$ and $(5, 3)$. The sequence is $-3, 2, 1, -2, 3$.

3.   See Figure 3.

n	1	2	3	4	5	6	7
a_n	0	2	6	12	20	30	42

Figure 3

4.   Row 5 of Pascal's triangle is 1, 4, 6, 4, 1. $(2x-1)^4 = (2x)^4 - 4(2x)^3(1) + 6(2x)^2(1)^2 - 4(2x)(1)^3 + (1)^4 \Rightarrow$

$$(2x-1)^4 = 16x^4 - 32x^3 + 24x^2 - 8x + 1$$

5.   The sequence is arithmetic. The common difference is -3.

6.   The sequence is geometric. The common ratio is -2.

7. $a_n = 2 + (n-1)(-3) \Rightarrow a_n = 2 - 3n + 3 \Rightarrow a_n = -3n + 5$

8.   Note: $2 \cdot r^2 = 4.5 \Rightarrow r^2 = \dfrac{4.5}{2} \Rightarrow r = 1.5$, thus $a_n = 2(1.5)^{n-1}$

9.   Yes, the common ratio is 2.5.

10.   No, there is no common ratio.

11. $9\left(\dfrac{-1+23}{2}\right) = 9(11) = 99$

12.   Here $r = -\dfrac{2}{3}$ and $n = 7$ thus $S_9 = 1\left(\dfrac{1-\left(-\frac{2}{3}\right)^7}{1-\left(-\frac{2}{3}\right)}\right) = \left(\dfrac{\frac{2315}{2187}}{\frac{5}{3}}\right) = \dfrac{463}{729}$

13. $3(2) + 3(3) + 3(4) + 3(5) + 3(6) + 3(7) \Rightarrow 6 + 9 + 12 + 15 + 18 + 21$

14. $\displaystyle\sum_{k=1}^{60} k^3$

15. $\dfrac{7!}{4!3!} = \dfrac{1 \cdot 2 \cdot 3 \cdot 4 \cdot 5 \cdot 6 \cdot 7}{(1 \cdot 2 \cdot 3 \cdot 4)(1 \cdot 2 \cdot 3)} = 5 \cdot 7 = 35$

16. $_5C_3 = \dfrac{5!}{3!2!} = \dfrac{1 \cdot 2 \cdot 3 \cdot 4 \cdot 5}{(1 \cdot 2 \cdot 3)(1 \cdot 2)} = 2 \cdot 5 = 10$

17.   This is an arithmetic series with $a_1 = 50$ and $d = 7$. The total number of seats is

$$S_{45} = \dfrac{45}{2}(2(50) + (45-1)(7)) = 22.5(408) = 9180 \text{ seats.}$$

18.   Symbolic: $a_n = 159,700(1.04)\wedge(n-1)$ for $n = 1, 2, 3, \ldots 7$. This is a geometric sequence.

Numerical: See Figure 18a. Graphical: See Figure 18b.

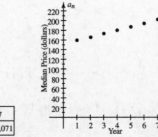

n	1	2	3	4	5	6	7
a_n	159,700	166,088	172,732	179,641	186,826	194,299	202,071

Figure 18a   Figure 18b

19.   (a) $a_n = 2000(2)^{n-1}$

(b) The sequence is geometric. The common ratio is 2.

(c) $a_6 = 2000(2)^{6-1} \approx 64,000$; After 30 days, there are 64,000 worms.

(d) See Figure 19.

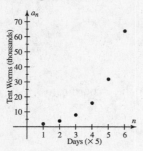

Figure 19

Chapter 11 Extended and Discovery Exercises

1.   Each term of the sequence is obtained by adding the two previous terms. 1, 1, 2, 3, 5, 8, 13, 21, 34, 55, 89,

144

2.   (a) Table $\mathrm{u}(n) = 2.85\mathrm{u}(n-1) - 0.19\left(\mathrm{u}(n-1)\right)^2$ with TblStart = 1 and ΔTbl = 1. See Figure 2a.

Note: be sure to set nMin = 1 and u(nMin) = {1}.

(b) Graph $\mathrm{u}(n) = 2.85\mathrm{u}(n-1) - 0.19\left(\mathrm{u}(n-1)\right)^2$ in [0, 22, 2] by [0, 11, 1]. See Figure 2b.

Note: be sure to set nMin = 1 and nMax = 20 in the WINDOW settings.

The moth population increases and then oscillates until it settles to a constant number (about 9.737

thousand).

[0, 22, 2] by [0, 11, 1]

   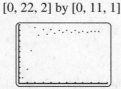

Figure 2a   Figure 2b

3.   (a) $\pi \approx \left[90\left(\dfrac{1}{1^4} + \dfrac{1}{2^4} + \dfrac{1}{3^4} + \dfrac{1}{4^4}\right)\right]^{1/4} \approx 3.138997889$

(b) $\pi \approx 3.141590776$; This is correct to 5 decimal places.

4.   (a) $S = \dfrac{2}{1-\left(-\frac{1}{2}\right)} = \dfrac{2}{\frac{3}{2}} = \dfrac{4}{3}$

(b) $S = \dfrac{1}{1-\left(\frac{1}{3}\right)} = \dfrac{1}{\frac{2}{3}} = \dfrac{3}{2}$

(c) $S = \dfrac{0.1}{1-(0.1)} = \dfrac{0.1}{0.9} = \dfrac{1}{9} = 0.\overline{1}$

(d) $S = \dfrac{0.12}{1-(0.01)} = \dfrac{0.12}{0.99} = \dfrac{4}{33} = 0.\overline{12}$

Chapters 1-11 Cumulative Review Exercises

1. This equation illustrates a distributive property.

2. The domain corresponds to the x-coordinates and the range corresponds to the y-coordinates.

   $D = \{-6, -2, 0, 2\}; R = \{0, 1, 3, 5\}$

3. $\dfrac{x^{-2}y^3}{\left(3xy^{-2}\right)^3} = \dfrac{x^{-2}y^3}{3^3 x^3 \left(y^{-2}\right)^3} = \dfrac{x^{-2}y^3}{27x^3 y^{-6}} = \dfrac{1}{27} x^{-2-3} y^{3-(-6)} = \dfrac{1}{27} x^{-5} y^9 = \dfrac{y^9}{27x^5}$

4. $\left(\dfrac{3b}{6a^2}\right)^{-4} = \left(\dfrac{b}{2a^2}\right)^{-4} = \left(\dfrac{2a^2}{b}\right)^4 = \dfrac{\left(2a^2\right)^4}{b^4} = \dfrac{2^4 \left(a^2\right)^4}{b^4} = \dfrac{16a^8}{b^4}$

5. $\left(\dfrac{1}{z^2}\right)^{-5} = \left(z^2\right)^5 = z^{10}$

6. $\dfrac{8x^{-3}y^2}{4x^3 y^{-1}} = 2x^{-3-3} y^{2-(-1)} = 2x^{-6}y^3 = \dfrac{2y^3}{x^6}$

7. The function is defined for all values of the variable except 8. The domain is $\{x \mid x \neq 8\}$.

8. The slope is $m = \dfrac{1-5}{0-(-2)} = \dfrac{-4}{2} = -2$. Since $f(x) = 1$ when $x = 0$ the y-intercept is 1. Here $f(x) = -2x+1$.

9. Horizontal lines have equations of the form $y = k$. The equation of the horizontal line passing through (2, 3) is $y = 3$.

10. The equation is in the form $f(x) = mx + b$. The slope is -3 and the y-intercept is 5.

11. Since the line is perpendicular to $y = -\dfrac{2}{3}x - 4$, the slope is $m = \dfrac{3}{2}$. Using the point-slope form gives

    $y = \dfrac{3}{2}(x-1) + 4 \Rightarrow y = \dfrac{3}{2}x - \dfrac{3}{2} + 4 \Rightarrow y = \dfrac{3}{2}x + \dfrac{5}{2}$

12. Since the line is parallel to $y = 2x - 7$, the slope is $m = 2$. Using the point-slope form gives

    $y = 2(x-5) + 2 \Rightarrow y = 2x - 10 + 2 \Rightarrow y = 2x - 8$

13. $\dfrac{2}{5}(x-4) = -12 \Rightarrow x - 4 = -30 \Rightarrow x = -26$

14. $\dfrac{2}{5}z + \dfrac{1}{4}z > 2 - (z-1) \Rightarrow 8z + 5z > 40 - 20(z-1) \Rightarrow 13z > 40 - 20z + 20 \Rightarrow$

$13z > 60 - 20z \Rightarrow 33z > 60 \Rightarrow z > \dfrac{60}{33} \Rightarrow z > \dfrac{20}{11}.$ The interval is $\left(\dfrac{20}{11}, \infty \right).$

15. First divide each side of $-3|t-5| \le -18$ by -3 to obtain $|t-5| \ge 6.$

 The solutions to $|t-5| \ge 6$ satisfy $t \le c$ or $t \ge d$ where c and d are the solutions to $|t-5| = 6.$

 $|t-5| = 6$ is equivalent to $t-5 = -6 \Rightarrow t = -1$ and $t-5 = 6 \Rightarrow t = 11.$

 The interval is $(-\infty, -1] \cup [11, \infty).$

16. $\left| 4 + \dfrac{2}{3}x \right| = 6 \Rightarrow 4 + \dfrac{2}{3}x = -6 \Rightarrow \dfrac{2}{3}x = -10 \Rightarrow x = -15$ or $4 + \dfrac{2}{3}x = 6 \Rightarrow \dfrac{2}{3}x = 2 \Rightarrow x = 3$

17. $\dfrac{1}{4}t - (2t+5) + 6 = \dfrac{t+3}{4} \Rightarrow \dfrac{1}{4}t - 2t + 1 = \dfrac{t+3}{4} \Rightarrow t - 8t + 4 = t + 3 \Rightarrow -8t = -1 \Rightarrow t = \dfrac{1}{8}$

18. $-3 \le \dfrac{2}{3}x + 5 < 11 \Rightarrow -8 \le \dfrac{2}{3}x < 6 \Rightarrow -12 \le x < 9.$ The interval is $[-12, 9).$

19. By substitution, $(3, -2)$ is a solution to the given system of equations.

20. See Figure 20.

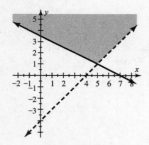

 Figure 20

21. Multiply the first equation by 2 and add the equations to eliminate the variable x.

$$2x - 4y = 2$$
$$-2x + 7y = 4$$
$$\overline{3y = 6}$$

 Thus, $y = 2.$ And so $x - 2(2) = 1 \Rightarrow x = 5.$ The solution is $(5, 2).$

22. $\begin{bmatrix} 1 & 1 & 1 & | & 5 \\ -2 & -1 & 1 & | & -10 \\ 1 & 2 & 8 & | & 1 \end{bmatrix} \begin{matrix} \\ R_2 + 2R_1 \rightarrow \\ R_3 - R_1 \rightarrow \end{matrix} \begin{bmatrix} 1 & 1 & 1 & | & 5 \\ 0 & 1 & 3 & | & 0 \\ 0 & 1 & 7 & | & -4 \end{bmatrix} \begin{matrix} R_1 - R_2 \rightarrow \\ \\ R_3 - R_2 \rightarrow \end{matrix} \begin{bmatrix} 1 & 0 & -2 & | & 5 \\ 0 & 1 & 3 & | & 0 \\ 0 & 0 & 4 & | & -4 \end{bmatrix}$

 $\begin{matrix} \\ \\ (1/4)R_3 \rightarrow \end{matrix} \begin{bmatrix} 1 & 0 & -2 & | & 5 \\ 0 & 1 & 3 & | & 0 \\ 0 & 0 & 1 & | & -1 \end{bmatrix} \begin{matrix} R_1 + 2R_3 \rightarrow \\ R_2 - 3R_3 \rightarrow \\ \end{matrix} \begin{bmatrix} 1 & 0 & 0 & | & 3 \\ 0 & 1 & 0 & | & 3 \\ 0 & 0 & 1 & | & -1 \end{bmatrix}$

 The solution is $(3, 3, -1).$

23. From the graph of the region of feasible solutions (not shown), the vertices are (0, 0), (0, 2.5), (2, 2), and (2.5, 0).

 The maximum value of R occurs at one of the vertices. For $(0, 0)$, $R = 3(0) + 8(0) = 0.$

For $(0, 2.5)$, $R = 3(0) + 8(2.5) = 20$. For $(2, 2)$, $R = 3(2) + 8(2) = 22$. For

$(2.5, 0)$, $R = 3(2.5) + 8(0) = 7.5$.

The maximum value is $R = 22$.

24. $\det A = 4(2) - 3(-3) = 8 + 9 = 17$

25. $2x^3\left(4x^4 - 3x^3 + 5\right) = 8x^7 - 6x^6 + 10x^3$

26. $(2z - 7)(3z + 4) = 6z^2 + 8z - 21z - 28 = 6z^2 - 13z - 28$

27. $4x^2 - 9y^2 = (2x)^2 - (3y)^2 = (2x - 3y)(2x + 3y)$

28. $2a^3 - a^2 + 8a - 4 = a^2(2a - 1) + 4(2a - 1) = \left(a^2 + 4\right)(2a - 1)$

29. $4x^2 - x - 3 = 0 \Rightarrow (4x + 3)(x - 1) = 0 \Rightarrow x = -\dfrac{3}{4}$ or $x = 1$

30. $x^4 - 10x^3 = -24x^2 \Rightarrow x^4 - 10x^3 + 24x^2 = 0 \Rightarrow x^2(x - 4)(x - 6) = 0 \Rightarrow x = 0, 4,$ or 6

31. $\dfrac{x^2 - 7x + 10}{x^2 - 25} \cdot \dfrac{x + 5}{x + 1} = \dfrac{(x - 2)(x - 5)(x + 5)}{(x - 5)(x + 5)(x + 1)} = \dfrac{x - 2}{x + 1}$

32. $\dfrac{x^2 + 7x + 12}{x^2 - 9} \div \dfrac{x^2 - 5x + 6}{(x - 3)^2} = \dfrac{x^2 + 7x + 12}{x^2 - 9} \cdot \dfrac{(x - 3)^2}{x^2 - 5x + 6} = \dfrac{(x + 4)(x + 3)(x - 3)(x - 3)}{(x - 3)(x + 3)(x - 3)(x - 2)} = \dfrac{x + 4}{x - 2}$

33. $\dfrac{2}{x + 5} = \dfrac{-3}{x^2 - 25} + \dfrac{1}{x - 5} \Rightarrow 2(x - 5) = -3 + 1(x + 5) \Rightarrow 2x - 10 = x + 2 \Rightarrow x = 12$

34. $\dfrac{2y}{y^2 - 3y + 2} = \dfrac{1}{y - 2} + 2 \Rightarrow 2y = 1(y - 1) + 2(y - 2)(y - 1) \Rightarrow 2y = y - 1 + 2y^2 - 6y + 4 \Rightarrow$

$2y^2 - 7y + 3 = 0 \Rightarrow (2y - 1)(y - 3) = 0 \Rightarrow y = \dfrac{1}{2}$ or $y = 3$.

35. $R = \dfrac{3C - 2W}{5} \Rightarrow 5R = 3C - 2W \Rightarrow 5R - 3C = -2W \Rightarrow \dfrac{5R - 3C}{-2} = W \Rightarrow W = \dfrac{3C - 5R}{2}$

36. $\dfrac{\dfrac{1}{x^2} + \dfrac{2}{x}}{\dfrac{1}{x^2} - \dfrac{4}{x}} = \dfrac{\dfrac{1}{x^2} + \dfrac{2}{x}}{\dfrac{1}{x^2} - \dfrac{4}{x}} \cdot \dfrac{x^2}{x^2} = \dfrac{1 + 2x}{1 - 4x}$

37. $\sqrt[3]{x^4 y^4} - 2\sqrt[3]{xy} = \sqrt[3]{(xy)^3 \cdot xy} - 2\sqrt[3]{xy} = xy\sqrt[3]{xy} - 2\sqrt[3]{xy} = (xy - 2)\sqrt[3]{xy}$

38. $\left(4 + \sqrt{2}\right)\left(4 - \sqrt{2}\right) = 4^2 - \left(\sqrt{2}\right)^2 = 16 - 2 = 14$

39. $8(x - 3)^2 = 200 \Rightarrow (x - 3)^2 = 25 \Rightarrow x - 3 = \pm\sqrt{25} \Rightarrow x - 3 = \pm 5 \Rightarrow x = -2$ or 8

40. $3\sqrt{2x + 6} = 6x \Rightarrow \sqrt{2x + 6} = 2x \Rightarrow 2x + 6 = 4x^2 \Rightarrow 4x^2 - 2x - 6 = 0 \Rightarrow 2(2x - 3)(x + 1) \Rightarrow$

$x = \dfrac{3}{2}$ or $x = -1$. The value $x = -1$ does not check. The only solution is $\dfrac{3}{2}$.

41. $(-3+i)(-4-2i) = 12 + 6i - 4i - 2i^2 = 12 + 2i + 2 = 14 + 2i$

42. $\dfrac{2-6i}{1+2i} = \dfrac{2-6i}{1+2i} \cdot \dfrac{1-2i}{1-2i} = \dfrac{2-4i-6i+12i^2}{1-4i^2} = \dfrac{2-10i-12}{1+4} = \dfrac{-10-10i}{5} = -2-2i$

43. $-\dfrac{b}{2a} = -\dfrac{8}{2(3)} = -\dfrac{8}{6} = -\dfrac{4}{3}; \ f\left(-\dfrac{4}{3}\right) = 3\left(-\dfrac{4}{3}\right)^2 + 8\left(-\dfrac{4}{3}\right) + 5 = -\dfrac{1}{3}$. The vertex is $\left(-\dfrac{4}{3}, -\dfrac{1}{3}\right)$.

   The minimum value is $-\dfrac{1}{3}$.

44. $y = 2x^2 + 8x + 17 \Rightarrow y = 2\left(x^2 + 4x + 4\right) + 17 - 8 \Rightarrow y = 2(x+2)^2 + 9$. The vertex is $(-2, 9)$.

45. $x^2 - 4x + 13 = 0 \Rightarrow x^2 - 4x + 4 = -13 + 4 \Rightarrow (x-2)^2 = -9 \Rightarrow x - 2 = \pm\sqrt{-9} \Rightarrow x = 2 \pm 3i$

46. $z^2 - 4z = 32 \Rightarrow z^2 - 4z - 32 = 0 \Rightarrow (z+4)(z-8) = 0 \Rightarrow z = -4$ or $z = 8$

47. (a) The graph intersects the x-axis at -3 and 1.

   (b) Because the parabola opens upward, $a > 0$.

   (c) Because there are two real solutions, the discriminant is positive.

48. $x^2 + 2x + 3 = 0 \Rightarrow x^2 + 2x + 1 = -3 + 1 \Rightarrow (x+1)^2 = -2 \Rightarrow$ no real solutions. This parabola does not cross the

   x-axis and it opens upward. The solution is $(-\infty, \infty)$.

49. (a) $g(-2) = 3(-2) - 2 = -8$, then $(f \circ g)(-2) = f(g(-2)) = f(-8) = (-8)^2 + 1 = 65$

   (b) $(g \circ f)(x) = g(f(x)) = g(x^2 + 1) = 3(x^2 + 1) - 2 = 3x^2 + 3 - 2 = 3x^2 + 1$

50. $f(x) = \dfrac{3x+1}{2} \Rightarrow y = \dfrac{3x+1}{2}$, interchange x and y and solve for y.

   $x = \dfrac{3y+1}{2} \Rightarrow 2x = 3y + 1 \Rightarrow 3y = 2x - 1 \Rightarrow y = \dfrac{2x-1}{3} \Rightarrow f^{-1}(x) = \dfrac{2x-1}{3}$

51. $\ln\left(x^3\sqrt{y}\right) = \ln\left(x^3 y^{1/2}\right) = \ln x^3 + \ln y^{1/2} = 3 \ln x + \dfrac{1}{2} \ln y$

52. $2 \log x - \log 4xy = \log x^2 - \log 4xy = \log \dfrac{x^2}{4xy} = \log \dfrac{x}{4y}$

53. $8 \log x + 3 = 17 \Rightarrow 8 \log x = 14 \Rightarrow \log x = \dfrac{7}{4} \Rightarrow 10^{\log x} = 10^{7/4} \Rightarrow x \approx 56.23$

54. $4^{2x} = 5 \Rightarrow \log_4 4^{2x} = \log_4 5 \Rightarrow 2x = \log_4 5 \Rightarrow 2x = \dfrac{\log 5}{\log 4} \Rightarrow x = \dfrac{\log 5}{2 \log 4} \approx 0.58$

55. See Figure 55. The vertex is $(1, 3)$, and the axis of symmetry is $y = 3$.

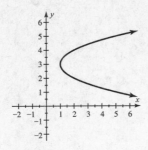

Figure 55

56. $x^2 - 6x + y^2 + 2y = -6 \Rightarrow \left(x^2 - 6x + 9\right) + \left(y^2 + 2y + 1\right) = -6 + 9 + 1 \Rightarrow \left(x - 3\right)^2 + \left(y + 1\right)^2 = 4$

The center is $(3, -1)$, and the radius is 2.

57. See Figure 57.

58. See Figure 58.

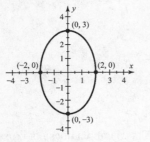
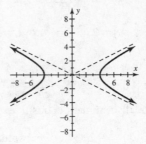

Figure 57   Figure 58

59. Vertical transverse axis with vertices $(0, \pm 2)$ and asymptotes $y = \pm \dfrac{2}{4}x \Rightarrow \dfrac{y^2}{4} - \dfrac{x^2}{16} = 1$

60. Horizontal major axis with vertices $(\pm 4, 0)$ and minor axis endpoints $(0, \pm 2) \Rightarrow \dfrac{x^2}{16} + \dfrac{y^2}{4} = 1$

61. Substitute $y = x^2 + 1$ in the second equation and solve for x.

$x^2 + 2\left(x^2 + 1\right) = 5 \Rightarrow 3x^2 - 3 = 0 \Rightarrow 3\left(x^2 - 1\right) = 0 \Rightarrow 3(x + 1)(x - 1) = 0 \Rightarrow x = -1$ or $x = 1$

When $x = -1$, $y = (-1)^2 + 1 = 2$. When $x = 1$, $y = (1)^2 + 1 = 2$. The solutions are $(-1, 2)$ and $(1, 2)$.

62. See Figure 62.

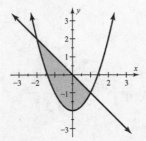

Figure 62

63. This sequence is arithmetic, the common difference is -2.

64. This sequence is geometric, the common ratio is 0.2.

65.　This sequence is geometric, the common ratio is 4.

66.　This sequence is arithmetic, the common difference is 6.

67.　Note: $d = 5 - 2 = 3$, thus $a_n = 2 + (n-1)(3) \Rightarrow a_n = 2 + 3n - 3 \Rightarrow a_n = 3n - 1$

68.　Note: $r = \dfrac{12}{4} = 3$, thus $a_n = 4(3)^{n-1}$

69.　$9\left(\dfrac{3+35}{2}\right) = 9(19) = 171$

70.　Here $r = -2$ and $n = 11$ thus $S_{11} = 1\left(\dfrac{1-(-2)^{11}}{1-(-2)}\right) = 1\left(\dfrac{2049}{3}\right) = 683$

71.　$(2x+3)^4 = (_4C_0)(2x)^4 + (_4C_1)(2x)^3(3) + (_4C_2)(2x)^2(3^2) + (_4C_3)(2x)(3^3) + (_4C_4)(3^4) \Rightarrow$

　　$(2x+3)^4 = 16x^4 + 96x^3 + 216x^2 + 216x + 81$

72.　$(2a-5b)^3 = (_3C_0)(2a)^3 - (_3C_1)(2a)^2(5b) + (_3C_2)(2a)(5b)^2 - (_3C_3)(5b)^3 \Rightarrow$

　　$(2a-5b)^3 = 8a^3 - 60a^2b + 150ab^2 - 125b^3$

Applications

73.　$r = \sqrt{\dfrac{14}{\pi}} \approx 2.11$ inches

74.　See Figure 74.

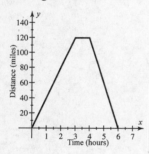

　　　　Figure 74

75.　(a)　$f(x) = \dfrac{170}{2} + 0.4x \Rightarrow f(x) = 0.4x + 85$

　　(b)　$0.4(90) + \dfrac{W}{2} = 130 \Rightarrow 36 + \dfrac{W}{2} = 130 \Rightarrow \dfrac{W}{2} = 94 \Rightarrow W = 188$ pounds

76.　Let x and y represent the speed of the airplane and the speed of the wind respectively. Then the system
　　needed is $x - y = 360$ and $x + y = 400$. Adding the two equations will eliminate y.

　　$x - y = 360$
　　$\underline{x + y = 400}$
　　$2x = 760$　　　Thus, $x = 380$. And so $(380) + y = 400 \Rightarrow y = 20$.

　　The speed of the airplane is 380 mph and the speed of the wind is 20 mph.

77.   Let x represent the width of the tent floor. Then $2x - 6$ represents the length.

$$x(2x-6) = 108 \Rightarrow 2x^2 - 6x - 108 = 0 \Rightarrow x^2 - 3x - 54 = 0 \Rightarrow (x+6)(x-9) = 0 \Rightarrow x = -6 \text{ or } 9$$

Since the value $x = -6$ has no physical meaning, the solution is 9. The dimensions are 9 feet by 12 feet.

78.   Let x represent time required to weed the garden if they worked together. Then $\dfrac{x}{60} + \dfrac{x}{90} = 1$.

$$180 \cdot \left(\dfrac{x}{60} + \dfrac{x}{90} \right) = 1 \cdot 180 \Rightarrow 3x + 2x = 180 \Rightarrow 5x = 180 \Rightarrow x = \dfrac{180}{5} = 36 \text{ minutes}$$

79.   (a) $xy = 96$ and $3x - y = 12$

(b)   Solve the second equation for y and substitute the result in the first equation. $3x - y = 12 \Rightarrow y = 3x - 12$

$$x(3x-12) = 96 \Rightarrow 3x^2 - 12x - 96 = 0 \Rightarrow x^2 - 4x - 32 = 0 \Rightarrow (x+4)(x-8) = 0 \Rightarrow x = -4 \text{ or } x = 8. \text{ Since}$$

the numbers must be positive, the solution is $x = 8$. The numbers are 8 and 12.

80.   Find the sum of the series $1 + 3 + 5 + \cdots + 23$. The sum is $12\left(\dfrac{1+23}{2}\right) = 12(12) = 144$ musicians.

Critical Thinking Solutions for Chapter 11

Section 11.1

•   Each year there are r times as many female insects as there were the previous year. The annual growth factor is 1.04.

Section 11.2

•   No, since we can not determine the sign of r. Note that $r = \pm \sqrt[4]{\dfrac{a_5}{a_1}}$.